Essentials of
World Regional
GEOGRAPHY

Essentials of
World Regional
GEOGRAPHY

third edition

George W. White

South Dakota State University

Joseph P. Dymond

The George Washington University

Elizabeth Chacko

The George Washington University

Justin Scheidt

Georgia Southern University

Michael Bradshaw

College of St. Mark and St. John, Plymouth, U.K.

McGraw Hill Education

ESSENTIALS OF WORLD REGIONAL GEOGRAPHY, THIRD EDITION

Published by McGraw-Hill Education, 2 Penn Plaza, New York, NY 10121. Copyright © 2014 by McGraw-Hill Education. All rights reserved. Printed in the United States of America. Previous editions © 2011 and 2008. No part of this publication may be reproduced or distributed in any form or by any means, or stored in a database or retrieval system, without the prior written consent of McGraw-Hill Education, including, but not limited to, in any network or other electronic storage or transmission, or broadcast for distance learning.

Some ancillaries, including electronic and print components, may not be available to customers outside the United States.

This book is printed on acid-free paper.

1 2 3 4 5 6 7 8 9 0 DOW/DOW 1 0 9 8 7 6 5 4 3

ISBN 978–0–07–336935–8
MHID 0–07–336935–7

Senior Vice President, Products & Markets: *Kurt L. Strand*
Vice President, General Manager, Products & Markets: *Marty Lange*
Vice President, Content Production & Technology Services: *Kimberly Meriwether David*
Managing Director: *Thomas Timp*
Brand Manager: *Michelle Vogler*
Director of Development: *Rose Koos*
Development Editor: *Jodi Rhomberg*
Director of Digital Content: *Andrea Pellerito*
Marketing Manager: *Matthew Garcia*
Director, Content Production: *Terri Schiesl*
Content Project Manager: *Kelly A. Heinrichs*
Senior Buyer: *Laura Fuller*
Designer: *Tara McDermott*
Cover Image: © *Georgina Bowater/Corbis*
Lead Content Licensing Specialist: *Carrie K. Burger*
Compositor: *Lachina Publishing Services*
Typeface: *10/12 Times Roman*
Printer: *R. R. Donnelley*

All figures except where noted are either original, or reproduced/adapted with permission from Contemporary World Geography, 4th ed. by Bradshaw et al. Copyright © 2012 McGraw-Hill Global Education Holdings, LLC.

All credits appearing on page or at the end of the book are considered to be an extension of the copyright page.

Library of Congress Cataloging-in-Publication Data

White, George W., 1963-
  Essentials of world regional geography / George White, South Dakota State University, Joe Dymond, George Washington University, Elizabeth Chacko, George Washington University, Michael Bradshaw, College of St. Mark and St. John, Justin Scheidt, Georgia Southern University. — Third edition.
      pages cm
  Includes index.
  Rev. ed. of : Contemporary world regional geography / Michael Bradshaw, George W. White, Joseph P. Dymond, c2004.
  ISBN 978–0–07–336935–8 — ISBN 0–07–336935–7 (hard copy : alk. paper) 1. Geography. I. Bradshaw, Michael J. (Michael John), 1935- Contemporary world regional geography. II. Title.
  G116.B72 2013
  910–dc23
                                                                      2013026117

The Internet addresses listed in the text were accurate at the time of publication. The inclusion of a website does not indicate an endorsement by the authors or McGraw-Hill Education, and McGraw-Hill Education does not guarantee the accuracy of the information presented at these sites.

www.mhhe.com

In Dedication

— — — — — —

Maureen, Madison, and José

Thomas, Rebecca, and Abraham

Papa

Valerie, Paul, and John

Brief Contents

Table of Contents

About the Authors

George W. White

George W. White grew up in Oakland, California. He pursued graduate work in Eugene, Oregon, completing a Ph.D. at the University of Oregon. He then moved to Frostburg, Maryland, where he met his wife. George served as a faculty member for 15 years at Frostburg State University, attaining the rank of full professor and serving for a time as department chair. He then moved to South Dakota State University, where he now serves as head of the Department of Geography. Political geography and Europe are two of his primary interests. He authored a book titled *Nationalism and Territory: Constructing Group Identity in Southeastern Europe* (2000) and a second titled *Nation, State, and Territory. Vol. 1. Origins, Evolutions, and Developments* (2004).

After meeting Michael Bradshaw, George was impressed by Michael's long and distinguished career of teaching, research, and publication. He accepted the opportunity to join Michael in his plans to write a new world regional geography text. He initially took lead authorship of the chapters on Europe and Russia and Neighboring Countries, later adding Northern Africa and Southwestern Asia.

George became a geographer because he believes that the field of geography is alive and dynamic, attuned to our ever-changing world and its great diversity. The world regional approach represents the breadth of the field of geography, and world regional geography texts are the epitome of the geographer's art. George White chose to collaborate with Michael Bradshaw on this project because the text combines local practices with global processes, and explains interactions between the two as they shape each other.

Joe Dymond

Joe Dymond earned a master of science degree from The Pennsylvania State University in 1994 and a master of natural sciences degree from Louisiana State University in 1999. He taught world regional geography courses for the Louisiana State University Department of Geography and Anthropology from 1995 through 2000. During Joe's six years at LSU, he instructed thousands of students and was recognized in the spring of 1997, fall of 1999, and fall of 2000 for superior instruction to freshman students by the Louisiana State University Freshman Honor Society, Alpha Lambda Delta. Joe currently lives in suburban Washington, D.C., with his wife and children, and is an assistant professor and lead undergraduate major advisor in the Department of Geography at The George Washington University (GWU). In the fall of 2006, Joe was honored by GWU as a recipient of a Morton A. Bender Teaching Award. The George Washington University recognized him again in 2010 with a GWU Service Excellence: Student Choice Award. In the spring of 2012, Joe was celebrated by GWU as the recipient of the 2011–2012 GWU Writing in the Disciplines Distinguished Teaching Award.

Joe is the secondary author for Chapter 1 and is the lead author for the regional chapters on Australia, Oceania, Antarctica, and Latin America. Joe is interested in providing students with the geographic tools that will help them to better understand the human and environmental patterns present in their world. His greatest concern for geography students is that they obtain a comprehensive and fair perspective when learning about the people and places comprising the regions of the world. The style of this text attempts to tell the regional geographic story from many perspectives. This structure permits students to better analyze geographic characteristics, connections, and relationships around the world and to think critically about important global issues. *Essentials of World Regional Geography* **teaches** rather than **lectures**.

Elizabeth Chacko

Elizabeth Chacko was born and raised in Kolkata, India. She received her undergraduate degree in geography (with honors) from the University of Calcutta. Moving to the United States for further study, she earned a master's degree in geography from Miami University, Ohio. She also obtained a graduate degree in public health and a Ph.D. in geography from the University of California, Los Angeles (UCLA). Elizabeth taught geography at the college level at various institutions, including Loreto College, Kolkata; UCLA, and The George Washington University, where she is Chair and Associate Professor of Geography and International Affairs.

Elizabeth was selected as Professor of the Year from the District of Columbia in 2006 by the Carnegie Foundation for the Advancement of Teaching and the Council for the Advancement and Support of Education (CASE). She teaches courses on South Asia, globalization, medical and population geography, and development. Elizabeth's research interests include women's health and the role of culture in health and health care. She is currently engaged in research on transnationalism, the African immigrant community in the United States, and the return migration of Asian Indian professionals to India. Elizabeth is on the editorial board of the *Journal of Cultural Geography* and the *Professional Geographer*, and she is a member of the Board of Trustees of the Population Reference Bureau.

In this edition, Elizabeth is the lead author for the chapters on South Asia and East Asia. She is delighted to be part of an author team of committed geographers. She enjoys helping students understand the dynamic interactions between humans and the earth's surface and comprehend the interplay of economic, sociocultural, and political forces that impact globalization and the spatial variations that result at local, regional, and global scales. She hopes that this book will raise students' appreciation of the relevance and significance of geography in their lives.

Justin Scheidt

Justin Scheidt is an assistant professor of geography and geology at Georgia Southern University. Previously he served as a faculty member at Delta College and Ferris State University. He is currently working on his dissertation in geography, examining the regional development and planning of passenger high-speed rail systems in North America, and he completed coursework toward the Ph.D. degree at Florida State University from 2010 to 2011. He has a master of science degree in geological sciences from the University of South Carolina (2005), and a highest honors B.A. in geography from the University of Florida (2000), with minors in geology and environmental science. Justin brings experience to this textbook project in the fields of North America and Sub-Saharan Africa.

Justin has worked extensively on research and planning for passenger high-speed rail efforts in the United States, co-authoring publications on this topic with Dr. Joseph Schwieterman of DePaul University in the *Journal of Regional Planning* (2007, 2008). He was a guest speaker at the North American High Speed Rail Summit in Ottawa, Ontario, in November 2009, and he has given similar presentations at the Association of American Geographers National Conventions (2010, 2011). He currently lives in Canadian Lakes, Michigan, and is originally from Tampa, Florida.

Michael Bradshaw

Michael Bradshaw lives in Canterbury, England, and has two sons and three grandchildren. Michael taught for 25 years at the College of St. Mark and St. John, Plymouth, as Geography Department chair and dean of the humanities course. He has written texts for British high schools and colleges since the 1960s. In 1985, he was awarded a Ph.D. from Leicester University for his study on the impacts of federal grant-aid in Appalachia. His book *The Appalachian Regional Commission: Twenty-Five Years of Government Policy* was published in 1992. Since 1991, he has written for U.S. students and has been responsible for two physical geography texts and the successful world regional geography text, *The New Global Order*. Michael believes that we should all be better equipped to live in the modern, increasingly global world. Understanding of geographic differences should make us more able to assess crucial issues and value other people who bring varied resources and who face pressures that we find difficult to imagine.

Preface

As the authors of *Contemporary World Regional Geography*, we are pleased to bring you the third edition of *Essentials of World Regional Geography*. We created this text because many instructors teach under circumstances that make a shorter, streamlined text more desirable than the standard world regional text. It has its own unique features. In preparing this text, we adopted a fresh approach that combines fundamental geographical elements, internal regional diversity, and contemporary issues. These allow serious discussion of cultural and environmental issues along with political and economic changes, for example, in Russia and China. The shorter length and distinctive approach were received enthusiastically by nearly all our reviewers.

The main innovations are in the ordering of the text. Each of the 11 regional chapters opens with a full-page (or larger) map of the region, short accounts of people or events to provide a personal flavor of the region, an outline of the chapter contents, and a short section placing the region in its wider global context. Each chapter has three further sections. The first summarizes the distinctive physical and human geographies of the region; the second explores the internal diversity of the region at sub-regional, selected country, and local scales. These take forward our commitment to the comparative nature of world regional geography. The third section focuses on a selection of contemporary issues that are important to the people of each region and frequently have implications for the rest of the world. Many of these issues are highly contested, with opposing factions having dramatically differing viewpoints. We have outlined these views in debate formats to help students understand them. Reviewing instructors were enthusiastic about the teaching value of this overall approach.

The opening chapter contains a discussion of the basis and value of world regional geography and overviews of the main relevant aspects of physical and human geography. Students are introduced to the maps and diagrams that are features of each chapter to encourage comparative study and familiarization with the set of illustrations chosen. Maps take a prominent role, and photos throughout the text provide windows on the elements of the regional geographies. We know that students are interested in how geography can be used in the workplace, and so each chapter ends with a personal example of work in progress.

Chapter Highlights

Chapter 1, Essentials of World Regional Geography, defines geography and regional geography and introduces the concept of globalization-localization tensions. This is followed by an overview of physical geography with a focus on Earth's interior forces, climate, ecosystems, and human impact on the physical environment. The human geography overview is led by a consideration of how culture influences regional character, followed by major aspects of population geography, political geography, and economic geography. These aspects are connected in a summary of approaches to human development and human rights. At the end of the chapter, the world regions are defined with a summary of their main distinctive characteristics.

Chapter 2, Europe, begins a world region tour where many modern global processes and innovations began, often building upon previous African, Asian, and Arab achievements. The contemporary issues include the development and future of the European Union and an exploration of the growing multicultural nature of European society.

We next move eastward in **Chapter 3, Russia and Neighboring Countries**. This is a study of the geographical impacts of European-origin communist principles adopted by governments for most of the 1900s, followed by massive changes as the Soviet Union broke up in the early 1990s. The Russian "Empire" remains a political and economic reality in the region. Contemporary issues include questions of human rights and environmental problems, together with the region's wealth derived from oil and natural gas.

In **Chapter 4, East Asia**, we enter a region of cultural contrast to Europe, but one that contains the world's most significant emerging countries: Japan, China, South Korea, and Taiwan. These contrast with North Korea and Mongolia. Contemporary issues include the emergence of China as a world power and its distinctive population policy, human rights, local multinational corporations, and globally connected cities.

Chapter 5, Southeast Asia, deals with a region at a global crossroads where the impacts of commerce, cultural exchanges, conquest, and globalization are evident. Most countries in the region have coastal access and a history of population movements between the mainland and islands. A geographic transition zone between East and South Asia, Southeast Asia has incorporated demographic and cultural elements from both these regions. Among the contemporary issues examined in this chapter are regional cooperation through ASEAN (Association of Southeast Asian Nations), conflicts over ocean space and piracy, and the rise of Singapore as a significant financial and trading hub in the region.

Moving westward, we reach **Chapter 6, South Asia**, with its distinctive cultural background, including the origins of Hinduism and Buddhism, and colonial experiences. After gaining independence from the British Raj in 1947, the new countries attempted self-sufficiency, avoiding close relations with other regions and leading a group of nonaligned countries. This policy was partly successful, but switched in the 1990s to a more global outlook. The contemporary issues include ethnic conflicts and environmental problems, alongside considerations of population and urban growth.

Further westward we enter **Chapter 7, Northern Africa and Southwestern Asia**, at the junction of Asia, Europe, and Africa. Although a mainly arid region, its people initiated, influenced, and passed on many cultural and technical innovations to the surrounding regions. Today it is the world's center of the Islamic religion and has the world's largest oil resources. However, its fragmented and conflicting peoples tend toward political instability. The contemporary issues include the Israeli-Palestinian conflict, the Iraq situation, and aspects of human rights.

Southward is the subject of **Chapter 8, Sub-Saharan Africa**, the world's poorest region despite its leading role at the outset of human history. After major migrations of African peoples, Muslim and European influences took control of much of the region. Most countries gained independence, mainly in the 1960s, but struggled through internal political conflict and poverty. The contemporary issues include the role of HIV/AIDS, the culture shocks of global elements, exploding city populations, and the question of this century's challenge to Africans.

The tour then reaches **Chapter 9, Australia, Oceania, and Antarctica**. Though distant from many of the other world regions, countries in this world region were first brought into the global system through European colonization and territorial claims. Continued globalization is increasingly connecting this world region to other world regions.

Crossing the South Pacific Ocean to **Chapter 10, Latin America**, we find a world region where many indigenous peoples remain but enjoy little political or economic power in contrast to the descendants of European colonists and those of mixed ethnic groups. Contemporary issues include the deforestation of Amazon rain forest, the international drug trade based in the northern Andes, and the growth of huge cities in Mexico and Brazil.

Finally, we reach **Chapter 11, North America**, with the world's most affluent societies in the United States and Canada. The United States in particular sets the conditions of globalization, although not all the impacts offer better livelihoods to all Americans. Contemporary issues include the impacts of immigration, the North American Free Trade Agreement, and the role of French-speaking Québec in Canada.

What's New to This Edition

- **New information.** The world in which we live is constantly changing. This edition features new information on population changes, migration, the global economy, gender (in)equality, conflict, and the environment. Globalization is causing some societies to become increasingly connected as they share products, styles, and ideas. At the same time, others remain disconnected, and often by choice. In either case, the world's connections are continually changing.

- **New and improved physical features maps.** Each new chapter opener is now a redrawn and enlarged full-page physical features map, or a two-page map in the case of Chapter 3: Russia and Neighboring Countries and Chapter 7: Northern Africa and Southwestern Asia. The new maps show major physical features, country boundaries, capital cities, and other major cities that students can easily reference while reading the chapter. Every map has been evaluated for size, labeling, and color consistency.

- **New climate maps.** Regional climate maps have been rendered and placed in each chapter, where they are more easily accessed by students and instructors.

- **Learning objectives** have been added to each chapter to help guide both the instructor and students through the key content of the chapter.

- **Updated data tables.** Throughout the text, tables have been updated to reflect the most current data.

- **Global economic crisis.** This text incorporates coverage of the 2008–2009 global economic crisis. Details are provided in most regional chapters about the local, regional, and global effects of the crisis. Further coverage incorporates documentation and analysis of the post-crisis recovery for many areas.

- **Natural disasters and human-environment issues.** Unfolding crises are seamlessly integrated into the regional discussions of the human dimension of physical geography and environmental issues. Disaster coverage is also integrated into relevant political and economic discussion. The scale of impacts from local to global is analyzed to help students to appreciate how events in one part of the world can affect everyday life in others. Examples include earthquakes in Haiti and New Zealand, an oil spill in the Gulf of Mexico, and an earthquake, tsunami, and nuclear disaster in Japan.

- **Economic crisis in Europe.** The recent world recession hit Europe particularly hard, and it has changed the relationships between the members of the European Union.

- **Russia as a world power.** The changing global economy has changed Russia's ability to act as a world power. Discussions include Russia's invasion of South Ossetia.

- Chapter 1 includes an expanded section on "**deglobalization**," which takes into account the rising costs of fuel and transportation and the rise of new technologies that use new materials.

- Chapter 2 discusses actions taken in Europe to address **climate change**, which includes new information on investments in **green energy** and new developments in Europe's auto industry. It also offers new information on the continued development of **Russia's economy**, especially its agricultural sector.

- Chapter 4 offers new discussions on the **globally connected cities of East Asia** (Hong Kong, Beijing, Tokyo,

Seoul) and an expanded section on **environmental problems in China**.

- Chapter 5 contains a more current discussion of **ASEAN** and **APEC** as well as a section on the tensions surrounding the **South China Sea**, and the impacts of human rights violations and economic sanctions on **Myanmar**.
- Chapter 6 highlights the **changing urban landscape** of South Asia and the growing urban population of India. It also touches on manufacturing and infrastructure in India.
- Chapter 7 contains a new section on the **Arab Spring** as well as new information on resources such as oil and water, and on **human rights**, especially concerning women in society.
- Chapter 8 includes significant updates, including current information on **South Sudan** and new maps and tables with up-to-date data.
- Chapter 9 offers updated discussions on recent geologic events, Aboriginal communities of Australia, current and projected **water issues in Australia**, the impact of regional resources on local, regional, and global trace/economies of Australia and New Zealand.
- Chapter 10 highlights include more detailed discussions on **Mexico City**—the expansion of the metropolitan area as well as squatter settlements, impoverished neighborhoods, and crime. It also includes a section on the **Andes Mountains**; the significant components of the economy, crime and drug trafficking and efforts to combat it, urban geography, and societal and cultural developments like the "mall/shopping culture" and the "bicycle culture."
- Chapter 11 contains a new section on **water issues and potential solutions** in the 21st century, with a focus on the process of desalination and its applications in North America. It also has a new agricultural regions map, which highlights the wide spatial diversity of land use across North America.

A Text for Students

Students are encouraged to think about what it means to be part of a global community and to develop their geographical understandings of world events. This text features:

- **Accessibility.** Reviewers commented on the clarity of writing, clear definition of terms, and up-to-date illustrations.
- **Consistent structure.** The clear and consistent structure within each chapter encourages readers to compare world regions.
- **Superior illustrations.** Straightforward maps and diagrams with styles that are repeated in each chapter allow students to easily compare regions.
- **An efficient and economic option.** A book of fewer pages encourages student participation.

DIGITAL RESOURCES

McGraw-Hill offers various tools and technology products to support *Essentials of World Regional Geography*, third edition.

McGraw-Hill's Connect Plus™

 McGraw-Hill's Connect Plus (*www.mcgrawhillconnect .com*) is a web-based assignment and assessment platform that gives students the means to better connect with their coursework, with their instructors, and with the important concepts that they will need to know for success now and in the future. The following resources are available in Connect:

- Auto-graded assessments
- LearnSmart, an adaptive diagnostic tool
- SmartBook, an adaptive reading experience
- Powerful reporting against learning outcomes and level of difficulty
- McGraw-Hill Tegrity Campus, which digitally records and distributes your lectures with a click of a button
- The full textbook as an integrated, dynamic eBook that you can also assign
- Instructor Resources such as an Instructor's Manual, PowerPoint slides, and Test Banks
- Image Bank that includes all images available for presentation tools

With Connect Plus, instructors can deliver assignments, quizzes, and tests online. Instructors can edit existing questions and author entirely new problems; track individual student performance—by question, assignment, or in relation to the class overall—with detailed grade reports; integrate grade reports easily with Learning Management Systems (LMS); and much more. By choosing Connect, instructors are providing their students with a powerful tool for improving academic performance and truly mastering course material. Connect allows students to practice important skills at their own pace and on their own schedule. Importantly, students' assessment results and instructors' feedback are all saved online, so students can continually review their progress and plot their course to success.

LearnSmart™

LEARNSMART®

No two students are alike. Why should their learning paths be? LearnSmart uses revolutionary adaptive technology to build a learning experience unique to each student's individual needs. It starts by identifying the topics a student knows and does not know. As the student progresses, LearnSmart adapts and adjusts the content based on his or her individual strengths, weaknesses, and confidence, ensuring that every minute spent studying with LearnSmart is the most efficient and productive study time possible.

LearnSmart also takes into account that everyone will forget a certain amount of material. LearnSmart pinpoints areas that a student is most likely to forget and encourages periodic review to ensure that the knowledge is truly learned and retained. In this way, LearnSmart goes beyond simply getting students to memorize material—it helps them truly retain the material in their long-term memory. Want proof? Students who use LearnSmart are 35 percent more likely to complete their class; 13 percent more likely to pass their class; and have been proven to improve their performance by a full letter grade.

Visit *www.learnsmartadvantage.com* to discover for yourself how the LearnSmart diagnostic ensures students will connect with the content, learn more effectively, and succeed in your course.

SmartBook™

SmartBook is the first and only adaptive reading experience available for the higher education market. Powered by an intelligent diagnostic and adaptive engine, SmartBook facilitates the reading process by identifying what content a student knows and doesn't know through adaptive assessments. As the student reads, the reading material constantly adapts to ensure the student is focused on the content he or she needs the most to close any knowledge gaps.

Tegrity™

Tegrity Campus is a service that makes class time available all the time by automatically capturing every lecture in a searchable format for students to review when they study and complete assignments. With a simple one-click start and stop process, you capture all computer screens and corresponding audio. Students replay any part of any class with easy-to-use, browser-based viewing on a PC or Mac. Educators know that the more students can see, hear, and experience class resources, the better they learn. With Tegrity Campus, students quickly recall key moments by using Tegrity Campus's unique search feature. This search helps students efficiently find what they need, when they need it, across an entire semester of class recordings. Help turn your students' study time into learning moments immediately supported by your lecture. To learn more about Tegrity, watch a two-minute Flash demo at *http://tegritycampus.mhhe.com*.

Customizable Textbooks: Create™

Create what you've only imagined. Introducing McGraw-Hill Create—a new, self-service website that allows you to create custom course materials—print and eBooks—by drawing upon McGraw-Hill's comprehensive, cross-disciplinary content. Add your own content quickly and easily. Tap into other rights-secured third-party sources as well. Then, arrange the content in a way that makes the most sense for your course. Even personalize your book with your course name and information. Choose the best format for your course: color print, black and white print, or eBook. The eBook is now viewable on an iPad! And when you are finished customizing, you will receive a free PDF review copy in just minutes! Visit McGraw-Hill Create at *www.mcgrawhillcreate.com* today and begin building your perfect book.

CourseSmart eBook

CourseSmart is a new way for faculty to find and review eBooks. It's also a great option for students who are interested in accessing their course materials digitally and saving money. CourseSmart offers thousands of the most commonly adopted textbooks across hundreds of courses. It is the only place for faculty to review and compare the full text of a textbook online, providing immediate access without the environmental impact of requesting a print exam copy. At CourseSmart, students can save up to 50 percent off the cost of a print book, reduce their impact on the environment, and gain access to powerful Web tools for learning, including full text search, notes and highlighting, and e-mail tools for sharing notes between classmates.

To review comp copies or to purchase an eBook, go to *www.coursesmart.com*.

Additional Teaching/Learning Tools

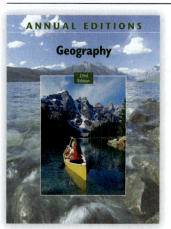

Students of geography and other disciplines, as well as the general reader, will find these unique guides invaluable to their understanding of current world countries and events.

The **Annual Editions** series is designed to provide students with convenient, inexpensive access to current, carefully selected articles from the public press. They are updated regularly through continuous monitoring of over 300 periodicals. Organizational features include an annotated listing of selected websites, an annotated table of contents, a topic guide, a general introduction, and a brief overview for each section.

Each title offers an instructor's resource guide containing test questions and a helpful user's guide called "Using Annual Editions in the Classroom."

Annual Editions: Developing World, **13/14 by Griffiths (ISBN 9780078135910; MHID 0078135915)**
Annual Editions: Geography, **06/07 by Pitzl (ISBN 9780073515519; MHID 0073515515)**
Annual Editions: Global Issues, **13/14 by Jackson (ISBN 9780078135989; MHID 0078135982)**
Annual Editions: World Politics, **13/14 by Weiner (ISBN 9780078135996; MHID 0078135990)**

The **Taking Sides** volumes present current issues in a debate-style format designed to stimulate student interest and develop critical thinking skills. Each issue is thoughtfully framed with an issue summary, an issue introduction, and a postscript, or challenge questions.

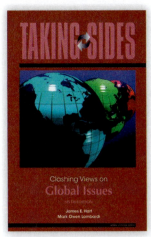

The pro and con essays—selected for their liveliness and substance—represent the arguments of leading scholars and commentators in their fields. **Taking Sides** readers feature annotated listings of selected websites. An instructor's resource guide with testing materials is available with each volume. To help instructors incorporate this effective approach in the classroom, an excellent resource called "Using Taking Sides in the Classroom" is also offered.

Taking Sides: Clashing Views on Global Issues Expanded, **seventh edition by Harf/Lombardi**
(ISBN 9780078050442; MHID 0078050448)
Taking Sides: Clashing Views in World Politics, **sixteenth edition by Rourke**
(ISBN 9780078139543; MHID 0078139546)

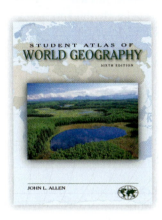

The **Student Atlas** series combines full-color maps and data sets to introduce students to the importance of the connections between geography and other areas of study, such as world politics, environmental issues, and economic development. In particular, the **Student Atlases** combine over 100 full-color maps and data sets to give students a clear picture of the recent agricultural,

industrial, demographic, environmental, economic, and political changes in every world region. These concise, affordable resources provide the most recent geographic data for geography students.

Student Atlas of World Geography, **eighth edition by Sutton**
(ISBN 9780073527673; MHID 007352767X)
Student Atlas of World Politics, **tenth edition by Allen/Sutton**
(ISBN 9780078026201; MHID 0078026202)

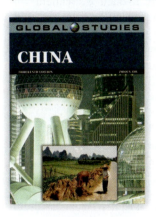

Global Studies is a unique series designed to provide comprehensive background information as well as vital current information regarding events that are shaping the cultures of the regions and countries of the world today. Each **Global Studies** volume features country reports in essay format and includes detailed maps and statistics. These essays examine the social, political, and economic significance of each country. In addition, relevant and carefully selected articles from worldwide newspapers and magazines are included to further foster international understanding.

Global Studies: Africa, **fourteenth edition by Krabacher/Kalipeni/Layachi**
(ISBN 9780078026232; MHID 0078026237)
Global Studies: Japan and the Pacific Rim, **eleventh edition by Collinwood**
(ISBN 9780078026249; MHID 0078026245)
Global Studies: Latin America and the Caribbean, **fifteenth edition by Goodwin**
(ISBN 9780078026263; MHID 0078026261)

Acknowledgments

The authors wish to express special thanks to McGraw-Hill for editorial support through Michelle Vogler and Jodi Rhomberg; the marketing expertise of Matthew Garcia; and the production team led by Kelly Heinrichs, Tara McDermott, Carrie Burger, Judi David, and Laura Fuller.

We would like to thank the following individuals who wrote and/or reviewed learning goal–oriented content for **LearnSmart**.

Sylvester Allred, Northern Arizona University
Tristan J. Kloss, University of Wisconsin–Milwaukee
Arthur C. Lee, Roane State Community College
Trent McDowell, University of North Carolina at Chapel Hill
Jessica Miles

Input from instructors teaching this course is invaluable to the development of each new edition. Our thanks and gratitude go out to the following individuals who either provided market feedback for White et al., *Essentials of World Regional Geography*, third edition, or completed a detailed chapter review of an earlier edition of this text.

Alabama State University, David A. Iyegha
Alvin Community College, Johanna Hume
Appalachian State University, Christopher Badurek
Appalachian State University, Kathleen Schroeder
Arizona State University, J. Duncan Shaeffer
Austin Peay State University, Peter P. Siska
Ball State University, Reuben Allen
Baptist Bible College of Pennsylvania, Susan Cagley
Bay Mills Community College, Rick Elder
Bellevue University, Gabrielle Collins
Bemidji State University, Jeff Ueland
Bergen Community College, Keith Kyongyup Chu
Blinn College, Rhonda E. Reagan
Blinn College, Susan Slowey
Blinn College, Angie E. Wood
Bluegrass Community & Technical College, Clovis Perry
Bowling Green State University, Michael Anthony Kimaid
Bowling Green State University, Kefa M. Otiso
Broward College, Susan Oldfather

California Poyltechnic State University, William Preston
California State University–Bakersfield, Vondana Kohli
California State University–East Bay, Dianne E. Meredith
California State University–Northridge, James Craine
California State University–Northridge, Ronald A. Davidson
California State University–Northridge, Edward L. Jackiewicz
California State University–San Marcos, Laura Martin Makey
Cameron University, Edris J. Montalvo, Jr.
Central Connecticut State University, Richard W. Benfield
Central Oregon Community College, Sonja Porter
Central Piedmont Community College, David Privette
Central Washington University, Cameron McCormick
Chaffey College, Peter Konovnitzine
Clinton Community College, Dean B. Stone
Coastal Carolina University, Daniel A. Selwa
College of DuPage, Keith Yearman
Collin County Community College, Irina Vakulenko
Columbus State University, Amanda Rees
Community College of Denver, Peg Clover-Stipek
Concordia University–Nebraska, Jack Kinworthy
Craven Community College, Robert Scull
Defiance College, Don Buerk
Delta State University, Mark Bonta
Duquesne University, Charles Wilf
Eastern Kentucky University, Bruce E. Davis
Eastern Kentucky University, Elizabeth J. Leppman
Eastern Iowa Community College District, John Lindberg
Edinboro University, Wook Lee
Edison State Community College, Katherine Clifton
Elmhurst College, Ralph Feese
Endicott College, Michael Kilburn
Florida International University, Patricia L. Price
Fort Hays State University, Paul E. Phillips
Front Range Community College, Colorado,
     Uwe R. Kackstaetter
Gadsden State Community College, George W. Terrell, Jr.
George Mason University, Patricia Boudinot
George Mason University, Sinclair A. Sheers
Georgia College, Amy R. Sumpter
Georgia College & State University, Chuck Fahrer
Georgia Southern University, Jason Dittmer
Grand Valley State University, Jim Penn
Greenville Technical College, Nick Hill
Hannibal-LaGrange University, Mark S. Quintanilla
Hardin-Simmons University, Tiffany M. Fink
Heartland Community College, Mark McBride
Hutchinson Community College, Femi Ferreira
Indiana State University, Cyril Oluyomi Wilson
Indiana University–South Bend, Gabriel Popescu
Iowa Lakes Community College, Sharon Hackenmiller
Ivy Tech Community College, Milan Andrejevich
Jackson Community College, Glenn M. Fox
Jacksonville College, Patricia Richey
Kansas State University, L. Scott Deaner
Kansas State University, Matthew J. Gerike
Kansas State University, Chris Laingen
Kansas State University, Max Lu

Kansas State University, Sumanth G. Reddy
Kansas State University, Jeffrey S. Smith
Kansas State University, Jacob Sowers
Kennesaw State University, Ulrike Ingram
Kennesaw State University, Lynn M. Patterson
Kennesaw State University, Harold R. (Harry) Trendell
Kent State University, Weronika Kusek
Kent State University, Amy Rock
Kirkwood Community College, Jeremy J. Brigham
Lake Erie College, Darlene Hall
Las Positas College, Thomas Orf
Long Beach City College, Ebenezer Kofi Peprah
Long Beach City College, Ray Sumner
Louisiana State University, Kent Mathewson
Marshall University, Joshua Hagen
Marshall University, James M. Leonard
Martin Luther College, Earl Heidtke
Marygrove College, Tal Levy
Metropolitan State College of Denver, Kenneth W. Engelbrecht
Michigan Technological University, Bradley H. Baltensperger
Middle Tennessee State University, Denis A. Bekaert
Middle Tennessee State University, Douglas Heffington
Middle Tennessee State University, Paul J. O'Farrell
Mississippi State University, Dalton W. Miller, Jr.
Missouri State University, Deborah Corcoran
Missouri State University, Ju Luo
Missouri State University, Paul Rollinson
Missouri State University–West Plains, John H. Fohn II
Montclair State University, Danlin Yu
Moorpark College, Andrea Ehrgott
Morehead State University, Royal Berglee
Morris College, Patricia Ali
Mt. San Antonio College, Elizabeth Lobb
Muskingum University, Stephen Van Horn
National-Louis University, Frank Scruggs
New England College, Wayne Lesperance
North Country Community College, William F. Price
Northern Illinois University, Sarah A. Blue
Northern Illinois University, Xuwei Chen
North Hennepin Community College, Karen Johnson
Northwestern State University, Dean Sinclair
Northwest-Shoals Community College, Selina Pearson
Ohio Mid-Western College, Bill Dykes
Oregon State University, Cub Kahn
Ozarks Technical Community College, Susan Siemens
Palm Beach Community College, Shari L. MacLachlan
Pittsburg State University, Tim Bailey
Red Rocks Community College, Laura A. Zeeman
Rochester College, Mark Manry
Rowan University, New Jersey, Denyse Lemaire
Salt Lake Community College, Robert Adam Dastrup
Samford University, Jennifer Rahn
San Jacinto College–Central Campus, Michael Modica
Santa Fe College, Heidi L. J. Lannon
Shawnee State University, Anthony Dzik
Sitting Bull College Library, Mark Holman
South Dakota State University, James Peterson

Southern Arkansas University, Natalia G. Murphy
Southern Utah University, Paul R. Larson
Southwestern Illinois College, Jeff Arnold
Tarleton State University, Greg Arkinson
Tarleton State University, Robert Atkinson
Tennessee State University, Gashaw Bekele
Texarkana College, Janet G. Brantley
Texas A&M University, Erik Prout
Texas Tech University, Cynthia L. Sorrensen
Three Rivers Community College, Peter Patsouris
Towson University, R. D. K. Herman
Trinity International University, Linda Fratt
United States Military Academy, Francis A. Galgano, Jr.
United States Military Academy, Steven Oluic
University of Akron, Robert Barrett
University of Central Oklahoma, Robert M. Kerr
University of Colorado–Colorado Springs, Emily Skop
University of Delaware, Yda Schreuder
University of Kansas, Joshua Long
University of Kentucky, Stanley D. Brunn
University of Louisiana–Lafayette, Dennis Ehrhardt
University of Louisville, Carol L. Hanchette

University of Nebraska–Kearney, Brett R. Chloupek
University of North Carolina–Wilmington, W. Frank Ainsley
University of North Carolina–Wilmington, Robert Argenbright
University of Northern Colorado, Phil Klein
University of North Texas, Donald I. Lyons
University of Regina, Ben Cecil
University of South Carolina–Aiken, Linda Q. Wang
University of Vermont, Cheryl Morse Dunkley
University of Wisconsin–Green Bay, Marcelo Cruz
University of Wisconsin–Manitowoc, Melvin Johnson
University of Wisconsin–Milwaukee, Linda McCarthy
University of Wisconsin–Stevens Point, Jia Lu
University of Wisconsin–Stevens Point, Ismaila Odogba
University of Wisconsin–Whitewater, Margo Kleinfeld
Utah State University, Cliff B. Craig
Western Illinois University, Yongxin Deng
Western Kentucky University, John All
Western Kentucky University, Amy T. Nemon
Western State Colorado University, Phil Crossley
Westfield State College, Julie Urbanik
West Virginia University, Brent McCusker
Youngstown State University, Craig S. Campbell

Market in Moshi, Tanzania. *This town is situated 30 km (18 mi.) from Mt. Kilimanjaro. Fertile volcanic soils combined with orographically enhanced rainfall make the region south of Kilimanjaro, including Moshi, one of the most agriculturally productive in Tanzania.* Photo: Joseph P. Dymond.

Essentials of World Regional Geography

LEARNING OBJECTIVES

After reading this chapter, you should be able to:

- Describe the study of world regional geography, and its importance to a changing world.

- Define globalization and localization.

- Understand the concept of region, and describe how regions change over time.

- Describe the broad physical geographic concepts of atmosphere and water interactions, lithosphere movement and relief, ecosystems and biomes, and natural hazards and resources. Explain how these phenomena contribute to defining world regions.

- Identify human impacts on natural environments, both historic and modern.

- Define cultural geography and explain how culture defines regions. Provide examples of divisions based on language, religion, ethnicity, class, and gender.

- Understand population change in terms of birth and death rates, fertility rates, and migration.

- Interpret a population pyramid, consumer goods chart, and income distribution chart. Explain how each of these highlights differences among and within world regions.

- Define nationalism, and explain how the concept of the nation-state contributes to the modern world map. List different types of governance, and describe the scale at which these operate.

- Explain how economic development is measured, and connect wealth, economic development, and type of economic system to differences among and within world regions.

- Discuss what is meant by the global economy, and identify important elements that make up the global economy.

- Explain what is meant by human development, and recognize the difference between the HDI and the GDP. Describe modern theories of development.

- Trace the evolution of the concept of human rights, and explain how recognition of these rights can vary from region to region.

- Identify the 10 major world regions and locate these on a world map. Compare and contrast the geographic characteristics that define these regions.

Physical Map — North and South America, Atlantic and Pacific Oceans

Grid references (top and bottom): A 160° B 140° C 120° D 100° E 80° F 60° G 40° H 20°

Row numbers (left/right margins): 1–10

ARCTIC OCEAN

80° 60° 40° 20° 0° 20° 40° 60° 80°

Beaufort Sea
Point Barrow
Queen Elizabeth Is.
Sverdrup Is.
Parry Is.
Ellesmere I.
Devon I.
Victoria I.
Baffin Bay 6,660 ft 2,030 m
Greenland
Gunnbjørn Fjeld 1,214 ft 3,700 m
Nuuk
6,952 ft 2,119 m
Reykjavik
Iceland
Denmark Str.
Greenland Str.
Arctic
North Atlantic Drift
British Isles

Mt. McKinley 20,321 ft 6,194 m
Alaska
Alaska Range
Anchorage
Kodiak I.
Gulf of Alaska
Aleutian Is.
Yukon
Yellowknife
Edmonton
Mackenzie
Hudson Bay
Labrador
Labrador Sea
Cape Farewell
10,111 ft 3,082 m

Vancouver
Seattle
Coast Mts.
NORTH AMERICA
Winnipeg
Canadian Shield
Montréal
Newfoundland
Cape Race
London
Land's End

Rocky Mountains
Great Plains
Toronto
Chicago
St. Louis
Missouri
Mississippi
Appalachian Mts.
Boston
New York
15,420 ft 4,700 m
C. Finisterre
20,751 ft 6,325 m
Madrid

Sa. Nevada
Mt. Whitney 14,495 ft 4,418 m
Denver
Mt. Mitchell 6,683 ft 2,037 m
Washington, D.C.
Cape Hatteras
Gulf Stream
ATLANTIC
Azores
Lisbon
Casablanca
Tubqal Atlas 13,665 ft 4,165 m

San Francisco
Los Angeles
Dallas
Atlanta
Bermuda Is.
Madeira
Canary Is.
Algiers

New Orleans
Florida
3,632 ft 1,107 m

Sa. Madre Occ.
Monterrey
Gulf of Mexico
Havana
Cuba
Bahama Is.
22,949 ft 6,995 m
Canary Current
Nawadhibu
Sa

Tropic of Cancer
Honolulu
Hawaiian Is.
19,373 ft 5,905 m
C. San Lucas
Mexico City
Pico de Orizaba 18,405 ft 5,610 m
Greater Antilles
Hispaniola
Lesser Antilles
North Equatorial Current
Cape Verde
C. Verde
Dakar
Bamako

13,796 ft 4,205 m
Hawaii
North Equatorial Current
Guatemala City
Caribbean Sea
Punta Gallinas
18,947 ft 5,775 m
23,924 ft 7,292 m
Conakry

PACIFIC
Equatorial Counter Current
Galápagos Is.
Caracas
Llanos
9,219 ft 2,810 m
Guiana Highlands
Bogotá
São Paulo
SAU
Abidjan
Accra
Guinea Current
Gulf of Guinea

Equator
Kiritimati
Jarvis I.
Line Islands
South Equatorial Current
Quito
Chimborazo 20,577 ft 6,272 m
9,888 ft 3,014 m
Manaus
Amazon
Belém
Fernando de Noronha
C. de São Roque
South Equatorial Current
Benguela

17,549 ft 5,349 m
19,199 ft 5,852 m
Punta Negra
Selvas
SOUTH AMERICA
Recife
Ascension

Marquesas Is.
Huascarán 22,204 ft 6,768 m
Lima
Campos
Salvador
Brazil Current
St. Helena
OCEAN

Cook Islands
Society Is.
Tuamotu Is.
La Paz 21,463 ft 6,542 m
Brasília
Brazilian Highlands
9,482 ft 2,890 m
Pico da Bandeira
Rio de Janeiro
São Paulo
20,075 ft 6,119 m

Tubuai Is.
Pitcairn Is.
Desventurados Is.
Andes
Asunción
Porto Alegre

Tropic of Capricorn
Easter I.
Sala-y-Gómez
8,399 ft 2,560 m
Pampas
Montevideo
Rio de la Plata
Tristan da Cunha
Gough I.

Juan Fernández Is.
Santiago
22,841 ft 6,962 m
Buenos Aires

OCEAN
Patagonia
Humboldt Current
Antarctic Circle

13,313 ft 4,058 m
20,380 ft 6,212 m

Falkland Is.
Falkland Current
South Georgia
27,113 ft 8,264 m

Tierra del Fuego
Str. of Magellan
Cape Horn
Drake Passage
Scotia Sea
South Sandwich Is.

17,355 ft 5,290 m
South Shetland Is.
South Orkney Is.

Antarctic Pen.
Bellingshausen Sea
Peter I I.
Mt. Jackson 10,446 ft 3,184 m
Antarctic Circle

Amundsen Sea
Alexander I.
Weddell Sea
Berkner I.
Ellsworth Land

ANTARCTICA

Elevation (legend):

13,124 ft	4,000 m
6,562 ft	2,000 m
3,281 ft	1,000 m
1,640 ft	500 m
656 ft	200 m
0 ft	0 m
	Depression

656 ft	200 m
6,562 ft	2,000 m
13,124 ft	4,000 m
19,686 ft	6,000 m
26,248 ft	8,000 m

Inland ice, glaciers
Ice shelf
Pack ice
→ Cold current
→ Warm current
vvvvv Limit of drift ice

Figure 1.1 World: physical features, country boundaries, and capital cities.

1.1 Contemporary Geography

What Is Geography?

Geography is a discipline that studies spatial patterns in the human and physical world. Geographers examine where and how the human and natural features of Earth's surface are distributed, how they relate to each other, and how they change over time. The distribution of language, religion, human and natural resources, and vegetation are among the many spatial phenomena examined by geographers. They attempt to explain such patterns, how they are changing, and what they might become. Many jobs require a geographic understanding. For example, urban planners need to be aware that cities contain people with varied preferences, traditions, fears, and desires; they are constantly moving around, interacting with one another, and having impacts on their urban environment. Disciplines such as history, sociology, economics, political science, and environmental science increasingly view spatial differences as crucial to understanding the Earth's human and physical conditions. They give significance to the research carried out by geographers.

Geography is a unique discipline encompassing both the physical and social sciences. **Physical geography** includes natural environmental processes across Earth's surface that result in the distribution of climate zones, plant ecologies, soil types, mountain formation, and fresh water distribution, among other patterns. In addition, physical geography increasingly examines the impacts of human actions on Earth's natural environments. **Human geography** is the study of the distribution of people (Figure 1.1) and their activities (economies, cultures, politics, and urban changes). World regional geography is an integrative analysis of the relationship between the human and the physical.

Geographers study places on Earth's surface as the environments or spaces where people live and through which they make life meaningful. Geography thus provides a place- or space-related **spatial view** of the human experience.

Place matters. Consider what it would be like if you grew up in a different country. How might that affect the language you speak, your family's religious preference, the food you eat, the music you listen to, or the schools you attend? How might the weather and other environmental conditions affect

you? What might be different about you? Might your views of world issues and possible solutions differ from what they are now? What might be the same? These are the sorts of questions geographers ask about people and the places where they live. Geographic literacy is increasingly essential because people are connected and interacting on unprecedented and still increasing levels.

Maps and Geographic Information Systems

In their approach to understanding and solving human problems, geographers use maps to present information about location, distance, direction, flows, and other characteristics of places. **Maps** are relatively small representations of much larger areas of Earth's surface. On maps, a scale implies a mathematical relationship between features on a map and on the ground area it represents. Map scales vary with the purpose of the map (Figure 1.2). Small-scale maps show areas at frac-

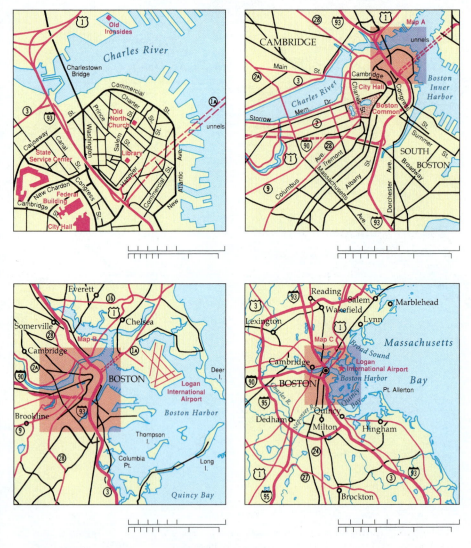

Figure 1.2 Map scale and detail. Maps of Boston from large scale (1:25,000) to small scale (1:1,000,000). The larger scale maps show more detail (streets, buildings) than the small scale maps (only major roads) for the same size of map.

tions of 1:250,000 or smaller (e.g., 1/1 million or a ratio of 1:1 million). Large-scale maps have map-to-ground ratios ranging from 1:10,000 to 1:250,000 (e.g., 1:50,000). The world maps used in this text (see, for example, Figure 1.5b) are examples of small-scale maps, in which the scale along the equator is approximately 1:120 million. Large-scale maps cover smaller areas, as in town maps, but can include more details. Not everything can be drawn to scale on maps, or else features would be too small to be seen: for example, roads, rivers, and buildings are denoted by symbols.

Geographic information systems (GIS) combine maps and aerial and satellite images with data relevant to the area (Figure 1.3). Such systems bring together a range of information that can form the basis for finding answers to complex questions. GIS became a significant tool in geography in the 1970s. The technology, utility, and application of GIS has increased substantially from its early use in the discipline, and today the majority of published large-scale maps now relate to satellite images.

Latitude and Longitude

Absolute location is the precise position of places on Earth's surface. The most universally accepted means of determining absolute location is by calculating latitude and longitude.

Latitude and longitude form the framework of an internationally accepted, location-based reference system that pinpoints absolute location (Figure 1.4). **Latitude** describes how far north or south of the equator a place is, measured in degrees. The north pole is at 90°N and the south pole at 90°S. Although the equator is an imaginary line, its position has a direct physical relationship between Earth and the sun as the line along which the most direct radiation from the sun reaches Earth. The equator encircles the globe midway between the north and south poles and is the 0° (zero degree) line of latitude. The almost spherical Earth's circumference is around 40,000 km (25,000 mi.) at the equator. A circle that joins places of the same latitude on Earth's surface is called a **parallel of latitude**. The ground distance from one degree

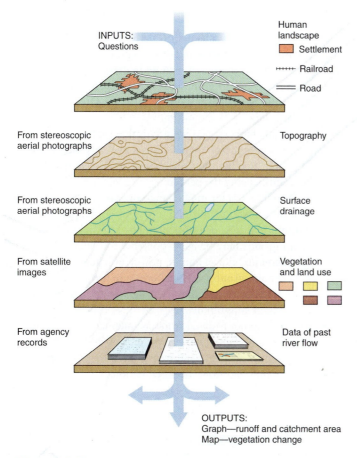

Figure 1.3 Geographic information system. The layers of information in this example are used to monitor a river system feeding a reservoir. Outputs from the system could include graphs of seasonal streamflow per unit of drainage basin area, or maps of dominant vegetation types or changing land use.

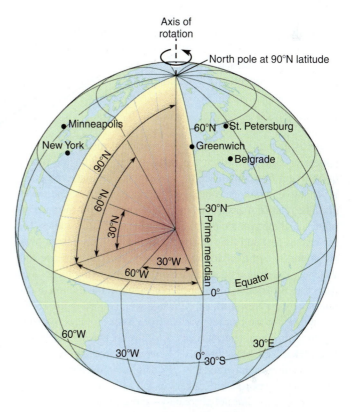

Figure 1.4 Location: latitude and longitude. Locating places on Earth's surface is made possible by the coordinate system of latitude and longitude. The degrees of latitude and longitude are formed by angles focused at Earth's center. Define the differences between 60°N and 30°N and between the 0° meridian and 60°W in terms of angles at Earth's center. Give the latitude and longitude of the cities marked on the globe.

of latitude to the next is approximately 110 km (69 mi.) on Earth's surface. For a long time, latitude was found by measuring the angle of the sun above the horizon at noon.

Longitude measures position east or west of an imaginary line drawn from the north pole to the south pole—a half circle—that passes through the former Royal Observatory at Greenwich, London, United Kingdom. Lines joining places of the same longitude are called **meridians of longitude**. The position of the prime meridian passing through Greenwich (0°) was chosen by an international conference in 1884, when London was the world's most powerful decision-making city. Methods of determining longitude, especially when charting the position of a ship, were more complex and took longer to evolve than latitude measurements. Longitude lines are not parallel like their latitude counterparts, and thus they do not provide equal measurements of ground distance. Longitude lines are farther apart at the equator and closer together at the poles. Instead of measuring sun angle at noon, people had to create tables of planetary positions, and they needed accurate clocks (chronometers first made in the late 1700s). In the late 1900s, radio beacons and satellites provided standard reference points by emitting radio pulses that could be timed and interpreted rapidly in computerized navigation systems to give accurate position fixes in global positioning system devices. Air travel, ocean transport, and increasingly, automobile use all rely on latitude and longitude through global positioning systems (GPS). Engineers on construction sites, real estate developers, and city planners also depend on GPS technology.

Distance and Direction

Physical **distance** between places is usually measured in kilometers or miles. However, the **relative distance**, or the time it takes to get from one place to another, may be much more significant to commuters, commercial shipping interests, and travelers. The relative measures of time-distance and cost-distance are often substituted for measured distance in geographic studies. The increasing cost and time to cover distance between places gives rise to the idea of the **friction of distance**. As costs increase with distance, interactions decrease. However, improving technology decreases the friction of distance. For example, the friction of distance between New York and Chicago was reduced in the 1800s, when time for the journey was cut from weeks to days by building the Erie Canal and railroads. Air travel between these places today takes a couple of hours. The increasing availability of rapid transportation facilities and the "global information highway" (the Internet) bring people into easier contact with each other, making them relatively—but not absolutely—closer. Friction of distance is still affected by other factors that slow or increase contacts among people. These factors include natural obstacles such as mountains and oceans, political factors such as country boundaries or conflict, and cultural factors such as language differences.

Geographers give directions by the cardinal points: north, east, south, west. Direction and distance help to define the locations of places relative to each other.

The Regional Approach to Geography

Our world of closer connections among peoples and places and increasingly rapid changes focuses attention on regional geography. A **region** is an area of Earth's surface with physical and human characteristics that distinguish it from other places. Regions vary in geographic scale from major divisions of the world (world regions) to single countries and parts of countries; metropolitan regions, for example, are areas that focus on large cities and their suburbs. **Regional geography** evaluates differences and similarities within and between defined areas, or regions, of Earth's surface.

Regional studies combine consideration of the systematic physical (rocks and landforms, climates, natural vegetation, soils) and human (culture, population, politics, economy) features that give a region geographic character. Regional linkages and boundaries change over time, or they may be perceived and defined differently by their inhabitants or different geographers. The range of regional scales between the global and local includes world regions, subregions, countries, and smaller regions within countries.

This text adopts the regional geography approach. Regions form and change as different systematic aspects interact and influence one another in specific areas of the world.

- People create regions over time by interacting with natural environments and other regions. People in some regions may view people in other regions as friendly or as different, and perhaps even as hostile. Perceptions of the region's identity are often helped by propaganda.
- Regional character and identity influence the actions of people living there. There is thus a two-way interaction of people creating their region and of the region's character influencing the people.
- Regional character may change over time at varied rates as a result of introducing new people, products, cultural features, or political control. Individuals and small groups may be as responsible for change as governments. For example, powerful leaders, inner government groups, or local pressure groups may dominate the outcome of issues leading to changed or unchanged geographies.
- Regions are not isolated from one another. Ties among them became more intense over the last 200 years, causing more far-reaching and frequent changes of internal geographies and flows between regions.

1.2 Globalization and Localization

Two geographic trends help us understand what makes regions unique as a result of the increasing flows among them: globalization and localization.

Globalization

Globalization, in its simplest form, is the increasing level of interconnection among people and places throughout the world. Economic globalization involves the integration and exchange of capital, technology, and information across country borders. It also affects society, culture, politics, and the natural environment. It could be argued that globalization began with the European discoveries in the late 1400s or even earlier, but the speed and intensity of globalization, especially in terms of world trade and the flow of financial investments, increased markedly beginning in the 1990s. Few people used the term "globalization" before 1990, but it is now mentioned daily in the media. It is an essentially geographic phenomenon.

Global connections include the spread of ideas, technologies, crimes, and diseases; flows of goods and services; long-term migrations of people for work, seekers of political asylum, and family consolidation; short-term flows of people for business purposes or tourism; impacts of dominant ideologies, both religious and political; and the spread of images and messages through the media of film, TV, the Internet, and print. Today a few languages, notably English, Chinese, French, Spanish, and Arabic, act as a basis for global communication. Although many worldwide flows continue to be controlled by country-based legislation and policing, others, such as the trafficking of arms, drugs, or slaves, are difficult for individual countries to control.

Localization

Localization stems from established local identities that existed prior to the intrusion of globalizing forces. Such identities respond to globalization as an increasing or a strengthening of local traditions in resistance to the global diffusion of human practices. The "local" scale is any place less than global in size, including countries, regions within countries, and wider regions that include groups of countries.

The rapid and widespread acceptance of the "globalization" theme by politicians, members of the media, special interest groups, and academia, among many others, created substantial confusion over precise meanings. Some people assume pervasive globalization is already overriding country and smaller community boundaries and interests. They consider globalization as either a great opportunity for a more cohesive world or a danger to cultures, economies, politics, and environments. Others think the term is overused and unjustified when country governments remain the dominant political entities and when cultural awareness and identities remain strong.

1.3 Regions and Natural Environments

Physical geography is the study of natural environments and their world distributions. In world regional geography, the inter-actions between people and natural environments are important. Major physical features, such as oceans, seas, mountain ranges, and rivers, have directed patterns of human movement. For many centuries, human activities relied on natural resources. The length of the growing season, the amount of water available, soil types, and mineral-bearing rocks influenced the locations of people. Natural environments played important parts in the locations of early culture hearths and concentrations of people. From the 1800s, physical geography also played important roles in the locations of new human-engineered urban environments. Today, in a reversal of the idea that physical geography determines human affairs, human beings increasingly influence many of the ways in which natural processes function. For example, natural vegetation has been removed to make way for agriculture, rivers are straightened or diverted, slopes are graded, and mountains are tunneled. We now believe that human actions are increasingly altering weather and climate.

Powerful Natural Systems

The Earth contains four major natural environments, which are powered by energy from Earth's interior and the sun. They work together to produce a dynamic and interacting system. The **solid earth environment** (lithosphere) receives its energy from the Earth's interior, which causes huge sections of the crust to collide with one another, slide past one another, or move apart. These interactions create earthquakes, volcanoes, mountain chains, and continental movements. **Earth-surface environments** come about from the interactions of the atmosphere and hydrosphere with the lithosphere. These interactions produce rain, glacier ice, wind, and ocean waves and shape landforms such as hills, valleys, cliffs, and beaches. The **atmosphere-ocean environment** receives incoming solar energy, which controls the circulations of the atmospheric gases and oceanic waters (atmosphere and hydrosphere). This solar energy produces weather, climate, and longer-term climate changes. In the **ecosystem environments** (biosphere), plants convert solar energy into food for animals. Living organisms (plants, animals, and humans) respond to and modify local climates, landforms, and soils. It is important to examine these environments in greater detail.

Solid Earth Environments

Earth is a multilayered planet with a hot molten core surrounded by the solid mantle and a relatively thin outer crust. Earth's interior provides the energy that forces large blocks of surface rock, known as **tectonic plates**, to crash into one another, slide past one another, or move apart. Earth's surface is comprised of about a dozen major and several minor plates (Figure 1.5), each up to thousands of kilometers across and around 100 km (65 mi.) thick. Most earthquakes and volcanic eruptions of molten rock from beneath Earth's surface occur along plate boundaries. The volcanic "ring of fire" and earthquake zone around the Pacific Ocean was created by plates clashing there.

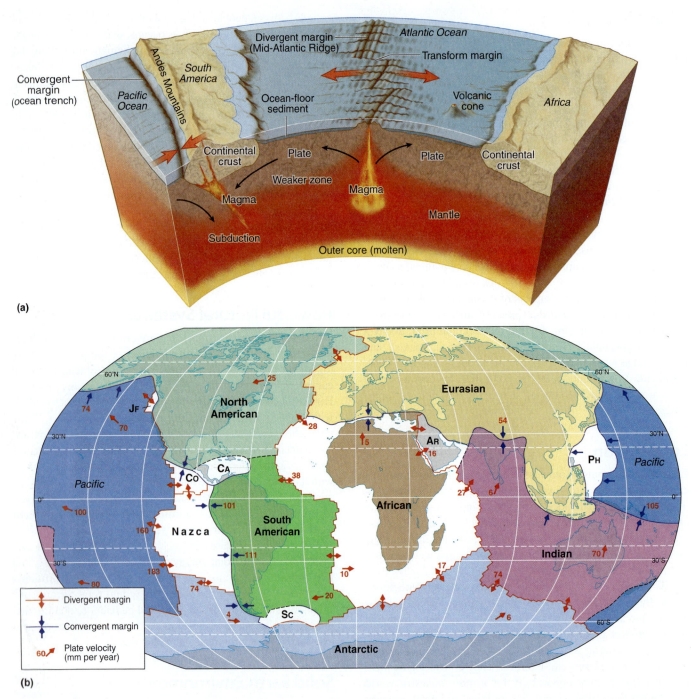

(a)

(b)

Figure 1.5 **Plate tectonics, ocean basins, and mountain ranges.** (a) A diagram that simplifies the plate tectonics relationships of the South Atlantic Ocean area. The large red arrows depict the plate movements: diverging from the Mid-Atlantic Ridge where molten rock (magma) rises and solidifies as new plate; converging where the Pacific Ocean and Andes Mountains meet, where plate material is destroyed and rises as molten magma to form volcanoes. A mid-ocean ridge is associated with diverging plates, and an ocean trench and high fold mountains with convergence. Where a plate is broken horizontally, a transform margin forms. All plate margins are associated with earthquakes and volcanic activity. (b) World map of major and the main minor plates. The minor plates include Nazca, Cocos (Co), Caribbean (CA), Juan de Fuca (JF), Arabian (AR), Philippine (PH), and Scotia (Sc). *Source: (b) Data from NASA.*

The plate movements cause the opening and closing of ocean basins and the raising of mountain systems. Where plates move apart, or diverge, fissures open and rock erupts as molten lava, adding to the edge of a plate where it solidifies. Such **divergent plate boundaries** include the Mid-Atlantic Ridge, where the eruption of molten lava builds Iceland.

Collisions occur along **convergent plate boundaries** and often force one plate upward to form mountain systems such as the Andes Mountains of South America. A plate that is forced beneath another plate is said to be **subducted**. The subducted solid rock melts under high temperatures generated by burial and friction. The molten rock produced rises toward the surface under pressure and erupts to create volcanic mountains and huge lava fields such as those forming the Columbia Plateau in the northwestern United States. Japan also lies along a number of convergent boundaries that result in subduction (Figure 1.6a).

Along transform plate boundaries, plates move horizontally against each other. They create earthquakes but usually without volcanic eruptions, such as those that occur along the San Andreas Fault in California (Figure 1.6b).

Earth Surface Environments

Earth's surface is 71 percent covered by oceans and only 29 percent occupied by the continents and islands on which people live. Weather and the action of the sea interact with Earth's internal continent-building forces to produce differences in the height and shape of the land, or its **relief**, in features such as mountains, valleys, and plains.

Once land emerges above sea level, atmospheric processes and ocean waves etch the details of surface relief. The temperature and chemical composition of the atmosphere, together with water from rain and snowmelt, react with the rocks exposed at the surface and dissolve them or break them into fragments. Such changes are called **weathering**. The broken and dissolved rock material forms the mineral basis for soil. On steep slopes, weathered material moves downhill under the influence of gravity. Such movement may be rapid in slides, flows of mud, or avalanches, or slow in local heaving and downslope creep of the surface. The mobile fraction of this broken rock material often enters rivers or glaciers and is moved toward the ocean.

The concentrated flows of water or ice and rock particles in rivers and glaciers gouge valleys in the rocks—a process called **erosion**. Glaciers formed of ice move slowly, helped by meltwater lubrication in the summer. When the flows reach a lowland, lake, or ocean, the rock particles drop to the valley, lake, or ocean floor in the process of **deposition**. Wind blows fine (dust-size) rock particles long distances, while sand-sized particles are moved around deserts or across beaches to form dunes. Along the coast, sea waves and tides erode cliffs and deposit beach materials, often moving the rock particles supplied by rivers, glaciers, or wind (Figure 1.7). The combination of these internal and external forces shapes the land surface features of regions. These features also are determined by the presence of plate margins, the nature of the rocks, the climate, and how long natural forces operated without catastrophic changes. Human activities also explain the character of surface land features.

(a)

(b)

Figure 1.6 **Internal and external earth forces mold the landscape.** (a) Workers inspect a caved-in section of a prefectural road in Satte, Saitama Prefecture. On March 11, 2011, it was damaged by one of the largest earthquakes ever recorded (9.0 magnitude by some measures) in Japan. (b) The gash across the Carrizo Plain in southern California is caused by the transform San Andreas Fault splitting a section of Earth's crustal rocks. Two plates move against each other with the western side moving northward. Rain and rivers etched out the line of weakness along the fault. Continuing movement along the fault offsets the lines of stream valleys on either side. *Photos: (a) © AP Photo/Saitama Shimbun via kyodo News; (b) © BrandX/Punchstock RF.*

Figure 1.7 **Eroding the land.** Ocean waves attack the cliffs and form beaches in Oregon. Inland, valleys are formed by a combination of river action and slope processes. *Photo: © Corbis RF.*

Soils form as broken rock matter interacts with weather, plants, and animals. Rock materials supply or withhold nutrients. Rainwater and snowmelt make any nutrients present available to plants. Decaying plant and animal matter release the nutrients back to the soil in mineral form. Soil fertility is based on rock structure, nutrient content, heat, and moisture.

Atmosphere-Ocean Environments

The **climate** of a place is based on long-term averages of the weather conditions (mainly temperature and precipitation). Differences in climatic conditions result from the transfers of heat and moisture through the atmosphere and oceans, and how they interact with the surface conditions. The transfers are powered by energy from the sun.

Heating the Atmosphere-Ocean System

Mostly visible light rays from the sun reach Earth's surface. Earth's atmosphere filters out other elements of solar-ray energy that harm living organisms, including ultraviolet rays, x-rays, and gamma rays. Absorption of the light rays at Earth's surface causes rock, soil, and ocean water to be heated and to radiate heat upward. This heat is then absorbed in the lower atmosphere by water vapor and carbon gases, raising the temperature of the air. This natural process in Earth's atmosphere is known as the **greenhouse effect**. An approximate balance between the incoming solar radiation and radiation from Earth to space reduces temperature fluctuations over time in Earth's atmosphere.

Important geographic differences in solar heating cause climatic differences from place to place. Earth rotates on its axis every 24 hours, creating day and night. It revolves in orbit around the sun once a year, producing seasonal changes. The seasonal progression of the overhead sun north and south of the equator brings summers of warmer weather and long days to each hemisphere, while the winter hemisphere with low sun angles has cooler weather and longer nights. Earth's axial tilt causes the sun to be directly overhead at noon at the Tropic of Cancer (Northern Hemisphere summer) between June 19 and June 23 and at the Tropic of Capricorn (Southern Hemisphere summer) between December 19 and December 23.

Because the sun is more directly overhead for most of the year in tropical regions, it produces higher temperatures there (Figure 1.8). Tropical areas have an excess of incoming energy over that which is radiated back to space. The polar regions, however, have a deficit of energy: in winter, they have several months of almost complete darkness, losing energy to space without any coming in. Flows of air and ocean water transfer the heat from the tropics toward the polar regions. Tropical oceans are huge reservoirs of heat, which ocean currents move poleward to heat the atmosphere of temperate and high latitudes. The air and waters cool by releasing heat in higher latitudes and then return as cool flows to the tropics, where they are reheated. This system makes human habitation possible outside the areas of greatest solar radiation.

Air and Water Circulating in the Atmosphere-Ocean System

Oceans are the major sources of water that evaporates into the atmosphere, condenses into clouds, and produces precipitation as rain, hail, or snow. The areas of the world with the highest rainfalls are near the equator (Figure 1.9), where warm, humid airstreams collide, force the air to rise, and produce frequent rainstorms. High precipitation totals also occur where moisture-laden winds from the ocean meet tropical islands or the coasts of temperate continents. **Orographic lifting** ("mountain-related") occurs when moisture-laden winds are forced upward over mountains (as in Canada and southern Chile). As the air lifts, it is forced to cool and loses its ability to hold moisture. Clouds form and rain occurs. As the air descends down the other side of mountains, it warms and releases little if any of its remaining moisture. The leeward sides of mountains often are dry.

Earth's rotation affects winds and ocean currents across the surface. The effect of rotation increases away from the equator toward the poles, bending winds to form circulating weather systems, including cyclones (counterclockwise wind circulation in the Northern Hemisphere, clockwise in the Southern Hemisphere) and anticyclones (clockwise circulation in the Northern Hemisphere, counterclockwise in the Southern Hemisphere).

World Climate Regions

The transfers of energy and moisture and the circulation within the atmosphere produce distinctive seasonal changes and weather systems that distinguish climate regions (Figure 1.10). **Tropical climates** experience high temperatures (around 30°C

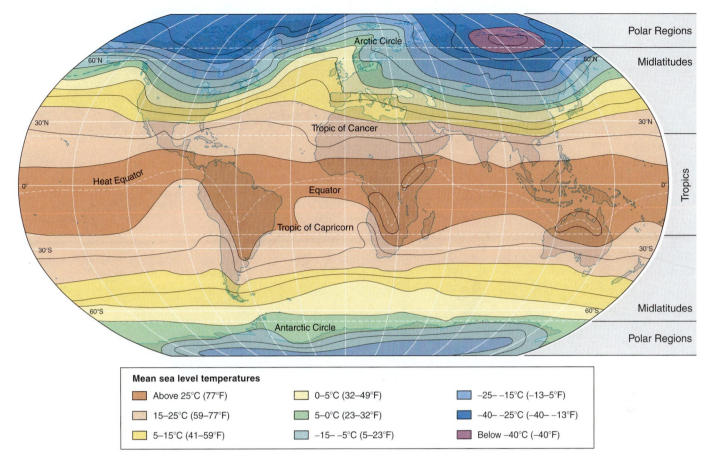

Figure 1.8 **Temperatures at ground level: January.** Isotherms (lines joining places of equal temperature) during the Northern Hemisphere winter. The heat equator connects points of highest temperature at each meridian of longitude. Compare the Northern and Southern Hemispheres for the extent of very cold temperatures and the position of the warmest band of temperatures. What effect do the oceans and continents have on the air temperatures above them?

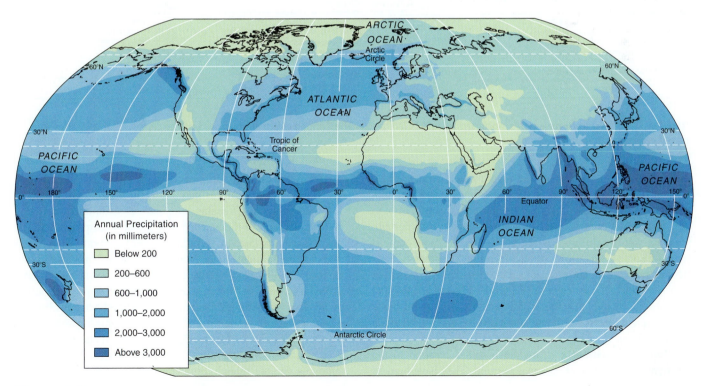

Figure 1.9 **World precipitation (rain, hail, snow).** Locate the main areas of low and high precipitation. Areas with less than 200 mm of precipitation in a year are classed as arid; those with more than 2,000 mm are nearly all in the tropical ocean areas. Warm air (which can hold more moisture than cold air) from the northern and southern tropics converges at the equator, rises, and forms clouds that precipitate rain.

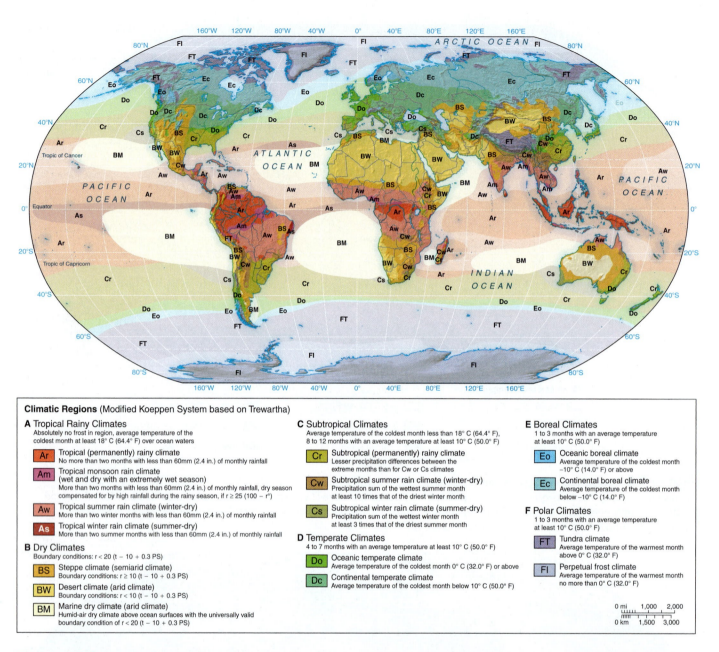

Climatic Regions (Modified Koeppen System based on Trewartha)

A Tropical Rainy Climates
Absolutely no frost in region, average temperature of the coldest month at least 18° C (64.4° F) over ocean waters

Ar Tropical (permanently) rainy climate
No more than two months with less than 60mm (2.4 in.) of monthly rainfall

Am Tropical monsoon rain climate
(wet and dry with an extremely wet season)
More than two months with less than 60mm (2.4 in.) of monthly rainfall, dry season compensated for by high rainfall during the rainy season, if r ≥ 25 (100 − r°)

Aw Tropical summer rain climate (winter-dry)
More than two winter months with less than 60mm (2.4 in.) of monthly rainfall

As Tropical winter rain climate (summer-dry)
More than two summer months with less than 60mm (2.4 in.) of monthly rainfall

B Dry Climates
Boundary conditions: r < 20 (t − 10 + 0.3 PS)

BS Steppe climate (semiarid climate)
Boundary conditions: r ≥ 10 (t − 10 + 0.3 PS)

BW Desert climate (arid climate)
Boundary conditions: r < 10 (t − 10 + 0.3 PS)

BM Marine dry climate (arid climate)
Humid-air dry climate above ocean surfaces with the universally valid boundary condition of r < 20 (t − 10 + 0.3 PS)

C Subtropical Climates
Average temperature of the coldest month less than 18° C (64.4° F), 8 to 12 months with an average temperature at least 10° C (50.0° F)

Cr Subtropical (permanently) rainy climate
Lesser precipitation differences between the extreme months than for Cw or Cs climates

Cw Subtropical summer rain climate (winter-dry)
Precipitation sum of the wettest summer month at least 10 times that of the driest winter month

Cs Subtropical winter rain climate (summer-dry)
Precipitation sum of the wettest winter month at least 3 times that of the driest summer month

D Temperate Climates
4 to 7 months with an average temperature at least 10° C (50.0° F)

Do Oceanic temperate climate
Average temperature of the coldest month 0° C (32.0° F) or above

Dc Continental temperate climate
Average temperature of the coldest month below 10° C (50.0° F)

E Boreal Climates
1 to 3 months with an average temperature at least 10° C (50.0° F)

Eo Oceanic boreal climate
Average temperature of the coldest month −10° C (14.0° F) or above

Ec Continental boreal climate
Average temperature of the coldest month below −10° C (14.0° F)

F Polar Climates
1 to 3 months with an average temperature at least 10° C (50.0° F)

FT Tundra climate
Average temperature of the warmest month above 0° C (32.0° F)

FI Perpetual frost climate
Average temperature of the warmest month no more than 0° C (32.0° F)

0 mi   1,000   2,000
0 km   1,500   3,000

Figure 1.10 **World climate types.** Included are letters referring to the Köppen classification based on the climate characteristics of natural vegetation zones.

or 80–90°F) through the year. The main tropical climates have a north-south distribution, from the equatorial climate with rain at all seasons (Ar), through wet-dry seasonal climates (Aw), to very arid (BW) climates with almost no appreciable rainfall. Seasonal wet-dry differences are greatest in the monsoon climates of South Asia (Am). Distinctive tropical weather systems include the frequent and widespread development of clusters of thunderstorms near the equator (Figure 1.11a). Intense tropical cyclones (also called hurricanes and typhoons) tend to originate in the tropical waters north and south of 10 degrees of latitude (Figure 1.11b).

Temperate climates in middle latitudes respond to the sun's changing overhead seasonal position. They have marked seasonal temperature differences between summer and winter. Summer temperatures may be warm to hot, while temperate winters include freezing temperatures as latitude increases north and south of the equator. The greatest temperature range from average winter to average summer temperatures exists in the continental interiors of North America and in northern and central Asia (Ec, Dc). Territories receiving persistent air flow from large bodies of water, such as Western Europe's prevailing winds from the Atlantic Ocean, have modified oceanic climates which reduce the range from coldest to warmest average temperatures (Eo, Do). Most precipitation falls on coastal hills and windward, west-facing sides of mountain ranges. Precipitation amounts decline inland. Temperate climates have a west-east distribution, with mild and moist west coasts (Do, Eo, Cs), drier continental interiors with significant ranges in winter-summer temperatures (BS, Dc, Ec), and east coasts with moderate

(a)

(b)

Figure 1.11 **Tropical thunderstorms and hurricanes.**
(a) Thunderstorms over Africa, seen from a space shuttle. Several cloud tops amalgamate in fibrous masses of ice that spread outward and downwind. (b) Hurricane Katrina over the Gulf of Mexico, August 28, 2005. The hurricane passed over Florida (on the right) and narrowly missed Cuba and Mexico (foreground) before turning north toward the Mississippi River delta and New Orleans. Several hundred kilometers across, this hurricane shows the cloud-free eye of descending air in the center, surrounded by swirling clouds affected by strong winds. Once over the Gulf of Mexico, it picked up energy from the warm surface waters, intensifying the winds and surge of ocean water. *Photos: (a) NASA; (b) NOAA.*

contrasts in summer-winter temperature and rainfall amounts (Cr, Do, Eo).

Frontal cyclone systems with low atmospheric pressures bring rain and high winds across vast areas in the temperate latitudes. Temperate west coasts lacking persistent year-round westerly winds have transitional climates characterized by long, warm, dry summers and mild, wet, windy winters. They are typical of lands around the Mediterranean Sea, the south-central Californian coastlands, the south-central Chilean coastlands, and the southwestern-facing coasts of South Africa and Australia. On eastern coasts, air drawn in from the ocean produces summer rains, while cold winter winds from the interior bring frost and snow.

Polar climates are extremely cold through the year, seldom rising above freezing temperatures. Winter conditions dominate the Arctic climates (FT, FI), although short summer spells may melt some of the snow and ice. Truly polar climates are frozen all year. Temperate cyclones occasionally invade polar regions, bringing high winds and precipitation.

Global Climate Change

Changes in the path of Earth's solar orbit and the planet's axis angle alter the intensity of solar radiation and are significant, naturally occurring factors in long-term climate change cycles. Evidence exists for cyclic patterns of warming and cooling phases.

An example of a long-term cooling phase was the intensive freeze that was the latest part of the Pleistocene Ice Age. It lasted for most of the last 100,000 years, ending around 10,000 years ago with a period of warming and ice melt that led to a sea surface rise of around 100 m (300 ft.) to its present level. During the freeze, huge ice sheets dominated the northern parts of North America and Europe, and sea levels were lowered around the world.

After the ice cover retreated, smaller fluctuations of climate brought the warmest conditions around 5,000 years ago. The "Little Ice Age," from approximately AD 1430 to 1850, caused upland glaciers to advance several kilometers down valleys and cultivation to retreat from higher areas in temperate countries. Climatic warming since the early 1800s resulted in a reversal of those trends, with glacier melt and higher temperatures. This phase coincided with the Industrial Revolution, which spread from Europe and North America.

Human reliance on fossil fuels (carbon-based fuels such as oil, natural gas, and coal) for both energy production, such as electricity generation, and energy consumption, such as fueling our cars or heating our homes, produces greenhouse gases (GHGs). The anthropogenic GHGs are emitted into the lower atmosphere surrounding Earth. Earth has a natural greenhouse effect, whereby GHGs in the lower atmosphere prevent some of the heat energy created when the sun warms Earth's surface from escaping into the outer atmosphere. The naturally occurring GHGs thus help to produce temperatures warm enough for habitation. The human-generated GHGs in the lower atmosphere are producing an enhanced greenhouse effect. In other words, the steadily increasing levels of carbon-based gases in the lower atmosphere make the atmosphere trap more radiation, or heat energy, and thus keep warmer temperatures present at Earth's surface. One of the many responses to increasing surface temperatures is the melting of mountain glaciers and ice in Antarctica and the Arctic.

Ecosystem Environments

Plants and animals live in communities in which they share the physical characteristics of heat, light, water availability, and nutrients. Most plants produce the food that animals require by capturing and storing the sun's energy in chemical form. An **ecosystem** is the total environment of such a community and its physical conditions. Ecosystems exist at all geographic scales, but for the purposes of this text they are discussed in relation to the largest scale, or **biome**. The five main types of biomes are forest, grassland (Figure 1.12), desert, polar, and ocean.

Hazards and Resources

Natural hazards and resources are distributed unevenly on Earth and contribute to the differences between world regions. The study of them involves both physical geography and the part of human geography that considers the cultural perceptions of what is useful or harmful to human populations.

Figure 1.12 **World biome types: savanna grassland in the Ngorongoro Conservation Area, Tanzania.** Tropical savanna grassland is common in Eastern Africa. The grasses grow in areas which receive moderate, seasonal amounts of average annual rainfall. Tropical savanna regions are characterized by grasses, bushes, and scattered trees. The grasses support large numbers of grazing animals, such as zebras, which then become a food source for lions and other carnivores at the top of the food chain. The abundance of vegetation varies in tropical savanna areas depending on the season and on surface features such as streams and creeks. *Photo: © Joseph P. Dymond.*

Natural resources are Earth's materials that human societies use to maintain their living systems and built environments. They include fertile soils, water, and minerals in the rocks. However, resources valuable to one society or technology are not always rated highly by other groups. For example, Stone Age peoples used flint and other hard rocks that flaked with sharp edges to make tools and weapons, but such rocks have few uses today. The clay mineral bauxite was ignored until it was found that refining it produced the strong, lightweight metal aluminum. Among energy resources, emphasis shifted over time from wood to wind, running water, coal, oil, natural gas, and nuclear fuels.

Renewable resources replenish naturally. The best example is solar energy, which provides a constant stream of light and heat to Earth. Water is a renewable resource that is recycled from ocean to atmosphere and back to the ground and oceans. All renewable resources are, however, ultimately finite in quantity and quality or limited by human ability to exploit them. For example, the limits of water supply affect irrigation-based development in arid countries. In fact, as the world's population increases, water is becoming increasingly scarce.

Nonrenewable resources are not replenished after they are extracted and used. They include the fossil fuels (e.g., oil, natural gas, and coal) and metallic minerals available in rocks. Though these are exhaustable, technological advances or new and increased demands drive our continued efforts to find new sources, to extract sources that were once thought to be uneconomical, and to recycle. Such technologies extend the lifetime usefulness of nonrenewable resources.

Natural hazards such as volcanic eruptions, earthquakes, hurricanes and other storms, mudslides, river and coastal floods, and coastal erosion pose difficulties and challenges for humans. They interrupt human activities but seldom deter humans from settling or developing a region if its resources are attractive. For example, people are drawn to living in California or the major cities of Japan despite earthquakes. Similarly, people continue to live and work in areas prone to hurricane damage or river flooding. In areas such as the Mississippi River valley in the United States and the lower Rhine River valley in the Netherlands, protective walls are designed to cope with all but the worst river floods. Extreme weather events may produce levels of flooding that overflow these walls, as in 1995 along the lower Rhine and in 2005 in New Orleans.

Hazards cause loss of life and destruction of property, but the costs of protection against hazards are also high. Most protection is provided in economically wealthier countries and succeeds in preventing high death tolls, though property and broader economic damage tend to be very high. In contrast, economically poorer countries have few resources available to construct protective structures against natural hazards. Consequently, they often suffer major losses of life after floods, hurricanes, or earthquakes. For example, Hurricane Katrina struck Louisiana in 2005 and an earthquake set off a tsunami that hit Japan in 2011. Because of Katrina, just over 1,800 people died and around US$100 billion in property damage resulted with increases to as much as US$250 billion in total economic damage when lost businesses and disrupted supply chains are included. Similarly, the Japanese earthquake and tsunami in 2011 killed 28,000 people and likewise caused US$100 to $250 billion of property and broader economic damage. In contrast, the Indonesian tsunami in 2004 resulted in at least 230,000 deaths and US$10 to 15 billion in economic damage. Similarly, the earthquake in Haiti in 2010 caused 200,000 to 250,000 deaths and about US$8 to 9 billion in economic damage.

Human Impacts on Natural Environments

Natural environments operate largely outside human controls, being powered by energy from the sun or Earth's interior. However, specific human activities have altered the functioning of natural systems as they change rates of erosion, remove natural vegetation cover, or emit pollutants into the atmosphere and hydrosphere. Even early humans changed natural environments, for example, by burning vegetation around forest edges to expand grasslands. This in turn increased populations of grazing animals such as bison in North America that provided meat and hides for Native Americans. As humans have developed technologies from farming and mining to industry, human activities have had increasing impacts on natural systems.

Farming, Forestry, and Fishing

The first farmers settled the lighter soils where there was not much vegetation to clear. From around 1000 BC, new iron implements made it possible to fell trees on a larger scale and extend farmland into heavy clay soil areas. Phases of woodland clearance increased soil erosion in uplands, deposition in lowlands, and changed the species composition of animals and plants. Drainage of marshes and protection of coastal lands also altered ecosystems. In recent times, the widespread use of fertilizers and the accumulation of mining and nuclear wastes as well as other toxic emissions have contaminated soil and water, which in turn have made land unusable and uninhabitable for many years.

Another environmental alteration is **desertification**, which is the destruction of the productive capacity of an area of land. It can occur naturally with climate change, but human stripping of vegetation cover or soils has been a major cause. For example, deserts, such as the Sahara, are expanding in area (Figure 1.13). The southern margin of the Sahara has experienced desertification since the 1970s after commercial agriculture extended into less humid areas and then years of drought followed (see Chapter 8). It resulted in the deaths of cattle and vegetation and forced many communities to move away from their traditional lands.

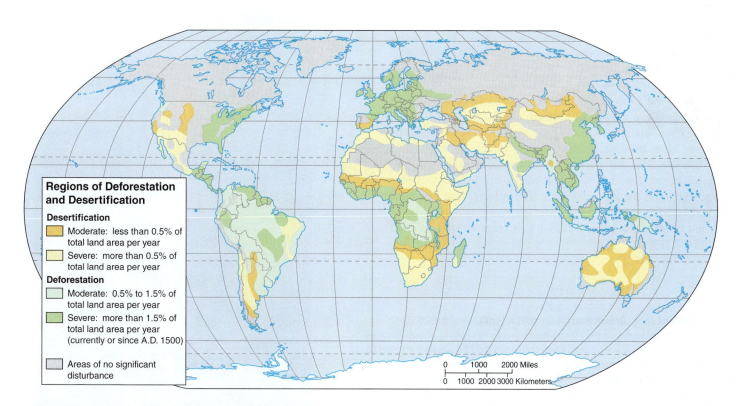

Figure 1.13 **Desertification and deforestation.** Climate change and human activities lower the productive capacity of the land in many areas.

As farming has expanded and the demand has grown for wood products, the world's forests have been cut (see Figure 1.13). Today, about half of the Earth's forests are gone, with only about 22 percent of its original, old growth forests remaining. Until about 100 years ago, most forest removal occurred in Europe, North America, and North Africa and Southwest Africa. Deforestation has reversed in the last 100 years as tree planting has led to the increase of second growth forests. About half of the world's forests are now in tropical areas such as Brazil, central Africa, and Southeast Asia. In the 1990s, approximately 16 million hectares of forest were cut every year, but this declined to 13 million hectares per year in the 2000s. Aggressive tree planting programs in countries such as the United States, China, India, and Vietnam reduced the net loss of forests from 8.3 million hectares per year in the 1990s to 5.2 million hectares per year in the 2000s.

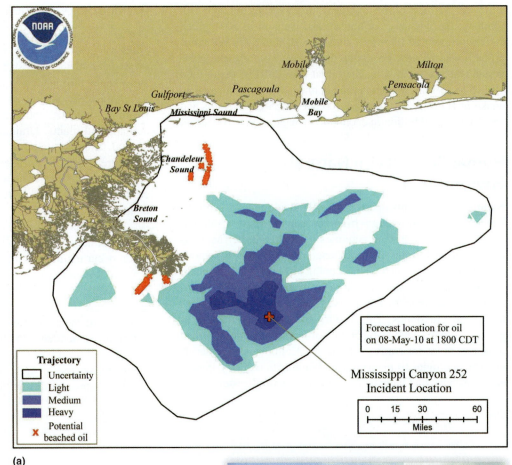

(a)

The demand for ocean fish has been met in the last several decades with the introduction of new technologies. However, increased fishing is now removing more than 70 percent of the world's fish stocks faster than they can be replenished. For example, a number of northern cod fisheries collapsed in the 1990s, leaving less than 1 percent of their previous populations. Many haddock, salmon, Pacific halibut, Alaska pollock, Atlantic redfish, Atlantic mackerel, shrimp, and king crab fisheries are overfished. Bluefin tuna are so depleted that they may go extinct in the next few years. The Atlantic bluefin tuna is particularly vulnerable because its spawning grounds are in the Gulf of Mexico where the British Petroleum oil spill occurred in 2010 (Figure 1.14).

Figure 1.14 **British Petroleum (BP) oil spill.** On April 20, 2010, an explosion occurred on the Deepwater Horizon drilling rig about 40 miles southeast of Louisiana's coast. The explosion killed 11 workers and led to an oil spill which continued until the oil well finally was capped on September 20, 2010. It was the largest oil spill in U.S. history. (a) The extent of the oil spill on May 8, 2010. (b) Oily water in the Gulf of Mexico. The Gulf of Mexico is a critical spawning area for bluefin tuna and important to yellowfin as well. With an abnormally large number of dead baby dolphins washing ashore starting in February 2011, the impact on marine life was only in the early stages. *Map: NOAA. Photo: U.S. Coast Guard, photo by Jaclyn Young.*

(b)

Industry and the Global Atmosphere

Though humans have been modifying Earth's natural systems for centuries, the industrial revolutions that began in the 1750s marked the beginning of new technological developments that have allowed humans to modify natural environments on unprecedented scales (Figure 1.15). With the building of modern factories, wastes have been poured into the atmosphere and rivers. Consequently, environmental stresses have built up rapidly over large areas of the Earth's world regions.

The carbon gases emitted by the burning of fossil fuels such as coal, oil, and natural gas may very well exceed the amounts that natural systems can absorb. Certainly, carbon dioxide and other gases in the atmosphere have been rising for the last 200 years. These gases trap solar energy in the lower atmosphere, enhance the greenhouse effect, and likely contribute to **global warming**. Today emissions from the burning of fossil fuels are increasing rapidly in newly industrializing countries such as India and China (Figure 1.16).

Rising temperatures melt ice sheet margins, and the meltwater pours into the oceans. The level of the oceans is forecast to rise a meter or so within the next 50 years. The low-lying coasts of coral islands, wetlands, port areas, and hundreds of millions of people will be at risk. This is by far the most important global environmental issue today and will increasingly affect people's lives around the world.

The ozone layer, which is comprised of large concentrations of ozone gas (O_3), forms a protective shield in the atmosphere and blocks incoming ultraviolet rays from the sun. In the 1980s, it became clear that chlorine gases, including the human-made chlorofluorocarbons used in refrigeration systems, gradually permeate upward and destroy ozone by chemical reactions, and they have created an "ozone hole" over Antarctica. The reactions are most intense during the cold polar winters over Antarctica, which makes the hole largest at this time of year. The loss of the protective ozone shield increases people's risk of skin cancer. Governments agreed to end the use of ozone-depleting gases, and the policy is reducing ozone destruction. However, the potential dangers will last for several decades. Indeed, ozone depletion is now a concern in the Arctic, where it was once thought that it did not become cold enough to allow the chemical reactions that destroy ozone. However, with increased greenhouse gases being trapped in and then warming Earth's lower atmosphere, the upper atmosphere is no longer heating to its previous levels. The atmosphere above the Arctic is now notably colder. During the winter of 2011–2012, the first ozone hole above the Arctic was discovered.

While global warming and ozone depletion may have worldwide effects, acid deposition affects areas up to several hundred kilometers downwind of major urban-industrial areas. It is mainly caused by sulfur and nitrogen gases from power stations and vehicle exhaust, which react with sunlight in the atmosphere and return to the ground as acids. Soils and lakes that are close to the pollution sources and are already somewhat low in plant nutrients suffer first. The phenomenon is a growing menace downwind of new industrial areas in developing countries.

Figure 1.15 **Human modification of the earth.** When it was built in the 1930s, Hoover Dam was the largest dam in the world. It created Lake Mead, which extends for 112 miles (180 km) behind it, and it has changed the ecosystems of the Colorado River. By providing drinking water and electricity, it also made it possible for millions of people to inhabit the dry Southwest, which includes cities such as Las Vegas, Phoenix, Los Angeles, and San Diego. *Photo: U.S. Bureau of Reclamation, photo by Andy Pernick.*

Rio and Kyoto

International conferences have been called to address and limit the amount of human-generated emissions that affect global and regional natural systems. Notable conferences took place in Rio de Janeiro, Brazil, in 1992 and in Kyoto, Japan, in 1997. The United Nations Conference on Environment and Development in Rio de Janeiro, Brazil, in 1992, is now commonly known as the "Rio Earth Summit." The Rio conference produced a number of conventions, namely the UN Framework Convention on Climate Change (UNFCCC), the Convention on Biological Diversity (CBD), and the United Nations Convention to Combat Desertification (UNCCD). The UNFCCC established an overall policy framework for addressing climate change and laid the foundation for combating global warming. It went into effect in March 1994 after 50 countries ratified it.

The **Kyoto Protocol** to the UNFCCC was adopted on December 11, 1997, and addressed specific targets for reducing the six primary greenhouse gases: carbon dioxide (CO_2), methane (CH_4), nitrous oxide (N_2O), hydrofluorocarbons (HFCs), perfluorocarbons (PFCs), and sulphur hexafluoride (SF_6). Industrialized countries, primarily in Europe but also including the United States, Canada, Japan, Australia, and New Zealand, were known as Annex I countries. According to the Kyoto Protocol, these countries needed to reduce their greenhouse gas emissions so that their average yearly emissions for the years 2008 to 2012 were 5 percent less than their emissions in 1990. The protocol finally was ratified by enough countries by the end of 2004 so that it went into effect on February 16, 2005. The United States did not ratify the protocol though it was by far the world's largest emitter of carbon dioxide (CO_2) at the time (36.1 percent of total world emissions). In the meantime, many developing countries have increased their emissions with continued industrialization. As China surpassed the United States to become the biggest emitter of greenhouse gases sometime in 2007 or 2008 (see Figure 1.16), it has become clear that developing countries need to assume new responsibilities in collective global actions.

Other international meetings have been convened to develop new initiatives to follow after the Kyoto Protocol's 2012 end date. Examples include conferences held in Bali, Indonesia, in 2007, and in Copenhagen, Denmark, in 2009. The most recent conference was in Rio de Janeiro, Brazil, in 2012 and was called Rio 2012 or Rio+20. The conference focused on seven critical issues that ultimately affect the natural environment: jobs, energy, cities, food, water, oceans, and disasters. Despite much discussion, these conferences made little progress in building on the Kyoto Protocol's ambitious goals.

Multiple methods for reducing greenhouse gases exist, and most of them will be employed. Examples include biofuels such as ethanol based on corn, sugarcane, or switchgrass. Some automobiles run on compressed natural gas, while further research is being conducted on hydrogen fuel cells and better batteries for electric cars. More generally, investments in solar and wind energy have increased dramatically. Some countries are turning to nuclear power, though it creates other environmental problems. Some are imposing carbon taxes on their industries. Others use cap-and-trade systems that place limits on emissions but allow companies to sell their unused allotments to other companies. This system supposedly encourages companies to see a profit in emitting less.

Figure 1.16 **Air pollution: Beijing.** As China's economy rapidly expands, so do its industrial emissions. *Photo: © AFP/Getty Images.*

1.4 Regions and Human Geography

Human geography (see p. 4) is concerned with the spatial variations of humans and their characteristics. Specific examples include cultural, political, economic, population, and urban geographies. **Cultural geography** is concerned with such phenomena as material traits, social structures, and belief systems. Specific examples include language use, religious beliefs and affiliations, dress, food, recreation, building practices and styles, and social organization. Culture explains how peoples create and re-create the distinctiveness of their regions, which in turn shapes the lives of regions' inhabitants and their descendants. People often behave according to their culture's norms and find comfort and security through identification with a group that shares common cultural characteristics.

The most important thing to understand about culture is that it is learned behavior. A combination of traditions and behavior practices is transmitted from generation to generation along with adaptations, variations, and new ideas or innovations more recently acquired or accepted by various groups of people. Cultural identification is not mutually exclusive. One may be part of many different culture groups at the same time. Biology and genetics do not determine any element of culture, but some cultures focus on gender and racial characteristics when deciding who does and does not belong to their group.

Language

A **language**, which includes speech, writing, and signing, is a means of communication among people. It is an important factor in geographic diversity. Languages grow out of historic experiences and traditions and often provide a shared identity for a cultural group. Some groups make a point of using their language to enhance their identity and separate themselves from others. For example, the French-speaking Québecois people in Québec, Canada, emphasized their language when they wanted to achieve greater political recognition.

Some six thousand languages are spoken around the world, many by small groups of people in isolated environments such as South America's Amazon River basin. Related languages can be grouped in families (Figure 1.17). For example, the Indo-European family includes most of the languages of South Asia (e.g., Hindi) and many of the languages of Europe and Russia and its neighboring countries. Some of these languages, especially Spanish, Portuguese, English, French, and Dutch, spread with European colonization after around AD 1450. They dominated the Americas and the South Pacific and become important in many parts of Africa and Asia.

Many of these languages became **world languages** (lingua francas), which are spoken by people who otherwise do not share a mother tongue. They are among the world's 12 most widely spoken languages, each having over 100 million speakers. Six of these (English, French, Spanish, Russian, Arabic, and Mandarin Chinese) are official languages of the United Nations, mostly (with the exception of Arabic) chosen from the victorious allies at the end of World War II. Other significant world languages are Hindi/Urdu, Portuguese, German, Persian, and Swahili. The dominance of these languages has been brought about by globalization. As borders have been opened, some forcibly, economic competition has allowed some languages to flourish and expand at the expense of others.

Religion

Religions also generate geographic variations in culture through strong group loyalties and exclusive attitudes. Each **religion** is an organized set of practices that professes to explain our existence and purpose on Earth. Many also have value systems and faith in and worship of a divine being or beings. Religious belief often influences social and legal practices, shapes views of the natural environment, and reflects itself in the cultural landscape, such as in building designs (Figure 1.18). Religions play a significant role in transferring cultural values and practices from one generation to the next.

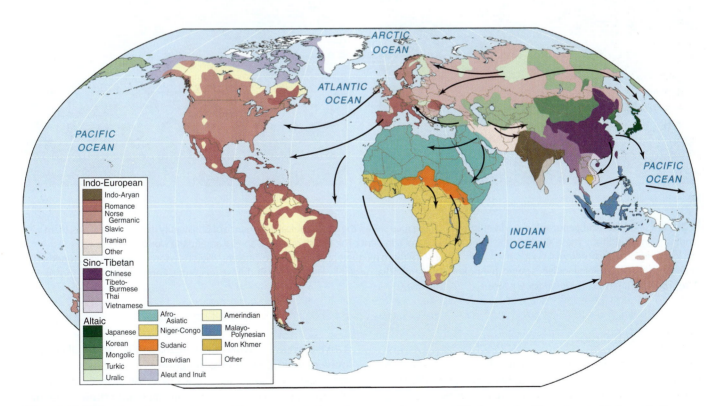

Figure 1.17 **Cultural geography: world language families.** The arrows show some diffusion routes. The map records the majority languages; there are often many different languages, including those of minority groups, in each area.

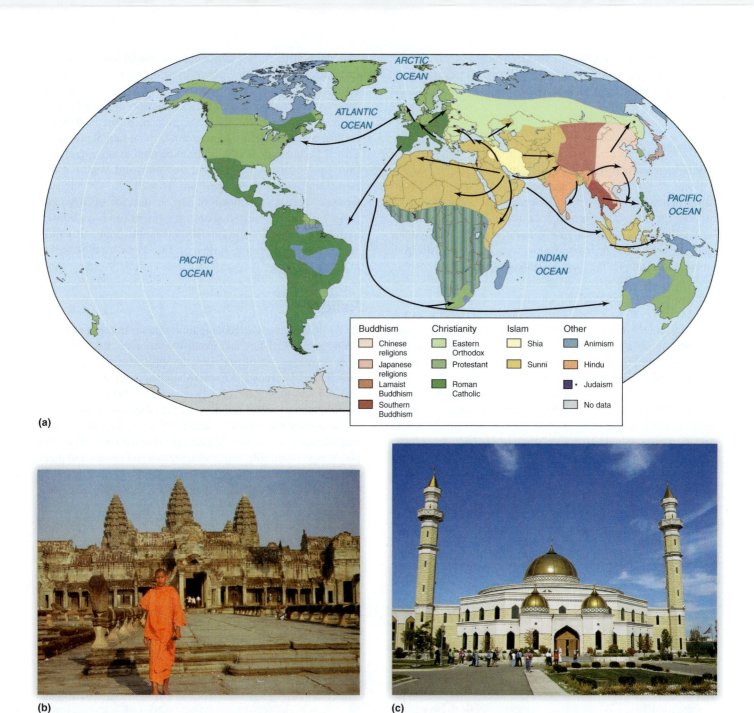

(a)

Buddhism
- Chinese religions
- Japanese religions
- Lamaist Buddhism
- Southern Buddhism

Christianity
- Eastern Orthodox
- Protestant
- Roman Catholic

Islam
- Shia
- Sunni

Other
- Animism
- Hindu
- Judaism
- No data

(b)

(c)

Figure 1.18 Cultural geography: world religions. (a) The map shows the geographical dominance of Christianity in its various forms, Islam, Hinduism, and Buddhism. These also have the highest numbers of adherents. The Jewish religion is spread around the world, mainly in cities. Older and local religious practices remain in localized areas. Many places have mixtures of allegiances to different religions—sometimes working together, other times in conflict. Religious landscapes: (b) Angkor Wat, Cambodia, originally Hindu, now Buddhist; (c) Islamic Center of America, mosque in Dearborn, Michigan. *Photos: (b) © Ronald Wixman; (c) © George White.*

Major World Religions

The religions claiming the largest numbers of adherents are Christianity (2 billion), Islam (1.4 billion), Hinduism (900 million), and Buddhism (500 million). Judaism has a smaller number of adherents (14 to 15 million) but an important place among world religions. Christianity, Islam, and Buddhism are religions that can be joined by anyone in any country, and they actively seek to expand their memberships: this makes them **universalizing**, or global, **religions**. Hinduism and Judaism are considered to be mainly a matter of birth by adherents, closely tied to family and region: they are **ethnic religions**. People practicing ethnic religions do not actively seek converts because they see their religions as only appropriate for their own ethnic groups.

These major religions also can be categorized into western and eastern groupings. The western religions, namely Judaism, Christianity, and Islam, all originated in Southwestern Asia (see Chapter 7). Christianity and Islam were created from the already existing structure of Judaism. All three believe in the same God and share many religious texts and figures. Because they all consider Abraham to be the earliest patriarch of their faiths, they often are called the "Abrahamic religions." They do not share texts or religious figures with Hinduism and Buddhism, which are considered eastern religions. These religions both originated in South Asia (see Chapter 6) and are connected because Buddhism grew out of Hinduism. As Buddhism spread into East Asia, it encountered and interacted with other religions such as Daoism in China and Shintoism in Japan.

Religion and Society

Religious adherence often motivates believers to advocate certain social, economic, and political policies because religions have very specific views on such issues as abortion, sexuality, family life, gender roles, education, the natural environment, crime, punishment, and the propriety of certain activities and professions. For example, many Roman Catholic and Muslim leaders officially oppose birth control, although individual families increasingly make their own choices. Some religious views argue that humans are either part of or stewards of the natural environment and, therefore, should conserve it. Their views conflict with those that argue that Earth was created for human use.

Most Jews, Christians, Muslims, Hindus, Buddhists, and peoples of other faiths live together peacefully. However, religious differences between peoples may result in conflicts, especially when one group's actions violate or offend another group's beliefs. Those in or seeking political power sometimes deliberately stereotype and emphasize differences to fan ignorance and exacerbate conflict for their own political gain. These actions encourage violence and religious extremism. Since the end of the Cold War in 1991, the growth of Christian, Hindu, Jewish, Islamic, and Buddhist extremism in some countries is one sign that religious ideologies have replaced the political ideologies of the Cold War period.

Cultural Status: Race, Class, and Gender

The concept of meritocracy, which grants status based on achievement, only is practiced to certain limits around the world. Inherited monetary wealth, political allegiance, media prominence, or sporting performance can be more important in defining a person's status. In many cultures, however, one's position in society is more profoundly determined by one's ethnicity, race, class, and gender. Because these concepts have such an influence on everyone's lives and geographic diversity, it is important to discuss what they mean.

Ethnicity and Race

An **ethnic group** is a cultural group whose members are defined by such characteristics as common origin (real or imaginary), religion, language, customs, or physical features. The term is often used synonymously for tribe, clan, or segregated minority though they are not the same.

Individuals become so accustomed to behaviors and standards of their own ethnic groups that they judge others as "wrong" or "backwards" when they do not behave and act the same. For example, in some cultures, people eat hamburgers, French fries, and pizza with a knife and fork rather than with their hands. Others drive on the left side of the road instead of the right side. These reflect preferences that have become standards for groups and are not caused by a lack of morality or development. Nevertheless, the term **ethnocentrism** refers to those who judge other ethnic groups harshly because they do things differently.

The concept of **race** is assumed by many to be based on essential biological differences, but characteristics such as skin color, eye shape, or hair type vary as much within so-called "races" as between them. Although race and racism are often at the center of human conflicts, the most basic human biological features, DNA and blood type, demonstrate little variation across the human species, which is a single reproducing group. However, racial characteristics frequently are chosen to divide people into groups of "us" and "them."

Examples of cultural racism include South Africa, where the minority white population operated a supremacist apartheid ("apartness") policy until 1994, separating whites, blacks, and other "colored" peoples into segregated neighborhoods. The country is now reversing this policy (see Chapter 8). In the United States, many African-Americans still struggle against discrimination from people of "whiter" European origin. Even in Brazil, where Native American, European, African, and Asian peoples mix freely, the upper socioeconomic classes are of European origin and lower socioeconomic classes consist of people of African or Native American origin.

In many countries, ethnicity decides the membership of opposing political groups and may lead to armed conflict. The promotion of ethnic groups as "superior nations" and the demotion of others as "inferior" had much to do with the Holocaust associated with World War II and other ethnic cleansings in Europe and Asia. Before that, colonial powers often highlighted and exploited ethnic and tribal differences by playing one group against another. After the former colonial countries in Africa, Asia, and Latin America achieved independence, many new rulers established one-party rule and often emphasized the perceived differences among ethnic groups by rewarding those in their own ethnic group and penalizing others. For example, Saddam Hussein in Iraq favored the Sunni Muslims over the more numerous Shia Muslims. Such favoritism created a hostility and distrust that continues today.

Class Distinctions

Class hierarchies within societies are created by an emphasis on such criteria as ethnicity, race, religion, material wealth, education, perceived birthright, and other social characteristics. In the United Kingdom, the royal family is on the top of a hierarchy of hereditary dukes, earls, and knights with commoners at the bottom. In India, the caste system and, increasingly,

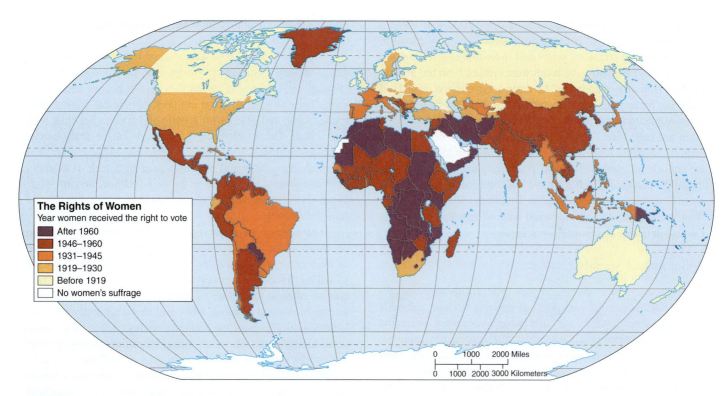

Figure 1.19 **Women: the right to vote.** This map illustrates that women only recently have gained the right to vote in many countries. Are there any geographical patterns to when this right was obtained?

The map legend reads:

The Rights of Women
Year women received the right to vote
- After 1960
- 1946–1960
- 1931–1945
- 1919–1930
- Before 1919
- No women's suffrage

education define position in society. In Communist countries, the expectation of a classless society is contradicted by the status and privileges given to members of the Communist Party. Most Americans identify general class differences by material possessions and appearances largely based on income: a lower-middle class of factory and shop workers, an upper-middle class of managers and professionals, and a group of exceptionally wealthy financiers, property owners, and sports and media stars.

Gender Inequalities

Gender—the cultural implications of one's sex—also highlights differences and inequalities of opportunity within and among societies. Males dominate most societies and have a history of denying full rights to women. Although major changes occurred in the 1900s, particularly in extending voting franchises to women (Figure 1.19), some countries still deny women the human rights defined by the United Nations. Many cultures prefer and favor males. Consequently, abuse, aborted female fetuses, and denial of adequate food, shelter, and access to health care has taken its toll on women in what is called "gendercide." As many as 100 million fewer women are alive today simply because of the poorer treatment they receive for being women.

In Afghanistan, girls and women risk having their noses and ears cut off or acid thrown in their faces for trying to go to school. In Africa, female genital mutilation, or "cutting," continues in some societies. In Iran and northern Nigeria, women—but not men—can be stoned to death for adultery. Illiteracy

among women is still much higher than among men in most of Africa and Asia. Few European countries allowed women to vote in elections until the 1900s, with Switzerland delaying it until 1971 and one of its cantons until 1990.

Women, even in the world's wealthier countries, commonly receive lower wages than men for the same jobs and constitute a minority of doctors, engineers, corporate executives, and elected politicians. Some jobs such as nursing, secretarial work, elementary school teaching, and sales clerking have been widely regarded as "women's work" and often have lower status. Further, women commonly take a major role in running the household and child care, which often affects their career opportunities.

Some aspects of women's inequality are gradually being tackled. The high rates of female illiteracy in poorer countries affect population growth rates and resource consumption. Better education gives women confidence, enables them to take jobs, improves their self-esteem, and increases their roles in decision making. Women, especially educated ones, commonly want fewer children than men. However, male dominance exercised through cultural traditions often still prevents women from controlling how many children they have.

Women hold as many as half of the jobs in countries like the United States. In the United States, they also possess roughly half of the master's degrees but are only 2 percent of CEOs of Fortune 500 companies and comprise less than 13 percent of board members in American companies. Overall, women's salaries are on average 83 percent of those of their male counterparts. Though sex discrimination is not allowed in many countries, most workplaces are male-structured, providing few childcare facilities, flex-time or part-time work schedules, or extended leaves to take care of family matters. Where these shortcomings exist, women still find it difficult to advance.

To measure gender inequalities in and among countries, the United Nations developed the Gender Inequality Index (GII) for its Human Development Report 2010. The GII is "a measure that captures the loss in achievements due to gender disparities in the dimensions of reproductive health, empowerment and labour force participation. Values range from 0 (perfect equality) to 1 (total inequality)." Actual values for countries range from 0.049 to 0.769. In the most recent 2011 report, Sweden ranked as the most gender-equal country and was followed by the Netherlands, Denmark, Switzerland, and Finland. The lowest-ranked country was Yemen. It was closely followed by Chad, Niger, Mali, and Democratic Republic of Congo. The rankings indicate that gender inequality is linked to economic prosperity and political stability.

According to the Human Development Report 2011, gender inequality also impacts the natural environment. For example, in countries where inequality is high, women have less control over their reproduction and less access to family planning. It means that women have more children than they personally desire. "Meeting unmet need for family planning by 2050 would lower the world's carbon emissions an estimated 17 percent below what they are today." Also, the Gender Inequality Index indicates a causal link between gender inequality and deforestation in more than 100 countries between 1990 and 2000. Less deforestation occurs in democracies, but democracy requires more gender equality.

Though gender inequalities exist, they were much greater 50 years ago. The gender gap has closed dramatically over the last few decades for two reasons: the contraceptive pill and higher education. The contraceptive pill has allowed women to choose when to have children. Many have chosen to put off having children to pursue a higher education, which in turn has increased their incomes. At the same time, women have changed their career paths. For example in 1966, 40 percent of American women who received bachelor degrees majored in education and 2 percent majored in business. By 2010, 12 percent majored in education while 50 percent majored in business. There are also 3 million more women than men pursuing university degrees.

Women are also better able to compete in the modern global economy as manufacturing jobs have declined and knowledge-based service sector jobs have increased. More men lost jobs in the world recession from 2007 to 2009. In the European Union, women obtained 6 million of the 8 million new jobs created from 2000 to 2010. In the United States, the Bureau of Labor Statistics estimates that women hold two-thirds of the jobs in 10 of the 15 job categories that are growing the fastest. Increasing female employment compared to that of males will tremendously boost the GDPs of countries. According to a study by Goldman Sachs cited in the *Economist*, growing female employment will increase Italy's GDP by 21 percent, Spain's by 16 percent, Japan's by 9 percent, and those of the United States, France, the United Kingdom, and Germany each by 8 percent.

1.5 Regions and Population

From the time that humans first walked the Earth, it took until 1804 for the world's population to reach one billion people. Just over 200 years later, in 2011, the world's population passed the seven billion mark (Figure 1.20). This tremendous change in population has altered resource use, environmental impacts, economic systems, political relations, and some of the differences among world regions. To appreciate these changes and their impacts, it is important to understand population dynamics such as growth, settlement patterns, and migration.

Population Distribution and Dynamics

Where People Are

With countries and other political units varying in land area, meaningful comparisons between places are commonly measured by looking at **population density**—the numbers of people per given area (e.g., square kilometer or square mile). Population densities vary greatly around the world (Figure 1.21) and can be used to map out **population distribution**. The

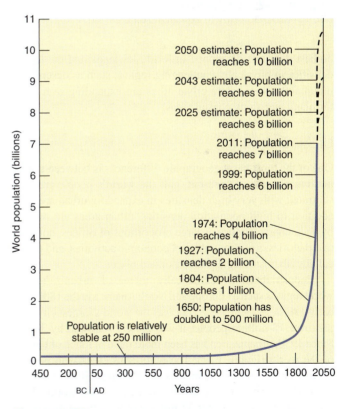

Figure 1.20 **World population growth.** For most of the human occupation of Earth, population growth was slow compared to the last 300 years. The population took 1,300 years to double from 250 to 500 million, then doubled again in 254 years to reach 1 billion. The next doubling took 123 years to reach 2 billion, followed by a doubling time of just 47 years to reach 4 billion. It is estimated the world's population will reach 8 billion in 2025, which would make the doubling time 51 years. *Source: United Nations.*

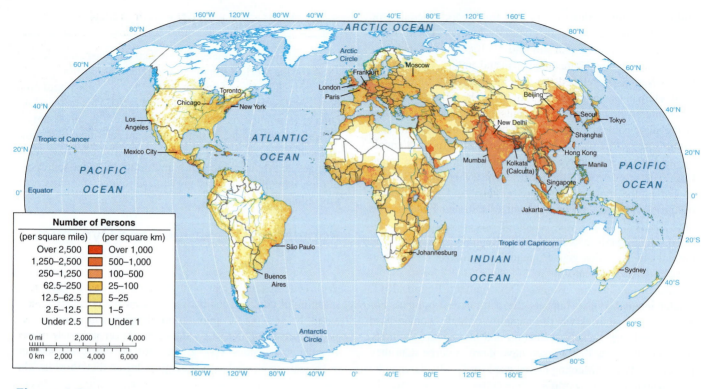

Figure 1.21 **World population distribution.** Which world regions have the highest and lowest densities of population? As you read through this chapter, try to explain the differences.

highest densities developed in fertile lowlands and the lowest densities in physically challenging regions such as deserts and mountains. Now, many urban areas are expanding over good farmland and along coastal areas.

Urban Growth

One of the most basic geographic differences is between **urban** and **rural areas**. At present, half the world's people are concentrated with very high densities in expanding urban areas. In addition to high population densities, urban areas are marked by such characteristics as large proportions of offices, housing, and factories, a concentration of economic activities, and political boundaries of legally incorporated places.

Industrialization dramatically increased urbanization in world regions such as Europe and North America in the 1800s and early 1900s. In 1950, New York was the world's largest city with just over 12 million people; it was closely followed by Tokyo. Since 1950, urbanization has been more rapid in much of the rest of the world (Figure 1.22). The number of cities with populations over 10 million inhabitants, called **megacities**, increased from two in 1950 to 23 in 2011. Tokyo is now the world's largest urban area with almost 37 million people. It is followed by Delhi (India), Mexico City, New York–Newark (20 million), São Paulo (Brazil), Shanghai (China), Mumbai (Bombay, India), Beijing (China), Dhaka (Bangladesh), and Kolkata (Calcutta, India). Listings of major city populations and their projected growths are in each regional chapter. Where a country has a single very large city, often several times the population of its other cities, it is known as a **primate city**.

Population Ups and Downs

Demography is the study of population numbers, densities, growth and decline, migration, and their relationships. The following terms illustrate how they are measured.

- The **crude birth rate (CBR)** refers to the number of live births per 1,000 inhabitants per year in a given population. It is related closely to the **total fertility rate (TFR)**—the average number of births per woman in her lifetime. Total fertility rates of 6 to 7 are typical of many developing countries, while industrialized countries have rates of 2 or below.

- The **crude death rate (CDR)** is the number of deaths per 1,000 inhabitants per year in a given population. It is often broken down into age groups. **Infant mortality** (deaths per 1,000 live births in the first year of life) and child mortality (deaths per 1,000 live births in the first five years of life) are examples. Infant mortality rates below 10 (i.e., 10 infant deaths per 1,000 live births) in industrialized countries compare with those above 100 in many developing countries. The crude birth and death rates are labeled "crude" because they do not account for the age and gender structure of a population.

- The crude birth rate minus the crude death rate equals the rate of natural population increase or decrease.

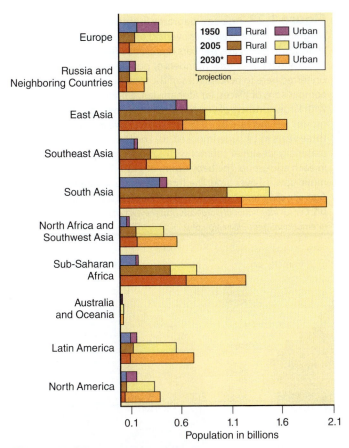

Figure 1.22 Growth in population and urbanization. Identify those world regions that have grown the most in total population and urbanized the most since 1950.

The Migration Factor

Migration is the long-term movement of people into or out of places. Immigration ("into") and emigration ("out of") can dramatically change the population of a country despite natural increase. For example, immigration is a major cause of population growth in the United States, Canada, Australia, and some European countries, as natural increase is slow. In general, major migration flows are tied to globalization.

By 2010, 215 million people lived outside their country of birth, making up 3.1 percent of the world population. However, migrants are not evenly distributed. More than 20 percent of the populations of Australia and Switzerland are foreign born. Canada, Germany, the United States, Sweden, Ireland, the United Kingdom, and France have foreign-born populations of more than 10 percent.

With travel costs high and entry difficult, about 78 million migrants have not journeyed to wealthy countries but instead only were able to trek to neighboring countries. This has been particularly true for refugees fleeing war-torn countries such as Iraq, Somalia, and Afghanistan. About 70 percent of migrants from Sub-Saharan Africa stay in their region. Some migrants seek more moderate-income countries like Argentina, India, Russia, and South Africa.

Immigrants play significant roles in their host and home countries and the global economy. For example, migrants from South and Southeast Asia have moved to the Persian Gulf oil countries and Europe where their labor is needed. Also, educated men and women move to major economic urban centers for better-paying jobs. On the one hand, these migrants represent a "brain drain" from their home countries. On the other hand, along with many of the world migrants, they send their paychecks home as remittances. These money transfers provide a high proportion of national income in some poorer countries. For example, remittances from abroad make up over half the national income of Haiti, and 40 percent of Zimbabwean households receive money from abroad.

Overall Population Change

A change of 1 or 2 percent in a country's annual population growth rate will have a dramatic effect over time. This effect is illustrated by **population doubling time**, the time in years taken to double the number of people in a place. A population increase of 1 percent growth leads to a population doubling time of 70 years. An increase of 2 percent means a doubling in 35 years; 3 percent growth means a doubling in 23 years. Developed countries today commonly have below 0.5 percent population increase, while developing countries have rates of 2 to 3 percent. Countries with high emigration, low birth rates, or high death rates may experience population losses.

The composition and recent history of a country's population characteristics are often summarized in an age-sex diagram, also termed a "population pyramid" (Figure 1.23). Migrations into the country or baby booms show up as expansions in particular age and gender groups; deaths in major wars may be reflected in a narrowing of specific cohorts. Long life spans result in larger groups of older people, which are illustrated at the tops of the pyramids.

Is the World's Population Growth Sustainable?

Population growth puts greater demands on Earth's resources. In 1798 the English economist Thomas Malthus argued that without moral restraint in choosing the number of children that we have, world population growth would exceed the growth of food production, creating overpopulation. This would lead to widespread famine, disease, and war. Overpopulation and its consequences have occurred locally, but new technologies have prevented it from happening globally. Technology has expanded resource production through the development of genetically modified foods, fertilizers, additives, plastics, synthetics, and alternative fuels. However, the mathematics of population increase make it questionable whether technology alone will provide the solutions. Some argue for conservation and the slowing, stopping, or reversing of population growth. Certainly, our lifestyle choices concerning transportation, food, and clothing, among many others, will impact future patterns of human and physical geography.

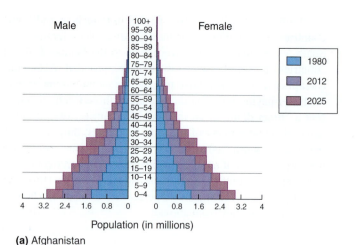

(a) Afghanistan

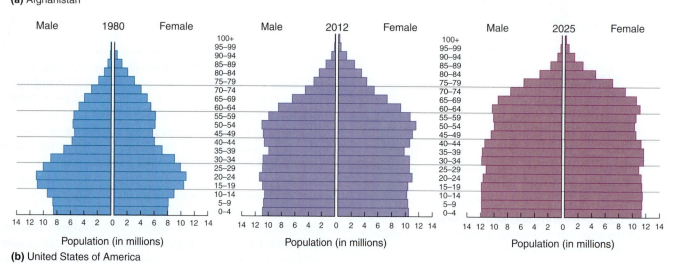

(b) United States of America

Figure 1.23 **Age-sex diagrams (population pyramids).**
Diagrams for three years are overlaid or set side-by-side to show changes. In each case, the bars represent a five-year age group (male and female). Total numbers of people in each group are preferred here to percentages of the total population to allow comparisons of places. (a) Afghanistan, a materially poor country with large numbers of young people and fewer old; middle-age groups will increase by 2025. The progressive increases allow the three years to be superimposed. (b) The United States, a typical materially wealthier country with a more even spread of numbers in each age group. The baby boom of 1950–1965 is seen to move upward over time. The three years are separated for clarity. Age-sex diagrams can be found in Chapters 2 through 10. *Source: U.S. Census International Database: http://www.census.gov/population/international/data/idb/region.php.*

1.6 Regions and Politics

Political geography is the study of how governments and political movements (e.g., nongovernmental organizations, labor unions, political parties) influence the human and physical geography of the world and its regions. Cultural phenomena (e.g., language and religion) and physical features (e.g., the distribution of fresh water) influence governments, political movements, and their relationships.

The world is primarily divided up into **countries** (also called states), which are bordered territories with governments that have political control, or sovereignty, over the internal and external affairs of the territory's contents and inhabitants. The number of self-governing countries increased from 62 in 1914 to 74 in 1946 and 195 today. The world's newest country, South Sudan, came into being in July 2011. Each country ideally is recognized by other countries, but such is not always the case. For example, Taiwan is not recognized by the People's Republic of China, which claims Taiwan as its territory. Taiwan once had the recognition of more than 100 countries, but the list has declined to just 23. One of the newer countries, Kosovo, had only received the recognition of 91 countries by mid-2012.

Country governments promote and protect their peoples in world affairs and may join other country governments in mutual trading or defense agreements. Countries tax their citizens to provide public services, including military defense, and encourage economic and social welfare. Countries often have systems of regional, state, or local government that carry out some of the governmental responsibilities at different geographic levels. In world regional geography, countries provide the main subunits of study within the world regions.

Nations and Nationalism

Countries often are called nations, but these two terms do not have the same meaning. A **nation** is a group of people who share a common identity, a sense of unity, and a desire for self-governance. Each nation defines its common identity in its own way and is thus called an "imagined community." Typical imaginations of commonality are shared language, religion, history, ideology, ethnicity, and race. An ethnic group also is defined by a set of shared characteristics, but many ethnic groups have no sense of unity and do not wish to govern themselves. When a group sharing one or more of the aforementioned characteristics becomes politicized, it becomes a nation. Thus, some nations are comprised of a single ethnic group, such as the English, Serbs, and Japanese. However, other nations are multiethnic, such as Americans. Many Americans emphasize their different ethnic backgrounds by combining their ethnic and nation identities with a hyphen (e.g., Irish-American, Japanese-American, and African-American).

Nationalism is pride in one's national identity and the belief that one's national interests are more important than all other interests. To protect and advance each nation's interests, nationalists believe that each nation must be able to govern itself, which is best ensured with a separate state (or country). Thus, nation and state are linked together into the concept of the nation-state. However, not all nations have their own states, though they may wish to. For example, Basques, Kurds, and Palestinians are all stateless nations and live as minorities in other groups' nation-states. Consequently, the nationalists in many such groups try to create their own nation-states, but such attempts often result in violence and oppression. Not all violence is begun by the minorities; often it begins with the dominant national groups who see the minorities as threats.

Indigenous peoples are the first inhabitants of any given area. Today indigenous groups often remain as minority populations within individual countries, subject to some denial of human rights and development opportunities. The numbers of indigenous peoples often dramatically declined during the colonial period and world wars, and some groups were exterminated. Other groups evolved their own cultural and political aspirations as minority "nations." For example, Native Americans initially fought back against the taking of their lands by European colonists. Since then, they have gained rights and varying degrees of autonomy in the United States. In Canada, they recently won major rights as "First Nations."

Governments

Government functions are concentrated in **capital cities**, where the heads of state live and administrative and government offices are situated. Many capital cities are the largest cities in the country, like London (United Kingdom), Tokyo (Japan), and Nairobi (Kenya). In some countries, new capital cities were built as a gesture to replace colonial choices, to expand economic development of the interior, or to provide a more central location. Washington, D.C., Brasília (Brazil), Abuja (Nigeria), and Canberra (Australia) are examples.

Global Governance

No global government exists with the same powers as country governments. Any vision of a worldwide government remains a long way off. However, the term **governance** is increasingly being applied to entities other than country governments. These entities are categorized as either **intergovernmental organizations (IGOs)** or **nongovernmental organizations (NGOs)**. In various ways, they seek to legislate and regulate human activities and set new governing standards, often based on universal principles. Together with their networks that function across country borders, often with minimal government consultation, they are increasingly challenging and eroding the sovereignty of countries.

Intergovernmental Organizations (IGOs)

The United Nations (UN), which includes almost all of the world's countries, is the largest IGO. Founded in 1945 at the end of World War II, its goal has been to prevent and stop wars between countries by serving as a forum for dialogue and by promoting cooperation in international law and security, social progress, human rights, environmental protection, and economic development. Member countries pay dues, and these are used to fund various programs and specialized agencies. Examples of programs are UNICEF (UN Children's Fund) and WFP (World Food Program). Examples of specialized agencies are the International Monetary Fund (IMF), the World Bank, and the WHO (World Health Organization).

The UN has few specific programs in the security field apart from the Security Council and the groups of military peacekeepers that are drawn from member countries. Although the UN has had difficulty in preventing all civil wars, nuclear testing, or drug, weapons, and slave trafficking, its role in world affairs continues to grow. For example, it has completed almost 50 missions and is currently undertaking 16 others. A criticism and challenge for the UN is that its wealthier donor countries like the United States try to use the organization to advance their own foreign policies. At the same time, poorer countries like Bangladesh, Pakistan, and India obtain much-needed funds at the expense of having their soldiers serve on dangerous UN missions.

Not all IGOs are global in intent like the UN. Some are associations of countries with cultural and historical connections. The Commonwealth of Nations (formerly the British Commonwealth) is comprised of countries that were once part of the British Empire, while Islam unites members of the Organization of the Islamic Conference. Other IGOs are closely connected to specific world regions such as the European Union (EU), the African Union (AU), and the Organization of American States (OAS).

Country Groupings for Trade or Defense

Countries make agreements with other countries to foster security through common trading and defense interests. Governments influence world trade patterns whenever they encourage their people to export goods and whenever they control certain imports by charging taxes, or tariffs, on them. The General Agreement on Tariffs and Trade (GATT) was established in 1948 to increase world trade by encouraging countries to lower their tariffs. In 1995, the World Trade Organization (WTO) took over GATT's role of trying to prevent discrimination among trading partners, but many believe that its rules favor the wealthier countries at the expense of the weaker ones.

Most progress on liberalizing trade has been made at the world regional level in free-trade areas, the members of which impose common tariff rates on imports. The largest trading group at present is the European Union (EU). Other examples are the North American Free Trade Agreement (NAFTA) for Canada, the United States, and Mexico, and the Asia-Pacific Economic Cooperation Forum (APEC). Such groupings of regional interests are considered in each chapter of this text.

The Cold War period that began after World War II generated defense agreements on both sides. The North Atlantic Treaty Organization (NATO) linked North America and

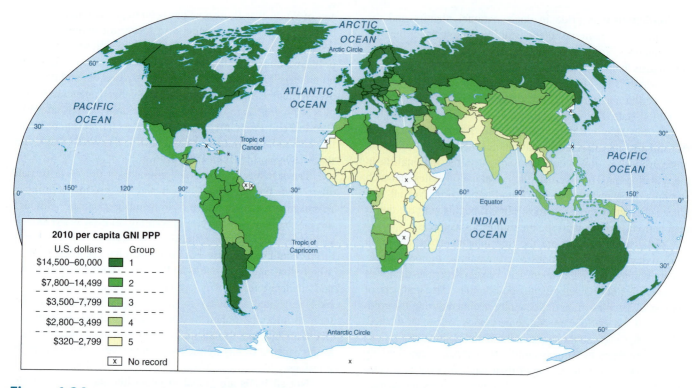

Figure 1.24 **Major income groups of countries.** A World Bank division based on GNI PPP per capita for each country. How do the five categories relate to the major world regions? *Source: Data (for 2010) from* World Development Indicators, *World Bank.*

western Europe in a common response to a perceived military threat from the Soviet Union and still exists today despite the disintegration of the Soviet Union.

The Association of South East Asian Nations (ASEAN) began with political objectives during the Cold War, opposing Communist countries such as Vietnam, but later shifted to increasingly economic objectives and admitted Vietnam as a member.

Nongovernmental Organizations

Increasingly, nongovernment organizations (NGOs) such as aid bodies have been assuming responsibilities for government-like activities. They include any group engaging in collective action of a noncommercial, nonviolent manner that is not on behalf of a government. Some NGOs are locally based, others are associated with particular countries, and the largest engage in international activities. Examples of the last group include the International Red Cross, Green Crescent, Oxfam, Save the Children, Amnesty International, Greenpeace, and Médécins Sans Frontières (Doctors Without Borders). Some are better known than many smaller countries. Each has a particular concern, such as human rights or the environment. Many international NGOs are contracted by governments and international agencies to supply aid irrespective of country borders. The number of NGOs working or consulting with the United Nations rose from under 500 in 1970 to over 2,000 today.

1.7 Regions and Economics: Wealth and Poverty

Economic geography is concerned with the spatial patterns of material wealth and poverty, natural resource use, goods' production, distribution, and consumption, and labor and capital flows. The spatial distribution of material wealth depicted on the World Bank map demonstrates the pattern of economic inequality (Figure 1.24). According to the World Wealth Report, the number of dollar millionaires (or HNWI: High Net Worth Individuals) in the world in 2011 was 10 million (up from 5.2 million in 1997). For the first time since the world recession, the collective value of their material wealth declined. At US$42 trillion, it was a decrease of almost 1.7 percent from the previous year. Although comprising only 0.14 percent of the world's population, they control one-third of the world's wealth. Significantly, the geographic distribution of the world's extremely wealthy also is shifting. In 2010, the number of dollar millionaires in Asia surpassed those in Europe, and then in 2011 overtook those in North America for the first time. China, Japan, South Korea, Thailand, Malaysia, and Indonesia are largely responsible for Asia's growth in millionaires. Their growth offset 20 percent decreases in the number of millionaires in India and Hong Kong. Nevertheless, the North American HNWI wealth at US$11.4 trillion still remained greater than Asian HNWI of US$10.7 trillion.

Measuring Wealth and Poverty

To give more precise meanings to material wealth and poverty, specific indicators have been chosen to compare and understand the differences between countries and regions. The ownership of consumer goods and access to piped water and energy resources are vivid indicators of differences in material wealth among countries (Figure 1.25). Poor people's luxuries such as better drinking water, food, clothing, and shelter are often wealthier people's normal expectations.

The economic development of countries is commonly measured by two statistics of income. **Gross domestic product** (GDP) is the total value of goods and services produced within a country in a year. Gross national product (GNP), now called **gross national income** (GNI), adds the role of foreign transactions to GDP. Per capita figures of a country's total annual income are averages of GDP or GNI per person in the population. They do not indicate personal incomes. The divisions shown on the World Bank map (see Figure 1.24) are based on GNI per capita, with countries divided into five income groups: high, upper middle, middle, lower middle, and low.

GDP and GNI are informative, but they may not reflect the costs of living in a country. The **purchasing power parity** (PPP) estimates of GNI and GDP are more meaningful comparisons of living costs among countries and are used extensively in this text. Because prices in India, for example, are much lower for equivalent items you might buy in the United States, US$440 will buy as much in India as US$2,230 does in the United States. To illustrate this idea, *The Economist* devised a "Big Mac index" based on exchange rates against the U.S. dollar. In January 2012, the burgers that sold for an average of $4.20 in the United States would cost $6.81 in Switzerland, $5.86 in Brazil, $2.70 in Mexico, $2.55 in Russia, $2.44 in China, and $1.62 in India. Countries with high incomes and high living costs have a lower PPP estimate of income than the GDP or GNI based on exchange rates; poorer countries often have higher estimates. For example, in 2007 Switzerland had a GNI per capita income of US$60,820 but a GNI PPP per capita estimate of US$44,410; Mexico had comparable values of US$9,400 and US$13,910.

Figure 1.26 shows that the distribution of wealth is geographically uneven: in 2010 approximately 61 percent of world income was accrued by countries having 20 percent of the world's population, while 21 percent of world income was accrued by countries having 60 percent of the population.

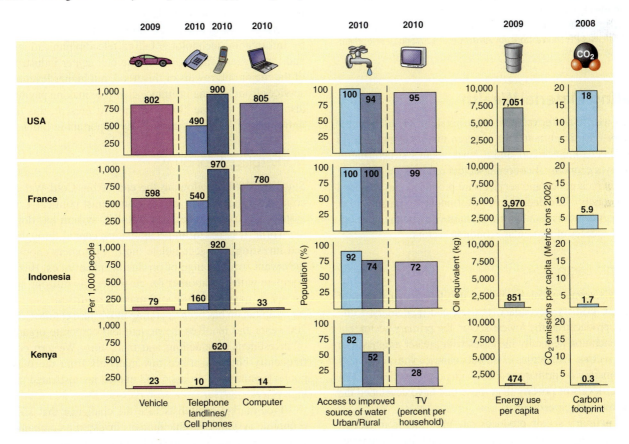

Figure 1.25 **Consumer goods, water access, and energy use.** The ownership of consumer goods is shown as the number of goods per thousand people (e.g., 802 people per thousand have a vehicle in the United States). Access to clean water is given as a percentage of the urban and rural populations. TV ownership is given as a percentage of homes. Energy use is annual kilograms of oil equivalent per capita. Carbon dioxide emissions are in annual metric tons per capita. How do these items demonstrate differences in affluence among the United States, other materially wealthy countries (France), middle-income countries (Indonesia), and materially poor countries (Kenya)? This type of diagram occurs in each of the regional chapters 2 through 11, enabling comparisons. *Source: Data for 2008–2010 from 2012 World Development Indicators, World Bank.*

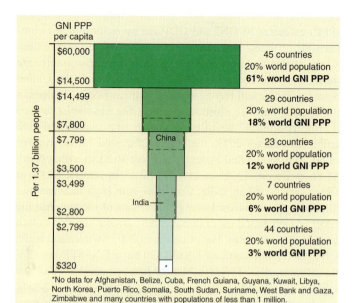

GNI PPP
per capita

$60,000 — $14,500	45 countries / 20% world population / **61% world GNI PPP**
$14,499 — $7,800	29 countries / 20% world population / **18% world GNI PPP**
$7,799 — $3,500 (China)	23 countries / 20% world population / **12% world GNI PPP**
$3,499 — $2,800 (India)	7 countries / 20% world population / **6% world GNI PPP**
$2,799 — $320	44 countries / 20% world population / **3% world GNI PPP**

Per 1.37 billion people

*No data for Afghanistan, Belize, Cuba, French Guiana, Guyana, Kuwait, Libya, North Korea, Puerto Rico, Somalia, South Sudan, Suriname, West Bank and Gaza, Zimbabwe and many countries with populations of less than 1 million.

Figure 1.26 **Distribution of world incomes.** The 2010 world population was divided into five groups of 1.4 billion people each. Gross national income purchasing power parity (GNI PPP) forms the basis of comparison of these groups. Each country's position is shown in the regional chapters, 2 through 11. The United States is in the top group and has 4.5 percent of world population and 20 percent of GNI PPP. *Source: Data (for 2010) from* 2012 World Development Indicators, *World Bank.*

Creating Material Wealth

Material wealth and poverty in countries stems from a number of factors. Two important ones concern the development of a country's economic sectors and a government's structuring of its country's economy. Economic sectors refers to the differing stages at which goods and services are produced and provided within a country. The decision of governments to pursue either free-market capitalism or central planning affects the development and success of particular economic sectors.

Economic Sectors

A country's economy can be divided into four major groups of production: primary, secondary, tertiary, and quaternary. Little wealth is produced in the lowest sector, the primary sector, but wealth generation generally increases with greater emphasis on the other sectors. Exceptions exist, but countries with economies focusing mainly on primary production tend to have less wealth and higher levels of poverty than those that have developed the other sectors of their economies.

The **primary sector** produces raw materials from natural sources including minerals, oil, gas, timber, and fish. Farm products come from domesticated plants and animals, subject to local soil and climate conditions. Many materially poor countries are poor because their economies, and thus the livelihoods of most of their citizens, focus heavily on this low-wage sector.

The **secondary sector** focuses on manufacturing and construction. Extra value and profit come from using raw materials

generated by the primary sector to produce clothes, furniture, food and drink products, pharmaceuticals, railroads, engines, trucks, cars, airplanes, consumer electrical goods, and many other products. The cost of raw materials is a relatively small part of total product costs, which include the costs of building factories and equipping them with machinery, the wages of factory workers, and the cost of getting the products to those who want to buy them. The value of the final goods brings greater profits than primary products.

The **tertiary sector** centers on the service industries, which grew on the backs of the manufacturing companies and from government decisions to provide health services, education, and a wide range of other services such as legal, financial, and media. This sector also includes retail and wholesale trade, and is mainly present in urban areas. Since the second half of the 1900s, all service industries have been experiencing huge employment growth.

As the tertiary sector grew and expanded, a new **quaternary sector** has developed. It focuses on information-based services such as legal, financial, and media, and their increasing use of the Internet and information technology (IT). As a specialized and highly sophisticated branch of the services sector, it first concentrated in the cities of materially wealthy countries. However, as these services have become more complex, the headquarters of many large corporations have sent many aspects of work to low-cost countries, encouraged and enabled by information technology, falling telecommunications costs, and low wages. The growth of the call center industry is one of the major outgrowths of this trend, but higher-level business services from logging insurance claims to making payments are also moving to these global centers. At present, this overseas movement of activity is particularly important in India.

Marketplace Economy

Since the 1990s, following the end of the Cold War and the collapse of the Soviet Union's Communist political–economic system, the **free-market**, or **capitalist, system** has dominated the world. It is based on accumulating capital through profits and investing it to accumulate more. Customers choose what they want from a range of products and services. Businesses compete with one another for customers by offering the lowest prices possible while still making profits.

Capitalism has been operating in Western countries for over 200 years and involves the private and corporate organization of investment, production, and marketing. Western countries frequently buy low-priced raw materials from the materially poorer countries to produce and market sophisticated goods at high prices.

Free-market capitalists face the challenge that affects all economic systems. Fallible humans invest, run companies, and generally perform roles to the best of their ability. Sometimes the investments produce profits, but not always. Sometimes managers take advantage of weaknesses in the system by fixing prices with their competitors or by dishonest accounting. Even in countries with well-regulated economies, major corporations may crash in scandal and create personal catastrophes for employees, suppliers, and customers.

In theory, governments and entities like the World Trade Organization intervene in free-market economies mainly to regulate the terms of trade and ensure fairness among producers. In practice, they decide on what trade should happen and what is fair, but their decisions are not necessarily carried out in the public interest. Politics means that some groups are favored when policies and laws are made. Consequently, some imported products are taxed, and some domestic goods are subsidized despite the goal of free trade. Furthermore, governments of many wealthier countries provide social services and build infrastructure (roads, airports, harbors, water supplies, waste disposal) that give businesses and people in those countries many cost advantages compared to the poorer countries. As a result of increased government intervention, capitalist countries are becoming less "free" market.

Central Planning Economy

Communists detested the free-market system because they believed that it allowed the rich to become richer and the poor to become poorer. To prevent this from happening, they designed centrally planned systems that were adopted by the former Soviet Union, its satellite countries, the People's Republic of China, and linked countries such as Cuba. This system places planning and decision-making responsibilities in the central government on the grounds that the whole country's interests come first and the central ministries know what is best for the people. They plan the production of goods considered essential—whatever the cost and whether or not the goods meet consumer demands. Central governments provide desired medical care and education and develop strong military defenses.

Those in command of centralized policymaking, however, often made large-scale mistakes, handicapped even more than in the free markets by a lack of information or by personal bias or interest. Many leaders were afraid to change policies, even if inefficient or oppressive, while regional bureaucrats often obeyed central commands despite knowing the policies would fail. Overproduction of some goods and underproduction of others led to these countries failing to produce the consumer goods available in most Western free-market countries. Incomes for most families remained modest, while members of the Communist Party hierarchy became relatively wealthy or privileged elites.

In the 1980s, dissatisfaction grew in the Soviet Union as the economic situation worsened for most. In 1991 the Soviet Union broke up and caused the collapse of economic relationships with the former countries of the Soviet bloc in East Central Europe and in other allied countries worldwide. These countries then entered the global free-market capitalist economic system, but it has been a traumatic change for most of them. The People's Republic of China avoided this when it began adopting enough capitalist practices as early as 1978 to bring about economic growth.

The Global Economy

The United States, European Union countries, and East Asia (Japan and China) represent the pillars of the global economy (Figure 1.27). Growing communications, greater free trade, and the movement of jobs and workers has resulted in greater "globalization," a process that is leading to the increasing economic interconnectedness of countries. Globalization has been encouraged and facilitated by multinational corporations, global financial institutions, trade organizations, and the rise of global city-regions. Intervals of recession and rising fuel costs create the opposite trend of "deglobalization" or localization as local goods and services become cheaper and countries engage in **protectionism** when jobs are being lost.

Multinational Corporations

Multinational corporations (MNCs) make goods or provide services for profit in several countries but direct operations from a headquarters in one country. The term *transnational corporation* (TNC) is often used instead to refer to corporations that are no longer rooted in a single country. The greater ease of travel and telecommunications contacts, together with the Internet transfer of information, encouraged MNCs to expand in numbers and operations. For example, in the early 1990s, an estimated 37,000 TNCs had 170,000 foreign affiliates. In 2008, the number had grown to 82,053 with 807,363 foreign affiliates. About 60 percent of all international trade is actually trade within MNCs, making MNCs a major force in globalization trends.

Multinational corporations place production facilities in countries outside their homelands to take advantage of cheaper labor, land, energy, and to avoid tariffs, stringent worker safety standards, and environmental laws in their home countries

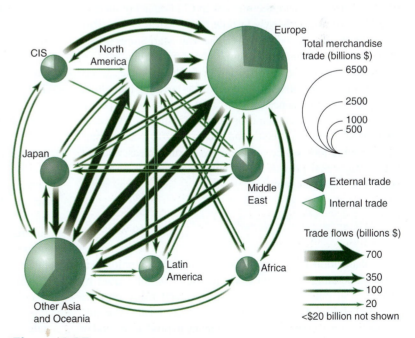

Figure 1.27 **The network of world trade, 2008.** *Source:* Global Shift *by Peter Dicken. Copyright 2007 by Guilford Publications, Inc. Reproduced with permission of Guilford Publications, Inc.*

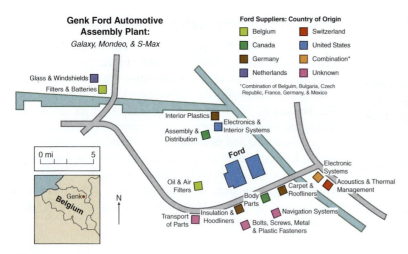

Genk Ford Automotive Assembly Plant:
Galaxy, Mondeo, & S-Max

Ford Suppliers: Country of Origin

- Belgium
- Canada
- Germany
- Netherlands
- Switzerland
- United States
- Combination*
- Unknown

*Combination of Belguim, Bulgaria, Czech Republic, France, Germany, & Mexico

Glass & Windshields
Filters & Batteries
Interior Plastics
Electronics & Interior Systems
Assembly & Distribution
Ford
Oil & Air Filters
Electronic Systems
Carpet & Roofliners
Acoustics & Thermal Management
Body Parts
Transport of Parts
Insulation & Hoodliners
Navigation Systems
Bolts, Screws, Metal & Plastic Fasteners

0 mi — 5

Genk
Belgium
N

Figure 1.28 **Multinational corporation's international linkages.** In the automotive industry, for example, automobiles are assembled from parts manufactured in many different places around the world. In the United States, General Motors (GM), Ford, and Chrysler receive about one-third of their parts from foreign companies, one-third from foreign companies in the United States, and one-third from American companies. Canada, Japan, and Mexico are the largest suppliers. Parts manufacturers are increasingly opening factories near automobile assembly plants as illustrated by this map of a Ford assembly plant and its foreign-parts suppliers in Genk, Belgium. The close geographical proximity of these factories not only illustrates "just in time" delivery but also shows how international an industrial park of just a few square miles can be. *Source: Adam P. Lewis and Ashley Wolfe.*

(Figure 1.28). For example, auto manufacturers spread the manufacture of components across several countries to ensure supplies during labor strikes and to react to local needs. Of the top Fortune 100 companies in 2012, 71 were from five countries: the United States (29), Japan (12), Germany (11), France (9), and China (9), with the number in China quickly increasing. The five largest MNCs in order were Royal Dutch Shell (the Netherlands), Exxon Mobil (U.S.), Wal-Mart Stores (U.S.), BP (British Petroleum) (UK), and Sinopec (China).

MNCs wield considerable power in the countries where they operate. Some MNCs are perceived as uncaring monolithic institutions without concern for the best interests of the people they employ in either home or adopted countries. However, other MNCs transfer wealth and technology to poorer countries, provide jobs where none existed in rural areas, and pay better wages and provide better employee benefits and prospects than local companies.

Outsourcing, Offshoring, and Offshore Financial Centers

As globalization is breaking down barriers, corporations are cutting operating costs and lowering their tax obligations by engaging in outsourcing and offshoring and using offshore financial centers. **Outsourcing** occurs when a business or government agency contracts with a company to produce a good or perform a service that it once produced or performed for itself. It occurs both inside and outside of countries. **Offshoring**

refers to the shifting of a job to another country, often overseas. A single job can be outsourced and off-shored at the same time. However, many companies simply move jobs to branch offices overseas, meaning that such jobs are offshored but not outsourced. In 1983, for example, American Airlines established Caribbean Data Services in Bridgetown, Barbados, to process the paperwork related to its tickets and boarding passes. It became the largest single employer in Barbados. U.S. insurance companies process claims in Ireland. India, with its large population of English speakers, is growing as one of the world's major call centers for multinational corporations. Some Indian companies even train their staff to respond in American accents.

One of the fears of globalization is that more and more jobs will be lost overseas. However, outsourcing hit a peak in 2004 and has generally declined since. Low wages are attractive, but local workers' knowledge of a local market is still important to maximizing profits, in turn minimizing the number of jobs that American companies are exporting. At the same time, foreign companies are employing American workers to produce and sell their products in the United States because American workers better understand American consumer choices and demands. Moreover, low wages is not the only goal of outsourcing. For example, as demand for environmentally friendly front-loading washing machines increased in the United States, Whirlpool found it more cost effective to pay $32 per hour to German workers than $23 to American workers to make these machines because German factories were already tooled for such machines and German workers already had the knowledge to build them. It shows that knowledge, experience, and infrastructure are as important as low wages in minimizing costs.

Individuals and businesses also save money by moving their financial assets to **offshore financial centers (OFC)**. The term comes from the fact that many of these places are small islands in the Caribbean and the South Pacific, although some of them are mainland countries like Switzerland, Luxembourg, Costa Rica, and Kuwait (Figure 1.29). OFCs attract individuals and businesses with low tax rates, few regulations, and secrecy. The United States loses about US$40 billion a year in taxes from monies moved to OFCs. Apple Inc. recently became best known for its "Double Irish with a Dutch Sandwich," whereby it logs large amounts of its profits at its two subsidiaries (Apple Operations International and Apple Sales International) in low-tax Ireland. From its Irish subsidiaries, other profits are moved tax-free to the Netherlands and the British Virgin Islands. Apple also has a subsidiary (iTunes S.à.r.l.) in low-tax Luxembourg where a little more than a dozen employees record more than one billion U.S. dollars in sales annually. These practices have lowered Apple's tax rates to less than 10 percent, which stands in sharp contrast to the 24 percent that Walmart pays in taxes.

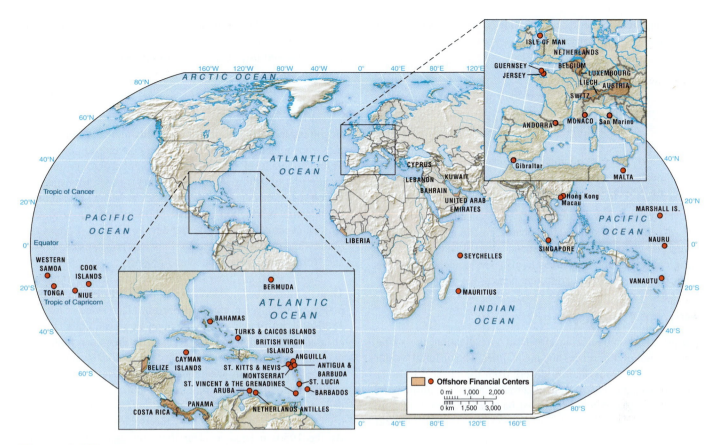

Figure 1.29 Offshore financial centers (OFCs). By offering low tax rates, few regulations, and secrecy, OFCs attract assets from individuals and businesses.

While many companies open up subsidiaries in OFCs, some subsidiaries have less than their nameplate in the lobby of another business. For example, fewer than 3 percent of companies in the Virgin Islands have any presence there. The loss of taxes and opportunities for criminals and terrorists to hide their monies has prompted the Organization for Economic Cooperation and Development (OECD) in 2009 to compile a blacklist of OFCs for the Group of 20, the world's largest economies. This has helped these countries collect almost US$19 billion in additional taxes. Similarly, after the government on the island of Vanuatu required banks to have a permanent office with at least one full-time employee, 30 out of 37 offshore banks there ceased to exist.

"Deglobalization" and the New Industrial Revolution

The world economy does not always become more globalized every day. For example, when prices of commodities such as oil rise, they dramatically increase transportation costs and erase the savings of low wages found far away. When oil prices hit record highs in spring 2008, it became cheaper for Americans to produce steel at home rather than buy it from China. Although Chinese wages were a fraction of U.S. wages, steel is heavy and expensive to transport. Subsequently, the U.S. steel industry prospered after years of decline. Such phenomena can be called "deglobalization" or localization. Deglobalization also often occurs during recessions, when countries place tariffs on foreign goods and services in an attempt to stimulate businesses at home.

Deglobalization is occurring with a new industrial revolution. The old industrial revolution emphasized the mass production of identical objects created by large numbers of workers doing simple, repetitive tasks. Profits depended greatly on locating factories where cheap labor could be found. Thanks to manufacturing becoming digital, the development of new materials like carbon-fibre composites and processes like 3D printing, robots, and online collaborative services, goods can be made cheaply in small quantities. Rather than being characterized by large, sprawling factories, the new industrial landscape is transforming into a multitude of small production facilities located near markets where customers can communicate with the production facility and receive their products quickly.

1.8 Geography, Development, and Human Rights

Geographers consider how some places experience greater material and personal well-being than others, and how improvements, or **development**, of poorer regions may occur. For each place or region they build a knowledge of:

- the complex interactions of people with the natural environment,
- the historic growth of population numbers and cultural expressions,
- the evolution of political systems and their present operation,
- the growth of economic output, and
- the effect that different views of human rights have on those living in different parts of the world.

In this chapter we highlighted geographically related analyses of natural environmental concerns, resources, cultural features, population, political governance, wealth and poverty, and the global economy. They affect human development and human rights in different ways in different parts of the world.

Human Development

Despite the homogenizing effects of globalization, the world remains full of differences and inequalities. The material wealth of many Americans sharply differs from the poverty of other Americans and the extreme material poverty of millions in Africa and Asia. Such differences have led to studies of how some regions and countries move ahead and others fall behind in terms of their levels of development. The process of enhancing human capabilities and improving quality of life by providing access to better incomes, education, health care, piped water, and energy supplies is known as **human development**. Concern is given to the possibilities of helping materially poor, "less developed" or "underdeveloped" countries and regions to catch up with the wealthier countries.

The United Nations' **human development index (HDI)** is a measure of human well-being, incorporating statistics calculated from life expectancy, education attainment, and health, as well as income. Materially poorer countries investing heavily in education and health care, such as Costa Rica and Sri Lanka, provide a better quality of life for their people and have a higher HDI than GDP (merely based on a country's income) rank. By contrast, many of the oil-rich Persian Gulf countries have high income rankings based on oil exports but lower HDI rankings because of poor improvements in schooling, especially for girls, and health care—although both are improving.

Sustainable human development involves economic growth that does not deplete renewable resources for the future. It thus links to both human and natural resources, drawing together studies of human and physical geography.

The United Nations Human Development Program and recent World Bank publications focus on the need to eradicate material poverty. The last 50 years saw major reductions of income poverty in large parts of the world, improvements in human development indicators—particularly in health and education—and the wider spread of law and fair administration of justice. However, the fact that so many people in the world remain materially poor is a challenge. In 2000, the United Nations and other global organizations formulated the Millennium Development Goals, to be achieved by 2010–2015 (Table 1.1). A 2012 report revealed that Goal 1 was likely achieved in 2010, but it will be a challenge to meet all goals by 2015.

An Unequal World

Beginning in the 1950s, the economically more developed countries encouraged the economically less developed countries to improve their levels of human development through modernization: the transformation of a country's economy from relying on traditional agriculture (the primary sector) to one based on industrial production and mass consumption (the secondary sector), eventually leading to the development of services (the tertiary and quaternary sectors). This was how Western Europe and North America developed and prospered.

Modernization, however, did not proceed as imagined. Though Europe began decolonizing in the late 1940s, the colonial economic system that had been in place since at least the 1800s did not fundamentally change. European colonial powers traded low-value raw materials with their own high-value manufactured goods. As colonizers and colonies were simply replaced by materially wealthier **core countries** and dependent **peripheral countries**, a similar economic relationship continued. It was difficult for the materially poorer peripheral countries to follow the same path as the wealthier core countries. By the 1950s it became clear that although most materially wealthier places got wealthier, few poorer places experienced improvements.

In the 1960s and 1970s, those living in materially poorer countries were concerned about the lag in economic development, and they also resented previous colonial domination. Many materially poorer countries strove to be self-sufficient, and they disengaged from involvement in the West-dominated world economy when they achieved political independence. They developed home-based manufacturing (**import substitution**) and created local trade barriers, placed restrictions on

Table 1.1	Millennium Development Goals

In 2000, many target dates were set for between 2010 and 2020, but lack of progress by 2006 suggested that these goals will take much longer to achieve.

Goal 1: Eradicate extreme poverty and hunger.

Goal 2: Achieve universal primary education.

Goal 3: Promote gender equality and empower women.

Goal 4: Reduce by two-thirds the under-five mortality rate.

Goal 5: Improve maternal health.

Goal 6: Combat HIV/AIDS, malaria, and other diseases.

Goal 7: Ensure environmental sustainability.

Goal 8: Develop a global partnership for development.

foreign corporations, and formed trading groups of countries with similar concerns. By the 1990s, however, it became clear that isolation from the world economy did not result in economic equality with the Western countries.

In many of the materially poor countries today, few jobs in the **formal economy** pay taxable salaries. The majority of people have to gain income as best they can in the **informal economy**. They cannot use their homes or land as collateral for loans, and this failure to extend formal property rights to most of the population is a major cause of poverty and a lack of enterprise opportunities.

In response to the lack of local opportunities and a positive view of the potential of individuals, groups of people set up microcredit banks, building on the experience of the Grameen ("Village") Bank in Bangladesh, which grew out of a program of small individual loans created by Nobel Peace Prize winner Professor Mohammad Yunus. Small amounts of money enabled craft workers and others to establish small businesses that helped them emerge from poverty. These methods are now applied in numerous countries, including materially wealthier countries such as the United States and Canada. Though the movement has had success, it also has its critics. With most of the borrowers being women, some argue that many women are forced to turn the money over to men but remain responsible for repaying the loans.

Responsible Growth and Emerging Countries

By 2004, "**responsible growth**" was replacing earlier terminology. Arising from the Johannesburg World Summit on Sustainable Development held in August 2002, a consensus emerged directing world leaders to new development paths that build on the UN Millennium Development Goals and connect economic growth, environmental sustainability, and social equity. In this view, directed at the initial period to 2015 and longer term toward 2050, poverty reduction is not an end in itself but a precondition for peaceful coexistence and ecological survival.

Meanwhile, terms for developing countries like *Third World* were falling out of favor. Not only did they have negative connotations, they applied to too many countries with vastly different circumstances to be meaningful. Antoine van Agtmael believed that his Third-World Equity Fund failed to attract investors because of its name: "Third World" was seen as stagnating. So he coined the term "emerging markets" to refer to the more economically successful developing countries. The use of the word *emerging* caught on and is seen in the broader term **emerging countries**. In 2001, Jim O'Neill, chief economist at Goldman Sachs, invented the acronym BRICs for Brazil, Russia, India, and China, which, along with countries like Argentina, Chile, Indonesia, Mexico, Poland, Singapore, Saudi Arabia, South Africa, South Korea, Taiwan, and Turkey, are more economically developed. Indeed, China's GDP surpassed Germany's in early 2008 to become the world's third-largest economy and then surpassed Japan's in mid-2010 to become the world's second-largest economy.

Historically, emerging economies were highly dependent on the industrialized world. As suppliers of raw materials, their economies suffered when industrialized countries went into recession. However, emerging economies grew so rapidly through much of the 2000s that the term "decoupling" was coined. It referred to the fact that many emerging economies had developed industry and were diversifying, perhaps to the extent of breaking their dependence on industrialized economies. Though this will take time, many emerging economies recovered more quickly from the 2007–2009 global recession than industrialized economies, especially those with oil and other raw materials to export.

The Growing Middle Classes

The middle classes include those who have a stable income and have a third of their income left over after obtaining basic food and shelter. This extra income allows middle-class individuals to invest in their health care and their children's educations. These are luxuries that the poor do not have. The actual income for middle class varies from country to country, but it typically falls between US$2 and US$60 per day, depending on location.

In 1820, during the early years of the Industrial Revolution, less than 3 percent of the world's population was middle class. Over the course of the 1800s and early 1900s, the first growth spurt for the middle class occurred in Western Europe. Still, by 1914, the world's middle classes had grown only to about 13 percent. A second growth period occurred from 1950 to 1980 in Western countries, bringing the total to about 33 percent. The most recent growth surge began in 1980. By 2005, the middle classes accounted for roughly half of the world's population. In China alone, the middle class grew from 15 percent in 1990 to 65 percent in 2005. India's middle classes were only at 5 percent in 2005, but they are expected to grow to 20 percent by 2015.

Surveys taken by the Pew Research Center indicate that the middle class is more concerned about free speech, fair elections, and eradicating government corruption than the poor, who are more concerned about the circumstances of poverty. In countries where the middle class has grown rapidly in recent years, there likewise has been a sharp rise in demands for greater political freedom and in protests against corruption. This shows that close connections exist between human development and human rights.

Issues of Human Rights

The concept of **human rights** emerged in the late 1700s, in part from revolutions in Europe and the United States. "Liberté, Égalité, Fraternité" (freedom, equality, brotherhood) was the slogan in France during the 1789 revolution, and similar ideas were enshrined in the U.S. Bill of Rights. Stressed again after World War II by the United Nations Declaration of Human Rights (Table 1.2), international agreements and actions often refer to these rights. However, few countries fully implement the UN list, and some groups claim that the imposition of human rights legislation impinges on their traditional rights.

Human rights fall into three different categories. The United States and Western Europe have long argued for **political rights** (the right to vote and participate in one's own

Table 1.2	United Nations Declaration of Human Rights
Freedom from discrimination because of gender, race, ethnicity, national origin, or religion.	
Freedom from want and a decent standard of living.	
Freedom to develop and realize one's human potential.	
Freedom from fear of threats to personal security in arbitrary arrest or violence.	
Freedom from injustice.	
Freedom of thought and speech to participate in decision making and forming associations.	
Freedom for decent work without exploitation.	

government). Others, including the former Communist governments, argued for **social rights** (the right to have a job and earn a living with basic material standards). People in materially poorer countries often argue for **cultural rights** (the right to protect one's cultural traditions, often developed in response to Western ideals and practices). An example of cultural rights involves Western medicine patents, where people in other countries must pay royalties to a Western corporation for a medicine they had discovered hundreds of years ago before the institution of patenting systems.

Agreement on the nature of human rights and justice is not global. Muslim countries in particular debate the largely secular basis of the United Nations list that conflicts with their religious beliefs and fierce legal provisions. Saudi Arabia refuses to attend UN conferences on such topics on the grounds that it has a God-given right to gender discrimination. As another example, Western countries criticized the former Taliban government in Afghanistan and other strict Muslim regimes for oppressing women and carrying out amputation punishments for "minor" crimes. And yet, Muslims claim that their strict rules protect women, while many Western governments allow men the freedom to mistreat and degrade women. Different cultural definitions of human rights contribute to differences among people's expectations in different places.

1.9 Major World Regions

In this text we identify 10 world regions (Figure 1.30) that are the subjects of Chapters 2 through 11. Each **world region** contains a group of countries which exhibit distinctive physical and human geographies combined with internal diversity in countries and local areas. Global and local connections contribute to regional identities.

Europe

Europe (Chapter 2) is the source of many Western trends in politics, culture, and economics. The region is defined as those countries that are members of the European Union or likely to be in the next few years. It is marked by temperate natural environments and advanced capitalist economies. Its cultures

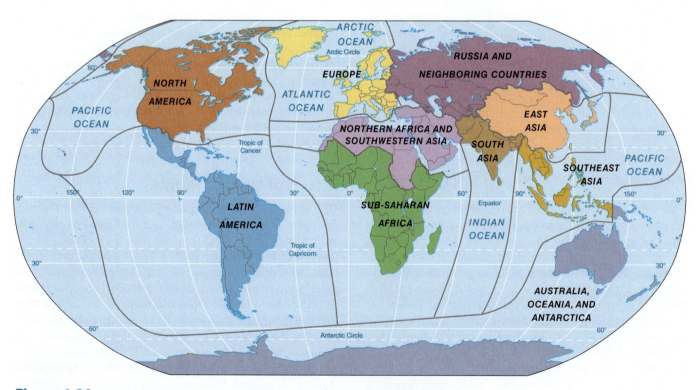

Figure 1.30 **Major world regions.** These regions are the subjects of Chapters 2 through 11.

continue to be affected by the past dominance of Roman Catholic, Protestant, or Orthodox Christian groups, increasingly challenged by secular ideologies and the religious demands of immigrants from other world regions. Europe is marked by a variety of spoken languages, while increased immigration from the late 1900s makes it home to a growing number of people speaking other languages from diverse world regions, contesting what it means to be European.

Russia and Neighboring Countries

Russia and Neighboring Countries (Chapter 3) includes all the countries that emerged from the breakup of the Soviet Union apart from the three small Baltic countries of Estonia, Latvia, and Lithuania (now part of Europe, Chapter 2). The region extends from easternmost Europe across northern Asia—a huge area that resulted from the expansion of the Russian Empire in the last four hundred years. European cultures extended into Asia, incorporating and interacting with a wide range of peoples, cultures, and languages in varied environments to produce today's geographic diversity. Through much of the 1900s, the centralizing Communist government of the former Soviet Union kept under control the historic clashes between the Orthodox Christians and Muslims and between the Russian and other peoples. These conflicts emerged again after the Soviet Union breakup, and they are part of the difficult transition from Communist to marketplace economies and open religious allegiance.

East Asia

East Asia (Chapter 4) includes Japan, the Koreas, Mongolia, Taiwan, and a resurgent China. The region includes the world's most successful countries in recent economic growth, challenging the economic dominance of the United States and Europe. In economic terms, China now has the second-highest total output by value (GNI PPP) in the world after the United States, and Japan is third. Neither North Korea nor Mongolia share in the economic growth. The historic and modern influences of the Chinese kingdoms and related cultures were less affected by European colonization than other world regions, and they bring special character to this region and to surrounding countries, where many Chinese families and business personnel live. Geographic diversity occurs in all countries.

Southeast Asia

Southeast Asia (Chapter 5) includes the independent kingdom of Thailand and the former European colonies of Myanmar (or Burma), Malaysia, and Singapore (all ex-British colonies); Indonesia (Dutch); the Philippines (Spanish and the United States); Vietnam, Laos, and Cambodia (French Indochina); and East Timor (Portuguese, taken over by Indonesia). Many countries in Southeast Asia engage in intense economic competition with one another while simultaneously participating in economic trade blocs to strengthen the region's global position.

South Asia

Diversity within *South Asia* (Chapter 6) is based on the major religions, which include Islam, Hinduism, Sikhism, Jainism, and Buddhism. Today, Hinduism (the dominant religion in India) counts 80 percent of India's population as its adherents; Islam is the national religion of Bangladesh and Pakistan; Buddhism is dominant in some smaller countries. The British Indian Empire, which included the whole region, ended with the partition into separate countries at independence in 1947. Since independence, partly successful policies of self-sufficiency and alignment with the Soviet bloc brought moderate economic growth before the rise in foreign trade and investment began in the 1990s. Growing middle classes and technocratic and wealthy elites exist alongside large numbers of poor people.

Northern Africa and Southwestern Asia

Northern Africa and Southwestern Asia (Chapter 7) has a distinctive and strategic position at the junction of Europe, Africa, and Asia. It includes the birthplaces of the world's three monotheistic (believing in one God) religions: Judaism, Christianity, and Islam. The Islamic religion is dominant today, partly paralleled in its extent by the Arabic language—although the two Muslim countries with the largest populations in this region, Iran and Turkey, have their own languages. The presence of Jewish Israel in the predominantly Muslim region creates cultural, economic, and political tensions, which result in almost continuous hostilities. The location of the world's largest oil reserves and water shortages in the largely arid natural environment of much of this region pose internal problems of uneven resource availability.

Sub-Saharan Africa

Sub-Saharan Africa (Chapter 8) was the cradle of the human race, and most of its current population consists of indigenous peoples and their many ethnic groups. It remains one of the most culturally diverse world regions. The region contains great mineral riches and has underdeveloped economic potential. European traders established commercial farming and mining, but colonial settlement by Europeans occurred much later and on a smaller scale than in the Americas or India. Decolonization began in the 1950s, but it was often followed by repressive dictatorship governments. The people of this region are among the world's most materially poor.

Australia, Oceania, and Antarctica

The region *Australia, Oceania, and Antarctica* (Chapter 9) covers a vast stretch of latitude and longitude and is defined by proximity to and the influences of water. Australia and New Zealand are the region's most materially wealthy and Western countries. The South Pacific islands of Oceania are distributed across great ocean distances and are challenged by a lack of locally held global political or economic influence. Antarctica is increasingly visited by tourists and scientists, yet it does not have a permanent resident population. Human-induced climate change may significantly alter the world's political and economic interest in the Antarctic continent.

Latin America

Latin America (Chapter 10) has many political and economic issues that relate to the region's history of colonization and its position as the nearest neighbor of the United States. The region's indigenous peoples were reduced in numbers by the Spanish and Portuguese colonization of Latin America, which began in the 1500s, but they still form major portions of the population in the Andes and Central American highlands. The Latin-based Romance languages and Roman Catholic religious culture brought by Spanish and Portuguese settlers influence most of this region. There are enclaves of other European languages, including French, Dutch, and English, particularly in the Caribbean.

North America

North America, comprising the United States of America and Canada (Chapter 11), is the world's materially wealthiest region, containing the only current world superpower (the United States). The two countries are dominated by cultures first brought by European settlers beginning in the 1500s. French and Spanish are spoken locally, but English is the most widely spoken language in North America. The region's indigenous peoples, whose numbers were significantly diminished by European diseases and conflict with the settlers, formed smaller and smaller proportions of the population, but some lands and facilities were restored to them in the later 1900s.

Geography at Work

AAG President in Iran: Reconciling Differences

Geography provides fundamental insights into the nuances of globalization, whether economic, political, or cultural. Geography can build bridges and help ease cultural tensions and misunderstandings. For example, Professor Alexander B. Murphy, from the Department of Geography at the University of Oregon, traveled to Iran in 2004 when he was president of the Association of American Geographers (Figure 1.31). He addressed the Second International Congress of Geographers of the Islamic World, and he spoke to students and faculty at two of Tehran's major universities.

Everywhere Alex went, people were extraordinarily nice, helpful, and friendly. He had traveled previously in other parts of the Muslim Middle East— especially Egypt, Jordan, and Palestine. Iran is clearly different—in language, in culture, in social norms, and much more. To visit Iran is to understand the fallacy of treating the "Islamic World" as a monolith.

Very few Americans go to Iran these days, and this opportunity allowed Alex to give Americans a human face in Iran. At the same time, he gained insight into the diversity of opinion in Iran on political and social issues. Iranians express views that span the political spectrum.

Figure 1.31 Professor Murphy, the tallest person, with colleagues at the Iranian geographer's conference. *Photo: Courtesy Alexander Murphy.*

What particularly impressed Alex, however, was how well informed Iranians seem to be—gaining information not just from government news sources but from the Internet and from friends and relatives in other parts of the world. He saw how much Iranian society has opened up in recent years, and his discussions with people and his media appearances highlighted a side of America that Iranians rarely see.

Frankfurt, Germany. *The Eurotower, which contains the headquarters of the European Central Bank (ECB), oversees the euro.*
Photo: © Peter Adams/The Image Bank/Getty Images.

Europe

LEARNING OBJECTIVES

After reading the chapter, you should be able to:

- Explain how European culture has influenced today's global economic and political systems, citing specific examples of European contributions.

- Describe Europe's diverse landforms, climates, and natural resources.

- Identify connections between industrialization and Europe's environmental problems, and describe Europe's policies on climate change.

- Discuss Europe's population growth trends in terms of aging and rural-urban relationships.

- Describe the formation of Europe's modern political map and the consequences of nationalism. In doing so, refer to the locations of Europe's major ethnic groups, languages, and religions.

- Compare the relative importance of manufacturing, energy sources, service industries, and agriculture in Europe.

- Identify the main subregions of Europe on a map and describe how these can be distinguished in terms of their physical, cultural, economic, and political geography.

- Describe the formation of the European Union and the EU's role in changing spatial patterns of agriculture and industry since World War II. Discuss supranationalism and devolution in the European context.

- Understand why immigrants are attracted to Western, Northern, and Mediterranean Europe, and how immigration has changed the cultural makeup of these subregions.

Figure 2.1 **Europe: physical features, country boundaries, and capital cities.** The region is defined by the Atlantic Ocean to the west, the Arctic Ocean to the north, the Mediterranean Sea to the south, and the land boundary with Moldova, Ukraine, Belarus, and Russia to the east.

(a)

(b)

(c)

Figure 2.2 **Diverse transformed landscapes of Europe.**
(a) Urban Leiden, low-lying Netherlands. (b) Rural Monterchi in hilly Italy. (c) Local outdoor market in Campo de Fiori, Rome, Italy. *Photos: (a) © Mike Camille; (b) © Alasdair Drysdale; (c) © Emily A. White.*

2.1 European Influences

European influences dominate studies of world regional geography in the twenty-first century. For almost 500 years, from the late 1400s to the mid 1900s, European empires colonized or forced the other world regions into trading relations with European countries. In doing so, they also forced their cultures on other peoples. For example, the English, French, Spanish, and Portuguese languages and the Roman Catholic and Protestant variants of Christianity spread around the world through trade, religious missions, and political control. Other ideas that originated in Europe and diffused include democracy, colonialism, imperialism, capitalism, the basis of scientific research, nationalism, fascism, communism, socialism, and genocide. At

the same time, many local ideas, technologies, and faiths from cultures in Asia, Africa, and the Americas were imported into various European cultures.

European discoveries and interactions abroad stimulated an interest in scientific and technological advancement at home. The scientific discoveries include the Earth being spherical and revolving around the sun, the laws of motion and gravity, the theory of evolution, genetic science, the causes of disease and pharmacology, radioactivity, and the theory of relativity. New technologies include movable type for the printing press, the microscope, telescope, steam engine, railroad, internal combustion engine, automobile, radio, antibiotics, orbital satellite, and digital computer. Many of these early discoveries led to and propelled the Industrial Revolution, which in turn increased the economic and political power of Europe.

As many Europeans used their technologies to explore and gain more material wealth from abroad, they also used their technology to transform Europe's natural environments (Figures 2.1 and 2.2). For 2,000 years farms and market towns replaced forests; in the last 200 years factories and cities replaced many farms. Rivers were straightened for navigation and dammed to prevent flooding and to provide power. The earth was mined for coal and ore and then drilled for oil. These activities improved the material standard of living and created new landscapes, although some had negative environmental impacts.

For better or worse, over the last 500 years, many Europeans used their local ideas to shape the modern global economy, structure the world's political system, and spread their culture around the world. Though they forced their ways on other cultures for most of that time, they became less aggressive in their interactions with other cultures after the trauma of World War II. Since then, many Europeans have been major proponents of human rights and human development. European cultures continue to interact with and react to cultures in other areas of the world, illustrating that globalization is not destroying geographic differences but is often resulting in different and alternative cultural practices. Indeed, though many Europeans are cooperating both economically and politically more than before, Europe is as culturally diverse as ever.

2.2 Distinctive Physical Geography

The regional physical geography of Europe is generally noted for closeness to the ocean and abrupt changes in the physical landscape over short distances (see Figure 2.1). Europe's natural environments have small-scale geologic provinces, mineral resources, extensive plains and valley routeways, long, indented coastlines, temperate climates, woodlands, and areas of fertile soil. These characteristics influenced human actions and have been shaped by human actions over time, contributing to the development of diverse modern geographies.

Geologic Variety

Within its relatively small area, Europe includes almost the world's entire range of geologic features. Examples include ancient shield areas around the Baltic Sea, the uplands of central Europe, the young folded mountains of the Alps, and the extensive plains in countries around the North and Baltic seas (see Figure 2.1). Volcanoes and earthquakes are active along the Mediterranean Sea.

The Mediterranean Sea is the remnant of a larger ocean that occupied the area between Africa and Europe but closed as the two continents clashed along a convergent tectonic plate margin (see Figure 1.5). Within this zone, the young, folded mountains of the Alps form the highest ranges, with many peaks rising above 4,000 m (13,000 ft.). The highest point is Mount Blanc (4,807 m, 15,771 ft.), found on the French-Italian border. Farther east, the ranges are not so high in Austria, where few peaks exceed 3,000 m (10,000 ft.). The high Alps are part of a series of ranges that includes the Sierra Nevada in southern Spain, the Pyrenees between France and Spain, the Apennines that form the Italian peninsula, the coastal ranges of the Dinaric Alps in Croatia, Bosnia, Macedonia, and Albania, and the Pindus in the Greek peninsula. The curve of the Carpathian Mountains in Slovakia and northern Romania, which continues in the Balkan Mountains of Bulgaria, forms a further extension.

Around the young, folded mountains are hilly plateaus (1,500 m, 4,500 ft.) of older rocks that once formed mountain ranges. The ranges were worn down by erosion and then raised again by faulting. Such areas include much of Spain and Portugal (the Meseta), France's Massif Central and Brittany areas, the Rhine Highlands of southern Germany, the Bohemian Massif of the Czech Republic, the uplands of western and northern Britain, and those of Norway. As the Atlantic Ocean opened along a divergent plate margin, volcanic activity and uplift occurred along continental margins. Rifting extended through the North Sea area, providing a downfaulted block of rocks that became oil reservoirs.

Lowlands, areas less than 300 m (1,000 ft.) above sea level, are extensive. The North European Plain alone stretches from southeastern England and northeastern France eastward across the Low Countries, northern Germany, Poland, and the Baltic countries into Russia.

Long Coastlines and Navigable Rivers

Peninsulas such as Scandinavia (Norway and Sweden), Jutland (Denmark), and Brittany (France) in the north, and Iberia (Spain and Portugal), Italy, and Greece in the south allow arms of the ocean, such as the Baltic, North, Mediterranean, Adriatic, and Aegean seas, to reach far inland. Islands in the Baltic and Mediterranean seas add to the length of coastline that encouraged many groups to engage in trade and develop ship technology at a time in history when water transportation was easier than land transportation. Many **estuaries**, where rivers meet the seas, facilitated port building. During the most recent Ice Age, glaciers excavated deep valleys of Norway's Kjöllen Range as they cut their way to the sea. When the ice melted, the sea rose and flooded inlets that are known as **fjords**.

The Rhine and Elbe rivers in Germany and the Danube River flowing from Germany through Austria and the Balkans served as particularly important trade routes. The Rhône, Seine, and Loire rivers in France, the Thames River in England, the Vistula River in Poland, and the Po River in Italy were also significant in movements of people and goods from early times. River valleys large and small also provided good soils. Sites for towns at river crossings and at the highest navigation points attracted and concentrated the settlement of growing populations. Modern investments in the Euroports at the mouths of the Rhine and Rhône rivers are important for maintaining Europe's continuing role in world trade.

The Danube River is longer than the Rhine, but the latter is used heavily for transportation and is the world's busiest waterway. Some of the main efforts at managing the Rhine waterway were devoted to making navigation possible for large barges from the international port of Rotterdam at its mouth up to Basel, Switzerland. The river was canalized to that point. Canalization straightened the rivers, producing faster flow with greater channel bed erosion in some sections and silt deposition in others. Channel deposition partly filled some sectors and caused flooding. Higher levees were constructed to protect from flooding the lands on either side that were used more and more intensively. At the Rhine mouth in the Netherlands, great efforts are also put into keeping out the sea, though both the river and sea periodically breach flood barriers.

The Rhine River and its tributaries became a very important transportation corridor as industrialization grew after the late 1800s. The coalfield and steelmaking areas of the Ruhr and Saar are now less productive and polluting, but some 20 percent of the world's chemical industry output occurs along the river, with major centers around Basel, Mannheim, and the Ruhr area. Pollution was at its worst in the 1970s. Since then, an international agreement has been reached to reduce the problem, helped by improved water treatment technology. Concerns remain over nondegradable chemicals and metals that are still at high levels in the river.

Temperate Climates

Europe's climates are mostly temperate and humid, but they include long, freezing winters in the north and dry summers in the south. Most of the region enjoys environments that support farm production and do not require high heating or air conditioning costs to be livable (Figure 2.3). No part of Europe is more than 500 km (320 mi.) from the coast, and mild and humid oceanic atmospheric influences affect the whole region.

Oceanic temperate climates extend from Norway near the Arctic Circle and the summer midnight sun to the southern warmth of Portugal. The southern coasts have mild winters (0–10°C, 30–50°F) and warm summers (18–25°C, 60–80°F) with precipitation through the year. The North Atlantic Drift current, an extension of the Gulf Stream, has a major impact, bringing warmer ocean waters across the North Atlantic Ocean. Inland, the far north of Norway, Sweden, and Finland and the eastern parts of the region have climates with long, very cold winters and months of snow cover.

In the far south along the Mediterranean Sea coastlands, the **subtropical winter rain climate** (Mediterranean climate) prevails. Summers are hot and dry with drought-like conditions.

In the winter, the midlatitude belt of cyclones moves southward, bringing rain and wind. In Central and Eastern Europe the **continental temperate climate** has severe winters as cold winds bring freezing temperatures from Russia and the northern countries. In summer, air over the land brings higher temperatures than on the coasts and thunderstorm rains.

Since the mid-1800s, Europe's climates have been getting warmer. Retreating glaciers in alpine valleys (Figure 2.4) have influenced the tourist industry and the generation of hydroelectricity. This retreat and decreasing extent of snow cover shortened the winter sports season but added warm days for summer vacations. Inlets to hydroelectricity projects are now sited higher up the valleys, increasing the elevation (and energy) of water falling on the electricity-generating turbines.

Many Europeans are concerned about the potential impacts of global warming. Rising sea levels could drown the extensive low-lying parts of Europe, including the many

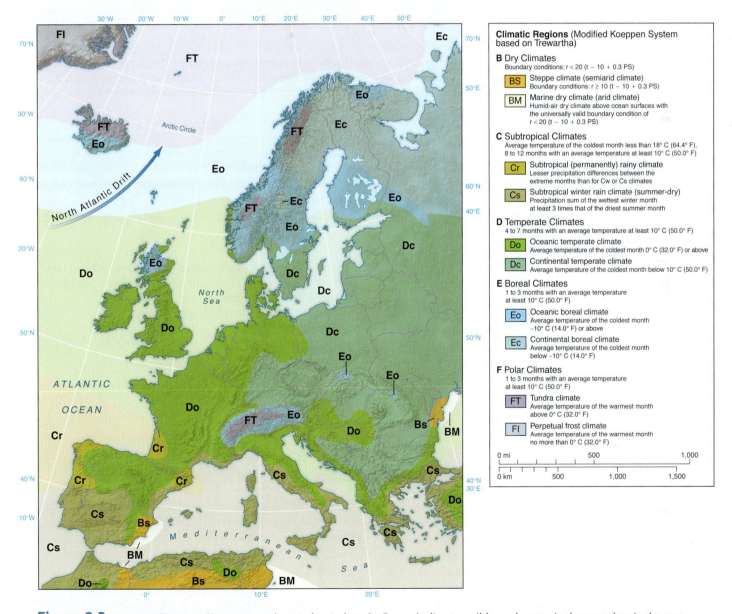

Figure 2.3 **Europe: climates.** The ocean and westerly winds make Europe's climates milder and wetter in the west than in the east.

(a)

(b)

Figure 2.4 **Europe: changing climate.** One hundred years has made a difference in Zermatt, Switzerland. (a) The glacier on the left side of the 1880 painting disappeared in the modern photo, (b), having retreated over a kilometer. Bare rock and icemelt deposits are exposed on the valley floor. Locate the church in both views and compare other 100-year differences. How might the changing climate have affected tourism? *Photos: (a & b) © Photo Klopfenstein-Adelboden.*

areas of former coastal wetland that have been reclaimed and intensively populated. The Netherlands has particular worries, but attention has also focused on the plight of Venice in northern Italy. Venice, a medieval trading city, is one of Europe's greatest architectural treasures and tourist attractions, but it is gradually subsiding (Figure 2.5). Increasingly frequent high tides weaken its foundations. The worst flood in Venice's history occurred in 1966. A south wind raised the tide level by 2 meters (6 ft.), covering St. Mark's Square with oil-polluted water that left stains on all the buildings. In 1990, St. Mark's Square was flooded nine times, and it now deals with flooding almost 200 times per year. In 2008, Venice experienced its fourth-highest flood in modern history. The industries and oil refineries of Venice's lagoon pollute the waters but also demand deepwater channel access, making protection of Venice difficult. As Venice has become increasingly less secure, the historic city's population has fallen from 175,000 in 1950 to 60,000 in 2012.

Forests, Fertile Soils, and Marine Resources

The natural vegetation of Europe is temperate forest. Deciduous trees (oak, elm, chestnut, beech) and associated brown earth soils dominated most lowland environments. Evergreen trees (mainly firs and pines) grew on the thin soils of the uplands and on sandy soils elsewhere, lowering their quality by acidification.

Because clearance of light woodland was easier, initial settlement (which occurred around 5,000 years ago by early Neolithic farmers) favored thinly vegetated uplands on limestone,

sandstone, and some granite rocks. Also favored were the easily cultivated **loess** (windblown glacial debris) soils found along the southern edge of the North European Plain from Poland to northern France and southern Britain. The loess soils also provided an easy route of diffusion for farming technology

Figure 2.5 **Venice.** The medieval port city accumulated huge wealth on its unique site in a lagoon. Modern Venice is prone to flooding, as seen here. *Photo: © Michael Bradshaw.*

originating around the Mediterranean Sea, via the Danube River valley.

After the introduction of Iron Age tools, humans rapidly cut into the denser forest on the lowlands, farming the exposed fertile brown earths. By the time of the Roman occupation of France and lowland Britain, a large proportion of the forest on the lighter lime-rich soils had been cut and some inroads made to the denser forest on heavier soils. Further expansion of the cultivated area onto the clay soils occurred during the Middle Ages. Huge demands for wood to construct ships and for charcoal used in smelting iron resulted in the removal of most natural woodland by around 1600. Later reforestation and the importing of exotic species added diversity to European woodlands.

As soon as humans cut the forest, soil washed down hillsides more rapidly and added silt to rivers. In the Middle Ages, for example, a combination of growing populations, rigid political systems, close grazing by sheep and goats, and climate change caused intense soil erosion on the hills of the Mediterranean peninsulas. In much of Europe outside the alpine area, however, the slopes are less steep, and cultivation methods were adopted that maintained the soils and their productivity over many centuries. In the 1800s, competition from cheap grain imported from newly opened and settled lands in North and South America resulted in once-plowed lands in Europe being sowed with grass for livestock

production, further reducing soil erosion and helping to maintain soil quality.

In the Mediterranean, Baltic, and North seas and the North Atlantic Ocean fishing, related ports, and ships grew in significance in the later medieval period. In the 1900s and 2000s, overfishing led to decreasing supplies and made ocean fishing allocations a source of contention and conflict among European countries.

Environmental Impacts

Today, European countries are concerned about maintaining environmental quality, spending billions of dollars protecting the natural environment (see "Geography at Work: Meeting the Challenge of Climate Change," p. 74).

Impacts of Industrialization

The Industrial Revolution and the spread of factory-concentrated production led to widespread pollution of the rivers and air (Figure 2.6). Many rivers lost their fish stocks. Occasional major pollution incidents still result in fish kills in major rivers such as the Rhine. Various governmental agencies are making great efforts to improve the quality of river water in European countries. In the Thames River of England, the reduction of pollution was so successful that fish stocks revived in the 1990s after decades of absence.

Winter smoke fogs (smog) that blighted the major industrial and urban areas of Europe in the 1950s have greatly declined due to legislation that outlawed domestic and industrial coal burning. Emissions of sulfur compounds from thermal power stations burning coal created **acid deposition**—of dry particles near the source or of wet "acid rain" farther downwind. It affected coniferous forest on thin soils and shallow lakes in the Alps and Scandinavia. The highest levels of acid pollution occurred during the Communist era along the Czech, Polish, and East German borders in an area known as the **Black Triangle**. Since the 1990s, acid rain and its effects have been greatly reduced, and alpine forests have recovered.

Figure 2.6 Environmental degradation in East Central Europe. Pollution levels dropped in many areas of East Central Europe with the end of communism in 1991, after many inefficient factories closed because they could not compete in the global market. UN funds were provided to shut down this heavily polluting factory in Copsa Mica, Transylvania, Romania. Considered the most polluted place in Europe, local people called it "black town" because everything was black, from the air and sky to laundry on clothes lines and children's faces. Shown a few years after operations ceased, the sky is now blue and the grass green again. *Photo: © George W. White.*

Global Environmental Action

The negative environmental effects of industrialization have motivated many Europeans to engage in global efforts to reduce pollution. Representatives from European governments were active participants in the Rio Earth Summit in 1992 and the Kyoto conference in 1997. European countries comprise most of the industrialized countries on Annex I of the Kyoto Protocol (see discussion in Chapter 1, p. 18), accounting for more than 32 percent of global carbon dioxide (CO_2) emissions. Their ratification was key to making the protocol go into effect. When they accepted the Kyoto Protocol, European countries agreed to reduce their greenhouse gas emissions by 2012. They later hosted an environmental conference in Copenhagen, Denmark, in 2009 and actively engaged in Rio 2012 or Rio+20, a conference in Rio de Janeiro, Brazil, in 2012 (see Chapter 1, p. 18).

In 2005, the European Union Emission Trading System was established. It provides economic incentives for companies to emit less carbon dioxide. Known commonly as a "cap and trade" system, the European Union sets a maximum amount of carbon dioxide that individual companies are allowed to emit. Those companies wishing to emit more can purchase allowances from companies that emit less than their caps and thereby can trade/sell their excess permits. In 2007, the European Commission created the "20/20/20 by 2020" plan, which seeks to cut emissions 20 percent below 1990 levels, increase energy efficiency by 20 percent, and obtain 20 percent of energy from renewable sources by 2020. Norway became the most ambitious country in the world when it declared in 2008 that it plans to be carbon neutral by 2030. It will achieve this in part by investing in reforestation projects around the world to offset its own emissions.

2.3 Distinctive Human Geography

Though Europe is small in area compared to other world regions, it has a complex mosaic of languages and religions. In comparison to much of the rest of the world, Europe's populations are rapidly aging and very urbanized, with major European cities serving as important nodes in the global economy. Over the last 500 years, Europeans developed new forms of identities and governance, in turn creating a template for modern nations and states. Economically, Europeans began creating new global trade routes in the late 1400s, garnering them great material wealth and technological advantages. Europe became the hearth of the Industrial Revolution and today continues at the forefront of modern agricultural practices, sophisticated manufacturing industries, and service industries such as tourism.

Cultural Diversity

Patterns of Language

Most Europeans speak languages in the Indo-European language family, which fall into one of three major groupings: Romance, Germanic, and Slavic (Figure 2.7a). The Romance languages stem from Latin, the language of the ancient Romans, and generally are still found in many of the lands that the Romans controlled longest. After Rome fell and its unifying influence was destroyed, Latin splintered and evolved into the differing Romance languages spoken today. The major ones are Portuguese, Spanish, French, Italian, and Romanian.

Spoken to the north and west of the Romance languages are the Germanic languages, which include English, Flemish, Dutch, German, Danish, Norwegian, and Swedish. English was heavily influenced by French after the Normans invaded England in AD 1066. Thus English has many Romance words, causing many to think that it is a Romance language though it is fundamentally a Germanic language. The influence of French has also given the English language a comparatively large vocabulary.

The Slavic languages dominate in East Central Europe and are divided into three major groups: western, southern, and eastern. The western group includes Polish, Czech, and Slovak. The southern group is comprised of Slovenian, Croatian, Bosnian, Serbian, and Bulgarian. The Slavic word for *south* is *yug,* hence the name Yugoslavia ("land of the south Slavs"). The eastern Slavic languages are Russian, Ukrainian, and Belarussian and are spoken in Russia and Neighboring Countries (see Chapter 3).

In addition to these major language groupings, other smaller groupings and individual languages are noteworthy. The Celtic language group once was spoken in large areas of Central and Western Europe but was driven to the northwestern fringes of Europe by the Germanic peoples. Today, Irish Gaelic (Erse), Scottish Gaelic, Welsh, and Breton are the most commonly spoken of this group. In southeastern Europe, Greek and Albanian are spoken. Other languages found in Europe are Latvian, Lithuanian, Estonian, Finnish, Hungarian, Rom (Gypsy), and Basque.

Patterns of Religion

Christianity predominates in Europe, though Islam and Judaism are important religions in the region (Figure 2.7b). Christianity became significant in Europe after the Romans adopted it as their empire's official religion in AD 381. The empire broke in two, leaving a Roman Catholic, Western Europe centered in Rome and an Eastern Orthodox, Eastern Europe centered in Constantinople (now Istanbul). The religious split became official in AD 1054. Eastern Orthodoxy organized along national lines and includes Greek Orthodoxy, Bulgarian Orthodoxy, and others. Roman Catholicism remained more homogeneous until the Protestant Reformation in the 1500s. At that time, the northern, mostly Germanic areas outside of or on the fringes of the former Roman Empire broke from Rome.

Jews have lived in Europe since Roman times, first along the Mediterranean and then eventually in small groups everywhere else, especially East Central Europe. Jews contributed greatly to European culture. Persecution of the Jews occurred through the centuries and culminated in the Nazi Holocaust in the 1930s and 1940s that sharply reduced their numbers. The *American Jewish Yearbook* estimates that in 1933, 7 million

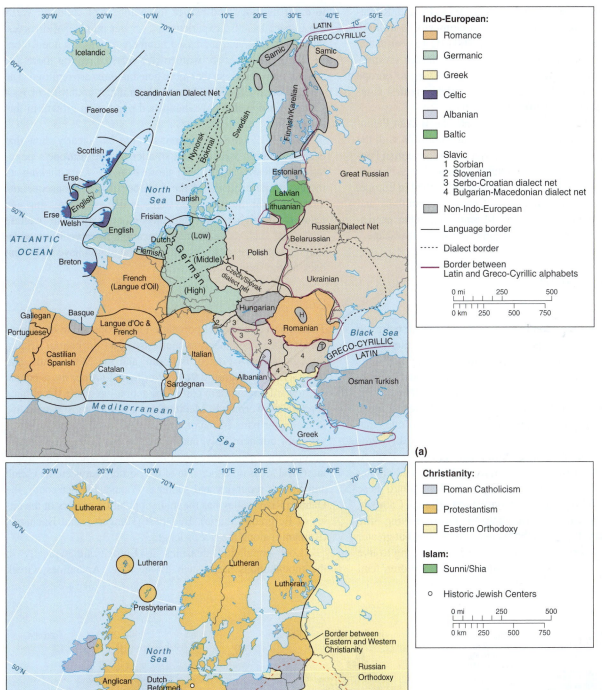

(a)

Indo-European:

Romance

Germanic

Greek

Celtic

Albanian

Baltic

Slavic
1 Sorbian
2 Slovenian
3 Serbo-Croatian dialect net
4 Bulgarian-Macedonian dialect net

Non-Indo-European

—— Language border

----- Dialect border

—— Border between
Latin and Greco-Cyrillic alphabets

0 mi   250   500
0 km   250   500   750

(b)

Christianity:

Roman Catholicism

Protestantism

Eastern Orthodoxy

Islam:

Sunni/Shia

○ Historic Jewish Centers

0 mi   250   500
0 km   250   500   750

Border between
Eastern and Western
Christianity

Figure 2.7 **Europe: languages and religious groups.** (a) Most European languages are in the Indo-European language family. The Latin alphabet, such as that used for English, is used by languages to the west of red line, and generally in Roman Catholic and Protestant Europe. The Greco-Cyrillic alphabets are generally used by Eastern Orthodox Christians to the east of the line. (b) Most Europeans are members of a Christian denomination though secularization is strong and Islam is growing. Are there any spatial correlations between language and religious groups? *Source: Data from Jordon-Bychkov.*

Jews lived in Europe as defined in this chapter. Poland, Romania, and Germany had the largest populations with 3 million, 980,000, and 565,000 respectively. In 2007, it was estimated that these countries had 3,200, 9,900, and 120,000 Jews respectively. Of the approximately 1.2 million Jews living in Europe in 2007, France and the United Kingdom had the greatest numbers with 490,000 and 295,000 respectively.

In the Middle Ages, Islam spread into the Balkan Peninsula with the Ottomans and into the Iberian Peninsula with the Moors. After 1492, Muslims were driven out of the Iberian Peninsula. Islam is still a significant religion in the Balkans. Following World War II, millions of Muslim migrants settled in Europe, especially in France (5 to 6 million), Germany (3 million), and the United Kingdom (1.6 million). Mosques and other expressions of Islamic culture are increasingly seen in the landscape.

Aging Populations

Dynamics

European countries are notable for their zero population growth rates. Europe's population is decreasing in most Mediterranean and East Central European countries and increasing only a little in most Western and Northern European countries (Table 2.1). Rates of natural population change ranged between −0.5 and 1.0 percent in 2012. Some countries, such as Germany, Italy, and many in East Central Europe, have death rates that are above birth rates. Throughout Europe, total fertility rates declined from as high as three births per female in 1965 to commonly less than two in 2008. Italy and Spain had among the lowest total fertility rates (1.4) in the world. Though these two countries and Portugal are predominantly Roman Catholic, the church's opposition to birth control is clearly having little effect. In Italy, few babies are born outside marriage, but greater access to careers for women, young people continuing to live with parents, and the end of pressures to have children tend

to defer marriage and reduce the numbers of children. Western Europe has also experienced a greater number of divorces, later marriages, and increased numbers of widows due to longer life expectancy for women (83 years), creating more and smaller households so that more housing units are required. The weak economies of the former Communist countries of East Central Europe also have dampened population growth in that subregion.

The 2012 age-sex diagrams (Figure 2.8) clearly show that each generation is becoming smaller in size. Europe's population is clearly aging, with those over 65 years making up increasing percentages of the total population. Consequently, the burden on the welfare systems is likewise increasing and resulting in higher taxes and the reduction of some welfare programs, such as health and education, after years of wide-ranging coverage. Many countries in the region have had to cut back on the services they provide.

Densities

The highest densities of population in Europe are in the urbanized industrial belt that runs southeastward from central Britain, through northern France, Belgium, and the Netherlands, and into Germany (Figure 2.9). In Northern Europe, most people live toward the southern part of the subregion where the climate is relatively warmer. In Mediterranean Europe most people live in the lower parts of major river valleys, such as the Po River valley of northern Italy, and along the coasts. In East Central Europe, the main concentrations of people are in the urban-industrial areas on either side of the borders between the Czech and Slovak republics and southern Poland. Farther south, the Danube River valley has the main population centers in the cities of Belgrade (Serbia), Budapest (Hungary), and Bucharest (Romania) along the river's length.

Low population densities in Europe are typically in the mountainous areas such as the Alps, Apennines, Greek

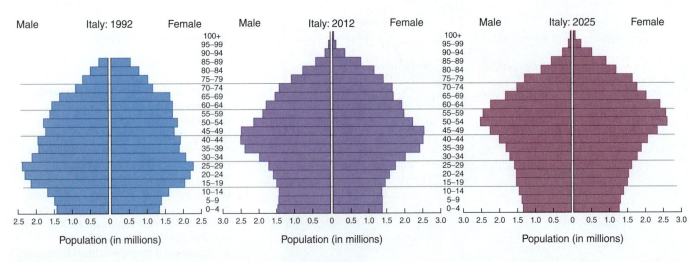

Figure 2.8 **Age-sex diagram for Italy.** The narrowing base of each successive pyramid is a feature of countries with declining birth and fertility rates. *Source: U.S. Census International Database: http://www.census.gov/population/international/data/idb/region.php.*

Table 2.1	**Europe:** Data by country, area, population, urbanization, income (gross national income purchasing power parity), ethnic groups						
		Population (millions)		**% Urban**	**GNI PPP 2010**	**2010**	
Country	**Land Area (km²)**	**Mid-2012 Total**	**2025 Est. Total**	**2012**	**Total (US$ billions)**	**Per Capita (US$)**	**Ethnic Groups (%)**
WESTERN EUROPE							
Austria, Republic of	83,850	8.5	8.9	67	333.9	39,790	German 99%
Belgium, Kingdom of	33,100	11.1	12.1	99	417.3	38,290	Flemish 55%, Walloon 33%
French Republic	551,500	63.6	67.4	78	2,254.9	34,750	Celtic/Latin with Teutonic, Slavic, Nordic, African, Indochinese, Basque
Germany, Federal Republic of	356,910	81.8	79.2	73	3,115.4	38,100	German 92%, Turkish 2%
Ireland, Republic of	70,280	4.7	5.6	60	150.1	33,540	Celtic, English
Luxembourg, Grand Duchy of	2,586	0.5	0.6	83	—	64,320	Celtic/French/German 75%, some Portuguese, Italian
Netherlands, Kingdom of	37,330	16.7	17.4	66	694.7	41,810	Dutch 96%, Moroccan, Indonesian, Turkish
Swiss Confederation	41,290	8.0	8.6	74	391.0	49,960	German 65%, French 18%, Italian 10%
United Kingdom of Great Britain and Northern Ireland	244,880	63.2	70.5	80	2,230.6	35,840	Celtic, English 94%, South Asian, black, other 6%
Totals/Averages	**1,421,726**	**258.1**	**270.3**	**77**	**9,588**	**37,148**	
NORTHERN EUROPE							
Denmark, Kingdom of	43,090	5.6	5.8	72	228.0	41,100	Danish, Inuit (Eskimo), Faroese, Greenlander
Finland, Republic of	338,100	5.4	5.8	68	198.9	37,070	Finn 93%, Swede 6%
Iceland, Republic of	103,000	0.3	0.4	94	—	25,220	Norwegian/Celtic descendants
Norway Kingdom of	323,000	5.0	5.9	80	286.3	58,570	Germanic (Nordic, alpine, Baltic)
Sweden, Kingdom of	449,960	9.5	10.2	84	372.6	39,730	Swedes, Finns, Danes, Norwegians, Greeks, Turks
Totals/Averages	**1,257,150**	**25.8**	**28.1**	**78**	**1,086**	**42,093**	
MEDITERRANEAN EUROPE							
Hellenic Republic (Greece)	131,990	10.8	11.1	73	312.7	27,630	Greeks 98%
Italian Republic	301,270	60.9	63.1	68	1,923.7	31,810	Italian, Sicilian, Sardinian, German, French
Portuguese Republic	92,390	10.6	10.7	38	261.6	24,590	Mediterranean
Spain, Kingdom of	504,780	46.2	47.3	77	1,465.2	31,800	Spanish 74%, Catalan 16%, Basque 2%
Totals/Averages	**1,030,430**	**128.5**	**132.2**	**69**	**3,963**	**30,842**	
EAST CENTRAL EUROPE							
Albania, Republic of	28,750	2.8	2.9	54	23.7	8,520	Albanian 95%, Greek 3%
Bosnia-Herzegovina, Republic of	51,130	3.8	3.7	46	33.5	8,910	Muslim 40%, Serb 38%, Croat 22%
Bulgaria, Republic of	110,910	7.2	6.7	73	101.2	13,440	Bulgarian 85%, Turkish 9%
Croatia, Republic of	56,540	4.3	4.1	56	83.4	18,890	Croat 78%, Serb 12%
Czech Republic	78,860	10.5	10.9	74	241.0	22,910	Czech 81%, Moravian 13%, Slovak 3%
Estonia, Republic of	45,100	1.3	1.3	69	26.5	19,810	Estonian 64%, Russian 29%
Hungary, Republic of	93,030	9.9	9.8	69	195.5	19,550	Hungarian (Magyar) 90%, Romany (gypsy) 4%
Kosovo	10,887	2.3	2.7	—	—	—	Albanian 92%
Latvia, Republic of	64,500	2.0	1.9	68	36.7	16,380	Latvian 55%, Russian 32%
Lithuania, Republic of	65,200	3.2	3.0	67	59.4	18,060	Lithuanian 80%, Russian 8%, Polish 8%
Macedonia, Former Yugoslav Rep. of	25,710	2.1	2.1	65	22.5	10,920	Slav 65%, Albanian 21%
Montenegro	13,812	0.6	0.7	64	—	—	Montenegrin 43%, Serbian 32%, Bosniak 8%, Albanian 5%
Poland, Republic of	312,680	38.2	37.4	61	731.5	19,160	Polish 98%
Romania	237,500	21.4	20.7	55	306.4	14,290	Romanian 89%, Hungarian 7%, Roma (gypsy) 2%
Serbia	102,170	7.1	7.0	59	80.8	11,090	Serb 62%, Albanian 17%, Montenegrin 5%
Slovak Republic	49,010	5.4	5.6	54	124.8	22,980	Slovak 86%, Hungarian 11%, Romany (gypsy) 2%
Slovenia, Republic of	20,050	2.1	2.2	50	54.4	26,530	Slovene 88%, Croat 3%, Serb 2%
Totals/Averages	**1,341,140**	**124.2**	**122.7**	**59**	**2,121**	**17,079**	
Europe Totals/Averages	**5,050,446**	**536.6**	**553.3**	**71**	**16,758**	**31,230**	

Source: *World Population Data Sheet 2008*, Population Reference Bureau; Microsoft Encarta 2005.

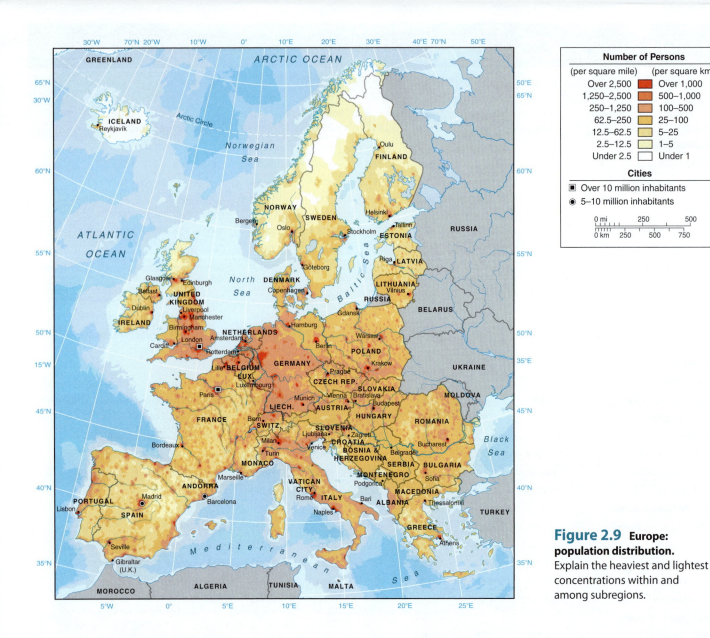

Figure 2.9 **Europe: population distribution.** Explain the heaviest and lightest concentrations within and among subregions.

mountains, the Pyrenees, the Carpathians and the Dinaric Alps or in the areas of extreme climates such the colder areas of northern Europe or the hot, dry interior of central Spain (compare Figures 2.1 and 2.3 with Figure 2.9).

Urban Pressures

European countries are among the most highly urbanized countries in the world. The high proportions reflect the economic focus on urban-based manufacturing and service industries. Europe has few extremely large cities, since functions are often spread among several cities (Table 2.2). Paris (France) and London (United Kingdom) are by far the largest urban centers. Many cities have populations in the 1 to 3 million range.

The geographic characteristics of the city landscapes of Europe result from centuries of making and remaking built environments. They record the consequences of change

(Figure 2.10). A number of Western and Mediterranean European cities contain within their centers relics of historic cultures such as ancient Greek, Roman, and Moorish. In medieval times, a network of towns grew all over Europe with walled defenses and market and administrative functions. Internal spatial differentiation by class occurred within buildings (with servants living in attics and basements and having to leave and enter from the back of buildings) rather than between central cities and suburbs.

With the Industrial Revolution of the 1800s, cities grew tremendously, and land uses differentiated into central business districts of shops and offices, industrial areas of large factories and worker housing, and suburbs in which the growing management classes lived. In the early 1900s, the development of public transportation in cities led to further differentiation of land use and the building of suburban housing linked by streetcar or bus to the city centers where retailing, commercial, and manufacturing activities concentrated.

During World War II, bombing and ground fighting reduced parts of many European cities to rubble, destroying older areas, including medieval and industrial buildings. The period from 1945 to 1970 was one of rehabilitation, expansion, and restructuring of cities. Nearly all cities expanded their functions and populations. City centers were rebuilt with utilitarian buildings (Figure 2.11), while large tracts of public housing catered to lower-income groups. New towns were built at a distance from the largest cities, separated by "green belts" of rural land in which new building was seldom allowed. A major result was decentralization—a movement away from the previous focus on the central business district and toward more economic activity in the extensive suburbs built since 1945.

Beginning in the 1970s, disillusionment grew in non-Communist Europe over the outcome of postwar reconstruction and the bleak nature of low-cost public housing. Large public housing tracts, especially where high-rise buildings were common, proved unpopular and often became centers of

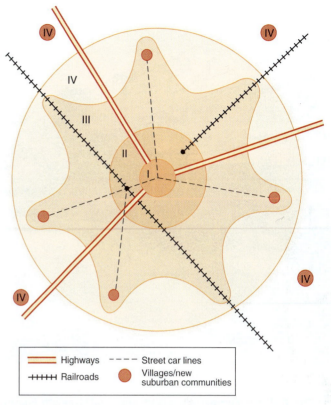

Legend:
— Highways
-- Street car lines
++++ Railroads
● Villages/new suburban communities

Figure 2.10 **Generalized morphology of non-Communist European cities.** I = Medieval core; II = Early growth area later penetrated by railroads and the location of early industrial factories and worker housing; III = Star-shaped pattern created by suburbanization along streetcar lines in the early 1900s; IV = Post–World War II suburbanization with automobiles.

Table 2.2	Populations of Major Urban Centers in Europe	
TOP 10 MOST POPULATED URBAN CENTERS		
City, Country	**2012 (millions)**	**2025 (millions)**
1. Paris, France	9.9	10.0
2. London, United Kingdom	8.6	8.6
3. Madrid, Spain	5.8	6.2
4. Barcelona, Spain	5.1	5.3
5. Berlin, Germany	3.5	3.6
6. Rome, Italy	3.3	3.3
7. Athens, Greece	3.2	3.3
8. Milan, Italy	2.8	2.9
9. Lisbon, Portugal	2.8	3.1
10. Vienna, Austria	2.4	2.6
NUMBER OF MAJOR URBAN CENTERS IN EUROPE, BY POPULATION CATEGORY		
Population Category	**Number of Urban Centers**	
10.00 million and greater	1	
5.00 – 9.99 million	3	
2.50 – 4.99 million	7	
1.00 – 2.49 million	31	

unemployment and crime. A significant move began toward public and private cooperation, particularly in efforts to make city centers and old dockland waterfronts more livable. Areas with buildings of historic interest were declared conservation areas, but this time the finances for pursuing such a policy came from private investors who saw the potential returns from encouraging tourism and building new accommodations for those wishing to move back near the city center.

Older couples whose children had become independent and younger professionals moved to neighborhoods near city centers. They bought and renovated poor and dilapidated housing, increasing the value of the housing stock and raising neighborhoods from low income to higher income. This process is known as **gentrification**. Other changes affected city centers, bringing pedestrian-only traffic, arcades of specialty shops, and the congregation of high-order services requiring support from people living in a wider area—theaters, opera houses, cinemas, and clubs. Areas around the city center that had become derelict, such as old railroad yards, factories, workshops, and worker housing, became sites for the new facilities.

(a)

(b)

Figure 2.11 **Europe: urban landscapes.** (a) Plymouth, England. The central shopping area was rebuilt after being bombed in World War II. It remains the largest retail district in the city despite the growth of out-of-town supermarket and warehouse shopping facilities since the 1980s. Some streets in this shopping area were made into pedestrian zones in the early 1990s. (b) Warsaw, Poland. The 38-story Palace of Culture was built with Soviet Union funds in the 1950s. It houses scientific and cultural institutions, theaters, and Congress Hall. *Photos: (a) © Michael Bradshaw; (b) © Jerzy Jemiolo.*

In the 1980s, the relaxation of planning constraints led to a further dispersal of economic activity from city centers to the suburbs and beyond. Manufacturing had already moved out from cramped inner city sites to suburban industrial estates. Now retail and office facilities followed to become suburban warehouse shopping, hypermarkets, shopping malls, and office parks. After communism ended in East Central Europe around 1990, East European cities began experiencing this same phenomenon. Though these changes parallel what has been happening in the United States, densities of European cities are still much higher than in the United States, and suburbanization is much less pronounced.

Global City-Regions

One way that Europe plays a leading role in the global economic system is through its cities, where multinational corporations site factories and where sophisticated service industries play important global roles. These European cities are no longer at the top of the list of the most populous cities in the world, but they are among the most globally connected cities. London and Paris, along with only New York and Tokyo, are the most important cities for accounting, adver-

tising, banking, and law. When we think of cities such as London and Paris, we should not simply consider their businesses as only serving Londoners and Parisians, respectively. The businesses within them serve people all over the world, whether in the Americas, Africa, or Asia. Many millions of dollars flow in and out of these cities from and to other countries around the world. After London and Paris, cities such as Frankfurt, Milan, Zürich, Brussels, and Madrid rank high in their global importance.

Evolving Politics

Modern Countries: Nation-States

Beginning shortly after AD 1000, the identities of Europeans changed as people began to shift their loyalties from more local feudal leaders to large social groups and emerging countries. With the French Revolution in 1789, the idea of the nation emerged. As noted in Chapter 1, a nation is an "imagined community" of people who believe themselves to share common cultural characteristics (e.g., language, religion, ideology). Soon many Europeans believed in the idea that each nation should be free to govern itself and can only do so if it has its own **state** (i.e., country), which is a politically organized territory with an independent government. Thus, nations become linked with states in what is known as the **nation-state**.

Many Europeans spread their ideas of the nation and the nation-state around the world. Today each country in the world is also called a nation-state, that is, the home of a single people (e.g., France as the home of the French, Japan as the home of the Japanese, etc.). However, few countries truly live up to the definition of the term. Even France, where the nation-state concept originated, contains peoples such as the Basques who regard themselves as a nation separate from the French. Thus, the European belief in linking nation to state has created more of a **nation-state ideal** than a reality. Nevertheless, since the age of nationalism began in the late 1700s, many dominant nations imposed their national cultures on other nations living within their borders in an attempt to make the nation-state ideal a reality. In response, many minority peoples resisted, and some have tried to establish their own nation-states. The aforementioned Basques are an example, and so are the Palestinians and Kurds (see Chapter 7). The nation-building projects of dominant peoples who try to homogenize their states and of minority peoples who try to declare their own nation-states have caused much violence, even war (see "Human Rights," p. 72).

Not all nations have been able to incorporate their nations' territories into their nation-states. Stronger nations hold onto some territory that other nations consider their own. These situations have caused the phenomenon of **irredentism**, which is the desire to gain control over lost territories or territories perceived to belong rightfully to one's group. With boundaries having changed so frequently in European history, irredentism is a major issue and a source of much conflict, more so than conflicts arising from simple cultural differences among groups.

Nationalism and World Wars

As capitalism and nationalism grew in Europe, so did competition. European armies that were created to conquer and colonize the rest of the world were turned toward one another in 1914 as war erupted between the European powers, later to be known as World War I. Though Germany and Austria-Hungary were decisively defeated in 1918, the trouble was not over. War costs and protectionism caused European economies to slump in the postwar 1920s and 1930s, bringing hardship to millions of individuals and families. Discontent and resentment grew in the defeated countries. The nationalist competition became more bitter and fed the more extreme but opposing ideologies of fascism and communism, two other concepts Europe gave to the world. A European war engulfed the rest of the world once again between 1939 and 1945, known as World War II. The intolerant side of nationalism under fascism led to the extermination of millions of people of specific groups such as Jews (see "Patterns of Religion," p. 46) and Roma (Gypsies), a phenomenon called **genocide**.

Europe Since 1945

After World War II, many Europeans seriously reevaluated their role in the world and their relationships with one another. The war had been so devastating that even the winners suffered destruction and huge financial, political, and cultural losses. In Western Europe, the United Kingdom, France, and the Netherlands were confronted with independence movements in their colonies at a time of weakness. Fueled by the ideologies of nationalism and communism that originally came from Europe, these movements ended four centuries of building colonial empires on which the "sun would never set."

At the same time, the politically and economically weak European countries were faced with the United States and the Soviet Union, which emerged as new world powers. At the end of World War II, the Soviet Union showed its strength when the Red Army moved into most of the countries of East Central Europe and fostered the establishment of **communism**. The Communists believed that capitalists used their riches to manipulate their governments in order to protect, even increase, their privileged positions in society and keep the majority of society, especially the working classes, powerless and in relative poverty. Instead, Communists argued for **democratic centralism**: the belief that the Communist Party, claiming to be the political party of the working class, was the only true representative of the people and, therefore, the only party with the right to govern. To keep capitalists and others from taking advantage of the people, Communists also believed in **state socialism**: governance by the Communist Party, which actively runs the political, social, and economic activities of the people. The state owned all the businesses and decided what was produced. The capitalist practice of competing companies producing similar products was seen as wasteful. Rather, large corporations owned by the state made each product. The state, not the free market of consumers, decided what needed to be produced through a **planned economy**.

As the Soviet Union imposed its Communist political and economic systems on East Central Europe, the rest of Europe (i.e., Western, Northern, and Mediterranean) immediately began to cooperate to counter further Soviet moves. In 1949, they formed the **North Atlantic Treaty Organization (NATO)** with the United States, which was seen as an ally against the Soviet military threat (Figure 2.12). To counter NATO, in 1955 the Soviet Union created a similar military alliance, known as the Warsaw Pact, among East Central European countries.

After the breakup of the Soviet Union in 1991, many questioned the need for NATO. Others feared that Russia would eventually become a formidable power again, though the Soviet Union no longer existed and Russia was weak. Having emerged from more than 40 years of Soviet domination, East Central European countries began joining NATO. In 2002, NATO formed a partnership with Russia, the country that NATO was created to defend against! Post–Cold War NATO is actually more focused on resolving or policing disputes within the expanded Europe and its immediate neighbors—as in Bosnia-Herzegovina, Kosovo, and Macedonia.

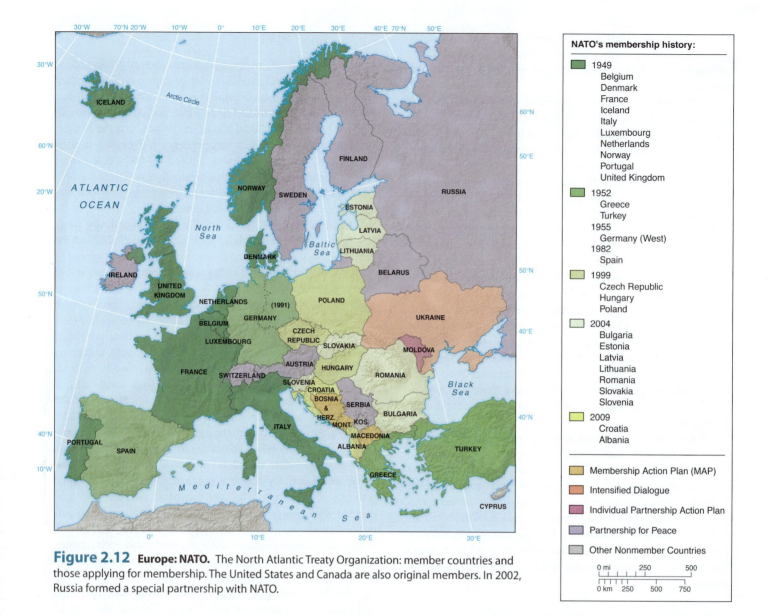

Figure 2.12 **Europe: NATO.** The North Atlantic Treaty Organization: member countries and those applying for membership. The United States and Canada are also original members. In 2002, Russia formed a special partnership with NATO.

NATO's membership history:

1949
Belgium
Denmark
France
Iceland
Italy
Luxembourg
Netherlands
Norway
Portugal
United Kingdom

1952
Greece
Turkey
1955
Germany (West)
1982
Spain

1999
Czech Republic
Hungary
Poland

2004
Bulgaria
Estonia
Latvia
Lithuania
Romania
Slovakia
Slovenia

2009
Croatia
Albania

Membership Action Plan (MAP)

Intensified Dialogue

Individual Partnership Action Plan

Partnership for Peace

Other Nonmember Countries

Global Economics

Capitalism and Imperialism

Of the European influences mentioned in the beginning of the chapter, capitalism characterizes today's global economy. Capitalism, the practice of individuals and corporations owning businesses and keeping profits, began in the late 1400s when merchants invested in trade expeditions that brought in precious metals (gold and silver). The best known of these voyages was made by Christopher Columbus in 1492. Funded by the Spanish crown, Columbus sailed westward in the belief he would get to India by a quicker route. He did not know that the Americas (supposedly later named after another Italian sailor, Amerigo Vespucci) lay in his path. At the same time, Vasco da Gama led the Portuguese explorations around the southern tip of Africa to India. The French, Dutch, and British followed the Spanish and Portuguese in the 1500s and 1600s. They all discovered and conquered new lands, radically altering and frequently destroying many local economies and cultures of indigenous peoples as they began a new era of **imperialism**. They also frequently pursued imperialism through **colonialism**, the settlement of their peoples in colonies in the new lands they conquered.

Industrial Revolution

European technological innovations in the mid-1700s led to the **Industrial Revolution**, which was characterized by the use of machines for the mass production of goods. The resulting need for workers led to the growth of large urban centers devoted to industrial work. Though machinery was first driven by waterpower, coal soon became the major energy source, leading to the growth of many urban-industrial centers near coalfields (Figure 2.13). With rivers, canals, and the sea as the initial forms of transportation used to transport raw materials and finished products, huge port facilities grew in the estuaries of Europe's major rivers. During the 1800s, railroads became more important and connected ports and coalfield industries.

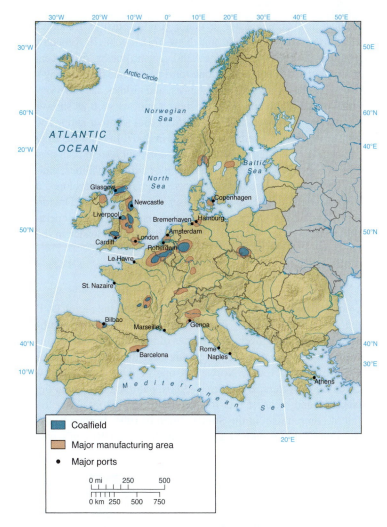

Figure 2.13 **Europe: major manufacturing areas in the early 1900s.** Note: Contemporary boundaries are drawn for reference.

The Industrial Revolution began in Great Britain and then spread across the English Channel to the Netherlands, Belgium, northern France, and the western areas of Germany. In the late 1800s and 1900s, the Industrial Revolution diffused further to central and eastern Europe and to other areas of the world. The European empires used their colonies to produce the raw materials needed in Western European factories. Cotton, wool, indigo, tobacco, and foodstuffs are a few examples. Forced to supply Western Europe with raw materials and having Europe as their only source of finished products, the colonies were pushed into economic dependency. Local economies and traditional ways of life were brought to an end as peoples in the colonies had to change their lives radically to produce exports for Europe.

Growing Economies

Two world wars and intervening economic depressions shattered the economies of European countries and their economic relationships with the rest of the world. After 1945, greater governmental intervention restored **productive capacity** (the amount of goods a country's businesses can produce) and now

ensures education, health care, unemployment benefits, and pensions for all. In non-Communist Europe, older, "heavy" (**producer goods**) industries, such as steelmaking, heavy engineering, and chemicals located on coalfields, were replaced in value of output and employment by motor vehicles, consumer goods, and light engineering products. The availability of electricity spread, so that new products could be manufactured wherever there was plentiful semiskilled labor, often in the larger cities that were also large markets for the consumer goods. This process led to greater material wealth in the cities compared to the older industrial and rural areas. Unemployment increased in the polluted old industrial centers.

New industrial areas grew at locations more suited to the needs of developing technologies and industries, although long-established production continued in the older areas. It was too costly to move old production facilities. The advantages of producing goods in an area having a trained labor force, assembly and distribution systems, and financial and other services, build up and reinforce the original locational advantages as **agglomeration economies**. Keeping production in an area despite higher costs than in possible competing areas creates **geographic inertia**. It is only when a substantial change in costs occurs, new products emerge, or the demand for the original product is reduced that new areas develop and older manufacturing centers decline.

Deindustrialization occurred when the numbers of jobs in manufacturing fell rapidly and factories became derelict in older industrial areas. A decline of 20 percent in European manufacturing jobs between 1970 and 1985 was more than balanced by a 40 percent rise in tertiary sector jobs. However, the skills of blue-collar miners and factory production-line workers were seldom convertible into the new white-collar office, hospital, or classroom jobs that were taken by younger and better-educated people. Those workers unable to retrain for the new jobs faced long-term unemployment. Older production workers, female workers, and poorly educated young people entering the labor force were particularly at risk.

Western Europe accounts for a large share of Europe's manufacturing, research and development, and high-tech industries. Germany in particular has one of the world's largest economies and the largest in Europe. It is often referred to as the "motor" of Europe, and it is responsible for much of Western Europe's economic strength. Total German exports are nearly equal to the combined exports of France, the United Kingdom, and the Netherlands. Its central position helped Germany to be the major trading partner for approximately 15 European countries as of 2011. In contrast, the United Kingdom was the main supplier of goods for only Ireland, and the Netherlands was the main supplier of only Belgium. France was not the main supplier of any other European country and was the secondary supplier of only Italy and Spain.

Sophisticated Manufacturing Industries

Two of Europe's major industries are automobiles and airplanes. Originally, the automobile industry was nationally based. For example, French companies were Bugatti, Renault, and Peugeot-Citroën; Italian producers were Fiat, Alfa Romeo, Ferrari, Maserati, and Lamborghini; German firms were Volkswagen, Porsche, BMW, Audi, and Daimler-Benz (Mercedes); and British companies were Rolls Royce, Jaguar, Aston Martin, Bentley, and MG Rover. Globalization has radically changed the European automobile industry. Foreign companies have made great inroads in Europe in recent years. However, European carmakers have extended their operations abroad and into one another's countries. (See Figure 1.28.) As automobile companies also are purchasing and investing in one another, it is no longer accurate to associate a company with a single country anymore. For example, Volkswagen now owns Bentley, Bugatti, Lamborghini, SEAT (Spain), and Škoda (Czech Republic). Renault purchased 44.4 percent of Nissan, and Nissan bought 15 percent of Renault to form the Renault-Nissan Alliance in 1999. Renault also acquired Dacia (Romania) in 1999, Samsung Motors (South Korea) in 2000, and a 25 percent share of AvtoVAZ (Lada) of Russia in 2008. In 2010, Renault-Nissan entered an alliance with Mercedes-Benz whereby both companies now provide engines to each other. In 2009, Fiat (Italian) began acquiring shares of Chrysler, reaching a 61.8 percent share by mid-2012. This gave Fiat the ability to reintroduce its cars into the United States after a 27-year absence. For a time in the 1990s, BMW owned MG Rover but then sold most of it. In 2007, SAIC (China) became parent company of MG Motor. In 2008, Tata Motors (India) took possession of the Rover brand, along with Jaguar Land Rover. No major British automobile companies exist anymore, although their names live on. Nevertheless, other European automakers are globally active by building automaking plants in Poland, Turkey, Brazil, India, and China. For example, Fiat's Italian factories are unprofitable compared to its factories in Latin America, Poland, and the ones that it contracts with in Russia. Fiat's new partnership with Chrysler will allow it to either withdraw from Italy or force Italian autoworkers to accept lower wages.

The aerospace industry is another key sector, with aeronautical research and manufacturing well developed. Defense was a reason for the national support of this industry, but commercial production receives more attention. The small size of European countries makes it difficult for them to compete commercially with a country as large as the United States and its airline manufacturers such as Boeing. To compete, a group of European airplane manufacturers formed the Airbus consortium in the late 1960s to pool the resources of several countries. Companies in France, Germany, the United Kingdom, and Spain all manufacture various components that are then shipped to assembly plants in either Toulouse or Hamburg (Figure 2.14). Toulouse is the bigger assembly plant and headquarters for Airbus. The first airplane rolled off the assembly line in 1972, and Airbus soon captured 10 percent of market share. Growth has been steady and rapid. Today, Airbus employs 55,000 and began capturing the biggest share of new passenger-jet orders worldwide in the early 2000s.

Competition between Airbus and Boeing is now fierce. With commercial airline traffic at a high and likewise airport congestion, the two companies are staking their market positions on two different scenarios of what the future may bring. Airbus believes that continued congestion will result in greater difficulties for

airlines in obtaining gates at major airports such as London's Heathrow, Tokyo's Narita, and New York's John F. Kennedy. Therefore, Airbus launched the A380, a plane that has 50 percent more floor space and at least one-third more seats than Boeing's 747. The plane flies greater distances, and airlines can have versions of the plane built to offer more comforts such as shower facilities, a piano bar, an exercise room, and a barber shop. The plane made its first commercial flight in 2007. In contrast, Boeing believes that greater airport congestion will lead to both new airport construction and the expansion of current facilities, calling for more mid-size and faster jets. Thus, Boeing developed a modest-sized but fast subsonic cruiser known as the 787 Dreamliner. It is difficult to predict the future, but European cooperation in aeronautics may make Europe the world's leader in the commercial airline industry.

Energy Sources

Domestic coal was the main source of energy in the region until the 1950s when cheaper imported oil, especially from Southwestern Asia, took over. Discoveries of natural gas in the Netherlands in 1963 led to the development of the North Sea basin (Figure 2.15a). Today, Norway is one of the world's largest crude oil exporters. By the mid-1980s, natural gas became the cheapest and cleanest fuel for electricity generators. After the fall of communism in the Soviet Union in 1991, Europe began receiving oil and natural gas via pipeline from Russia's vast

Figure 2.14 **Western Europe: cooperative European manufacture of Airbus aircraft.** The high levels of capital inputs and technological complexity required for constructing passenger aircraft make it impossible for the aerospace industry in one country of Europe to support the whole process. What are the political and economic implications of these movements?

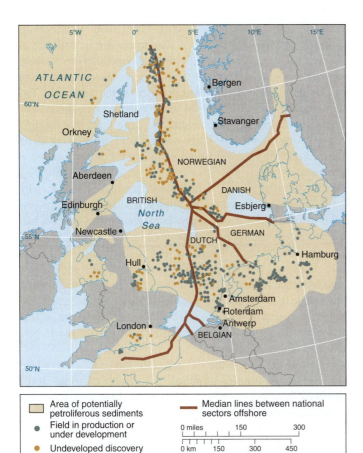

Area of potentially petroliferous sediments

● Field in production or under development

● Undeveloped discovery

Median lines between national sectors offshore

0 miles   150   300

0 km   150   300   450

(a)

(b)

Figure 2.15 **Energy sources.** (a) North Sea oil and natural gas fields (1995). (b) Wind-generated energy near Trekroner Fortress off the coast of Copenhagen, Denmark. *Sources: (a) Data from Pinder, David. The New Europe: Economy, Society and Environment. Copyright © 1998. Reprinted by permission of John Wiley and Sons, Ltd.; (b) © Loren W. Linholm.*

supplies, although its countries are concerned about the future level of control this might bring.

Of other sources of energy, nuclear-powered electricity generation is the most important. Europe generated 36 percent of the world's nuclear-powered electricity in 2006, more than any other world region. It also has 8 of the top 10 countries with the greatest share of their electricity derived from nuclear power. France ranks highest with 77 percent and is followed by Lithuania, Slovakia, and Belgium. Hydroelectricity is important in Alpine Europe. Wind power accounted for 3 percent of Europe's electricity but is growing rapidly, accounting for 36 percent of new electricity capacity installed in 2008. Germany

and Spain have the highest generation, but Denmark receives the greatest percentage of the energy from wind (Figure 2.15b).

Service Industries

After World War II, service jobs such as those in retailing, wholesaling, education, health care, and government grew in importance as the European countries became richer and instituted strong social welfare programs. By 2010, employment in the service sector made up over 75 percent of the total work force in most Western European countries and somewhat less in the other subregions of Europe.

The main growth areas in the service sector are in producer services and tourism. **Producer services** are involved in the output of goods and services, including market research, advertising, accounting, legal, banking, and insurance. They serve other businesses rather than consumers directly. Greater use of computers and information technology increased **productivity**, which is frequently measured by the amount of product generated or work completed per hour of labor. Producer services are closely related to global commercial developments and concentrate in the centers of major cities, where agglomeration economies are significant. As stated previously, agglomeration economies involve the clustering of businesses in a location, often near governmental agencies in the case of services, to save costs from the sharing of infrastructure, labor pools, market access, and transportation. London, Paris, Amsterdam, Frankfurt (Figure 2.16), and Munich have major shares of these industries. Some activities, such as processing paperwork, are decentralized in suburban office parks. However, major city centers retain the high-order functions in which face-to-face personal contact is important. By 2010, many back-office and call-center jobs went to cheaper locations in Ireland and overseas to India and the Caribbean.

Figure 2.16 **Western Europe: financial services.** Frankfurt, Germany, is one of Europe's financial centers. For example, the European Central Bank (ECB), which oversees the euro, is located here. *Photo: © Peter Adams/Image Bank/Getty.*

Agriculture

Beginning in mid-1700s, the Industrial Revolution transformed traditional agriculture in which peasants tilled the land with draft animals. Scientific crop rotation, which involved the planting of soil-enriching crops such as clover and turnips every four years, made it unnecessary to leave half or one-third of the land fallow every year. The use of tractors, fertilizers, and pesticides required fewer people to work larger farms. Particularly in the 1800s, people moved from the countryside to the cities, where they found jobs in industry and services. Although small farms of a few hectares continue to exist, many were taken over by more successful farmers, resulting in fewer but much larger farms, a trend known as **concentration**. Tractors, fertilizers, and pesticides also resulted in greater agricultural productivity per hectare, known as **intensification**. Today, agriculture still occupies almost as much land as before, but it accounts for only 2 to 7 percent of total employment in Europe.

Modern agriculture motivated farmers to switch from producing a variety of crops for their families to single crops that brought high profit, such as sugar beets or oil seeds. This is known as **specialization**. Specialization first occurred in basic grain and root crops. Then specialization developed in certain fruits and vegetables, more perishable and expensive to transport. These foods are associated with gardens, hence specialization in them is called **market gardening** (truck farming in the United States). For example, warm, dry summers and cool winters that seldom drop below freezing allow the Mediterranean countries to produce crops that are difficult to grow in the other subregions of Europe, such as olives, table and wine grapes, citrus fruits, figs, and specialized cereal grains for pasta. In addition, Portugal produces most of the world's cork for wine bottles, obtained from the bark of the cork oak (Figure 2.17).

Concentration, intensification, and specialization radically changed agriculture and the rural way of life over the last hundred years. Most large modern farms require farmers to have management skills. Farms also depend on industries that produce seeds, fertilizers, pesticides, and machinery to produce agricultural goods. They rely on food processing plants and marketing agencies to get their products to the consumer. The term **agribusiness** was coined to describe how commercially oriented farming has become and the close links that farming has with other industries. Thus, while fewer Europeans work on farms, agriculture is still a very significant sector of European economies.

Despite the rise of agribusiness, farming is still a risky prospect, with bad weather and pests still causing bankruptcy. To provide a stable food supply, European governments subsidize farmers. Within the European Union (EU) (see "Political Changes: European Union [EU]," p. 67), the Common Agricultural Policy (CAP) was created. It first provided a set of price supports that made farming almost risk-free. As technology advanced, it led to increased output. Mountains of grain and butter, and lakes of milk and wine resulted. Farming is now less risky, but agricultural subsidies to farmers remain the largest expense of the EU's funds.

During the 1980s, the EU began to change CAP to reduce the costly and unwanted surpluses. One result was to encourage livestock farmers to produce less from the same area, a process known as **extensification**. Despite cutbacks, the agricultural sector continued to produce surpluses. In the 2000s, the EU implemented more CAP reforms. Subsidies were "decoupled" from production, meaning that farmers now no longer receive more subsidies for producing more. Instead, the EU pays a flat amount that is tied to such concerns as rural development and environmental protection.

With agriculture a politically sensitive topic, cutting subsidies has been difficult. Countries like France, Spain, Ireland, and Portugal benefit greatly from agriculture and resist change. Protests and disputes also likely will emerge as the uneven geographies of CAP subsidies and preferences change. For example, France has historically benefited the most. In 2010, it received 17 percent of CAP funds, more than any other EU country. France has fought for years to protect its privileged position, but the situation is about to change. The newer EU members, particularly those in East Central Europe, only receive partial CAP subsidies and likewise find their agricultural sectors harmed by EU policies. This situation was scheduled to change in 2013 when more CAP subsidies would flow to East Central Europe and France would become a net contributor to CAP subsidies rather than a net beneficiary.

Tourism

Europe dominates the international tourist market as it received 373 million, or 40 percent, of the 942 million tourists worldwide in 2010. European countries held 5 of the top 10 places in terms of tourist arrivals: France was first (China with Hong Kong was second, the United States was third), and Spain was fourth with Italy fifth (Figure 2.18a; see Figure 2.5), United Kingdom sixth, and Germany eighth.

Figure 2.17 **A cork oak forest in Portugal.** Bottle corks are made from the bark of these trees. Little processing is required. Once the bark is stripped from the tree, bottle corks are cut. Numbers on the trees refer to a calendar year and indicate when the bark can be stripped from the tree again. *Photo: © Alexander B. Murphy.*

(a)

(b)

Figure 2.18 **Europe: tourist attractions.** (a) Florence, Italy, has cathedrals, art galleries, palaces, and the Arno River. (b) Mediterranean Europe: Olympia, Greece. Site of the classical Olympic games. Tourists walk through the arch to the stadium. Charred hills in the background are from devastating fires that swept across Greece in summer 2007 and killed more than 60 people. *Photos: (a) © Michael Bradshaw; (b) © Emily A. White.*

Most tourists went to the Mediterranean countries with their warm, sunny beaches, but the mountainous, rural, coastal, and historic urban areas elsewhere in Europe also increased their tourist trade (Figure 2.18b). While tourism is increasingly significant in Europe, tourist industries in other world regions have developed more rapidly. Europe still dominates the world in tourism, though its share of tourism receipts declined from 69 percent of the world total in 1975 to 38 percent in 2010.

Tourism is encouraged by the governments of each country, EU regional investment in infrastructure (roads, airports), and agreements on the sharing of health facilities, currency regulations, and customs. Individual countries promote their facilities through tourist boards. In the mid-2000s, tourism generated one EU job in eight, making it the largest industry in Europe. Much tourism employment, however, is seasonal, poorly paid, and female-dominated. The unskilled and low-paid nature of many jobs leads to them being taken up by migrants and thus having less impact on local economies than anticipated.

2.4 Geographic Diversity

Though Europe is smaller than many other world regions, it has considerable internal political, economic, and cultural diversity, which can be generalized into four distinct subregions (Figure 2.19 and see Table 2.1).

Subregion: Western Europe

Western European countries have been most significant in creating Europe's image as a global leader. The United Kingdom, France, and the Netherlands were three of Europe's most powerful colonial powers (Figure 2.20). They spread many European cultural characteristics around the world. Still today, English and French are two of the most commonly spoken world languages. Many Western Europeans were also the driving forces behind the spread of Christianity, expecially Protestantism. The British, French, and Dutch in particular carried it to North America, South Africa, Australia, New Zealand, and parts of Asia.

Western European countries were also among the first to experience the Industrial Revolution. The combination of colonialism and industrialization gave Western Europe the ability to establish many of the global trade flows that are still in effect today. The United Kingdom maintains ties through the British Commonwealth. In 1945, France created a currency known as the CFA franc (franc of the French Colonies of Africa) for use in its western and central African colonies. These colonies are now independent countries, but the CFA franc still exists and ties their economies closely to France. Supported by France and Canada, the Agency for Francophony is an organization that promotes the French language and culture around the world. Western European countries also still import raw materials from countries that were their former colonies and then sell finished products back to them. Colonialism helped to make Dutch cities such as Amsterdam and Rotterdam great trading centers, the latter being one of world's largest ports.

Today, Germany, United Kingdom, and France rank among the top 10 largest economies in the world. These three countries are members of Group of Eight (G8), an informal organization representing the world's most materially wealthy countries. (Other G8 countries are the United States, Canada, Japan, Italy, and Russia.) The political power of Western Europe is underscored by the UK and France occupying two of the five

Figure 2.19 Europe: subregions.

Figure 2.20 **Western Europe: Paris, France.** The Arc de Triomphe de l'Etoile stands on the hill of Chaillot and is the center of radiating avenues such as the Champs Elysées. In 1806 Napoleon conceived of a triumphal arch in the spirit of ancient imperial Rome that he could dedicate to the glory of his armies. Designed by Jean Francois Thérè Chalgrin (1739–1811), it was completed in 1836. Now a symbol of French patriotism, a huge French flag is hung from the ceiling of the arch on national holidays. *Photo: © Yann Arthus-Bertrand/Corbis.*

permanent seats of the UN Security Council, the most powerful organ of the United Nations. The human environment of this subregion ranks high in health, education, and income standards with all countries ranking within the top 20 on the Human Development Index.

German Reunification

Germany is Europe's largest economy but has not had the same political influence as France and the United Kingdom since the end of World War II. Following Germany's defeat, the Allies (the United States, the United Kingdom, France, and the Soviet Union) divided Germany into occupation zones (Figure 2.21a). Though allies during the war, the Soviet Union disagreed with the Western Allies over Germany's fate. In 1949, the Western Allies formed their three occupation zones into the Federal Republic of Germany (West Germany), a democratic/capitalist country. The Soviet Union then created out of its occupation zone the German Democratic Republic (East Germany), a Communist country. With Berlin similarly divided into four occupation zones, the three western zones became West Berlin and politically associated with West Germany, though it was deep inside East Germany. Likewise, the Soviet zone became East Berlin and part of East Germany.

The Cold War division of Europe between non-Communist West and Soviet Communist East ran through Germany and its capital Berlin. As the Cold War intensified, the division of Germany and Berlin deepened. For example, in 1955, West Germany joined the North Atlantic Treaty Organization (NATO) and East Germany the opposing Warsaw Pact. At first, the borders between the two separate German states were not closed. People freely traveled from one country to the other. In Berlin, thousands of people crossed the border every day. However, the differences between West and East Germany grew over time. The American Marshall Plan and Western investment allowed West Germany to experience an economic miracle, and citizens enjoyed political freedom. In contrast, immediately after the war the Soviets dismantled factories in its occupation zone and shipped them back to the Soviet Union as war reparations, stunting economic growth in East Germany; citizens were also denied their political freedoms. Soon, more and more people from the East moved to the West. In order to stop this exodus, the East German government, backed by the Soviet Union, erected a wall between East and West Germany and around West Berlin in 1961. Transportation routes were cut, the windows of houses facing the wall were bricked up, and families were separated. The wall disrupted the lives of Berliners, both East and West.

In the late 1980s, the weakening economies of Communist Europe and the Soviet Union began to undermine the Communist governments. Soviet leader Mikhail Gorbachev urged reform throughout the Communist countries, but East German leader Erich Honecker refused. Rising discontent among citizens led to mass demonstrations and people trying to flee East Germany. Erich Honecker was replaced by the more moderate Egon Krenz with promises of reforms. However, growing crowds in East Berlin pushed their way past the border crossings to enter West Berlin on November 9, 1989, marking the fall of the Berlin Wall. This event began the end of Germany's Cold War division. Less than a year later, the two Germanys once again became one country, and in 1991 Berlin was declared the capital of reunited Germany (Figure 2.21b).

Subregion: Northern Europe

Northern Europe is sometimes called "Norden." This subregion of cold climates is sparsely populated but rich in natural resources. From the 800s through the 1200s, its Viking inhabitants were very powerful and extended their control far beyond their subregion. Denmark and Sweden later became powerful empires. At the height its power, which lasted from 1610 to 1718 and was known as "the Great Power period," Sweden controlled the areas now known as Finland and the Baltics, and northern areas of Poland and Germany. The Swedish army was able to defeat Danish, German, and Russian forces, often simultaneously, before it was permanently weakened in 1721 and Sweden lost most of its possessions. Sweden and Denmark gave up imperial ambitions in the early 1800s and have since frequently taken positions of neutrality in international relations. Modern Norway was controlled either by Denmark or Sweden through the years and did not achieve self-governance until 1814. Finland was within the Swedish kingdom from the

(a)

(b)

Figure 2.21 **Western Europe: Germany.** (a) Divided among the Allied powers. (b) Berlin: The historic Reichstag building is the location of reunited Germany's parliament. The new glass dome symbolizes the transparency of German democracy. *Source: (a) Data from Jones 1994; data from Heffernan 1998; Photo: (b) © Heike Alberts.*

1100s to 1809, when it became a Russian possession. Finnish independence was declared in 1917. Today, the four largest countries in the subregion have some of the world's highest GDP per capita figures and standards of living with high levels of education and health care (Figure 2.22). These countries

are also great supporters of human rights. Their high standards make them reluctant to join any international organizations out of concern that membership will lower their standards. In the organizations that they join, they often do not fully participate in all activities for the same reason.

Fishing and wood products are generally important industries in Northern Europe. Both Sweden and Finland are major world producers of sawnwood, pulp for paper, and paper. Denmark is specifically known for Tuborg and Carlsberg beers and the toy company Lego, though Denmark has many other industries. Sweden excels in engineering, iron and steel, chemicals, and pharmaceuticals. Swedish inventions such as the ball bearing are indispensable to modern industry. Global companies include SKF, Ericsson, Volvo, Scania, and AB Electrolux. The Finnish company Nokia is world renowned for its mobile phone, Finland's most important export product.

Subregion: Mediterranean Europe

Mediterranean Europe played a major role in directing the early course of Western civilization, from ancient Greek and Roman ideas to those of the Italian Renaissance. In the 1400s, Portuguese and Spanish exploration launched the Age of Discovery and soon marked the beginning of European colonization of other lands and peoples. With colonization, the Portuguese and

Figure 2.22 **Northern Europe: Copenhagen, Denmark.**
Copenhagen's urban landscape combines modernity with
Scandinavian flavor. *Photo: © Loren W. Linholm.*

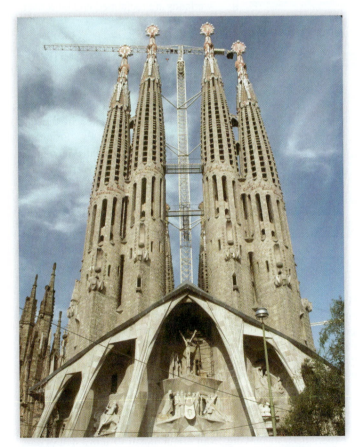

Figure 2.23 **Mediterranean Europe: Barcelona, Spain.**
Barcelona is world renowned for its Temple de la Sagrada Família.
Begun in 1882 and designed by the Catalan architect Antonio
Gaudi (1852–1926), this church is far from complete. *Photo: © Loren
W. Linholm.*

Spanish transplanted their languages and Roman Catholicism
around the world. By the Industrial Revolution of the 1800s,
however, Mediterranean Europe had lost a lot of its political
power, and Greece was part of the Ottoman Empire.

In the 1900s, differences among the Mediterranean
countries grew. Portugal and Spain were ruled by two dictators: Antonio Salazar, who came to power in Portugal in
1926, and Francisco Franco, who took control of Spain in
1938. Following the deaths of these dictators in the 1970s,
Portugal and Spain became more democratic and outward
looking, joining European organizations. Greece emerged
as a nation-state in 1832 but did not take its current shape
until the early 1900s when Ottoman control ended. Italy is
the youngest of the major Mediterranean nation-states, coming into existence in 1861 and experiencing some boundary
changes as late as the 1940s. Regionalism within Italy is
very strong, with most Italians considering themselves Sicil-
ians, Tuscans, Venetians, and so on first and Italians second.
Italy became the most industrialized of Mediterranean countries, followed by Spain (Figure 2.23). Portugal and Greece
still rely on agriculture, fishing, merchant marines, and tourism for much of their overseas income.

Industrialization began in Mediterranean Europe in the late
1800s in northern Italy and in the Catalonian region around
Barcelona in northeastern Spain. Most modernization occurred
after World War II and the incorporation of the Mediterranean
countries in the EU. Industrialization in Greece and Portugal
has been very modest. The two countries remained among the
poorest economically of the EU countries and received large
sums of regional development funds until countries of East
Central Europe joined the EU in 2004.

Italy's GDP is twice the total of the other Mediterranean
Europe countries and was the world's tenth largest in 2011.
The country's economic power comes mainly from northern-
based and high tech industries. The Po River valley between
the Alps and the Apennines is the largest center of manufacturing in Mediterranean Europe. For example, Fiat manufactures
automobiles in Turin. Venice, at the mouth of the Po River, is
famous for its glass manufacturing. Milan, the largest city in
northern Italy, is a major center of financial and other service
industries as well as a producer of a diverse group of manufactured goods including tractors, domestic electronic goods,
china, fashion, and pharmaceuticals. Milan and its surrounding
towns produce nearly one-third of Italy's GDP and form one of
the major growth areas of Europe.

Mediterranean Sea

Population growth in the 1900s, combined with crowding into
coastal locations, industrialization, and the great increase in
tourism, led to excessive pollution of the Mediterranean Sea—
an important issue for many prospective tourists (Figure 2.24).
The Mediterranean Sea is also one of the world's major shipping lanes (carrying 18 percent of the world's seaborne oil traffic) and has the world's highest level of oil pollution. It is an
almost closed sea with a single narrow connection to the Atlantic Ocean at Gibraltar; water mixing dilutes the pollution, but

(a)

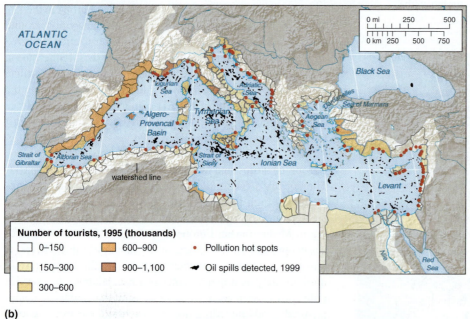

ATLANTIC OCEAN

Black Sea

Sea of Marmara

Aegean Sea

Levant

watershed line

Ligurian Sea

Adriatic Sea

Dardanelles

Algero-Provencal Basin

Tyrrhenian Sea

Strait of Gibraltar

Alboran Sea

Strait of Sicily

Ionian Sea

Nile

Red Sea

Number of tourists, 1995 (thousands)

☐ 0–150	☐ 600–900	● Pollution hot spots
☐ 150–300	☐ 900–1,100	➤ Oil spills detected, 1999
☐ 300–600		

(b)

Figure 2.24 Mediterranean Europe. (a) Marbella, Spain. Crowded beach on the Mediterranean coast welcomes millions of tourists, particularly from Northern and Western Europe, during the summer months. Many buy permanent accommodations or time shares here. (b) The growth of tourism and oil spills have contributed greatly to the pollution of the Mediterranean Sea and the emergence of "hot spots," environmentally endangered areas. *Photo: (a) © Alasdair Drysdale; Source: (b) Data from United Nations Environment Programme.*

urban-industrial areas, such as Barcelona, Marseilles, Genoa, Naples, and Athens.

Governments around the Mediterranean Sea met in 1975 and developed the Mediterranean Action Plan (MAP) to tackle pollution problems. Through MAP, the Barcelona Convention was adopted and required countries to reduce their pollution emissions. Sewage treatment improved in the wealthier countries such as France, but the poorer countries of North Africa, which also pollute the Mediterranean, have not been able to afford the necessary investment. Growing populations, tourist centers, and industrial projects have increased pollution, prompting six additional protocols and needed amendments to the Barcelona Convention in 1995. Even if pollution is reduced as countries get wealthier, the coastline and its delicate ecosystem are changed irrevocably when coastal wetlands are reclaimed and built over.

Subregion: East Central Europe

The countries of East Central Europe emerged as nation-states in the late 1900s and early 2000s. From the beginning of the nationalist idea in the late 1700s to the end of World War I in 1918, most of East Central Europe was dominated by four great empires—the Russian, German, Austro-Hungarian, and Ottoman empires (Figure 2.25). The subregion enjoyed a brief period of independence between World War I and World War II, but then most of its countries had Communist forms of government and economies imposed upon them (see "Europe Since 1945," p. 53). Most of these countries were directly controlled by the Soviet Union and were called Soviet satellite states. The Baltic countries were incorporated into the Soviet Union from 1945. Though Soviet domination ended by 1991, these countries share common experiences in moving from communism to more democratic forms of government and capitalist economies.

Prior to Soviet domination, East Central European countries also shared the common experience of being dominated by outside empires that treated their domains as colonies,

it takes 80 to 150 years for the Mediterranean to be completely renewed by the Atlantic Ocean. Confined within the Mediterranean, the chemicals and other nutrients from pollution decay and thereby deplete the sea's oxygen, killing sea life and creating large algae blooms that thrive in the anaerobic conditions. The Adriatic Sea has some particularly large algae blooms. The most polluted European areas of the sea are those near

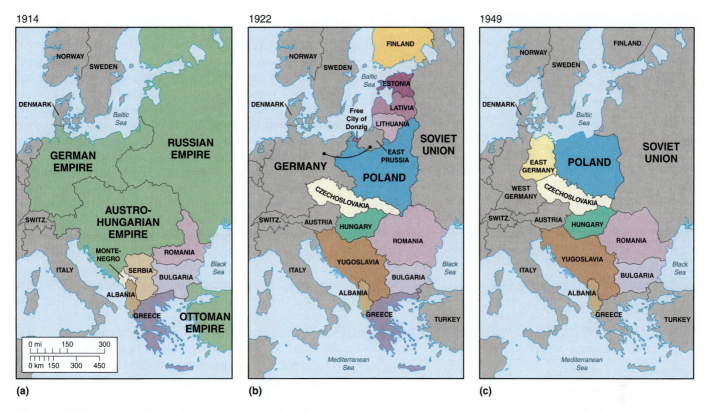

Figure 2.25 **East Central Europe: country boundaries in the 1900s.** Compare these with the contemporary boundaries shown in Figure 2.1. Note the sizes, shapes, and locations of Poland, Latvia, and Yugoslavia at each time. *Source: Data from Demko & Wood.*

preferring to extract resources and agricultural products rather than invest in industrialization. Thus the Industrial Revolution came slowly to East Central Europe. The Czech and Polish lands were somewhat exceptional in that a number of cities had factories. Hungary and Slovenia also developed industry. Nevertheless, a large number of people in the subregion still work in agriculture. Albania, for example, has 48 percent of its labor force engaged in farming, which is well above the 5 percent average for all of Europe.

After World War II, Communist economic policies were imposed on most countries in East Central Europe. The Communists improved the standard of living by encouraging industrialization, particularly in areas with coalfields, and by providing jobs for everyone, whether it made good economic sense or not. During Communist times, farming efficiency improved as industrialization provided tractors to replace horse-drawn plows. Great strides also were made in technology.

In terms of trade, Soviet Communists preferred not to be ensnared by capitalist practices, which they considered to be corrupting. Therefore, they kept trade among fellow Communist countries by setting up the Council for Mutual Economic Assistance (CMEA or COMECON), which also competed with the European Economic Community (EEC) (which is now the European Union [EU]). COMECON tied the subregions' economies with one another and with the Soviet Union, forcing greater dependency on the Soviet system and preventing any country from realigning itself with the West. Cheap oil

and natural gas from the Soviet Union also increased dependency, but polluted the natural environment. After the end of Soviet control in 1991, the countries of East Central Europe experienced economic crises as they reoriented themselves to the world economic system (see Chapter 3). Countries were challenged with breaking up business monopolies owned by the state, providing productive jobs, and cleaning up the environment.

After 1991, the difficulties in moving to capitalism initially led to an immediate drop in GDP and personal income. Eventually, Poland, the Czech Republic, Hungary, Slovenia, and the Baltic countries had the greatest successes in adopting capitalist practices with their new economic policies raising GDP again, often at faster rates than those of other European countries. These countries were also the most industrialized before and during communism, and they have had stronger traditions of democracy. Significantly, they also were closest to countries of the European Union (EU), making them easy beneficiaries of trade and investment from countries in the other subregions of Europe, especially from Germany.

Before Yugoslavia was engulfed in war beginning in 1991, it had one of the strongest economies and highest standards of living of Communist Europe. Not a Soviet satellite, Yugoslavia pursued its own course. It employed the Communist idea of centralized planning, but compared to the Communist countries in the Soviet sphere, it also allowed a more genuine practice of the Communist belief that workers should manage their

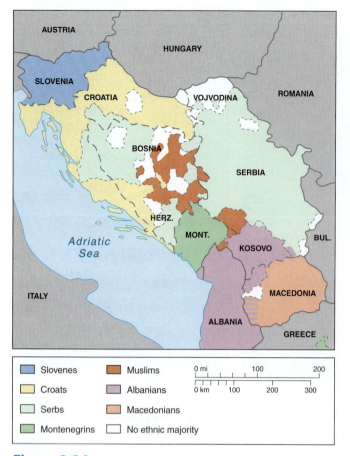

Figure 2.26 **East Central Europe: ethnic differences in former Yugoslavia.** Areas where a particular ethnic group comprises over 50 percent of the population. After 1991, Yugoslavia broke into five independent countries, although Serbia and Montenegro continued to call themselves "Yugoslavia" until 2002; Montenegro declared independence in 2006. Independence movements led to war in Croatia and Bosnia-Herzegovina. Violence in Kosovo (southern Serbia) in the late 1990s led to UN and NATO involvement in 1999. *Source: From* A Short History of the Yugoslav Peoples *(Cambridge University Press, 1989) by Fred Singleton. Copyright © 1989 by Cambridge University Press. Reprinted by permission.*

companies. Subsequently, productivity was high in Yugoslavia. Travel to non-Communist countries was not as restricted as in the Soviet sphere. As a result, thousands of Yugoslavs, especially Slovenes and Croats, sought work as guest workers in countries such as Germany. These workers sent millions of dollars back to family members in Yugoslavia.

Croatia may have been just as prosperous as Slovenia, but independence in 1991 was followed by the devastation of war and little foreign investment. However, in the 2000s, Croatia rebuilt, foreign investment returned, and Croatia's economy returned to normal. Having suffered great devastation, Bosnia-Herzegovina's economy is still very weak. Serbia's economy faces great difficulties following economic boycotts during the wars and NATO bombings. Though the war is over, the country receives little foreign investment. Macedonia escaped most of the ravages of war, but its landlocked position among unfriendly neighbors and its distant location from the wealthier countries of Europe attract little foreign investment to this largely agricultural country.

"Yugoslavia"

Most of East Central Europe transitioned peacefully from communism to democracy and market economies. The breakup of Yugoslavia in the 1990s was a major exception. In 1991 and 1992, all of Yugoslavia's republics except Serbia and Montenegro moved toward independence (Figure 2.26). The Serbs, who were the most numerous of Yugoslavia's ethnic groups, tried to keep the other republics from seceding from Yugoslavia, mostly because numerous Serbs lived in many of the seceding republics. Generally, the greater the number of Serbs that lived in a republic, the more problematic secession became. Thus Slovenia, which had very few Serbs, achieved independence relatively quickly and without much bloodshed. Secession was difficult for Croatia, where Serbs made up 12 percent of the population. After some warfare, Croatia achieved independence but lost control over its Serb-inhabited areas that resisted with the aid of the Serbian-dominated Yugoslav army. No ethnic group formed the majority in Bosnia- Herzegovina. The move toward independence was fiercely resisted by the large Serb population, which also received help from the Yugoslav army. International intervention in 1995 led to the division of the republic into two entities: a Muslim-Croat Federation and a Republika Srpska (Serb Republic) (Figure 2.27). While the war raged in Bosnia-Herzegovina, Macedonia avoided bloodshed, but to gain international recognition it had to officially change its name to the Former Yugoslav Republic of Macedonia (FYROM) to convince Greece that it would not attempt to annex the northern part of Greece, which is also called Macedonia. In 1999, UN forces took control of Serbia's southern province

Figure 2.27 **East Central Europe: Croatia and Bosnia.**

of Kosovo, which is inhabited by Albanians who desire independence. In 2002, Yugoslavia ceased to exist as the country's last two republics abandoned the name and called themselves Serbia and Montenegro. In 2006, Montenegro declared its independence from Serbia. In 2008, Kosovo declared independence from Serbia.

2.5 Contemporary Geographic Issues

Political Changes: European Union (EU)

After Soviet communism was firmly established in East Central Europe in the early 1950s, many Europeans felt that their nationalist notions and capitalist practices were threatened. Many European countries about to lose their colonies knew that they were too small to compete individually with the Soviet Union and United States. Moreover, the two world wars revealed the ugly sides of nationalist political competition coupled with capitalist competition and economic protectionism. Competition could lead to failure as well as success. On the other hand, cooperation in a non-Communist form was thought to result in everyone's success. Thus, to compete successfully again in the world economy over the long run, many Europeans in the non-Communist countries began to work together.

In 1949, not long after the end of World War II, Belgium, the Netherlands, and Luxembourg joined together in the Benelux customs union; the term **Benelux** is derived from **Be**lgium, the **Ne**therlands, and **Lux**embourg. In 1952, the Benelux countries joined together with France and West Germany to form the European Coal and Steel Community (ECSC). In 1957, the five ECSC countries plus Italy signed the Treaty of Rome to establish the European Economic Community (EEC), which was to evolve into the **European Union (EU)** known today (Figure 2.28). At the time, the EEC was expected to create a common market in which goods, capital, people, and services

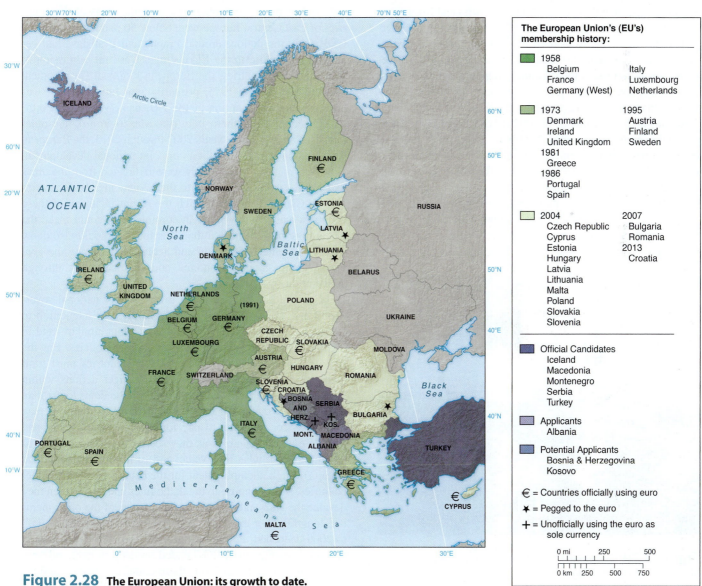

Figure 2.28 The European Union: its growth to date.

moved freely among countries. The European Commission became the executive arm of the community and was based in Brussels, Belgium. In 1967, the EEC changed its name to the European Community (EC) to emphasize the move from mostly economics to include social and political issues. Other countries were motivated to join. Denmark, the Republic of Ireland, and the United Kingdom became members in 1973, Greece in 1981, and Portugal and Spain in 1986. By then, the European Parliament, with elected members from all these countries, was created and located in Strasbourg, France. It forms the legislative branch of the EU along with the Council of the European Union (often known just as the Council of Ministers). The president of the Council rotates every six months among representatives from each of the EU's member countries. Other EU organs are headquartered in Brussels and Luxembourg City (Figure 2.29).

The Single European Act, signed in 1986 when Spain and Portugal joined the EC, set out the steps to complete a single market. The Treaty of Maastricht (1991) attempted to set a timetable for monetary and political union and changed the name of the organization to European Union in 1993. Austria, Sweden, and Finland joined in 1995. The Swiss and Norwegian referendum votes on joining were close but rejected membership at the time. The Swiss most likely did not want to give up their tradition of neutrality, and it is believed that the Norwegians' desire to protect their fishing grounds and their practice of whaling played a major role in their rejection of EU membership. Both countries were probably concerned that membership would lower their standards. In 2002, 12 of the 15 EU members adopted a common currency known as the euro. It replaced the traditional national currencies such as the German mark, the French franc, and the Italian lira. The common currency facilitates the free movement of labor, capital, and goods to strengthen businesses and allow greater opportunities for consumers to buy the best products at the lowest prices.

In 2004, 10 countries, many formerly part of the Soviet bloc in East Central Europe, joined the EU in the organization's largest expansion to date. Bulgaria and Romania joined in 2007. Turkey's application has long been under consideration, and Croatia applied for membership in 2003.

European cooperation has proceeded for almost 50 years. Some see it as successful and others do not. In either case, the idea of individual European countries competing alone in the global economy or attempting to exert political clout is still a daunting prospect. Though the Soviet Union no longer exists, countries like the United States, Japan, China, and Russia are still much larger than any one European country. For example, the three largest economies in the world, the United States, China, and Japan, have 314 million, 1.35 billion, and 128 million people respectively (mid-2012 data). In comparison, the most populous EU country is Germany with 82 million inhabitants. However, the 27 EU members have a combined population exceeding 500 million people and considerably more economic and political clout.

Though the EU is now very large with 27 members, the 12 new countries are mostly much poorer economically than the 15 older members and may be a financial drain. In 1998, before they became members, most of the areas of the countries of East Central Europe were below 50 percent of the EU's GDP. Only the regions around Prague and Bratislava had per capita GDPs close to the EU average. Until the admission of the 12 newest countries, the European Regional Development Fund, established in 1975, primarily sent funds to the western and southern margins of the EU, namely Ireland, Portugal, Greece, and southern Italy (the Mezzogiorno), and older industrial areas. For the years 2007 through 2013, the European Commission spent 336 billion euros on regional development. However, many of these funds were redirected from the areas that previously received them to the 12 new member countries of East Central Europe. Though the recent inclusion of these countries and possibly more in coming years could make the EU a more powerful force than it is today, it could also delay or thwart further integration and actually undermine the goal of achieving greater economic and political clout through cooperation.

Cooperation may also lead to the loss of national sovereignty and the erosion of national identity. For integration to succeed, the member countries have to adopt similar—and in many cases, common—laws and economic policies. For the euro to work, for example, all member countries must limit their spending and keep their annual budget deficits within 3 percent of their GDPs and their total public debts lower than 60 percent of their GDPs To do so, many EU countries may be unable, for example, to pay for their social programs and stimulate their economies as they see fit. They would have to give up a lot of what they value, and suffer through economic slumps of high unemployment for the benefit of other member countries. If they choose to break the rules and spend, then they devalue the euro and damage the economies of the other member countries. For example, as the world recession from 2007 to 2009 deepened, Portugal, Ireland, Italy, and Greece (which became known as the PIIGs) ended up deep in debt and with high budget deficits. In 2011, these countries' budget deficits ranged from 4 to over 13 percent of their GDPs, and each of

Figure 2.29 **European Union: Luxembourg.** The EU flag flies alongside the national flag in front of the Luxembourg City walls. *Photo: © David C. Johnson.*

their total public debts were over 100 percent of their GDPs. Previously these countries would have devalued their own currencies as a solution but, having adopted the euro, they cannot rely on this strategy anymore. They have to either abandon the euro or drastically slash their government budgets. Choosing the latter, these governments have become wildly unpopular with their citizens. The situation made people in countries planning to adopt the euro much less enthusiastic about the euro. These examples illustrate that many Europeans oppose the euro because of the restrictions that come with it. In fact, the current euro crises could end, even reverse, further European integration. Meanwhile, the United Kingdom, Denmark, and Sweden, which are wealthier countries, have not adopted the euro and continue to use their own national currencies.

Integration also requires the removal of barriers, including border controls between member countries. States will not be able to stop the entry of foreigners, whether from other EU countries or from abroad as they enter through other member countries. This means that these foreigners may take local jobs, demand cultural rights, and generally be a visible foreign presence. It also means that countries will not be able to stop the surge of cheap foreign goods and possibly contaminated food, driving businesses into bankruptcy and creating unemployment.

European integration generally is moving forward but may stall or reverse because many Europeans are concerned that further integration will lead to the loss of national sovereignty and the erosion of national identity. This concern is seen in the difficulties that many treaties and agreements have faced during the ratification process. For example, the aforementioned Maastricht Treaty (1991) initially was rejected by

Danish citizens before they later approved it. In 2001 and 2002, Irish citizens did the same with the Nice Treaty, which primarily was drafted to allow for the admission of 10 new members, mostly from East Central Europe. In 2005, French and Dutch citizens rejected the treaty that would have provided the European Union with a Constitution (Figure 2.30). In 2008, Irish citizens voted no to the Lisbon Treaty before voting to approve it in 2009. The Lisbon Treaty created a full-time EU president, eliminated the possibilities of individual countries being able to veto some issues, changed voting weights of countries to more accurately reflect their population sizes, and gave more power to the European Parliament. The rejections of these treaties likely stemmed from the fear of loss of national sovereignty. Others argue that "all politics are local," and thus it was also an opportunity for voters to express their discontent with their national leaders, especially those who were not protecting their nation's interests.

Though the European Union has its setbacks, nevertheless, it represents **supranationalism**, the idea that differing nations can cooperate so closely for their shared mutual benefit that they can share the same government, economy (including currency), social policies, and even military. During the Cold War, communism tried to offer a form of supranationalism, but after 1990 most Europeans abandoned the Communist experiment, leaving EU countries as the primary advocates of supranationalism. Members of the EU are still working out the details of their cooperation, but what they have accomplished is remarkable, considering that the more predominant nationalist idea, subscribed to by most of the world, holds that such cooperation is impossible between nations. Nationalism may still preclude ultimate political union in Europe.

(a)

(b)

(c)

Figure 2.30 **European Union: reactions.** Campaign posters in Ireland "for" and "against" the Lisbon Treaty in June 2008. (a) This "yes" poster expresses the desire for greater integration. (b) This "no" poster reflects general concerns among many Europeans that the European Union represents a high level of governance that is unconcerned about the needs of most citizens. (c) This "no" poster represents the concern of some Irish that membership in the European Union is economically exploitive and threatens Irish neutrality. *Photos: (a), (b) & (c) William Murphy.*

It remains to be seen if supranationalism will work or not (Table 2.3). In many ways, the European Union is a grand experiment, also called the "European Project." Interestingly, in 2012, the European Union was awarded the Nobel Peace Prize, an honor normally given to people rather than entities. The Norwegian Nobel Committee wanted to emphasize that the European Union achieved peace and reconciliation in a world region that was torn apart in two world wars. The chairman of the committee expressed a concern for an increase in extremism and nationalism that could cause Europe to disintegrate. Therefore, it was time to "focus again on the fundamental aims of the organization." Indeed, the European Union's emphasis on peace and economic prosperity along with NATO security during the Cold War has enabled Europe to regain a major place within the global economic and political system. For example, European countries now have many of the highest GDPs per capita in the world (Figures 2.31 and 2.32).

Devolution Within European Countries

As Europe moves toward greater economic and possibly political integration, it is simultaneously experiencing **devolution**, the process by which local peoples desire less rule from their national governments and seek greater authority in governing themselves. The desire for complete independence is known as **separatism**. One example of devolution is seen in the case of the United Kingdom of Great Britain and Northern Ireland. Great Britain is comprised of England, Scotland, and Wales. The rise of Scottish and Welsh nationalism over the last few decades has resulted in greater autonomy for Scotland and Wales, with both now having their own parliaments with limited powers. The people of Scotland may soon vote as to whether they want independence or not.

Northern Ireland's situation within the United Kingdom is more complex. All of Ireland was in the United Kingdom when it was created in 1801. In 1922, independent-minded Irish succeeded in separating all but six counties of Ireland and creating

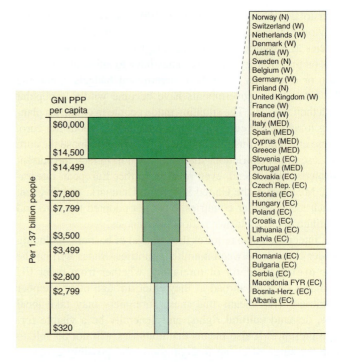

Figure 2.31 **Europe: national incomes compared.** The countries are listed in the order of their GNI PPP per capita incomes. Compare the relative material wealth of the countries in Western (W) and Northern (N) Europe with those in Mediterranean (MED) and East Central Europe (EC). Compare with Figure 1.26. *Source: Data (for 2010) from* 2012 World Development Indicators, *World Bank.*

the Irish Free State. Still part of the British Commonwealth, the Irish Free State had to swear allegiance to the British monarch until 1949, when it achieved its independence and adopted the name "Republic of Ireland." The six counties of the north remained part of the United Kingdom after 1922 as Northern Ireland. The violence in Northern Ireland often is depicted as a religious conflict between Roman Catholics and Protestants, but it is really a conflict between Irish nationalists and Unionists,

Table 2.3	**Debate:** The EU's future
Reasons for the EU's Success	**Reasons for the EU's Failure**
Economic union pools together the resources of member countries and thereby strengthens the economies of every member country.	Economic union undermines the ability of member governments to make economic decisions that are in their national best interests.
Economic union allows for the free flow of capital and labor, permitting capitalist tendencies to strengthen members' economies.	The free flow of capital and labor undermines attempts by member governments to protect their national economies.
The euro, or common currency, further facilitates the movement of capital and labor by doing away with the costs of converting currencies.	The euro forces member governments to have monetary and budgetary policies that may harm their national economies.
Political union increases influence in regional and global politics because it combines the political and military strength of member countries.	Political union forces member countries to adopt foreign policies that go against their national interests (e.g., forcing some, like Ireland, to give up their neutrality).
Political union results in common laws and standards for individuals and the environment in member countries. It makes for better social and natural environments.	Common laws and standards among member countries undermine national needs and traditions, often watering down social and environmental laws.

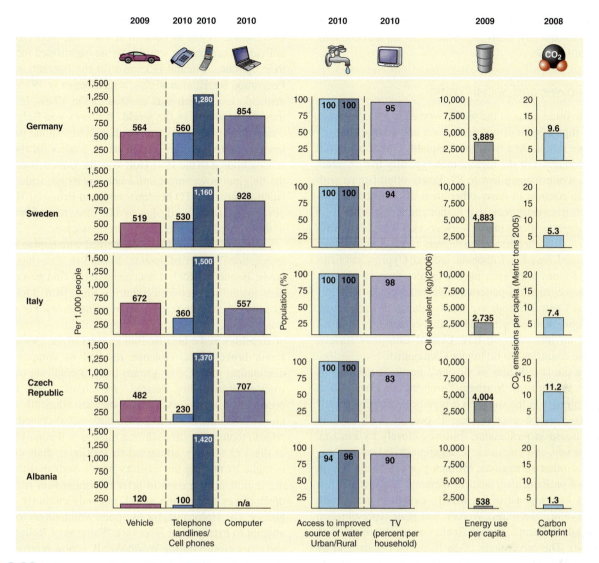

Figure 2.32 **Europe: ownership of consumer goods.** Note the diversity within Europe. Summarize the differences among the countries by subregion. *Source: Data (for 2007) from* World Development Indicators, *World Bank, 2009; Data for 2008–2010 from* 2012 World Development Indicators, *World Bank.*

also called Loyalists. Irish nationalists, feeling oppressed by the UK government, seek to unite the six counties of Northern Ireland with the Irish republic to the south. Because most Irish nationalists are Catholic and have the goal of uniting Northern Ireland with a Catholic country, and because Unionists/Loyalists tend to be Protestant and want to maintain a union with a Protestant country, the issue is often described as a religious conflict. As the conflict became progressively more violent, even ambivalent Catholics sought the protection of other Catholics, and Protestants sought refuge with other Protestants. The conflict has led to religiously segregated communities, though almost as many violent acts are committed by Catholics against Catholics and Protestants against Protestants as across religious lines. Violence declined in the late 1990s as the UK and Irish governments worked to create a shared government for Northern Ireland.

Europe has many other examples of devolution. Like the Northern Ireland case, the campaigns fought by the Basque peoples straddling the Spain-France border and the previously discussed disintegration of Yugoslavia are examples where

violence has played a major role in the process. Other cases of devolution progressed relatively peacefully such as Estonia's, Latvia's, and Lithuania's independence from the Soviet Union and the division of Czechoslovakia into separate Czech and Slovak republics. The Czechoslovak situation was referred to as the "Velvet Divorce," coined from the "Velvet Revolution" of 1989 that peacefully ended communism in that country. The fates of other cases of devolution are still unclear. Examples include the possibilities of Catalonian independence from Spain, the Northern League's desire for independence from Italy, and the dissolution of Belgium into a separate Flemish-speaking Flanders and French-speaking Wallonia.

Multicultural Societies

Following the end of European imperialism after World War II, Europe led the way in promoting women's rights and more generally human rights. At the same time, the need for labor to rebuild European economies attracted millions of immigrants, refugees, and asylum seekers, particularly to Western European countries. Subsequently, many European countries are becoming multicultural societies.

Women, Power, and Social Position

Compared to other areas of the world, women in Europe generally have a high degree of political power and a large share of the economic wealth. At a basic level, equality can be measured by life expectancy. For the region, men typically live 76 years on average, while women live to 82. Power often begins with education. In contrast to many other areas of the world, European women frequently receive a higher percentage of higher-education degrees than men. Norway ranks highest, with 56 percent of all higher-education degrees received by women. Ireland, Finland, Sweden, Lithuania, Iceland, Cyprus, and Estonia closely follow, with each having greater than 50 percent. The overall geographical pattern shows that women in Northern and East Central Europe are highly educated. In the latter subregion, communism, which advocated equality for women, clearly had an influence on East Central European society, though those countries are no longer Communist.

A more direct indicator of women's power is their representation in government. Northern European countries ranked among the highest in the world in this respect: Sweden placed fourth in 2012 with women holding 45 percent of the seats in the lowest house in parliament, followed closely by Finland, Iceland, Norway, and Denmark. The geographical pattern was varied in the other subregions. Women generally held a high percentage of parliamentary seats in Western Europe (the Netherlands was 14th highest in the world), though France (37th) and Ireland (89th) were low. Women had poor representation in Mediterranean Europe, though Spain was notably high (20th in the world). The percentages of parliamentary seats held by women in East Central Europe spanned the spectrum despite generally high education levels among women. Overall, women had greater political representation in the vast majority of European countries than women in the United States (81st).

Human Rights

Europe is the hearth for international organizations concerned with human rights. The International Movement of the Red Cross and the Red Crescent traces its origins to Switzerland. From the time of its Geneva Convention in 1864, it has helped to set the modern standard for the ethical treatment of wounded soldiers and civilians during wartime. Sweden and Norway are known for the Nobel Foundation and the Nobel Institute, founded in 1900 and 1904 respectively. Alfred Nobel, the inventor of dynamite, placed his fortune in a fund upon his death. The interest from the fund awards people who have worked to benefit humankind. The Nobel Peace Prize is perhaps the most well-known of these awards. Oxfam (short for the Oxford Committee for Famine Relief) began in Oxford, England, in 1942 to help starving people in Greece who were caught between Nazi occupation and an Allied blockade. Since that first mission, Oxfam has undertaken numerous operations around the world to bring food and medical supplies to those suffering from war and similar causes. Amnesty International began in London in 1961 and works today to free individuals around the world who are imprisoned for expressing their opinions. Finland is known for the International Helsinki Federation for Human Rights, which began in 1975. Belgium's genocide law, which was enacted in the 1990s, is one of the most far reaching in the world. It allows non-Belgians to file complaints for crimes committed anywhere in the world. When lawsuits were filed against American leaders for the deaths of schoolchildren in the 1991 Gulf War, the United States pressured the Belgium government into amending its genocide law, threatening to move NATO headquarters from Brussels. The Belgium government amended the law to refer cases to the home countries of the accused if such countries are democratic.

A significant institution concerned with human rights is the International Court of Justice, located in The Hague, Netherlands. Begun in 1899, it was designed to find peaceful resolutions to conflicts between countries. Since then, it has expanded its scope to include many human rights issues. For example, it was involved in assessing the responsibilities of those accused of atrocities in the horrific Balkan wars of the 1990s. As Yugoslavia broke apart, violence resulted as competing fervent nationalists of differing groups tried to eradicate other groups from territories they desired (see Figure 2.26). Eradicating people from a territory because they are ethnically different is known as **ethnic cleansing**. Serb nationalists coined the term to refer to their own actions during the war in Bosnia-Herzegovina in the 1990s. They attempted to legitimize their claim to the republic by making the territory purely Serbian. Ethnic cleansing is now a term used to refer to similar acts perpetrated in other places around the world, not only currently but also in history. The Holocaust, for example, which refers to the Nazis' attempt to exterminate Jews (see "Patterns of Religion," p. 46) and other groups during World War II, is now referred to as ethnic cleansing, although the term did not exist at the time.

Widespread atrocities, including rape, in Bosnia-Herzegovina led to changes in international law. To deal with the atrocities, the International Court of Justice set up a special war crimes tribunal, known as the International Criminal Tribunal for the former Yugoslavia (ICTY). The tribunal tried numerous cases and continues to do so. Prior to the Yugoslav conflict, international law viewed rape as an unfortunate by-product of war, but the tribunal set the precedent that rape is used as a weapon in war and deserves greater punishment than before. Rape is now officially recognized as a war crime. In one case, the tribunal convicted a man for the crime of not preventing a rape.

Refugees and Asylum Seekers

The relatively high level of economic health and political freedom found in Europe, particularly Western Europe, has been attractive to political refugees and asylum seekers from other areas of Europe and the world. Many arrive via Southern Europe from Northern Africa and Southwestern Asia. Portugal, Spain, France, Italy, and Greece have granted amnesty to more than three million illegal residents over the last 20 years. The numbers of individuals applying for asylum were at their peaks in 1992 and 2001, both coinciding with increased conflicts in other areas of the world. In 2010, the highest numbers of applications were to France, Germany, Sweden, Belgium, and the United Kingdom. These five countries accounted for more than two-thirds of all asylum-seekers in the European Union. Political refugees and asylum-seekers have contributed greatly to Europe's overall ethnic diversity.

Immigration and Evolving Identities

In addition to refugees and asylum seekers, Europe has received large numbers of immigrants looking for work. Immigration waves were created by World War II, for two reasons. First, the devastation of the war required massive rebuilding, but many young men of prime working age had been killed during the war. Second, the war weakened European countries with colonies, eventually forcing these countries to relinquish their colonies—a process known as **decolonization**. In short, European countries needed workers, and many were encouraged to come from the former colonies (Figure 2.33). The United Kingdom, France, and Germany have the largest numbers of immigrants. Yet all three have experienced and dealt with immigration differently over the decades and now face distinctly different problems.

The United Kingdom was the first to accept large numbers of migrants. In 1948, the British Nationality Act gave unrestricted entry to all subjects of the British Empire. The act also gave full political rights, which included the right to vote. Subsequently, waves of migrants began arriving in the United Kingdom in the late 1940s from colonies in the Caribbean and the Indian subcontinent (India and Pakistan). Racial Relations Acts were enacted in 1965, 1968, and 1976 to outlaw racial discrimination. Race relations, however, primarily were seen as both a local and an urban issue. Thus emphasis was placed on local governments and

urban redevelopment programs, particularly housing. As a result, racial minorities sought positions in local governments, but success in achieving these positions varied from locality to locality across the country. Eventually minorities rose to higher levels of government and now have representatives in the British parliament. The 1999 Immigration and Asylum Act shifted emphasis from the British Commonwealth to all countries. In 2000, the Immigration Minister stated that migration was a "central feature" of globalization "with potentially huge economic benefits for Britain if it is able to adapt to this new environment." She also stated that the country is "in competition for the brightest and best talents," and emphasized that "Britain has always been a nation of immigrants." The United Kingdom has not been without its ethnic and racial tensions, but its open attempt to build a multicultural society has resulted in far fewer tensions than in France.

France also received huge numbers of immigrants from its colonies: Algeria and other African countries, and those in Southeast Asia, especially Vietnam. However, the migration waves began about a decade after they did in the United Kingdom. *Liberté, égalité, fraternité* (liberty, equality, fraternity), a concept stemming from the French Revolution and reaffirmed in the 1946 and 1958 constitutions, summarizes the French view of migrants. To the French, all citizens of France are French regardless of gender, language, religion, race, and ethnicity. Thus the French government generally does not recognize ethnicity and does not publish any related statistics. In the name of equality, the idea of multiculturalism and affirmative action programs (which ensure minority access to education, employment, housing, and any other sector) are shunned because they are seen as favoritism and thus also forms of discrimination. In 2004, the French government banned the wearing of headscarves in schools in order to reinforce the separation between church and state. The ban angered many French Muslims, though many young Muslim women supported the ban as a way of helping them break from the traditions of their families so that they could more easily enter into French society. Ironically though, the French belief in equality has prevented legislation that would stop discrimination and in turn help minorities integrate into society. Despite French ideals, minorities feel highly discriminated against, and high unemployment has led to intense frustration and riots across the country that have resulted in the burning of thousands of automobiles. The French are holding firm to their view of what it is to be French, but that view is increasingly being challenged by France's minorities.

Germany has experienced and dealt with immigrants differently. Germany did not receive waves of immigrants from colonies because it did not have colonies. It initially recruited workers from other European countries like Portugal, Spain, Italy, Yugoslavia, and Greece. When not enough workers came from these places, workers were recruited from Turkey. Now more than 2 million Turks live in Germany (Figure 2.34). All these workers obtained the legal right to work but not citizenship and all the legal rights enjoyed by full citizens. The Germans coined a term for these foreign workers, *gastarbeiter*, or **guest worker**. Though allowed to stay as long as work was available, decades if necessary, the fundamental principle was that they would always be foreigners and would eventually go back to

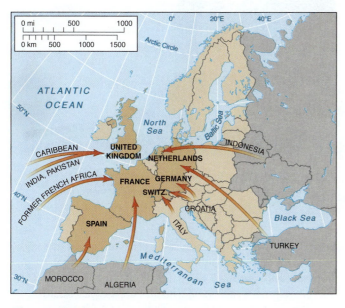

Figure 2.33 **Europe: sources of guest workers, 1970 to the 1990s.** Sources include people from the former colonies of the United Kingdom, France, and the Netherlands and countries around the Mediterranean Sea. What contributions do such workers make to the receiving and sending countries?

their respective homelands. The guest worker concept allowed Germans to hold to the idea that one was German according to ancestry and allowed Germany to preserve its "blood laws" concerning citizenship. By 1973 the labor shortage was over, and Germany stopped recruiting foreign labor. Soon, guest workers were encouraged to go back to their homelands, but many had been in the country for decades. They considered their original homelands "foreign" and would not leave. By the 1990s, the German population was aging, and Germany's economy then needed younger workers for business and to support the social security system. Yet without the same rights as Germans, guest workers felt like second-class citizens subject to discrimination. Germany's minorities became increasingly frustrated and tensions grew. Germans realized that the new global economy was a multicultural one and that Germany would have to open its arms to immigrants in order to survive. In 1999, the German government adopted the concept of "nation of immigrants" and changed its citizenship laws to allow immigrants to become Germans. Rules are strict. For example, a person is required to have irreproachably good conduct, show loyalty to the German constitution, sustain oneself and one's family without governmental support, and learn the German language. Tensions still exist within Germany because many immigrants have not been

Figure 2.34 **Europe: immigrant community.** Immigrants are changing European societies. This shop in Berlin, Germany, advertises in both Turkish and German. *Photo: © Helke Alberts.*

able to meet the standards. However, immigrants nevertheless have the ability to finally assimilate and show that Germany is becoming multicultural.

Geography at Work

Meeting the Challenge of Climate Change

Scientists across all disciplines share a concern that the world is close to reaching thresholds beyond which the effects of climate change will be irreversible. Creeping changes, like the gradual increase in temperature and the frequency and intensity of extreme events, expose society to new hazards and heightened risks that affect water and food supplies, power and transportation systems, and public health and safety. Socioeconomic changes happening at the same time affect how communities cope with and respond to those changes.

Lynn Rosentrater (Figure 2.35) is a geographer in Norway who specializes in methods for estimating climate risks. As an independent consultant, Lynn contributes to a number of international research programs investigating the social dimensions of climate change and also works with local governments to assess how they might be affected by climate change and what they can do to respond. In practice, local response strategies involve a mixture of tactics that include building climatic resilience, living with risks, and accepting losses.

There are two policy responses to climate change: mitigation—actions that address the causes of climate change—and adaptation, which deals with the consequences. Scientists agree that both measures are necessary, though Lynn's work focuses on issues that are particularly associated with adapting to climate change.

Within Europe, 12 countries out of a total of 32 have prepared National Climate Change Adaptation Strategies. Lynn's research suggests that adaptation is not an altruistic activity. Governments and institutions do it for their own benefit. Since adaptation anticipates the

Figure 2.35 **Lynn Rosentrater.** Using GIS to help local communities understand what climate change means for them. *Photo: Courtesy of Lynn Rosentrater.*

negative effects of climate change, early action saves on costs later. This contrasts with mitigation, where the benefits are deferred and shared across the planet.

Lynn sees herself as working at the interface between research and policy, and believes a key challenge is for decision makers across all levels of administration to understand the logic for swift action on climate change and to implement policies to ensure an optimal level of adaptation before unavoidable impacts are upon us.

Moscow, Russia. *A Russian woman stands in ice-cold water during the celebrations of the Epiphany Orthodox holiday in Moscow. For Orthodox Christians, Epiphany commemorates the baptism of Jesus in the Jordan River. Because this religious holiday occurs in January, the believers have the ice baths despite the extremely low temperatures—up to −32°C in Moscow.*
Photo: © Sergei Chirikov/epa/Corbis

Russia and Neighboring Countries

LEARNING OBJECTIVES

After reading the chapter, you should be able to:

- Explain why Russia and its neighbors are grouped together, and how the different subregions might identify with other world regions over time.

- Use a map to describe the changing climate, vegetation, and soil regimes from the southern to the northern lands of the world region, and relate this to land use and settlement patterns.

- Describe the cultural diversity of the region, from the Slavs to the peoples of the Southern Caucasus and Central Asia.

- Explain how Soviet urban development differs from that of Western Europe.

- Trace the political history of the region, from early Rus principalities to the Russian Empire, the Bolshevik Revolution, and the end of the Soviet Union. Compare the economic development of this world region to that of the European world region over the same time period.

- Name the political divisions of the Russian Federation and explain regional differences in terms of heartland and hinterland.

- Understand the basis for ethnic conflict in the countries of the Southern Caucasus and the Central Asia subregions.

- Explain how Russia is still a world power and how it is not.

- Identify major human rights issues and how they may hinder political and economic progress.

- Describe the region's major environmental problems.

Figure 3.1 Russia and neighboring countries: physical features, country boundaries, and capital cities. The region has mountains along its southern margin, with the Arctic Ocean to the north and the Pacific Ocean to the east. The land boundary with Europe in the west is marked mainly by lowland access. The interior lands are known for their climatic extremes: parched deserts in the south and harsh winters elsewhere.

ARCTIC OCEAN

UNITED STATES

Bering Strait

Chukchi Sea

CHUKCHI PENINSULA

Wrangel

St. Lawrence Island

Gulf of Anadyr

East Siberian Sea

CHUKCHI RANGE

Bering Sea

TAYMYR PENINSULA

Laptev Sea

KOLYMA RANGE

Karagin

NORTH SIBERIAN LOWLAND

Kolyma

Shelikhov Gulf

S I B E R I A

KAMCHATKA PENINSULA

CENTRAL SIBERIAN PLATEAU

EAST SIBERIAN UPLANDS

Lena

Sea of Okhotsk

Vilyuy

Lower Tunguska

Lena

Aldan

Shantar Islands

Kuril Islands

Angara

DZHUGDZHUR RANGE

Sakhalin

STANOVOY RANGE

Lake Baikal

YABLONOVYY RANGE

Amur

Tatar Strait

SIKHOTE ALIN RANGE

La Perouse Strait

Hokkaido

SAYAN MOUNTAINS

Irkutsk

Yenisey

Vladivostok

Sea of Japan (East Sea)

JAPAN

Honshu

MONGOLIA

Elevation (ft.)

below sea level

0   500   1,000   2,000   5,000   10,000

-10,000   -5,000   -500   0

0 mi   250   500   750   1,000

0 km   250   500   750   1,000   1,250   1,500

Lambert Azimuthal Equal-Area Projection

Yellow

CHINA

3.1 New World Regional Order

Until 1991, this entire world region was Communist and comprised of one country: the Union of Soviet Socialist Republics (USSR), more commonly called the Soviet Union. Of the Soviet Union's 15 republics, Russia was the largest and dominated politically and economically. The Soviet Union had many successes in its early years. However, in its last decades, its inflexible economic and political systems seriously weakened the country and allowed its discontented peoples to gain independence. The 15 republics of the Soviet Union became 15 independent countries. The three Baltic republics (Estonia, Latvia, and Lithuania), which had been part of the Soviet Union only since 1945, decided to seek greater cooperation with European organizations. The others, which comprise all the countries in this world regional chapter, share a more common political and economic legacy of the Soviet system. Recognizing the necessity of maintaining the trade flows established during Soviet times, the 12 former republics quickly formed the **Commonwealth of Independent States (CIS)**. Merely an organization that allows member countries to discuss and work out economic needs, CIS members express no desire to pursue any form of economic and political integration like that of the European Union (see Chapter 2). Indeed, though heavily tied to Russia (formally called the Russian Federation), most are trying to break their dependency on the Russian Federation by establishing connections to other world regions. However, the Russian Federation is larger than all the other countries in the region put together (Figure 3.1), nearly twice the size of the United States; the smallest country, Armenia, is only the size of Maryland and Delaware combined. Not surprisingly, Russia is still the dominant political and economic force of this diverse region (Figure 3.2). Thus, we find it appropriate to call the region "Russia and Neighboring Countries."

Extending from the eastern edges of Europe to the Pacific Ocean, Russia and Neighboring Countries is the world's largest region in terms of land area. However, with much of its land either in the frozen arctic environments of the north or the hot deserts of the south, the region is the least populated on Earth. Great distances, river systems that tend to flow toward the edges of the region such as to the Arctic Sea, and extreme climates of hot and cold that make it difficult to build rail and road have prevented governments from integrating the region with highly developed transportation and communication networks.

(a)

(b)

(c)

Figure 3.2 **Russia and Neighboring Countries: diverse landscapes and people.** (a) New church, Kharbarovsk. (b) Georgian boys dance group "Preserve Our Culture." (c) Russian village, south Siberia. *Photos: (a-c) © Ronald Wixman.*

3.2 Distinctive Physical Geography

The vast land area of Russia and Neighboring Countries results in a variety of natural environments. Vast plains and plateaus are bordered by mountains on the south. Winters are often long and harsh in some areas and summers hot and dry in other areas. Massive areas covered by a variety of natural vegetation types and soils provide a distinctive stage on which the development of human geographies occurred.

Plains, Plateaus, and Major River Valleys

Plains and low plateaus dominate most of the landscapes of this region. The North European Plain widens eastward from Poland into Belarus, Ukraine, and European Russia until it ends against the Ural Mountains (300–1,500 m; 1,000–5,000 ft.), which mark the line between Europe and Asia. Plains around the northern shores of the Black Sea and along the rivers leading to the Caspian and Aral seas extend into the nearly level, vast West Siberian Plains. Farther east, the relief again becomes hillier in the Central Siberian Plateau on ancient mineral-bearing rocks. The extensive areas of plains, interrupted by low hills, was a major factor in facilitating the invasions from eastern Asia and central Europe into Russia in medieval times and the later Russian imperial eastward expansion.

Across these low-lying landscapes flow some of the world's longest rivers. In the west, the Don River system flows into the Black Sea, which provides a strategic outlet to the Mediterranean; the Volga River flows into the Caspian Sea but is connected to the Black Sea via the Volga–Don canal. In Central Asia, the Amu Darya and Syr Darya rivers flow into the Aral Sea. The Ob, Yenisey, and Lena are the longest rivers of all and flow from the southern mountains of Central Asia and eastern Siberia northward to the Arctic Ocean; with their lower sections frozen for much of the year, meltwater frequently backs up and floods vast areas of wetland on both sides of their banks.

Southern Mountain Wall

The southern boundary of the region is marked by mountain systems. The Caucasus Mountains, between the Black and Caspian seas, are part of this line of mountain ranges that extends through the Elburz Mountains of northern Iran to the Tien Shan and the Pamir mountains along the southern borders of the Central Asian countries. These mountain ranges rise to over 7,400 m (24,000 ft.). They are snowcapped and in spring provide considerable meltwater for streams flowing through the dry areas of southern Russia and Central Asia. In the far east, the East Siberian Uplands and the volcanic peaks on the Kamchatka Peninsula parallel the Pacific coast. Among these high mountains are deep basins such as those filled by the Black Sea, Caspian Sea, and Lake Baikal.

Continental Interior Climates

Nearly all of Russia and Neighboring Countries lies north of 40°N and most of their area is north of 50°N. As such, these countries lie as far north as Alaska and Canada. The region contains places on Earth that are farthest from oceans, 500–2,000 km (320–1,250 mi.), and their moderating effects. They become extremely cold during winter and hot during summer, a situation described by the term **continentality** (Figure 3.3). The greatest temperature differences are in eastern Siberia, where large areas have January temperatures below −30°C (−22°F) and July averages of 12–16°C (56–60°F) (annual differences of 45°C, or 80°F). In the far east of Siberia, proximity to the Pacific Ocean results in more humid conditions, although winters remain long and very cold as Arctic winds sweep out from the continent's interior.

Western Russia also experiences continentality but not to such an extreme. Winter temperatures average −5 to −10°C (15–25°F), and summer temperatures average 15–20°C (59–68°F). With greater distance northward, the climate gets colder and the winters longer; northern Russia lies north of the Arctic Circle. The region's warmest climates are found around the Black and Caspian seas and in the Central Asian countries.

Precipitation declines with greater distances inland. Few parts of the region receive over 80 cm (32 in.) a year. Much of the region has moderate precipitation (40–80 cm, 16–22 in.) falling mainly in summer but producing a long-lasting snow and ice cover during the winter. Parts of southern Central Asia are arid because of high temperatures with high evaporation rates and distance from rain-bearing air masses. These areas rely on snowmelt from the mountains along their southern borders. Parts of far eastern Siberia have low precipitation and low temperatures.

Desert, Grassland, Forest, and Tundra

Latitude and climate (namely temperature and water availability) greatly influence the natural vegetation, which has also been greatly altered by human occupation. Hot deserts in the south change northward to steppe grassland, to deciduous and coniferous forest, and to tundra on the shores of the Arctic Ocean. These latitudinal distributions are interrupted only by mountain ranges (Figure 3.4a, b).

The deserts of the area east of the Caspian Sea contain few patches of grass and oasis vegetation and were once occupied by nomadic peoples. Trade routes via the region's oases provided long-established links between Europe and China.

North of the desert, the **steppe grasslands** grow on fertile **black earth soils (chernozems)**, which are similar to the soils of the North American prairies. The steppes extend from Ukraine into Kazakhstan, one of the world's major arable regions. In the Middle Ages, the steppe grasslands provided the grazing lands through which wave after wave of invaders moved westward into Europe.

Farther north, trees can grow where evaporation rates decrease, and a zone of wooded steppe gives way to deciduous

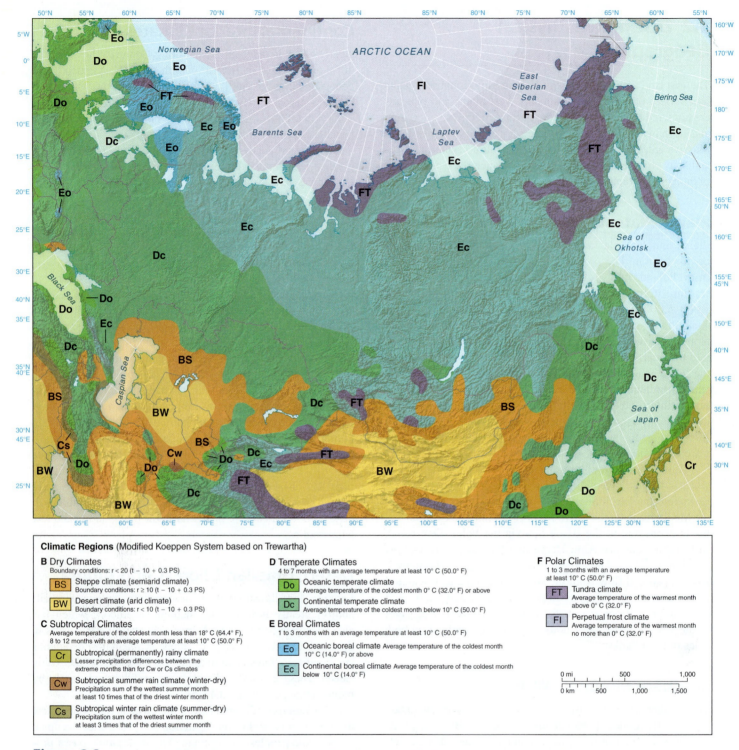

Figure 3.3 Russia and neighboring countries: climates. The dominant midlatitude continental interior climatic environments are characterized by harsh winter conditions, aridity in the south, and arctic conditions in the north.

forests, linked to fertile **brown earth soils**. The trees have been largely cut to extend farmland.

Northward and eastward—where temperatures and precipitation decline—the deciduous forest gives way to forests dominated by hardy birch trees and evergreen pine, fir, and spruce. This **northern coniferous forest**, or boreal forest (taiga), dominates vast areas of land from Moscow northward

and across most of Siberia. Pine and fir trees in the west give way eastward to larch. The taiga forms the world's largest forest area, covering 7.5 million km² (2.9 million mi.²; compare to 5.5 million km², or 2.12 million mi.², of rain forest in Brazil and 4.5 million km², or 1.7 million mi.², of related coniferous forest in Canada). It provides a huge reserve of biodiversity and is important in world climate change, absorbing one-third as much carbon dioxide as the tropical rain forest over a similar period. However, it is a very slow-growing forest system compared to other forests. Underlying the taiga are poor **podzol soils**, which have a gray sandy layer under a surface of slowly decaying leaves and above a layer where iron minerals accumulate, often becoming very hard and impeding drainage.

Around the shores of the Arctic Ocean and extending southward into the plateaus farther east is an area where trees will not grow. It is covered by grasses and low-growing shrubs. Such **tundra** vegetation is underlain by permanently frozen ground, or **permafrost**, that also extends southward beneath the taiga forest. The continuous and broken permafrost areas cover most of central and eastern Siberia and are up to 3,000 m (10,000 ft.) thick in parts of eastern Siberia. Farming is difficult if not impossible.

3.3 Distinctive Human Geography

Russia and Neighboring Countries remains ethnically diverse, reflecting three major influences: Christianity, penetrating the region from the southwest; Islam, entering from the south; and Mongol culture, sweeping in from the east. At the same time, this world region is currently experiencing some dramatic changes. The Slavic countries and the Southern Caucasus are experiencing dramatic population decline, and Central Asia is showing great growth. The decline of economic subsidies given to people during Communist times to work in harsher physical environments has given people the incentive to migrate to more amenable climates within the region. Ethnic minorities, particularly Russians, are migrating back to their home countries such as Russia. Capitalism is changing urban patterns, causing some cities to decline while others grow with modern shopping facilities and global connections. The peoples of Russia and Neighboring Countries have had to cope with a dramatic transformation from rigid and inflexible Communist political and economic systems to more dynamic and uncertain democratic and capitalist systems.

Cultural Diversity

Russia and Neighboring Countries is inhabited by over 100 ethnic groups (Figure 3.5), but many of its peoples have as much in common with people of neighboring regions as they do with one another.

Eastern Slavs

Russians, Ukrainians, and Belarussians are Eastern Slavs, offshoots of the broader Rus people and culture that emerged in the AD 800s and 900s. In 988, the Rus adopted Eastern Orthodox Christianity, which tied them to Constantinople. It also led to the adoption of the Cyrillic alphabet, which was derived from the Greek alphabet. All three Slavic groups are very closely related, as seen for example in the name *Belarussian*, which

translates as "White Russian." Differences between them developed after Poles, Lithuanians, and Austrians ruled the western lands of the Rus from the 1300s to the 1800s. During this time, Ukrainians and Belarussians developed separate identities from Russians. Polish, Lithuanian, and Austrian influence is seen in the fact that nearly 20 percent of Belarus's population is Roman Catholic. A number of Poles and Lithuanians live in Belarus today, the result of Poland's boundary being relocated in 1945. Though many eastern Slavs live in Moldova, native Moldovans are very closely akin to Romanians and are not related to the Slavs.

The name *Ukraine*, meaning "borderland," illustrates history's shifting boundaries and influences. To the Russians, Ukraine is Russia's borderland, but to many Ukrainians, who have stronger ties with Europe than the Russians, Ukraine is Europe's borderland. Russian ties with Ukrainians and Belarussians were reestablished in the early 1800s, when the Russian Empire extended west. **Russification** began soon afterward and was particularly strong during Soviet times. Today, 81 percent of Belarus's population is Belarussian and 11 percent is Russian, but 63 percent of the population regularly speaks Russian and not Belarussian. Not surprisingly, Belarus's leaders have expressed interest in uniting Belarus with the Russian Federation. In Ukraine, Russians, living primarily in the industrial areas of eastern Ukraine, account for 22 percent of Ukraine's population. Many Ukrainians, however, fear Russian domination and cultivate their ties with the West.

Peoples of the Southern Caucasus

The Southern Caucasus is a small area of the world but it is culturally very diverse (Figure 3.6), partly because it lies at the historic contact zone between the Turkish, Persian, and Russian empires. The many languages spoken in the subregion are of different language families, so most have very little in common. Georgian is in the Caucasian language family. Armenian is Indo-European but stands alone on its own branch. Azerbaijani is a Ural-Altaic language (see Figure 1.17).

The Georgians (see Figure 3.2b) and Armenians both accepted Christianity early in history, in the AD 300s. The Armenians claim their country was the first in the world to officially adopt Christianity. The Armenian Apostolic Church has been independent since the Middle Ages and expresses a unique view of Christianity. The Georgian Church is associated with Eastern Orthodox Christian churches.

Arabs introduced Islam to Azerbaijan in the 600s and 700s. In the 1500s, the Shia branch of Islam came to the country and now dominates. Despite close ties with Iran, Soviet secular policies greatly influenced Azerbaijanis. For example, Azerbaijani Muslims, unlike those in Iran, drink wine, and women are not veiled or segregated. In 1991, the Azerbaijani government also went against the wishes of Iran and adopted a modified Latin alphabet for Azerbaijani instead of the Arabic alphabet, the original alphabet of the Qu'ran used in many Muslim countries. The Latin alphabet is customarily used in Roman Catholic and Protestant countries but also in nearby Turkey, where the language is similar to Azerbaijani.

Central Asians

Modern Central Asians are settled, but many of their ancestors were nomadic, and their cultures still reflect the traditional ways of life. For example, in traditional Kazakh culture, it is customary to ask about the well-being of someone's livestock before inquiring about the person's health and that of his or her family. The dwellings for all these nomadic peoples is the yurt, a circular tent consisting of a willow wood frame covered in wool felt (Figure 3.7a). An opening at the top allows smoke to exit from the fire used for cooking and heating. Modern Central Asians no longer live in yurts, but they use them as decorative motifs for buildings or erect them in their yards and sleep in them during the summer. The yurt is an important symbol of national identity and appears on the national flag of Kyrgyzstan.

Though Central Asians share some cultural characteristics, they also have some distinct differences:

- Kazakhs emerged as a distinct people in the 1400s from a mixture of Turkic and Mongolian nomads of Central Asia.
- The ancestors of the Kyrgyz probably originated in Mongolia (Figure 3.7b). After the 800s, they mixed with Turkic tribes from the south and west and adopted their current name, Kyrgyz, which means "40 clans" in the Turkic languages. The 40 clans are represented on the national flag with a sun that has 40 rays. During czarist times, both Kazakhs and Kyrgyz were called Kyrgyz, illustrating that the languages of the two peoples are very closely related.
- Turkmens trace their ancestors back to Oghuz tribes that inhabited Mongolia and southern Siberia around Lake Baikal. In the 700s, these tribes migrated into Central Asia and assimilated Turkic and Persian tribes, giving rise to the Turkmen. "Turkmen" probably means "pure Turk" or "most Turklike of the Turks." Carpet making and horse breeding are important national traditions. Five traditional carpet designs are incorporated into Turkmenistan's national flag. Akhalteke, a breed of horse well adapted to the desert, is the breed of national significance, appearing as the central figure in Turkmenistan's national emblem.
- Uzbeks are a Turkic people who moved into the lands now known as Uzbekistan in the 1500s. They are closely related to Turkmens but also have close ties with Tajiks. The Uzbeks, who prospered greatly from the Great Silk Road, wear clothing made with fine fabrics, color, and ornamentation (Figure 3.8). All were expensive in earlier times, and the ability to incorporate as much as possible in one's dress showed one's wealth. This earlier cultural practice is evident today. On any given day, it is still common to see individuals elegantly dressed, though they have no special function to attend.
- Tajiks probably acquired their name from an old Arab tribe. The Tajik language is a Persian language and was

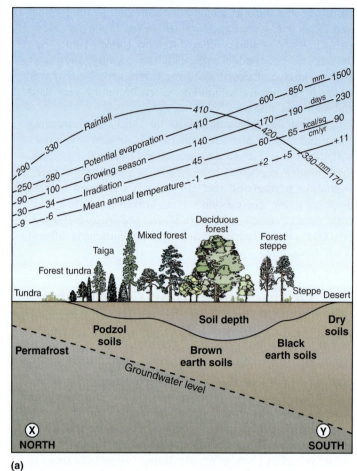

(a)

Figure 3.4 **Russia and neighboring countries: natural vegetation and soils.** (a) The section X-Y across the map illustrates the types of vegetation and soils in a north-south succession. (b) Map of natural vegetation and its relationship to soil types. *Source: Data from Brown et al. 1994.*

not distinguished from Persian (Farsi) until the Soviets designated Tajik as a unique language. At the same time, Tajiks had not differentiated themselves from Uzbeks, though the Uzbeks are a Turkic people. Both groups commonly spoke each other's languages.

Most Central Asians are Muslims. Those living closest to the Islamic heartland, such as the Uzbeks, converted to Islam as early as the 600s, soon after the rise of Islam. Those farther north, such as the Kazaks and Kyrgyz, converted to the religion only in the last 200 years. Some Central Asians seek ties with Iran (e.g., Azerbaijan and its Shia Muslims) or with other Sunni Muslim countries such as Turkey, Kuwait, and Saudi Arabia. These countries finance the building of mosques and other Islamic institutions in Central Asia. However, many Central Asians, like the Azerbaijanis, were heavily influenced by the secular ideas of Soviet communism and thus shun Islamic fundamentalists.

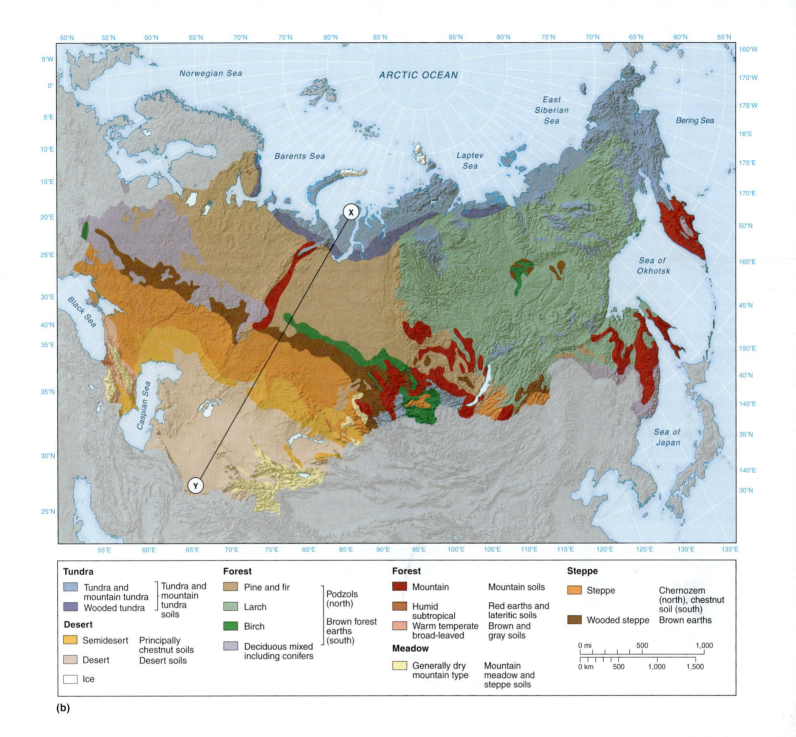

Tundra

Tundra and mountain tundra
Wooded tundra
} Tundra and mountain tundra soils

Desert

Semidesert — Principally chestnut soils

Desert — Desert soils

Ice

Forest

Pine and fir
Larch
Birch
Deciduous mixed including conifers
} Podzols (north)
Brown forest earths (south)

Forest

Mountain — Mountain soils

Humid subtropical — Red earths and lateritic soils

Warm temperate broad-leaved — Brown and gray soils

Meadow

Generally dry mountain type — Mountain meadow and steppe soils

Steppe

Steppe — Chernozem (north), chestnut soil (south)

Wooded steppe — Brown earths

0 mi   500   1,000

0 km   500   1,000   1,500

(b)

Women's Roles

Communist ideology professed that women were equal to men, and women began to achieve this equality following the revolution in 1917, before they were allowed to vote in the United States. During Soviet times, women in the subregion moved into positions traditionally held mostly by men in other societies (e.g., the United States), such as jobs in government, economics, medicine, or engineering. In some professions, women formed the majority. Most physicians, for example,

were women. During World War II, some of the most decorated fighter pilots and infantry were women. Though communism gave women great freedom in their career choices, it also idealized the factory worker, not the physician or engineer. Consequently, women were given the freedom to move into professions that lost much of their prestige compared to the same professions in the capitalist world. At the same time, however, women also were encouraged to do industrial and construction work, holding more than 60 percent of the construction jobs.

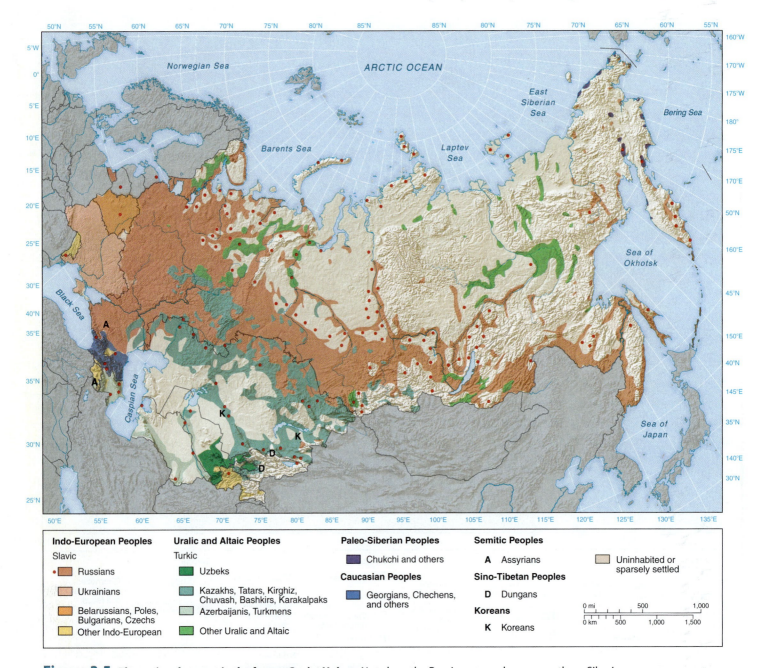

Indo-European Peoples

Slavic

- • Russians
- Ukrainians
- Belarussians, Poles, Bulgarians, Czechs
- Other Indo-European

Uralic and Altaic Peoples

Turkic

- Uzbeks
- Kazakhs, Tatars, Kirghiz, Chuvash, Bashkirs, Karakalpaks
- Azerbaijanis, Turkmens
- Other Uralic and Altaic

Paleo-Siberian Peoples

- Chukchi and others

Caucasian Peoples

- Georgians, Chechens, and others

Semitic Peoples

A   Assyrians

Sino-Tibetan Peoples

D   Dungans

Koreans

K   Koreans

- Uninhabited or sparsely settled

0 mi   500   1,000
0 km   500   1,000   1,500

Figure 3.5 **The national groups in the former Soviet Union.** Note how the Russians spread across southern Siberia.

Women achieved equality in their careers but usually not at home. After women completed a day's work at their jobs, many arrived home to undertake the traditional responsibilities of housework and child care. Men's attitudes and behaviors in regard to their wives did not change much during communism. As a result, between home and work, women ended up working long hours. As the region is transforming from communism to capitalism, the roles of women are changing. The idea that women are equal to men still exists, although more women than men have lost their jobs during difficult economic periods. Full and meaningful equality has yet to be realized.

Population

Decline and Growth

Not all of the countries of this world region are experiencing the same population dynamics. The Slavic countries and the Southern Caucasus are experiencing dramatic decline, and Central Asia is showing great growth.

Slavic populations may decline by several million by 2025 (Table 3.1). Growth rates typically ranged from 0.0 to −0.4 percent in 2012. Fertility rates were around 1.5, far below the fertility rate of 2.1 needed to maintain a population at its current number. The age-sex diagram for Russia (Figure

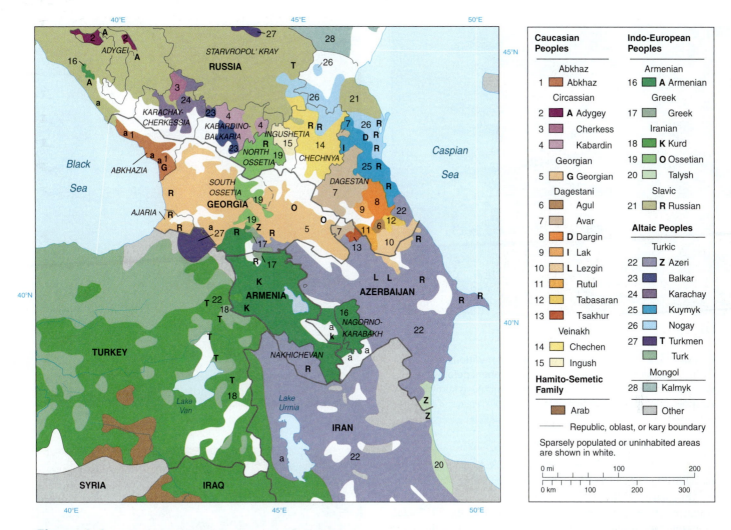

Figure 3.6 Ethnolinguistic groups of the Caucasus region. Compare the distribution of groups to the location of political boundaries. Do these comparisons help to explain conflict in the subregion?

3.9) shows an uneven pattern of age groups as a consequence of World War II and a baby boom from 1950 to 1970, when large families were encouraged by the state. Women dominate the over-60 age groups to a greater extent than in other parts of the world as a result of male deaths in World War II and Stalin's persecutions. In 2012, birth rates were 13 per 1,000 of the population, while death rates were 14 per 1,000. Though the total population is declining, it is at a much slower rate than just a few years ago. Nevertheless, the Slavic countries have some of the greatest rates of decline in the world. Ukraine is the highest in the region with 11 births and 14 deaths per 1,000 of the population.

For the Russian Federation, such figures hide variations within this huge country. In some parts, deaths outnumber births by two to one. The most significant geographic variation is between the western regions of the Russian Federation and the more recently settled regions of the north and Siberia. In western Russia, birth rates are low, death rates high, and the population is aging quickly. Yet the total numbers show less of a decline because other Russians are emigrating to western Russia from the north and Siberia and from the former Soviet republics.

Economic decline has been the major cause of population decline in the Slavic countries. Lack of government subsidies for industries in the north and Siberia is one factor. Elsewhere, many industrial cities have high employment because they have not been able to adapt to market economies. At times, particularly in the 1990s, many workers were not paid for months or years. Abortion rates are high, a legacy of Soviet times when abortion became a common means of birth control. Health care has also suffered greatly with economic decline. Hospitals have frequently run out of basic supplies such as drugs and anesthetics. The incidence of alcoholism is high, particularly among men. For many hospitals, 90 percent of emergency cases have been alcohol-related. Environmental contamination (e.g., nuclear and water pollution) is a major problem, too. The entire situation has given women little incentive to have children.

Some countries of the Southern Caucasus subregion experienced some decline after 1991. By 2012, growth rates ranged from 0.2 in Georgia to 1.3 in Azerbaijan. While the fertility rate

(a)

(b)

Figure 3.7 **Central Asia: Kyrgyzstan.** (a) Traditional nomads in the Suusamir Valley. Notice how the landscape is deforested, common where nomads live. (b) Kyrgyz man drinking kumya—an alcoholic drink made from fermented horse milk. *Photos: (a & b)* © *Ronald Wixman.*

of 2.3 in Azerbaijan indicates slow growth, Armenia's and Georgia's fertility rates are both 1.7, indicating declining populations.

In contrast to the rest of world region, the five Muslim countries of Central Asia anticipate growth from the 2012 total of around 64.6 million people to 77.2 million by 2025. In 2012, Kazakhstan, which is more than 30 percent Russian, had the lowest growth rate (1.4 percent per year) along with Turkmenistan; Tajikistan had the highest (2.2 percent). Death rates are very low—a formula for increasing population. Fertility rates fell from around 6 to just less than 3 between 1970 and 2012.

Distribution and Patterns

In the Russian Federation, the greatest concentration of people is in western Russia (Figure 3.10). Higher population densities continue eastward along the Trans-Siberian Railway to the southern end of Lake Baikal and Vladivostok on the Pacific coast. The extensive areas of mountains and desert in the south and of permafrost-ridden lands in the north and east contain few people. The more fertile lands of western Russia, Ukraine, Belarus, and Moldova have higher population densities and are punctuated with greater concentrations of people around industrial cities.

The Caucasus Mountains create uneven distributions of people in Georgia, Armenia, and Azerbaijan, with few people in the mountains and most in the plains. The distribution of population in Central Asia is marked by differences between areas of few people in the arid and mountainous zones and areas of higher densities in the irrigated lowlands.

During Russian imperial and Soviet times, all the peoples of this world region lived together in one country. To maintain

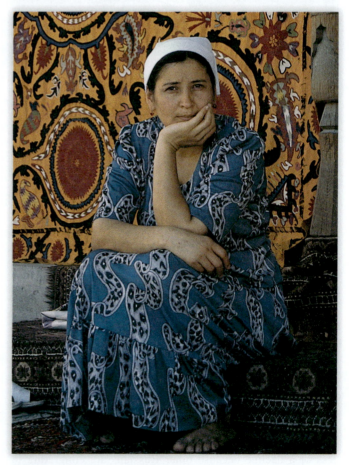

Figure 3.8 **Central Asia.** Uzbek woman in front of Suzanne embroidery. *Photo:* © *Ronald Wixman.*

Table 3.1	**Russia and Neighboring Countries:** Data by country, area, population, urbanization, income (gross national income purchasing power parity), ethnic groups

| Country | Land Area (km²) | Population (millions) | | % Urban 2006 | GNI PPP 2010 Total (US$ billions) | 2010 Per Capita (US$) | Ethnic Groups (%) |
		Mid-2012 Total	2025 Est. Total				
SLAVIC COUNTRIES							
Belarus	207,600	9.5	9.0	76	129.0	13,590	Belarussian 78%, Russian 13%, Polish 4%
Moldova	33,700	4.1	3.7	42	12.0	3,360	Moldovan/Romanian 65%, Ukrainian 14%, Russian 13%
Russian Federation	17,075,400	143.2	140.8	74	2,726.8	19,240	Russian 82%, Tatar 4%, Ukrainian 3%
Ukraine	603,700	45.6	42.4	69	303.8	6,620	Ukrainian 73%, Russian 22%
Totals/Averages	17,920,400	202.4	195.9	72	2,468	15,670	
SOUTHERN CAUCASUS							
Armenia	29,800	3.3	3.3	64	17.5	5,660	Armenian 93%, Azeri 3%, Russian 2%
Azerbaijan	86,600	9.3	10.4	53	83.9	9,270	Azeri 90%, Daghestani 3%, Russian 2%
Georgia	69,700	4.5	4.1	53	22.2	4,990	Georgian 70%, Armenian 8%, Russian 6%, Azeri 6%
Totals/Averages	186,100	17.1	17.8	55	123.6	7,228	
CENTRAL ASIA							
Kazakhstan	2,717,300	16.8	19.5	55	175.7	10,770	Kazakh 46%, Russian 35%, Ukraine 5%, German 3%
Kyrgyzstan	198,500	5.7	6.6	35	11.3	2,070	Kyrgyz 57%, Russian 18%, Uzbek 14%
Tajikistan	143,100	7.1	9.6	26	14.7	2,140	Tajik 65%, Uzbek 25%, Russian 3%
Turkmenistan	488,100	5.2	5.9	47	37.8	7,490	Turkmen 77%, Uzbek 9%, Russian 7%
Uzbekistan	447,400	29.8	35.6	51	87.7	3,110	Uzbek 80%, Russian 6%, Tajik 5%
Totals/Averages	3,994,400	64.6	77.2	44	327.2	5,065	
Region Totals/ Averages	22,100,900	284.1	290.9	65	2,820	12,750	

Source: *World Population Data Sheet 2012*, Population Reference Bureau; Microsoft Encarta 2005.

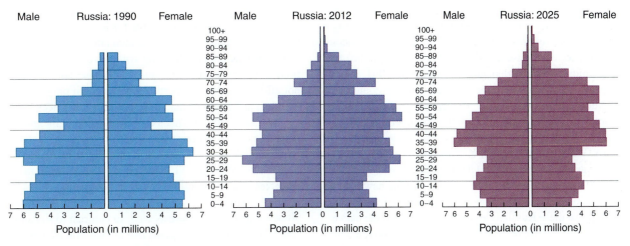

Figure 3.9 **Russia: age-sex diagram.** Account for the irregular shapes of Russia's population pyramids. How do they correspond to events of the last 50 years? *Source: U.S. Census International Database: http://www.census.gov/population/international/data/idb/region.php.*

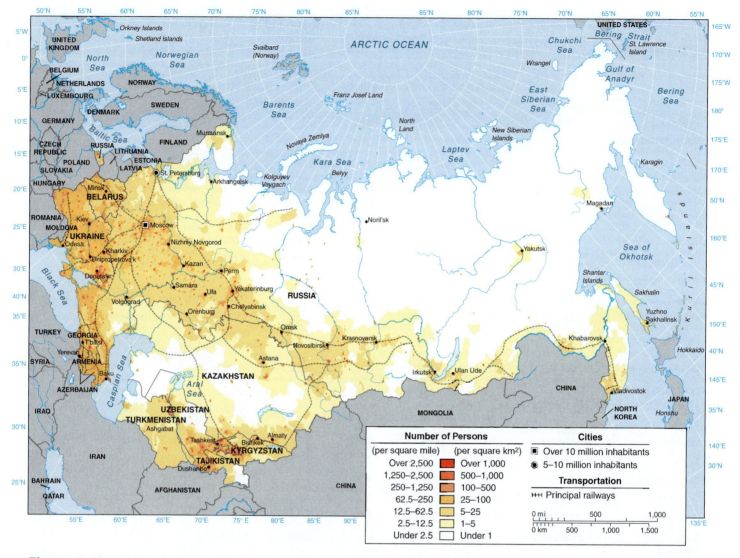

Figure 3.10 **Russia and neighboring countries: distribution of population.** For Russia, comment on the location of the highest and lowest densities of population. How do these reflect climatic conditions, initial historic superiority, the line of the Trans-Siberian Railway, and other factors?

control of this vast country, the government in Moscow encouraged Russians to move to other republics and the minority peoples, sometimes against their will, to move great distances from their territories. Consequently, when the 15 republics of the Soviet Union became independent countries in 1991, many people found themselves outside their designated countries. Twenty-five million Russians, or 17 percent of the Russian population of the Soviet Union, live in the "Near Abroad" (Figure 3.11). High concentrations live in Estonia, Latvia, northern and eastern Ukraine, and northern Kazakhstan. In many locales in the Crimean Peninsula (part of Ukraine), Russians constitute more than 90 percent of the population. Many of the new countries see their long-established Russian minorities as threats and have insisted that these Russians assimilate (e.g., learn the native language and way of life) or go back to Russia. By 2001, it was estimated that as many as 5 million Russians emigrated from the independent republics to the Russian Federation (Figure 3.12).

Time has not decreased hostility toward Russians in many countries. After making life miserable for Russians in Turkmenistan, the leader of that country ended dual citizenship for Russians in June 2003. Russians choosing Russian citizenship desperately tried to sell their property quickly because non-Turkmen were not allowed to own property. However, with only a few weeks' notice, many Russians had to sell at very low prices.

In the Southern Caucasus, most Armenians and Azerbaijanis do not live in their respective countries. Of the 6.3 million Armenians in 2001, only 3.8 million lived in Armenia. Many lived in Azerbaijan and Georgia. Of the 19 million Azerbaijanis, less than 8.1 million lived in Azerbaijan. Most still live in neighboring Iran. Azerbaijan is 90 percent Azerbaijani, but 90 other groups live there as well. In contrast to Armenians and Azerbaijanis, most Georgians live in Georgia. However, 30 percent of Georgia's population is composed of other groups such as Armenians, Russians, Azerbaijanis, Ossetians, and Abkhaz.

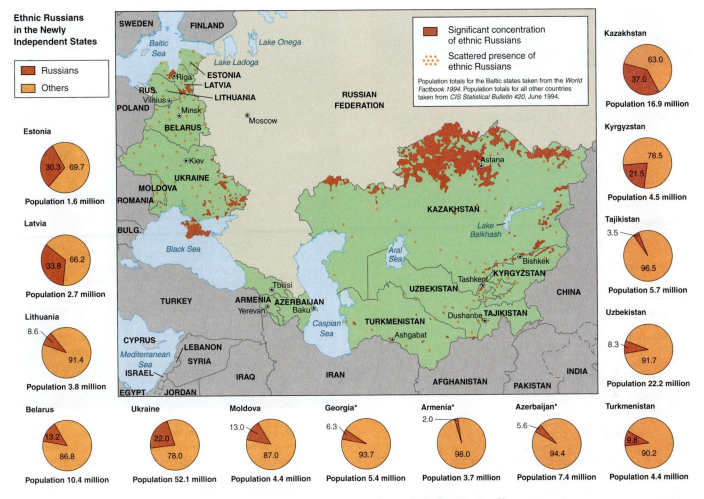

Ethnic Russians in the Newly Independent States

Russians
Others

Significant concentration of ethnic Russians

Scattered presence of ethnic Russians

Population totals for the Baltic states taken from the *World Factbook 1994*. Population totals for all other countries taken from *CIS Statistical Bulletin #20*, June 1994.

Estonia
30.3 / 69.7
Population 1.6 million

Latvia
33.8 / 66.2
Population 2.7 million

Lithuania
8.6 / 91.4
Population 3.8 million

Belarus
13.2 / 86.8
Population 10.4 million

Ukraine
22.0 / 78.0
Population 52.1 million

Moldova
13.0 / 87.0
Population 4.4 million

Georgia*
6.3 / 93.7
Population 5.4 million

Armenia*
2.0 / 98.0
Population 3.7 million

Azerbaijan*
5.6 / 94.4
Population 7.4 million

Turkmenistan
9.8 / 90.2
Population 4.4 million

Kazakhstan
63.0 / 37.0
Population 16.9 million

Kyrgyzstan
78.5 / 21.5
Population 4.5 million

Tajikistan
3.5 / 96.5
Population 5.7 million

Uzbekistan
8.3 / 91.7
Population 22.2 million

*Ethnic percentages for Georgia, Armenia, and Azerbaijan taken from the 1989 Soviet census; they may not accurately reflect present conditions.

Figure 3.11 **Ethnic Russians as minorities in neighboring countries.** In which countries are Russians the largest minorities? Do Russians outside of Russia enhance or hinder the political power of Russia in the region?

The peoples of the Central Asian countries also live in one another's countries. For example, the boundaries of Kyrgyzstan, Tajikistan, and Uzbekistan weave through the fertile and densely settled Fergana Valley, leaving many Kyrgyz, Tajiks, and Uzbeks in their neighbors' countries.

Urban Patterns and Linkages

Urbanization

This world region's highest rates of urbanization are in areas that emphasized urban-industrial development. The Russian Federation, Ukraine, and Belarus are the most urbanized, while Moldova and most of the Central Asian countries with their traditional ways of life are among the least urbanized. The Russian Federation has the greatest number of cities and the largest cities of the region (Table 3.2).

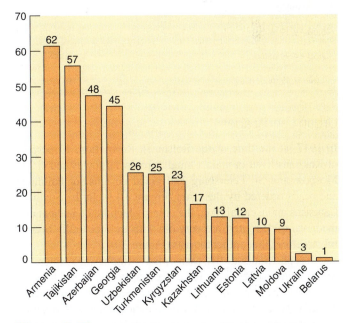

Figure 3.12 **Migration of ethnic Russians from the former Soviet republics, 1989 to 1998.** Each number represents the percentage of ethnic Russians who have emigrated from that country's total population of ethnic Russians.

| Table 3.2 | Populations of Major Urban Centers in Russia and Neighboring Countries |

TOP 10 MOST POPULATED URBAN CENTERS

City, Country	2012 (millions)	2025 (millions)
1. Moscow, Russia	10.5	10.5
2. St. Petersburg, Russia	4.6	4.6
3. Kiev, Ukraine	2.8	2.8
4. Tashkent, Uzbekistan	2.3	2.7
5. Baku, Azerbaijan	2.1	2.4
6. Minsk, Belarus	1.8	1.9
7. Kharkov, Ukraine	1.5	1.5
8. Novosibirsk, Russia	1.4	1.4
9. Almaty, Kazakhstan	1.3	1.4
10. Nizkniy Novgorod, Russia	1.3	1.3

NUMBER OF MAJOR URBAN CENTERS IN RUSSIA AND NEIGHBORING COUNTRIES, BY POPULATION CATEGORY

Population Category	Number of Urban Centers
10.00 million and greater	1
5.00 – 9.99 million	0
2.50 – 4.99 million	2
1.00 – 2.49 million	18

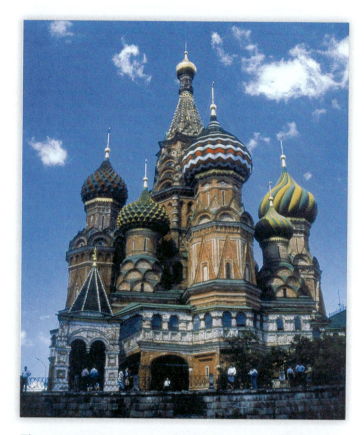

Figure 3.13 **Russian urban landscape.** The cathedral of the Intercession, commonly known as St. Basil's Cathedral, is located on Red Square, the center of Moscow. Ivan IV (The Terrible) had the cathedral built to commemorate Russia's victory over the Mongols. Originally intended as eight smaller, clustered cathedrals to represent Russia's eight victorious battles over the Mongols, the result was a single cathedral with eight domes and chapels signifying the victories. *Photo: © Ronald Wixman.*

Urban Landscapes

In 1917, at the time of the Bolshevik Revolution, 17 percent of Russians lived in mainly small, provincial towns and cities. Moscow and St. Petersburg had grand designs for buildings and roads ordered by the czars (Figure 3.13). In the Southern Caucasus and Central Asia, the cities, which lie on ancient and medieval trade routes, are very old with architecturally beautiful structures that attract tourists (Figure 3.14).

Joseph Stalin's five-year plans emphasized intensive centralization and rapid industrialization. Industrial centers expanded, and new specialist resource (e.g., oil, mining) centers emerged, often in new towns in remote regions. Urbanization increased to 48 percent of the region's population in cities by 1959 and to more than 70 percent in the 1990s. Stalinist urbanization produced distinctive, standardized cities throughout the Soviet Union. Housing for the workers consisted of

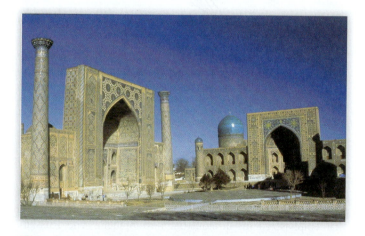

Figure 3.14 **Central Asia: Samarqand, Uzbekistan.** Historic Muslim buildings highlight the importance of Islam in Central Asia. Madrasah (theological seminary) buildings of the 1500s and 1600s border Registan Square. Typical features of Islamic art include arches, tilework, domes, and minarets. *Photo: © Ronald Wixman.*

numerous poor-quality, high-rise, concrete buildings with few shopping opportunities in their vicinity because Soviet planners did not foresee the need for much shopping.

One of the major trends to emerge since the breakup of the Soviet Union in 1991 is the building of suburbs, particularly around Moscow. The congestion and pollution of inner-city streets due to the growing use of cars and trucks, together with increased urban crime, caused richer families to move out.

A new focus of the market economy was not only the construction of new private housing but also the conspicuous appearance of suburban shopping malls with supermarkets, clothing chains, electronics stores, movie theaters, and foreign-owned hotels (Figure 3.15). In the 1990s, most were built in Moscow, followed by St. Petersburg. By 2003, 28 percent of Russia's retail sales were in Moscow though only 2 to 4 percent of Russians shopped in modern facilities. As a result of the fierce competition in Moscow, investors began searching for new cities, usually regional capitals with more than 1 million inhabitants. Kazan, Samara, Yekaterinburg, Krasnodar, and Kaluga have emerged as new cities of investment. For example, Marriott Renaissance opened a hotel in Samara in late 2003, and IKEA opened stores in Kazan (2004), Nizhniy Novgorod (2006), Yekaterinburg (2006), Rostov-on-Don (2007), Novosibirsk (2007), Adygea-Kuban (2008), Omsk (2009), and Samara (2009).

Evolving Politics

Russia has been the dominant country of this world region for almost a millennium. Prior to Soviet times, the Russian Empire was one of Europe's great empires. Unlike the European empires such as France and the United Kingdom, which conquered lands overseas, the Russian Empire conquered peoples and land adjacent to it, extending its borders as it annexed new territories with very diverse populations. The Soviet Union, founded in 1922, had many successes, but internal problems caused it to disintegrate in 1991. Though the 15 newly independent countries contain many peoples of ancient ancestry, many of the new countries scarcely had been independent before.

The Russian Empire

The Russian Empire traces its roots back to the principality of Muscovy, which began to expand in the 1400s. Muscovy's leader, Ivan III (the Great) married Sophie Paleologue, the niece of the last Byzantine emperor. After Constantinople fell to the Ottomans in 1453, Ivan was seen as the true inheritor of the Christian realm. He adopted the title "czar," derived from the Latin *caesar* for emperor. In 1613, Mikhail Romanov ascended the throne. The Romanov dynasty lasted until the last czar was deposed in 1917.

From 1480, Muscovy's territory continued to expand, especially to the east into Siberia and Central Asia but also to the west, bringing the Russians into contact with Central and Western European culture. The greatest territorial expansions were made under Peter the Great (1682–1725), Catherine the Great (1762–1796), and Alexander I (1801–1825). By the early 1900s, the Russian Empire achieved its greatest extent, stretching from Europe to East Asia and the Pacific Ocean (Figure 3.16).

Russia was a world power, but its strength was derived from its sheer size. Internally, it had many problems. Politically, the czars remained in total control and continued to claim to speak in the name of God, but absolute power inhibited the creation of an efficient government. Technologically, Russia was decades behind Western Europe. Most Russians remained enserfed peasants in a feudal system, farming with horses or oxen. At the same time, Russia had expanded its boundaries to include more than 100 different peoples. After the rise of nationalism in Europe in the mid-1800s, Russia's nationalities began clamoring for independence. This situation worsened as Russians themselves became nationalistic and began suppressing non-Russians. They developed Russification policies to force Russia's minorities to become more Russian ("Russified").

The Founding of the Soviet Union

World War I (1914–1918) exerted great stress on the Russian Empire. By 1917, starvation and a huge death toll worked together with long-standing opposition to czarist rule. Revolutionaries deposed Czar Nicholas II and set up a provisional government that proved to be weak. A number of national groups staked their claim to territory and declared independence. In late 1917, the Bolsheviks, a group of Communists also known as "Reds," overthrew the provisional government. Civil war ensued when anti-Communists, called "Whites,"

Figure 3.15 Moscow: IKEA. In recent years, European and American style super shopping centers have been appearing in the region. Notice how this IKEA has its name written in Cyrillic with the added expression of "для вашего дома" ("for your home"). *Photo:* © *Abramov Denis/ITAR-TASS Photo/Corbis.*

Figure 3.16 **Russian Federation: history of growth.** The expansion of the Russian Empire from the 1600s to the early 1900s. The Trans-Siberian Railway was built through territories that were regarded as safely Russian. It was rerouted after territory was lost. The Baikal-Amur Main Line Railway (BAM) was begun in 1974 and completed in 1988. It opened up more of the vast mineral and forest resources of eastern Siberia and the Russian far east to greater exploration.

tried to dislodge the Bolsheviks from power. By 1922, the Reds gained the upper hand, expelled their enemies, and reclaimed many of the lost territories.

In 1922, under the leadership of Vladimir I. Lenin, the Bolsheviks established the Union of Soviet Socialist Republics (USSR), commonly called the Soviet Union. This new country succeeded the Russian Empire, but the Bolsheviks changed the old system of government and economy. They abolished the monarchy and government by a privileged few and replaced them with "soviets." Soviets were workers' and soldiers' councils that drew their members from the common citizens. The Communists denounced religion because the clergy sanctioned the rule of political leaders who oppressed the common people. Consequently, the Bolsheviks attacked churches and mosques, dynamiting many and turning others into scientific centers such as planetariums to show the error of religious belief. After Joseph Stalin came to power in 1924, he purged millions of his enemies, suspected enemies, and potential enemies, including about half of the officers in the military (see the "Human Rights" section in this chapter, p. 106).

World War II

Stalin entered into a nonaggression pact with Germany's Adolf Hitler in 1939, allowing the Soviet Union to take control of territories that formerly belonged to the Russian Empire: the Baltic states, eastern Poland, and Bessarabia (now Moldova). Finland was also a target, but the Finnish forces successfully defended their country from the Red Army. The victory of the small Finnish military over the huge Red Army underscored the weakness of the Soviet Union and its military.

On June 22, 1941, Hitler launched Operation Barbarossa, the invasion of the Soviet Union. The campaign was initially devastating. The Soviet Union survived largely because of its sheer size. A vast amount of land was lost and millions of people were killed, but even larger areas were left unconquered. The Soviets moved war production farther east, out of the range of the Nazi military, including its air force. With help from the United States and the United Kingdom, production increased. Finally, the harsh Russian winter brought the German Nazi armies to a standstill. In the meantime, Stalin moderated many of his harsh policies, including the persecution of the Russian Orthodox Church. After a fierce battle at Stalingrad in 1943, Soviet forces began rolling back the Nazi invaders. The people of the Soviet Union had turned the tide of war and began winning what they call the "Great Patriotic War." By May 1945, the Red Army had swept through East Central Europe and occupied much of Germany, including the capital, Berlin.

Its victory in World War II allowed the Soviet Union to annex the Baltic countries and Moldova, former territories of the Russian Empire the Soviets invaded in 1940, and new territory in East Central Europe never before governed by the Russians (e.g., East Prussia, the northern half of which is now Kaliningrad, and areas of Poland). Victory also allowed Stalin to establish and support Communist governments in East Central European countries. To counter NATO and the Marshall Plan (economic aid from the United States) and later the European Economic Community, Stalin created the Warsaw Pact and Council for Mutual Economic Assistance (CMEA, also known as COMECON). Stalin also began persecutions again, accusing entire ethnic groups and nations of treason during the war. He moved millions from their homes, mostly to Siberia. After Stalin died in 1953, his successors were deliberately less harsh. However, Stalin's system of government and economy remained intact until the demise of the Soviet Union in 1991.

Glasnost

Stalin's policies allowed the Soviet Union to rebuild after World War II. However, rigid governmental structure and policies caused the economy to eventually decline by the 1970s. After Mikhail Gorbachev became the leader of the Soviet Union in 1985, he immediately set out to reform his country's political and economic systems. In doing so, he highlighted two concepts: *glasnost* (informational openness) and *perestroika* (economic restructuring) (see "*Perestroika*" in the next section). *Glasnost* referred to the policy of providing government information to citizens. It was intended to have a positive effect on the country by empowering citizens with knowledge. However, freedom of information also allowed citizens to learn about corruption, government abuse, forced labor, and many of the other problems of the Soviet government. Non-Russians vented their anger at the Soviet government for its Russification policies. In contrast, Russians complained that the Soviet government suppressed Russian culture, especially the Russian Orthodox Church, and distributed Russia's resources to the country's minorities. To avoid the wrath of citizens, many politicians echoed their anger, championed local and regional causes, and turned against the central government in Moscow. Even Boris Yeltsin, the elected leader of the Russian Republic in 1990, called for Russia's independence from the Soviet Union.

Mikhail Gorbachev believed in communism and tried to preserve the Soviet Union by implementing reforms. Ironically, he became the Soviet Union's last leader, as his policies led to the unraveling of the country in 1991. Led by Lithuania, the Soviet republics all declared independence by December.

Economically tied to one another, all the republics, except for Lithuania and the other two Baltic republics of Latvia and Estonia, formed the Commonwealth of Independent States, though Georgia waited a couple of years before joining. However, while Gorbachev's policies led to the rapid demise of his country, his reforms paved the way for further changes; without these reforms, the country would have probably disintegrated anyway, perhaps more violently.

Closed Economies Open Up

Prior to 1917, Imperial Russia was largely agricultural, with most people living as peasants on the land. Industrialization was only beginning in such cities as Moscow and St. Petersburg. When the Communists came to power, they were suspicious of capitalists whom they saw as exploiting the people. To prevent this from happening, they took complete control of the economy through the concept of a "planned economy" (see Chapter 2, p. 53). The distrust of capitalism also led the Communists to close off the Soviet Union's economy as much as possible from the global economy, which was primarily capitalist in nature. Communist economic policies were successful in the early decades of the Soviet Union but caused the country to disintegrate in its latter decades. This world region is now opening up to the global economy.

Five-Year Plans

With the Soviet economy initially largely agricultural, farmers could not relate to the urban-industrial ideology of communism. When Joseph Stalin came to power in 1924, he believed it necessary to forcefully industrialize the Soviet Union's economy to achieve the Communist ideal of a "workers' paradise." It was also necessary to overcome the technological advantages of the West.

To transform the economy, Stalin developed the idea of the **five-year plan**. The first one was launched in 1928 and called for collectivization and industrialization. Collectivization was a way of making farmers into factory-like workers and thus more sympathetic to the Communist way. Under collectivization, small family farms were merged together to create large farms, thousands of acres in size. The large farms were better designed to use modern farm machinery, then under production in the new factories. The government became the owner of the collectives and farmers became employees. With collectivization, farmers became more like factory workers, even living in tightly packed housing like urban factory workers.

Government owned all industries. In what is known as the **command economy**, the government set quotas favoring heavy industry over production of consumer goods. The five-year plans also established **central planning**. In contrast to capitalist economies, supply, demand, or profit making did not dictate what would be produced. With central planning, the government decided how many goods and services were needed by society, almost without cost considerations. Rapid and forced industrialization prevented the Soviet Union from experiencing the world economic depression of the 1930s, as the government kept investing in the economy and providing jobs. Significant amounts of the production came from the slave labor of the millions of people whom Stalin had purged.

Communism at an Economic Standstill

When Stalin died in 1953, his series of five-year plans had industrialized the Soviet Union's economy, despite the serious setbacks of World War II. The West, however, continued to develop economically so that the Soviet Union was still behind, despite spectacular scientific breakthroughs such as the space program. By the end of the 1950s, the Soviet Union had caught up only to the West of the 1920s, the time when Stalin began the five-year plans. In the meantime, industry in the West evolved to adapt to new materials such as plastics and other synthetics, and used new fuels such as petroleum and natural gas. Western industry also adapted to constantly changing consumer demands that required continual retooling and new locations for production facilities. The Stalinist Soviet economy, however, was so rigid that it could not adapt to change.

The weaknesses of the Soviet economic system compounded over time. For example, the military accounted for a large portion of the country's economy and saw little need to be efficient. With the government owning all businesses and with an absence of competition, managers saw no need to use fuel wisely or search for fuel alternatives in a country rich in natural resources. The Communist guarantee of a job for everyone meant that labor costs were high, though wages were low and no attempts were made to increase productivity by updating machinery and computers. Finally, the practice of central planning meant that government bureaucrats, not supply and demand, determined what was produced. The bureaucrats proved to be highly inefficient.

Resources such as oil, building materials, and equipment became scarcer because they could not be adequately exploited or delivered. The various ministries of the government hoarded them, attempting to become self-sufficient because other agencies could not supply them. This practice created further redundancies, shortages, and squandering of resources. Bureaucrats played it safe by locating new enterprises where they could best obtain supplies. In most cases, the locations were big cities. This led to excessive migration to cities, creating congestion, high living costs, and environmental problems. Ironically, though government bureaucrats contributed to the ruination of the Soviet economy, they became the unhappiest segment of society and greatly wanted change. As the privileged within the Soviet Union, they were displeased to see their standard of living drop below that of the poorer groups in the materially wealthy capitalist countries.

Perestroika

For *perestroika*, Gorbachev believed that it was necessary to divorce economics from politics, allow more local control, and introduce free market practices. Such policies went against Communist ideals and established interests. The bureaucrats fought Gorbachev and his policies, creating great political turmoil. As political battles raged, Gorbachev's economic policies ran opposite of what had been practiced since the 1920s. A crisis ensued as the economy tried to reverse direction. Companies, having existed in a noncompetitive environment, now had to meet their own costs and find their own customers. Without the ability to waste resources, they had to cut costs. To generate income, they cut production and raised the prices of their goods, but they soon found that they could not afford to buy supplies from one another. Furthermore, the lack of a capitalist banking and financial system meant that cash could not flow easily. Poor transportation and communication systems only exacerbated the situation. Production declined. To cut costs, companies laid off workers, increasing unemployment. The Soviet economy continued to spiral downward with frequent labor strikes and rampant crime.

Today's Economic Geographies

After the Soviet Union's disintegration, the new countries of this world region faced two economic problems. First, the heavy industry developed during Soviet times was far out-of-date. Since 1991, the independent republics have worked to update their factories, attract foreign investment, engage in the global economy, and develop more service industries. Second, Soviet economic policies tied the republics' economies very closely together. Many of the countries now prefer greater independence from the Russian Federation, but old connections are hard to break and new relationships difficult to form. For example, during Soviet times, eastern Ukraine was the most important iron- and steel-making region of the Soviet Union. Eastern Ukraine can provide for Ukraine's economic independence today, but the area's industries depend on imports of oil and natural gas, primarily from the Russian Federation, to meet 85 percent of their energy needs. As another example, during Soviet times, Moscow limited industrial development in the Southern Caucasus and encouraged agricultural production. Georgia supplied over 90 percent of the Soviet Union's tea and citrus fruits. Central Asia is a major supplier of cotton as well as mineral resources such as coal, iron, chromium, oil, and natural gas that are important to Russian factories. Many of the countries of the world region can find markets other than the Russian Federation for their products, but the transportation network leads to the Russian Federation.

After the demise of the Soviet Union, many people around the world wondered if Russia would remain a world power because it did not have a world power's economic strength. By 2000 Russia produced only 0.08 percent of the world GDP, and almost 40 percent of its people lived below the poverty line. The GDPs per capita within Russia and the other CIS countries were far below those of the capitalist countries of Europe and North America (Figure 3.17). Tremendous economic growth in the 2000s brought the economy back to where it was in 1990 and brought Russia's share of the world GDP to almost 2 percent. The percentage of Russia's people living below the poverty level fell to 14 percent in 2010; poverty was defined as those earning less than US$174 per month in 2010. The average share of household budgets spent on food dropped from 73 percent in 1993 to 44 percent in 2010. Nevertheless, long periods of economic distress have resulted in low levels of ownership of consumer goods (Figure 3.18). TV set ownership is high, as television provided mass communication from the government to the people during the Soviet era.

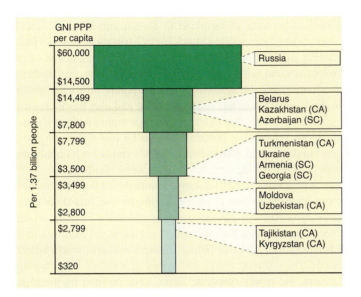

Figure 3.17 **Russia and neighboring countries: national incomes compared.** The countries are listed in the order of their GNI PPP per capita for 2010. CA = Central Asia, SC = Southern Caucasus. Compare with Figure 1.26. *Source:* 2012 World Development Indicators, *World Bank 2012.*

Agriculture

The Slavic countries' northerly latitudes and distant locations from moderating maritime influences result in dry conditions. Low-temperature climates in the north and high-temperature ones in the south make farming difficult almost everywhere. Even in the best locations, the late arrival of spring or the early appearance of winter can ruin a year's crops. The most productive farming areas are in the west of Russia, Ukraine, and Moldova in the former steppe and deciduous forest areas. Ukraine's vast plains of black soil produce abundant crops of wheat and make the country agriculturally self-sufficient. Moldova also has low relief and good soils like western Ukraine. Fruits, vegetables, wine, and tobacco are the main farm products. Belarus, however, is a country with low relief but mostly poor soils. Having been shaped by continental glaciation, the southern part of the country is covered by the Pripyat Marshes and the northern part by glacial moraines. The sandy glacial soils are good for potatoes. Toward the north in the Russian Federation, around Moscow, arable land gives way to livestock farming. East of the Urals, farming is restricted to areas that have sufficient water and length of summer growing season. The warm climates of the Southern Caucasus make it possible to grow citrus fruits, tea, tobacco, cotton, and rice.

In the 1950s, Soviet leader Nikita Khrushchev sought to increase agricultural production with his **Virgin Lands Campaign** to make the Soviet Union self-sufficient and independent of the global economy. This campaign promoted farming in lands where it had never been done before, but many of these lands had poor soil quality because they did not have either enough water or enough heat to grow crops. For example, many

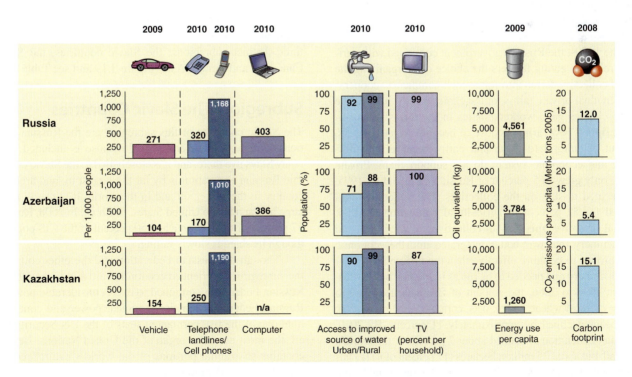

Figure 3.18 **Russia and neighboring countries: ownership of consumer goods, access to clean water, and energy usage.** What is the significance of relatively high television ownership? *Source:* 2012 World Development Indicators, *World Bank 2012.*

were in the semidesert and desert areas of Central Asia, especially in the Kazakh Republic and the adjacent dry areas of the Russian Republic, much of it stretching across southern Siberia. It was believed that these dry areas could be made productive by using new irrigation systems. From 1917 to 1987, irrigated land in the Soviet Union increased from 3.5 to 20 million hectares. At the same time, large drainage networks were built in the lands north of Moscow to remove water from the waterlogged soils. From 1956 to 1987, drained land increased from 8.6 to 19.4 million hectares.

The Virgin Lands Campaign proved to be a huge failure over the long run. Production increased, but many problems resulted. Massive amounts of water were diverted from rivers feeding the Caspian and Aral seas, wreaking havoc with their ecosystems. Pouring water on desert soils only led to soil salinization. Drainage projects were no more successful. Farm production fell on many of the newly established collectives. Eventually, the Soviet government stopped funding the Virgin Lands Campaign. Much of the land was slowly abandoned, though Kazakhstan is still one of the biggest producers of wheat and cotton.

During Communist times, most farmland was allocated to either state farms or collective farms. Typical state or collective farms consisted of thousands of acres and thousands of livestock with several hundred workers, each receiving salaries and benefits like other kinds of workers. Though state-owned farms provided job security, they were extremely inefficient. Private plots, making up only 3 percent of agricultural land, were much more efficient and accounted for approximately 25 percent of agricultural production in the 1980s.

Mikhail Gorbachev began agricultural reforms in 1988, but following the dissolution of the Soviet Union, the agricultural sector, like the rest of the economy, suffered and production fell, often to less than half for many products. The harvest in 1995 was Russia's worst since 1963.

In the 1990s, the Russian government continued with agricultural reforms, giving farmers the choice of reorganizing the state and collective farms into joint-stock companies, cooperatives, individual private farms, or the choice not to change. The first two choices gave workers shares in the farms and management rights. Few farmers chose to became private owners, partly out of cultural tradition and partly because the infrastructure and economic system do not support them. Private farmland only grew to 5 percent of Russia's farmland by 1995. Many decided to stay on the collectives. The collectives still accounted for 52 percent of agricultural production in 1996. Individual incentive, mixed with collectively owned land and farm machinery that was better integrated into the economic system, proved to be more effective than the private farms that lacked machinery and needed to be run as individual businesses.

From 1996 to 2006, the amount of farmland dedicated to the major crops of wheat, barley, oats, corn, rye, sunflower seed, and soybeans declined by approximately 11 percent. By 2010, the Russian Federation contained about 7 percent of the world's arable land, but about 35 million hectares (17 percent of it) lay fallow, which is more than the total arable land of France and Spain, the members of the European Union with the most arable land. However, from 1996 to 2006, greater yields in Russia increased total production by about 15 percent. The two largest crops, wheat and barley, grew by 35 and 28 percent, respectively. This allowed Russia to quadruple its grain exports (sevenfold for wheat specifically) in just 10 years, making it a major exporter of grain.

New laws enacted in the mid-2000s allowed foreign ownership of land. By 2010, leading buyers such as Alpcot Agro (Sweden), Hyundai Heavy Industries (South Korea), Morgan Stanley (United States), and Arab investors purchased almost 1 million hectares of farmland in Russia and Ukraine. Local investment groups such as Black Earth Farming also began purchasing land. While the earlier practice of breaking up the large collectives and selling them to individual farmers was ineffective, converting the collectives into corporate farms like those in the United States has been profitable and also has increased agricultural yields. For example, average grain yields in Russia are about 1.9 tons per hectare, which is far less than the 6.4 ton yield in the United States. Black Earth's yields for its fields are over 3.5 tons per hectare and are increasing steadily.

A heat wave in 2010 destroyed more than a third of Russia's crops, causing Russia to import grains for the first time since 2004. The Russian government responded by banning wheat exports, but because Russia had become the world's third-largest wheat exporter, the ban helped create a spike in world wheat prices. Though crop failure hurt many of Russia's small family farmers, the newer corporate farms have profited by the high prices and are likely to increase their share of Russian agriculture and increase Russia's agricultural productivity over time.

3.4 Geographic Diversity

Russia and Neighboring Countries has considerable internal political, economic, and cultural diversity that divide it into three distinct subregions: the Slavic countries, the Southern Caucasus, and Central Asia (Figure 3.19 and see Table 3.1).

Subregion: The Slavic Countries

The Slavic countries of this subregion are the Russian Federation, Ukraine, and Belarus. Moldova also is included because many Slavs live there, and it is closely tied to the Slavic countries. The Russian Federation is by far the largest in land area of any country in the subregion and in the world. It is nearly twice as large as Canada, the United States, or China. In 2008, Russia had 77 percent of the CIS area and 50 percent of the CIS population, though its share is slowly falling.

Though the Russian Federation and the other countries of the subregion experienced economic hardship in the 1990s, the Russian Federation continued to exert considerable power. The Russian Federation remains a nuclear power and continues to hold one of the five permanent seats of the UN Security Council, the most powerful organ of the United Nations. Because it contains substantial portions of the world's natural resources, the Russian Federation has considerable economic potential. In the mid-1990s, the Russian Federation joined the Group of Seven (G7), an informal organization representing the world's

Figure 3.19 **Russia and neighboring countries: subregions (the Slavic countries, Southern Caucasus, and Central Asia).** Note the distribution of the major cities and their concentration in the western part of the region.

most wealthy countries, changing the organization to the Group of Eight (G8). It held the presidency of the organization in 2006.

The Slavic countries seen on the map today became independent only in 1991 with the boundaries they had as Soviet republics. This situation is also true for the Russian Federation, though the Russians controlled the Soviet Union and its predecessor, the Russian Empire. Prior to 1991, Ukraine was only independent for a brief period after World War I until it became part of the Soviet Union in 1922. Before that, it was part of the Russian Empire. Belarus was never independent before 1991 and was usually part of either the Russian Empire or the Polish-Lithuanian Kingdom. Of the former Soviet republics, Belarus is the most closely tied to the Russian Federation. Belarussians have considered creating a Russian-Belarussian Federation since 1991.

Moldova, too, was never independent and only has been a distinct territory for fewer than 200 years. It was part of the Romanian province of Moldavia until the Russian Empire annexed the part east of the Dnieper River in the 1800s and named it Bessarabia. Romania annexed the territory after World War I, but the Soviet Union annexed it again after World War II. Many Romanians and Moldovans hoped to unite their two countries after the dissolution of the Soviet Union in 1991, but the Russian military stationed in the country prevented this. Worried about a union of Moldova and Romania, Russians and Ukrainians living in Moldova declared their own republic in the Transnistria region. Other ethnic minorities have made similar proclamations. The government has not been able to suppress these independence movements completely.

The Russian Federation

The Russian Federation is the modern political state representing the land known as Russia. To many Westerners, Russia is a mysterious land, hidden in cold, dark forests on the eastern and northern fringes of Europe. Europeans have regularly included the Russian heartland within Europe but at the same time have considered Russians too "Asiatic" to be European. For centuries, Europeans struggled to understand Russia, exemplified by Winston Churchill's remark that "Russia is a riddle wrapped in a mystery inside an enigma." Depending on their relationship with Europe and their desire to be within Europe, Russians themselves frequently alternate between emphasizing their European qualities and emphasizing their wider role spanning eastern Europe and northern Asia. At times, many Russians show a fear and hostility toward foreigners, known as **xenophobia**. After the fall of communism in 1991, Russia's internal political geography and its economic and social relationships all dramatically changed.

Political Divisions

The internal political geography of the Russian Federation includes a mixture of political units. To a large extent, the country's political geography was inherited from the Soviet system, though some changes have been made since 1991. The units fall into two categories: administrative and autonomous. Administrative units consist of 6 federal territories (*krays*), 49 regions (*oblasts*), and 2 federal cities (Moscow and St. Petersburg). Much like the states, counties, and municipalities of the United States, they were created to administer the large country. In contrast, the autonomous units, consisting of 21 republics (Figure 3.20), 1 autonomous

region (*oblast*), and 10 autonomous districts (*okrugs*), are able to craft many of their own laws and govern themselves somewhat differently than the rest of the Russian Federation.

Together, the 89 political units represent different levels of size, resources, and political power. Unlike the United States, where the federal government has the same relationships with lower levels of government (e.g., states, counties, etc.), Moscow has an asymmetrical relationship with its political units, particularly the autonomous territories, in which each political unit negotiates its own relationship with Moscow. The resource-rich republics tend to exercise the greatest authority over their own governance.

The autonomous territories were established by the Soviet Union to reflect the presence of ethnic minorities, such as the Tatars, Sakha (Yakuts), and Buryats (Figure 3.21). Soviet law protected minority languages, religions, and cultures. However, only 52 percent of the Russian Federation's approximately 30 million non-Russians live in the autonomous territories today. As a way of controlling their vast country, the Soviets drew boundaries for the republics that deliberately left many members of ethnic groups outside their intended territories, and included

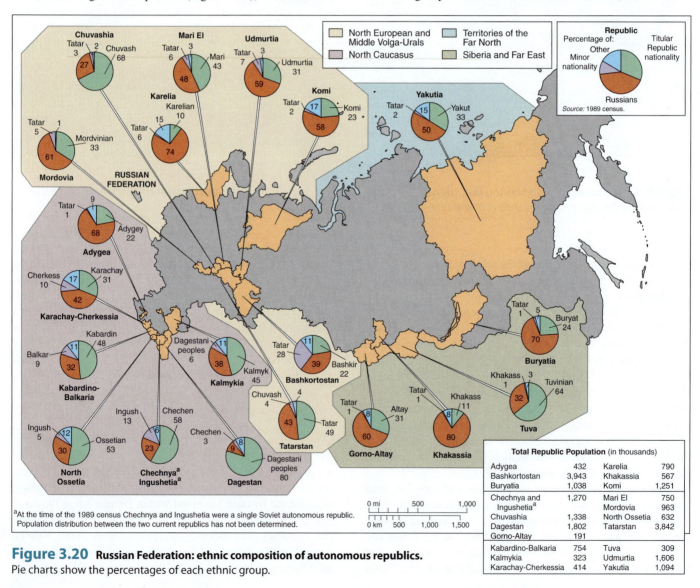

Figure 3.20 **Russian Federation: ethnic composition of autonomous republics.**
Pie charts show the percentages of each ethnic group.

Total Republic Population (in thousands)			
Adygea	432	Karelia	790
Bashkortostan	3,943	Khakassia	567
Buryatia	1,038	Komi	1,251
Chechnya and Ingushetia[a]	1,270	Mari El	750
		Mordovia	963
Chuvashia	1,338	North Ossetia	632
Dagestan	1,802	Tatarstan	3,842
Gorno-Altay	191		
Kabardino-Balkaria	754	Tuva	309
Kalmykia	323	Udmurtia	1,606
Karachay-Cherkessia	414	Yakutia	1,094

[a]At the time of the 1989 census Chechnya and Ingushetia were a single Soviet autonomous republic. Population distribution between the two current republics has not been determined.

Figure 3.21 **Russia.** Buryat singers at Datsan Temple, southeast of Irkutsk. *Photo: © Ronald Wixman.*

many Russians within them. Also, many recognized nationalities did not receive republic status. Of the 90 numerically significant groups recognized in the 1989 census, only 35 had a homeland. In addition to the earlier mentioned policies of Russification, boundary drawing was clearly a means that the Soviets used to divide and dilute non-Russian groups. The situation is particularly true in the North Caucasus, where the native peoples have most fiercely resisted Russian rule through history. For example, the Karbardians were grouped together with the Balkars, though they had more in common with their neighbors the Cherkessians.

Heartland and Hinterland in Russia

Another basis of regional differences in Russia is **heartland** and **hinterland**. The heartland lies west of the Urals and includes many of the original eastern Slav territories. It contains the greatest concentration of Russian people and accounts for much of the country's economic and political activity. It is also known as the Russian homeland. The Moscow and St. Petersburg urban regions, the Volga River valley, and the Urals contribute to the heartland's prominence. The Moscow region, approximately 400 km (250 mi.) square, is home to 50 million people and was the focus of Soviet central planning and transportation routes linking the entire country. Local manufacturing includes vehicle, textile, and metallurgical industries. St. Petersburg, a major Baltic port north of Moscow, is a smaller manufacturing center but still produces around 10 percent of total Russian output, including shipbuilding, metal goods, and textiles.

Southeast of Moscow, the Volga River is lined by a series of industrial cities that use the river transport, connected since the 1950s by a canal outlet to the Black Sea. Around 25 million people live in the Volga River region, which was developed for manufacturing during and after World War II—at a distance from advancing German armies and helped by the discovery of major local oil and natural gas fields. Manufactures include specialized engineering and the Togliatti car plant built by Fiat of Italy.

East of the Volga River basin, the Urals contain metal ores and are only a minor barrier to communications. Like the Volga region, the southern Urals were developed during and after World War II, principally as a metals center. Oil and natural gas fields are located to the south and east.

The Russian Federation's hinterlands include Kaliningrad *oblast* along the Baltic Sea, the Arctic regions around Murmansk, the mining areas east of the Urals, resource-rich Siberia, and the Pacific region inland from Vladivostok. Large expanses of the hinterland are virtually empty of people and economic activity and remain inaccessible to development. Siberia forms a huge area that can be divided into the more developed southern margins along the line of the Trans-Siberian Railway and the northern lands.

Siberia is an essential part of Russia, making up three-fourths of its land and providing a large proportion of its raw materials—a common feature of hinterland regions. In 1990, Siberia produced 73 percent of Russia's oil, 90 percent of its natural gas, 61 percent of its coal, all of its diamonds, and 30 percent of its timber and electricity. It also has a growing fishing industry on the Pacific coast.

Along the southern strip of Siberian Russia are a number of centers that were industrialized according to the Soviet planning processes and are separated by large tracts of sparsely settled land. In the Kuzbas region around Kuznetz, 2,000 km (1,200 mi.) east of the Urals, a major coalfield was developed initially to supply coal to the steel manufacturing centers of the Urals. This led to local industrialization, supported by iron ore and the return transport of bauxite from the Urals. Aluminum, steel, and products using them are major outputs. The world's largest aluminum plants at Krasnoyarsk, Bratsk, and Sayansk provided metal for the Soviet aircraft and missile industries.

Farther east, centers close to Lake Baikal, including Irkutsk, form a narrow belt of industrialization using hydroelectricity generated in the headwaters of streams flowing to the Yenisey and Lena river systems. Mining and lumbering are also important in isolated localities, while larger cities provide services to extensive areas in northern Siberia where population and mining activities are scattered.

The far east, flanking the southern Pacific coast of Russia, has ports such as Vladivostok and Nakhodk, and metal industries along the Amur River. In the 2000s, Japanese corporations began investing in the region's mines. The presence of a large population in neighboring China, coupled with a new Russian law in 2003 that allows foreigners to lease land in Russia for 49 years, has attracted many Chinese farmers.

Soviet centralized planning often chose locations for production facilities for political, rather than economic, reasons.

Special defense-related factories, for example, were built in remote locations where it was expensive to maintain them. Workers were paid more to work in these faraway facilities, and the cities that grew up around them received many subsidies. After 1991, the Russian Federation stopped giving financial support to these poorly located factories and cities. Without subsidies, many factories were not able to compete economically and had to shut down. In turn, unemployed people emigrated to the Russian heartland. Subsequently, regional differences and local characteristic lifestyles are now more pronounced within the Russian Federation.

The realities of such factors as proximity to consumer markets or ports with world trading connections are causing geographic shifts in regional production patterns. Moscow, for example, has had the most success in the transition to capitalism. As much as 20 to 25 percent of Moscow's population is middle class, the highest percentage in the Russian Federation. In other areas, less than 10 percent of the population has attained middle-class income.

Subregion: The Southern Caucasus

Georgia, Armenia, and Azerbaijan straddle the Caucasus Mountains and are frequently called the Transcaucasus, meaning "across the Caucasus" (Figure 3.22 and see Table 3.1). This term reflects a Russian ethnocentric view of the region as these countries are on the other side of the Caucasus Mountains from Russia and were once Russian colonies. The more neutral term *Southern Caucasus* is used to refer to these countries, and *Northern Caucasus* is applied to the part of the Caucasus in the Russian Federation.

Armenia and Georgia are both very mountainous with many peaks rising above 5,000 m (15,000 ft.). Azerbaijan is mountainous in the west along the borders of Georgia and Armenia, but the eastern areas of the country, formed by the Caspian Sea coast, are flat with areas below sea level. The origins of the word *Azerbaijan* are not clear, but one version derives from Persian words that mean "land of fire." Azerbaijan certainly contains great quantities of petroleum. "Land of fire" could refer to the surface deposits that burned naturally in the past or to the oil fires in Zoroastrian temples that once dominated the region. Zoroastrianism no longer exists, but it is a religious forerunner to Christianity and Islam, tracing its origins back to Azerbaijan. Zoroastrians believed that the Earth would be consumed in fire following judgment day. Interestingly, this belief developed in an area of the world where the land is oil-soaked and easily burns.

As former Soviet republics and current members of the CIS, Armenia, Georgia, and Azerbaijan struggle with establishing a viable political and economic existence. Ethnic conflicts ensued after 1991 and damaged these countries' economies. Since the late 1990s, their economies steadily have improved. All three countries have climates warmer than those of the other former Soviet republics and can produce agricultural products such as fruits, especially grapes, tobacco, cotton, and rice. They have also worked to reduce their dependency on Russia

Figure 3.22 **The Southern Caucasus: countries, cities, and major physical features.** Darkened areas within countries represent areas that have acted autonomously, often without the consent of their respective central governments.

by greatly altering and increasing their industrial and service sectors. Georgia, located on the sunny, warm, eastern shores of the Black Sea, has great tourist potential. In fact, a US$80 million Sheraton Hotel opened in Batumi in 2010, and a Radisson opened in 2012. Azerbaijan, located on the Caspian Sea, will become one of the world's leading oil producers if it is ever able to fully exploit its oilfields. Until recently, because most pipelines ran through the Russian Federation, the Russian government was able to control Azerbaijan's oil trading and limit Azerbaijan's desire to establish relationships with other countries, most notably with Iran, Turkey, the United States, and those in Europe. The recently opened pipeline to the Turkish port of Ceyhan is helping to break the dependence on the Russian Federation (Figure 3.23).

Armenia-Azerbaijan

The persecution of Armenians has had a lasting effect on this part of the world. In 1895, the Ottoman government massacred 300,000 Armenians within its realm. Again in 1915, during World War I, the Ottoman government tortured, exterminated, and deported its Armenian population, claiming that the Armenians were a threat. Somewhere between 600,000 and 2 million Armenians were exterminated out of a prewar population of about 3 million in what can be referred to as the "Armenian genocide." Many Armenians became refugees, migrating across their traditional homeland or leaving it altogether. By 1917, fewer than 200,000 Armenians remained in Turkey.

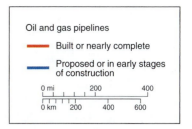

Figure 3.23 Russia and neighboring countries: pipeline outlets. Russia would like oil and gas exports from Central Asia and the South Caucasus to be piped to its Black Sea ports. The other countries and their international oil company partners are examining and building routes through Turkey to Ceyhan or through northern Iran. How does this situation illustrate the interdependence of countries in the world economic system and the importance of individual countries and their actions? *Source: © Ronald Wixman.*

It was not just this one period that inflicted a toll on Armenians. Over 1,000 years, foreign invaders wreaked havoc numerous times and scattered the population. Today, over half of the Armenian population lives outside of Armenia in a diaspora. About half of the diaspora community (e.g., those outside Armenia) lives in other CIS countries. The other half lives in communities from India across to Southwestern Asia, Europe, and North America, with a sizable number in the United States.

One of the largest conflicts among the former Soviet republics involved Armenia and Azerbaijan. In 1924, the Soviet government created an autonomous territory within Azerbaijan known as Nagorno-Karabakh. It was 94.4 percent Armenian. By 1979, Armenians represented only 76 percent of the region's population. Armenians began to fear their loss of numbers and objected to Azerbaijani laws that restricted the development of the Armenian language and culture. Clashes between the Armenians of Nagorno-Karabakh and Azerbaijanis began in the 1960s and developed into war by 1992. Armenian forces of Nagorno-Karabakh seized most of the territory and advanced westward to link their territory with Armenia. Afterward, they moved into Azerbaijan proper, but the Turkish and Iranian governments warned the Armenians to cease hostilities. Peace talks sponsored by the UN, Russia, Iran, and other countries ended the shooting war in 1994. In addition to Nagorno-Karabakh, Armenian forces continue to control approximately 20 percent of Azerbaijan. An official agreement on the governance and the political status of territories has not emerged.

Subregion: Central Asia

Central Asia's five countries were once part of a land called Turkestan. Turkestan was not a single, consolidated country but a loose confederation of tribes. Beginning in the 1800s, the Russian Empire expanded forcibly into Turkestan and took firm control of most of it by the end of the century. In the 1920s, Soviet authorities drew boundaries for new republics which emerged as the countries of Central Asia after the Soviet Union broke up (Figure 3.24 and see Table 3.1). With more than 40 percent of the population of the five Central Asian countries and as the only one that borders all the others, Uzbekistan is the dominant country.

The Central Asian countries share similar landlocked situations, arid or semiarid climates, and Muslim faith of many of their peoples. The subregion's fertile river valleys of the Amu Darya and Syr Darya are among the cradles of human civilization. They played important roles in the trading of goods and ideas from the time of the earliest civilizations in Mesopotamia, China, and the Indus River valley. Straddling old trade routes between east and west, most notably the Great Silk Road, great cities such as Bukhoro (Bukhara) and Samarqand (Samarkand) emerged. During the 700s and 800s, Bukhoro became one of the leading centers of learning, culture, and art in the Muslim world. Its grandness rivaled the other Muslim cultural centers of Córdoba (Spain), Baghdad (Iraq), and Cairo (Egypt). Some of Islam's greatest historians, geographers, astronomers, and other scientists came from the area. In the late 1300s, Timur (Tamerlane) emerged as the dominant leader in Central Asia

Figure 3.24 **Central Asia: countries, cities, and major physical features.** These countries have a mountainous southern border. Under the Russian Empire, they were carved out of the former Turkestan area. Only the cities of Almaty and Tashkent have populations of over 1 million people.

Figure 3.25). China's largest oil and gas provider also joined with a Kazakhstani oil and gas company to purchase MangistauMunaiGas, a Kazakhstani oil company.

All five countries produce similar commodities: oil, natural gas, and cotton. Thus, rivalry rather than cooperation is most common, especially in attempts to attract foreign trade. Kyrgyzstan and Tajikistan, however, do not have large fuel reserves like the other three. Consequently, the other three countries have shut off fuel supplies as a political lever against their neighbors.

The issue of water resources also encourages rivalry. Water is scarce in this dry area of the world but important for agriculture and power generation. Disputes occur over the allocation of water that flows in rivers passing through a number of the countries. Major rivers originate in the mountains of Kyrgyzstan and Tajikistan and flow down to the lowlands of Turkmenistan, Uzbekistan, and Kazakhstan. Each country frequently complains that the others withdraw an unfair share of the water from the rivers. Serious tensions have arisen between Kyrgyzstan and Uzbekistan over water in the Fergana Valley, where agricultural reform and land privatization programs are endangered by the water disputes.

and conquered lands far to the west, south, and east. A new flowering of culture began as numerous scholars and artisans came to reside in his imperial capital, Samarqand. Timur's grandson was one of the world's first great astronomers. Literature flourished and great religious structures and palaces were built (see Figure 3.14).

Following independence in 1991, GDPs dropped considerably before slowly improving. Soviet infrastructure has kept Central Asia dependent on the Russian Federation. Pipelines and transportation lines are of Soviet specifications and run primarily to the Russian Federation. The Russian Federation buys cheap raw materials from Central Asia and sends back more expensive finished products.

Central Asians are leery of Russia taking advantage of them but likewise worry about China to the east. Nevertheless, most of these countries joined new organizations with China, namely the Shanghai Cooperation Organization (SCO) and the Central Asia Regional Economic Cooperation (CAREC) Program. The CAREC Program began in 1997 when four of this subregion's countries (except Turkmenistan) joined with Afghanistan, Azerbaijan, China, and Mongolia. Focusing on the financing of transport, energy, trade policy, and trade facilitation, CAREC has increased trade between China and Central Asia. Trade increased from US$527 million in 1992 to US$26 billion in 2009. China also contributed US$10 billion in 2010 to help Central Asian economies cope with the world recession. As a result of this new cooperation, new oil and natural gas pipelines have been built between Turkmenistan, Kazakhstan, and China (see

3.5 Contemporary Geographic Issues

Is Russia Still a World Power?

As the Communist regime lost its grip on East Central Europe after the fall of the Berlin Wall in 1989 and then fell with the dissolution of the Soviet Union in 1991, Russians were confronted with profound change. On the positive side was the promise of political and economic reform, with greater freedom of expression and a better standard of living. On the negative side was the loss of much of their country, the open expression of great anti-Russian feelings from people within their world region, and the questioning and potential end of Russia as a world power. These experiences were a tremendous blow to the prestige of the Russian people.

When the 14 non-Russian republics declared their independence from the Soviet Union in 1991, it seemed that the peoples of these republics simply were expressing their right of self-determination. Many Russians, however, believed that the

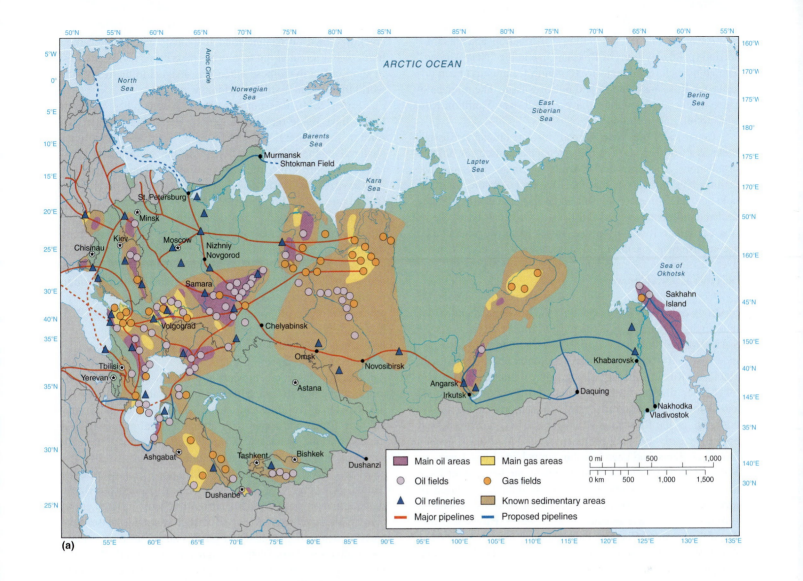

(a)

ARCTIC OCEAN

North Sea
Norwegian Sea
Barents Sea
Kara Sea
Laptev Sea
East Siberian Sea
Bering Sea
Sea of Okhotsk

Murmansk
Shtokman Field
St. Petersburg
Minsk
Kiev
Moscow
Chisinau
Nizhniy Novgorod
Samara
Volgograd
Chelyabinsk
Tbilisi
Yerevan
Omsk
Astana
Novosibirsk
Angarsk
Irkutsk
Daquing
Khabarovsk
Sakhahn Island
Nakhodka
Vladivostok
Ashgabat
Tashkent
Bishkek
Dushanzi
Dushanbe

▥ Main oil areas	▨ Main gas areas		
● Oil fields	● Gas fields		
▲ Oil refineries	▨ Known sedimentary areas		
— Major pipelines	— Proposed pipelines		

0 mi   500   1,000
0 km   500   1,000   1,500

(b)

Figure 3.25 Russia and neighboring countries: oil and gas.

(a) Assess the availability of oil and natural gas to Russia, the other neighboring countries, Europe, China, Japan, and the potential for developments in Siberia. (b) Moscow, Russia. New office building for LUKoil, financed by profits generated from high oil prices brought about by recent global demand. *Photo: © Mikko Stig/Rex USA, Courtesy Everett Collection.*

Soviet republics were Russian lands regardless of who lived in them. They saw the Soviet Union as rightfully theirs because it was created from the Russian Empire, lands that they struggled for and acquired over the course of centuries. Though these Soviet republics declared independence, many Russians do not see these 14 new countries as foreign. Instead, they see them as part of their "Near Abroad" and feel that they have an exclusive voice in both the internal and international relations of these countries.

As noted in the "Population: Distribution and Patterns" section (p. 86), millions of Russians live in the "Near Abroad." The adjustment from majority to minority status has been difficult for Russians, both in and out of Russia. The policies of these independent republics have caused over 5 million Russians to migrate to Russia.

Countries of the "Near Abroad" also are crucial to Russia's status as a world power. The Baltic republics housed key military installations. The Crimean Peninsula is a major Russian naval facility. It is no wonder that the Russian government tried to hold onto the Crimean Peninsula and its naval fleet after Ukraine declared independence. Kazakhstan was the center of the Soviet space program. Russians see the Soviet space program as their accomplishment, one that only Americans have matched. Control over the oil in the Caucasus and Central Asia is also key to Russia's role as a world power.

Coping with the loss of the "Near Abroad" is compounded by struggles by non-Russians within the Russian Federation for independence. Chechnya is the most vexing of them all. As the Soviet Union disintegrated in 1991, the Chechens moved toward independence from the Russian Federation and declared it in 1994. The Russian military responded by supporting a Chechen rebel group that sought to overthrow the Chechen government and keep Chechnya within Russia. The military campaign was not quick as the Russian government had hoped, but long and brutal. By early 1996, an estimated 40,000 to 100,000 people, mostly civilians, had been killed in a republic of just 1 million. Street-to-street fighting and bombings by the Russian air force made many of Chechnya's cities uninhabitable. Accused of gross human rights violations, Russia's international reputation suffered from the war in Chechnya. However, Russians feared that if they granted Chechen independence, then Russia's other minorities would also move toward independence. In addition, Chechnya, especially Grozny, is a major transit point for oil leaving the Caspian Sea region (see Figure 3.23).

Russia's ability to influence world affairs diminished in other ways in the 1990s (Figure 3.26). Soldiers in Russia's military are vastly underpaid, unprepared, and demoralized from their experience in Chechnya, which followed the earlier defeat in Afghanistan in the 1980s. Though Russia is a nuclear power, the United States, viewing itself as the sole victor of the Cold War, acted unilaterally in international affairs. At the same time, the North Atlantic Treaty Organization (NATO) expanded into East Central Europe to include countries formerly dominated by the Soviet Union. For both members and potential members of NATO, the expansion was a defensive move. For Russians, NATO's moves were provocative, particularly since they were made without Russian consent. NATO's expansion played on Russian fears that NATO was nothing more than another force, not unlike Napoléon or the German armies of the two World Wars, to menace Russia.

Closer relations with China led to the signing of a friendship treaty in 2001. The treaty is significant because it sets aside decades of tension caused mostly by border disputes and allows for cross-border trade. For example, Russia now sends oil and military supplies to China. Both China and Russia also are now cracking down on Islamic fundamentalists who straddle their borders and threaten the territorial integrity of their countries. In short, a Russo-Chinese alliance is a signal to the rest of the world that Russia is still a world power.

Since 2000, Russian's leaders, most notably Vladimir Putin, have made their country more authoritarian. Though foreign investment was welcomed after 1991, policies changed in the mid-2000s. The Russian government reasserted its control over the very profitable oil and natural gas industry, which it now uses as a major foreign policy tool (see Figure 3.25). For example, in 2005, Gazprom (Russia's state-controlled gas company) acquired the private oil firm Sibneft. Also, the state-controlled oil firm Rosneft obtained the main production operation of the privately owned oil company Yukos after the Russian government put Yukos's leader in jail and ordered Yukos's dismantling. Through political pressure and maneuvering, the Russian government also forced oil companies like British Petroleum (BP), ExxonMobil, and Royal Dutch Shell to forfeit much of their investment in Russian oil and natural gas companies, often at great financial loss.

Huge profits in recent years allowed Russian oil and natural gas companies to invest in their industry abroad. In the east, the Russian government has been negotiating deals to supply both China and Japan with oil and natural gas. If current trends continue, Russia will supply Europe with 94 percent of the oil and 81 percent of the natural gas that world region consumes by 2030. Poland, Slovakia, and Hungary already receive 100 percent of their oil from Russia, and the Baltic states, Slovakia, and Romania currently obtain 100 percent of their natural gas from Russia. Russian companies also have purchased significant shares of East Central European oil and gas companies. Also, Gazprom signed a deal with the German firm BASF in 2006. Still another plan to drill in the Shtokman gas field in the Barents Sea could lead to the sale of liquefied natural gas to the United States and open the door for Russian companies to gain shares in American gas distribution companies. However, the recent and tremendous growth of shale gas from "fracking" in the United States has led to the postponement of these plans.

This situation has allowed Russia to use oil and natural gas as a foreign policy tool to exert itself as a world power. For example, in 2004, gas supplies were temporarily shut off to Belarus during a dispute between Belarus's leader and the Russian government. In January 2006, during a very cold winter, gas supplies were also temporarily shut off to Ukraine following the Ukrainian government's resistance to paying higher prices for Russian natural gas. It also is believed that the Russian government was retaliating against Ukrainians for defeating a pro-Russian government in Ukraine and putting a pro-Western one in its place ("the Orange Revolution"). The shutting off of oil also has affected and angered Western Europeans. To prevent this in the future, the Russian government worked out a plan with the German government to build a pipeline known as "Nord Stream" under the Baltic Sea to Germany. Completely finished in late 2012, Russia now is able to stop the oil flow to its neighbors and East Central Europe without disrupting the flow to Western Europe.

Beyond the issue of oil and natural gas, Russia joined with Armenia, Belarus, Kazakhstan, Kyrgyzstan, and Tajikistan to form the Collective Security Treaty Organization (CSTO) in 2002. Uzbekistan joined in 2006. CSTO is very much like NATO in that an attack on one member is considered to be an attack on all. In 2003, the Russians opened an airbase in Kyrgyzstan, marking the return of the Russian military after all such forces had withdrawn from the country in 1998. In early 2009, Kyrgyzstan's government announced that it wanted the United States to shut down its military base at Manas, for which Kyrgyzstan received US$150 million per year. This decision came right after Kyrgyzstan received the promise of US$2.3 billion in investment and loans from Russia. However, Kyrgyzstan reversed its position six months later, after the United States tripled the rent it is paying for the base.

Russia's boldest move came in August 2009 when it intervened in a series of military clashes between Georgia's military and South Ossetian separatists. Siding with the South Ossetians, the Russian military entered not only South Ossetia, but also the breakaway region of Abkhazia and took control of key routes in Georgia, destroying Georgian military bases, sinking Georgian ships, and blockading its main port. Shortly afterward, the Russian government recognized the independence of South Ossetia and Abkhazia. Even prior to this military conflict, the Russian government had been issuing Russian passports to the inhabitants of these and other regions in Russia's near abroad.

The Russian government also has exerted its influence beyond the former boundaries of the Soviet Union. For example, it sold US$7.5 billion in military tanks, fighter aircraft, and antiaircraft missiles to Algeria. Russian weapons have also been used by Hezbollah in their fight against Israel. The Russian government also stood behind the Syrian government in the Syrian civil war, which contributed to the war's long duration and bloodiness. Perhaps Russian ties to Iran are most significant. Beginning under the leadership of George W. Bush, the United States began using sanctions and embargoes to force the Iranian government to end its nuclear research program for fear that Iran would produce not only nuclear energy but also nuclear weapons. In contrast, Russia is aiding the Iranian nuclear program and selling Iran military hardware. If successful, Russia will outmaneuver the United States and gain great investment opportunities and influence in one of the world's biggest oil-producing countries, while the United States will have neither.

(b)

(a)

Figure 3.26 **The Russian military presence.** (a) Russian soldiers sit atop an APC as a column of Russian troops moves toward Tskhinvali, the capital of South Ossetia, Georgia. The armed conflict between Russia and Georgia expanded on Saturday, September 4, 2010, with fighting spilling outside the Caucasus province of Ossetia, and both sides moving reinforcements into the region. The fiercest battles were in the South Ossetian city of Tskhinvali, where street fighting and artillery exchanges continued sporadically throughout the night. (b) Soldiers stand at the site of a suicide car-bomb attack on a Russian military base in the city of Buinaksk in the violence-plagued republic of Dagestan, Sunday, September 5, 2010. The attack resulted in the death of a number of soldiers and the wounding of dozens. A spokesman for the republic's Interior Ministry said the driver of the explosives-laden vehicle smashed through a gate of the base and headed for an area where soldiers were quartered in tents, but soldiers opened fire on him before he reached the center of the base. The spokesman said the driver rammed the car into a military truck where it exploded. *Photos: (a) © RIA Novosti/Topham/The Image Works; (b) © AP Photo/Kurban Labazanov.*

Since 2000, Russia has become a more full-fledged member of the informal organization representing the world's wealthiest countries, once known as the Group of Seven (G7) but, with Russia, known as the G8. In 2002 NATO gave Russia a voice in NATO affairs, though Russia is not a full member. With an improving economy, more effective political leadership, and a changing climate of international affairs, the Russian Federation has the opportunity to exert itself as a world power. However, Russia's political and military power has become highly dependent on commodities such as oil and natural gas, meaning that its power ebbs and flows with the price and availability of these commodities (Table 3.3).

Human Rights

In Imperial Russia, the czars sent their political opponents to Siberia, far from the political center of the empire. After the Bolsheviks came to power in 1917, they continued the practice of banishing their opponents. Joseph Stalin raised the practice to a new level when he created the "Main Directorate for Corrective Labor Camps," known in Russian as *Glavnoe uprav-lenie ispravitel no-trudovykh lagerei* or *gulag* for short. After Stalin's death in 1953, the number of people sent to the *gulag* dramatically decreased but did not stop. It is difficult to know exactly how many perished in the *gulag*, although estimates are in the tens of millions.

Since the end of communism, human rights abuses remain a concern. Human rights still are abused by police and in prisons. Court systems have yet to be reformed. Defendants sit in jail before their cases are heard, and low pay makes many judges prone to accepting bribes. Juries are rare. Judges frequently serve as prosecuting attorney as well as judge. Not surprisingly, the conviction rate has been 99.6 percent.

In some ways, human rights have worsened as some leaders rule like Stalin once did. For example, since Vladimir Putin came to power in 2000, wealthy businessmen, especially those who seek political power, have been accused of fraud, stripped of their wealth, and jailed if they do not first flee Russia. Journalists who investigate corruption or human rights abuses are found dead on city streets with bullets in their heads.

Yet human rights abuses are worse in Central Asia, such as under Saparmurad Niyazov, who ruled Turkmenistan from its independence in 1991 to his death in 2006. In 1999, he had himself elected president for life and built a huge statue of himself cast from pure gold. The statue revolves so that the sun is always behind it. Niyazov renamed the days of the week and the months of year, naming January after himself and April after his mother. Despite Niyazov's death, Turkmenistan still may have the most repressive regime, though freedom of the press is almost nonexistent across Central Asia. Political opposition groups are weak, and any of their leaders who do not operate

Table 3.3	Debate: Is Russia still a world power?
No Longer a World Power	**Still a World Power**
Russia lost possession of 14 republics (now independent countries) and with them sizable populations and resources.	Russia's heartland remains together, and Russia is still the largest country in the world, with a sizable population and considerable resources.
Russia's international power is undermined by the presence of 25 million Russians who now find themselves as ethnic minorities in other countries. Russia must be careful not to offend these other countries and thereby endanger the Russians who live in them. Russia may have to make economic concessions to these countries to protect Russian minorities in them.	Russians living as ethnic minorities in other countries increase Russia's political influence in those countries because they can vote and hold political office. Russia can also pressure those countries on behalf of Russians living in them. The Russian minorities also strengthen Russia's trading links with these countries.
The Russian government has been ineffective in controlling independence movements in its own republics. It has been condemned for human rights violations in Chechnya. Its military is underpaid, unprepared, and demoralized. Russia was unable to control the conflict in the former Yugoslavia and the expansion of NATO.	Russian soldiers are very patriotic, as in the past, and Russia is a nuclear power. With peacekeeping troops in the former Yugoslavia, Russia influences events in that area of the world. Russia also now has a voice in NATO's affairs. Russia helped to establish the Collective Security Treaty Organization (CSTO), which it uses to reestablish a military presence in Central Asia and pressure the United States to leave.
With rampant corruption, Russia's economy is weak and has floundered since the end of communism in 1991. Russia is unable to capitalize on selling its vast resources on the world market. Its control of Caspian Sea oil is questionable. Russia has little influence in global economic organizations.	Russia's economy is steadily improving, and the government is fighting corruption. Russia has signed agreements to supply oil to other countries and is able to control the flow of oil from the Caspian Sea. Russia has joined the Group of Seven (G7), making the organization the Group of Eight (G8).
Though Russia's military rapidly took control over South Ossetia and Abkhazia as it quickly defeated Georgia, Russia and its military are enormous compared to Georgia and its military. So the Russian victory was a given. However, the invasion showed that Russian military hardware is very old and in disrepair. Some Russian tanks could not turn left or right and were abandoned by their soldiers.	Russia's military strength was shown when the Russian army quickly defeated Georgia. It does not matter if Russia's military is old and inefficient as long as it can win wars.

outside their respective countries are most likely in jail, where torture is common.

Despite human rights abuses, democratic reforms since 1991 allowed groups working for human rights to operate, especially in the Slavic countries. Some human rights groups are outgrowths of international organizations, such as Amnesty International, though Putin has since limited their activities and made some illegal. Others are unique to the concerns of the region's people. The Committee of Soldiers' Mothers of Russia, the "Mother's Right" Foundation, and the Moscow Center for Prison Reform are a few examples.

Environmental Problems

Environmental damage to the tundra, forest, and desert areas are a serious issue for Russia and its Neighboring Countries today (Figure 3.27). For centuries, inhabitants of the region thought that the vast quantities of resources in their lands were inexhaustible. Later, communism preached that nature should be transformed to serve human needs. Rapid industrialization led to the massive exploitation and extraction of minerals and the construction of enormous factories, often built quickly with little thought to environmental protection. Emphasis on chemicals, steel, and the nuclear industry coupled with faulty storage of toxic rocket fuel and oil, the testing of weapons, and the mining of metallic minerals laid waste huge areas, many of which remain unreported and unreclaimed. A few examples illustrate the range of environmental problems facing Russia and Neighboring Countries.

- One of the major legacies of the Soviet Union is frequent oil pipeline breaks and leakages. Literally, hundreds occur every year. Some of the resulting oil spills are small, but others cover hundreds of square kilometers. Inefficient construction practices are the major cause. However, one

Figure 3.27 **Russia: areas of environmental degradation.**

incident in 2004, which resulted in a 540 km² (209 mi.²) oil spill in the Samara region, was caused by an incision carelessly made to illegally siphon oil.

- The city of Norilsk in western Siberia has the most polluted environment in Russia. The region contains some of the world's richest deposits of mineral ore: 35 percent of the world's nickel, 10 percent of the world's copper, significant cobalt reserves, and 40 percent of the platinum group of metals. The Norilsk Metallurgical Combine releases millions of tons of pollutants into the atmosphere each year. The stinking acidic haze blackens snow, kills trees for miles around, and poisons the river. Many locals, most of whom work in the factories, believe that the noxious fumes inoculate them against disease, but local physicians report high incidents of respiratory illnesses and shortened life expectancy, as low as 50 years.
- In addition to industrial pollution, nuclear contamination has caused major environmental damage. The most famous is the 1986 Chernobyl nuclear reactor explosion, which became a symbol for such disasters. It affected most of the area immediately around it north of Kiev, but the fallout was worst to the north in Belarus (Figure 3.28). Nuclear dumps on islands, peninsulas, and in seas, especially along the Arctic Ocean, have also released radioactive pollution.
- One of the greatest environmental disasters in the world occurred in the lands east of the Aral Sea (Figure 3.29a).

Although Central Asia is arid, snow on the high mountains to the east melts in spring, providing water to such rivers as the Amu Darya and Syr Darya that flow into the Aral Sea. This water provided the basis of local irrigation farming and small urban settlements throughout history, but the Soviet Union adopted ambitious plans to use the water on irrigated cotton farms inside and outside the main river basin (Figure 3.29b). Supported by growth-oriented, unbending bureaucrats who often acted against the advice of government scientists, the project was taken too far, and so much water was extracted that the rivers stopped flowing into the Aral Sea. The sea is now less than half its previous size, all transportation on it has ceased, and the supply of moisture for the mountain snows has dried up. Glaciers in the mountains are in retreat, groundwater levels have fallen, accompanied by ground salinization (Figure 3.29c), and dust storms now affect areas that had seldom experienced them. In 2005, a US$86 million grant from the World Bank provided for a 13 km (8 mi.) long dam to regulate the flow of the Syr Darya. The dam is allowing the northern Aral to rise 42 m (138 ft.), bringing the sea within 13 km (8 mi.) of the town of Aralsk.

- The Caspian and Black seas are threatened by oil, toxic waste, ships' ballast water, and nutrients from fertilized fields. Only five of the 26 fish species caught in the 1960s in the Black Sea were still found in 1992.

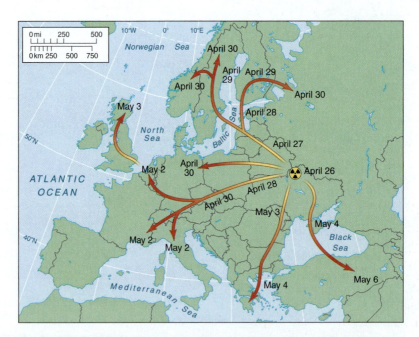

Figure 3.28 **Russia and neighboring countries: impact of a nuclear disaster.** The areas affected by the Chernobyl nuclear power plant explosion that released clouds of radioactive gases in 1986. The various winds at the time carried the radioactive gases in numerous directions over the days that followed, as seen on the map. In Ukraine, over 1.5 million people were affected, 10 percent of whom were severely radiated. Only two workers were killed in the blast, but with the explosion releasing 400 times more radiation than was released by the atomic bomb dropped on Hiroshima, Japan, local death rates are expected to increase. Like the atomic bombings of Japan, it will take decades to determine the real effects. In 2011, which marked the 25th anniversary of the explosion, the Union of Concerned Scientists estimated that excess cancer deaths will be close to 25,000. Also of concern is the deteriorating concrete shell that covers 200 tons of highly radioactive material at the reactor site. A new US$1.58 billion 18,000-ton sarcophagus is scheduled for completion by 2015.

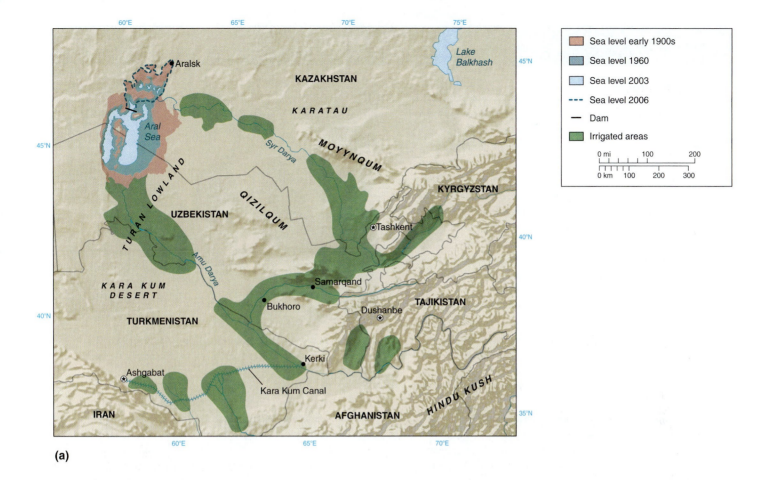

(a)

(b)

(c)

Figure 3.29 Central Asia: Aral Sea environmental disaster. (a) The Aral Sea, Kazakhstan, and its surroundings. The sea is a basin of inland drainage in an arid region. It is supplied by meltwater from the mountains to the southeast. Irrigation projects such as those supplied by the Kara Kum canal have been steadily diverting water from the Aral Sea, causing it to dry up. (b) Cotton farming. (c) Salinization: the white substance on the fields are salts brought to the surface by water poured on the land. *Photos: (b & c) © Ronald Wixman.*

Policy Analysis

During Soviet times, social sciences such as political science and international affairs generally were neglected. After Georgia regained its independence in 1991, geographers largely filled the need in these areas. This meant that human geographers primarily became the first generation of Georgia's diplomats in the 1990s. Since then, many geographers have been pursuing successful careers in the Georgian Foreign Service, serving as ambassadors in many countries (among them the United States) and as directors of various departments and units of the ministry. Among them is David Abesadze, who is head of the Policy Analysis Division of the Georgian Ministry of Foreign Affairs. At the same time, he is assistant professor of human geography on the faculty of Social and Political Studies at Tbilisi State University.

Dr. Abesadze (Figure 3.30) graduated from Tbilisi State University and holds B.A. and M.A. degrees in human geography and a Ph.D. in political science. He also studied political science at the Central European University (Hungary), where he earned an M.A. degree. He has been a fellow of nondegree programs at Stanford and Georgetown universities, and a visiting scholar at the George Washington University and Johns Hopkins University.

Dr. Abesadze has many responsibilities at the Ministry of Foreign Affairs of Georgia. For example, he regularly monitors and analyzes regional and functional issues, predicts future situations, and recommends alternative courses of action to the ministry's leadership. He also works on certain analytical issues as tasked by the ministry's leadership and cooperates with the academic community, nongovernmental organizations, and think tanks to exchange expert views on matters relevant to Georgia's foreign policy.

Figure 3.30 Dr. David Abesadze, Head of Policy Analysis Division of the Georgian Ministry of Foreign Affairs. *Photo: Courtesy David Abesadze.*

Generally speaking, Dr. Abesadze's daily work at the ministry requires in-depth analysis and prognosis of developments in world politics. He finds that being a geographer is highly advantageous to the performance of his duties. The extremely complex and dynamic nature of international politics makes it difficult for an observer to properly understand processes. Geography's ideas and principles remain highly applicable to a rapidly changing world. Therefore, Dr. Abesadze finds knowledge of geography and the skills of spatial analysis useful tools in pursuing his analytic work.

Beijing, China. *Soldiers march in formation outside the Meridian Gate, the southern entrance to the Forbidden City in Beijing. Built in 1420, the gate is now the only entrance to the palace complex for tourists. Photo: Elizabeth Chacko.*

East Asia

LEARNING OBJECTIVES

After reading the chapter, you should be able to:

■ Recognize the global impact of the huge population and growing economic output of the East Asian world region.

■ Use a map to locate and describe the diverse climates, landforms, vegetation, and natural resources of this region.

■ Discuss the human impacts on the natural environment, and the natural hazards that affect the people of East Asia.

■ Compare the population characteristics and dynamics of the major countries of the region. Link this to the high level of urbanization in the region, and identify global city–regions.

■ Locate the main subregions and countries of East Asia on a map, and describe how they can be distinguished on the basis of their physical, cultural, economic, and political geographies.

■ Describe Japan's postwar development and its transition from a manufacturing economy to a more diversified one.

■ Contrast the economic and political systems of North and South Korea.

■ Trace the political and economic changes in China since World War II, and describe China's move toward privatization and its emphasis on exports since 1976. Explain the changes in China's role in today's global economy.

■ Describe Taiwan's economy and the country's relationship to China.

Figure 4.1 **East Asia: mountains, lowland plains, major rivers, countries, and capital cities.** The region includes China and Japan, contrasting North and South Korea, Mongolia, and the island of Taiwan. The land boundaries are mainly mountains, but with significant breaks in the northwest, northeast, and southwest. Island chains separate the China and Japan (or East) seas from the Pacific Ocean.

4.1 East Asia's Global Influences

East Asia (Figure 4.1) includes the countries of China, Japan, North and South Korea, Taiwan, and Mongolia: some of the world's materially wealthiest and some of its poorest. The region's present character builds on millennia of cultural and technological development and interactions of regional and global powers. From the early 1800s, European colonial powers used force to draw East Asian countries in, but all countries in this region resisted European colonization.

After 1945, U.S. postwar reconstruction in Japan, South Korea, and Taiwan included the installation of democratic governments and connections to the U.S. economy. After its establishment in 1949, the People's Republic of China was governed by the Communist Party with central direction. Led by Japan, China, South Korea, and Taiwan, the region reemerged as a world political, economic, and cultural force in the 1990s. These East Asian countries not only interact with external globalization trends, but also contribute to those trends (Figure 4.2). By contrast, Mongolia is a much smaller country between Russia and China that follows policies it believes will placate its neighbors, although the end of the Soviet regime in 1991 led to greater political freedom. North Korea remains in the thrall of an inward-looking ruler who continues to lead the country into poverty as an international outcast.

Today, East Asia, led by Japan, China, South Korea, and Taiwan, refocuses global economic growth and political power on the countries situated around the Pacific Rim. The explosive growth of China's economy over the last three decades has catapulted it to the world's second-largest economy after the United States. Its global ties, ranging from investments in Africa to the sales of Chinese products in the United States, are far-reaching. Chinese, Koreans, and Japanese share many cultural characteristics, including Buddhism, the inclusion of Chinese characters in their scripts, close family life and kinship links, and a focus on communal organization (Figure 4.3). These common elements contributed much to modern attitudes toward the global economy. Today's business phenomenon of the "Asian Way" built on this common culture and is marked by a hierarchy of relationships that define how people work and live with each other. While the similarities are important, ideological, economic, and political differences have led to recent tensions between Japan and China, between the Koreas, and between Japan and North Korea.

4.2 Distinctive Physical Geography

The natural environments of East Asia have affected the cultural history and have added distinctive local features to the region. Large numbers of people crowd into relatively small areas of well-watered lowlands. They are much affected by natural events, especially in rural areas, and they themselves have major impacts on the physical environment by modifying natural features and processes.

Mountains and Major Rivers

Mountain systems and relatively small areas of lowland (see Figure 4.1) form the diverse relief of East Asia. Although the high proportion of rugged terrain suggests a difficult environment for human occupation, the region supports some of the highest densities of population in the world. High mountains extend eastward from the Himalayan Mountain ranges and the Tibetan Plateau, occupying over one-third of China. On the border with Nepal and India, the Tibetan Himalayas are the world's highest mountains.

The collision and subduction of tectonic plates have produced some of the most striking features of East Asia. The Tibetan Plateau, the highest and largest plateau on earth, was produced due to the collision of the Indian and Eurasian

Figure 4.2 **Pudong: China's new commercial hub.** The Pudong area across the Huang Pu from central Shanghai. Some believe the TV tower completed in 1998 is too *yang* (male), and that later buildings around it are more rounded and modest *yin* (female) to give it balance. *Photo: © Michael Bradshaw.*

Figure 4.3 East Asia: cultural features. A temple beside Dongting Lake, lower Chang Jiang, China. People come to pay homage to the Buddha and/or local deities, and it is a tourist stop. *Photo: © Michael Bradshaw.*

plates. Located on the Pacific Ring of Fire, the volcanic and earthquake-prone islands of Japan experience major shocks and eruptions as tectonic plates clash. Natural diasasters such as the May 2008, 7.9 magnitude earthquake that affected China's western Sichuan Province and the March 11, 2011, underwater 9.0 earthquake that caused a killer tsunami in Japan are consequences of this seismic activity. The Sichuan province earthquake left nearly 90,000 people dead or missing and 5 million homeless. Among the buildings that collapsed were schools, resulting in the deaths of thousands of students. In Japan, the 2011 tsunami caused some 14,000 deaths and the displacement of about 136,000 people. Damage to local and regional economies in Japan was compounded by radiation leaks from the tremor- and wave-damaged Fukushima Daiichi nuclear power plant.

The high flows of some of East Asia's most active rivers are supplied by summer meltwater from the high mountain snowfields to the west in combination with intense rains toward the coasts. The three major rivers of China are, from north to south, the Huang He (Yellow River), the Chang Jiang (Yangtze or Long River), and the Xi Jiang (West River) with its wide lower section, the Zhu Jiang (Pearl River). Inland, these major rivers carve deep valleys with steep slopes and boulder-strewn streambeds. The rivers carry large quantities of water toward the sea, together with eroded mud, sand, and rock fragments. As they near the coast, they drop their load of sand, silt, and clay to form wide, fertile plains and deltas. The tidal range and wave activity are both low around these coasts, allowing the large loads of silt and clay to build deltas at river mouths.

Subtropical and Temperate Climates

East Asia has a variety of climatic environments centered on the continental temperate regime (see Figure 4.4) that resembles the eastern parts of North America, from Cuba and Florida in the south to New England and Newfoundland in the north.

Southern China has a **subtropical rainy** climatic environment. In summer, southeasterly winds are drawn into central Asia as the continent heats up and rising air leads to low pressure. These winds bring moisture from the tropical oceans and heavy rains. In winter, the winds blow outward from the high pressure over the cold continent and are cooler and drier. In mid- to late summer, the heating of the Pacific Ocean makes the region subject to typhoons (the Asian equivalent of hurricanes) with strong winds and rain. Temperatures in Guangdong Province average 13°C (55°F) in January and 28°C (81°F) in July, with annual rainfall of 1,500 mm (60 inches), most of it in the summer.

Farther north, coastal China, the Koreas, and Japan have continental temperate climatic environments with summer rains and drier winters. Although summers are as warm as those farther south, icy winds blowing from Siberia make winters much colder. Japan receives more winter precipitation than the continental countries; winds blowing from the continental interior pick up moisture on crossing the Sea of Japan and precipitate snow on the west-facing mountains.

The western parts of China and the whole country of Mongolia have arid or semiarid climatic environments because of their distance from the ocean: humid ocean air loses its moisture by raining before reaching the interior. Winters are extremely cold. In the westernmost parts of China, high altitudes in the mountains and Tibetan Plateau cause even greater summer-winter and daily ranges of temperature. Lhasa in Tibet has an average temperature of 0°C (32°F) in December and 17°C (60°F) in summer.

Forests, Grasslands, and Desert

The range of climatic and relief environments in East Asia produces a variety of natural vegetation types, although human land uses later replaced or modified these ecosystems. The subtropical broadleaf forests in southern China covered extensive areas dominated by a great variety of species, including teak forests. Farther north, temperate deciduous and evergreen forests were largely cleared from the lower, more cultivable areas but remain on steep slopes in northeast China, the Koreas, Taiwan, and Japan. Inland, the increasing aridity causes the forests to give way to grassland and desert in northwest China and Mongolia. About 20 percent of China is desert. The Gobi Desert occupies much of Mongolia and large areas of northern China. Desertification due to overgrazing, deforestation, poor farming practices, and drought is on the rise in northern and western provinces of the country, which regularly experience sandstorms due to encroaching deserts.

Natural Resources

The major natural resources of East Asia are surface water flows, fertile soils, and minerals. The mineral resources of

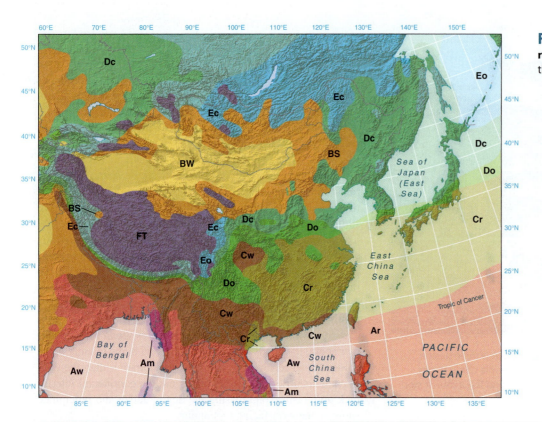

Figure 4.4 **East Asia: climate regions.** Note the contrasts within the region.

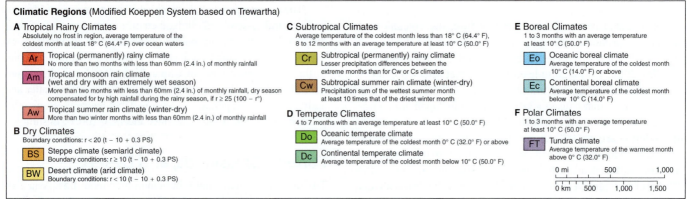

Climatic Regions (Modified Koeppen System based on Trewartha)

A Tropical Rainy Climates
Absolutely no frost in region, average temperature of the coldest month at least 18° C (64.4° F) over ocean waters

Ar Tropical (permanently) rainy climate
No more than two months with less than 60mm (2.4 in.) of monthly rainfall

Am Tropical monsoon rain climate
(wet and dry with an extremely wet season)
More than two months with less than 60mm (2.4 in.) of monthly rainfall, dry season compensated for by high rainfall during the rainy season, if r ≥ 25 (100 − r°)

Aw Tropical summer rain climate (winter-dry)
More than two winter months with less than 60mm (2.4 in.) of monthly rainfall

B Dry Climates
Boundary conditions: r < 20 (t − 10 + 0.3 PS)

BS Steppe climate (semiarid climate)
Boundary conditions: r ≥ 10 (t − 10 + 0.3 PS)

BW Desert climate (arid climate)
Boundary conditions: r < 10 (t − 10 + 0.3 PS)

C Subtropical Climates
Average temperature of the coldest month less than 18° C (64.4° F), 8 to 12 months with an average temperature at least 10° C (50.0° F)

Cr Subtropical (permanently) rainy climate
Lesser precipitation differences between the extreme months than for Cw or Cs climates

Cw Subtropical summer rain climate (winter-dry)
Precipitation sum of the wettest summer month at least 10 times that of the driest winter month

D Temperate Climates
4 to 7 months with an average temperature at least 10° C (50.0° F)

Do Oceanic temperate climate
Average temperature of the coldest month 0° C (32.0° F) or above

Dc Continental temperate climate
Average temperature of the coldest month below 10° C (50.0° F)

E Boreal Climates
1 to 3 months with an average temperature at least 10° C (50.0° F)

Eo Oceanic boreal climate
Average temperature of the coldest month 10° C (14.0° F) or above

Ec Continental boreal climate
Average temperature of the coldest month below 10° C (14.0° F)

F Polar Climates
1 to 3 months with an average temperature at least 10° C (50.0° F)

FT Tundra climate
Average temperature of the warmest month above 0° C (32.0° F)

0 mi   500   1,000
0 km   500   1,000   1,500

East Asia (Figure 4.5) include coal, oil and gas, iron, gold, rare earth minerals, and precious stones. China accounts for over 90 percent of the world's supply of rare earth minerals. Coal deposits are widespread in China, while major deposits of oil and natural gas occur in western China and in many offshore locations that are still being explored. Japan has few mineral resources and relies on imported raw materials for its industries. In 2012 China and Japan had a standoff over contested uninhabited islands in the East China Sea that are known as Senkaku in Japan and Diaoyu in China. Claimed by China, the islands were purchased by Japan in September 2012 for US$26.2 million from a private landowner. The tensions are in part due to claims on natural resources such as fishing grounds and oil and natural gas that may be in and around these islands.

Although the initial Chinese civilization near Xi'an was based on millet, the growing of wheat in the north and wet rice in the south made it possible to feed much larger numbers of people and maintain high population densities. The water and the annually renewed fertile **alluvial soils** of the lower valley floors enabled abundant crops to be harvested.

4.3 Distinctive Human Geography

In 2012, East Asia was home to 1.6 billion people, 22 percent of the world's population. Despite an emphasis on population control measures, this could rise to 1.6 billion by 2025 (Table 4.1). Since the population lives on only 11 percent of the world's land, East Asia is marked by some of the world's highest population densities. The combination of large populations, difficult environments, and involvement in the world economy leads to the growth of massive cities (Table 4.2).

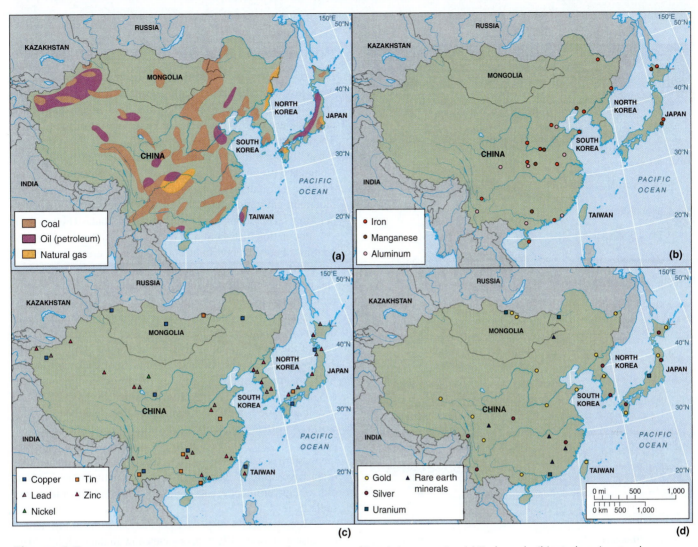

Figure 4.5 **East Asia: mineral resources.** Compare the resources of East Asian countries. (a) Fuels: coal, oil (petroleum), natural gas. (b) Metals: iron, manganese, bauxite (aluminum ore). (c) Metals: copper, lead, nickel, tin, zinc. (d) Precious metals: gold, silver, uranium, and rare earth minerals.

		Population (millions)		% Urban	GNI PPP 2010	2010	
Table 4.1		**East Asia:** Data by country, area, population, urbanization, income (gross national income purchasing power parity), ethnic groups					
Country	**Land Area (km²)**	**Mid-2012 Total**	**2025 Est. Total**	**2012**	**Total (US$ billions)**	**Per Capita (US$)**	**Ethnic Groups (%)**
Japan	378	127.6	119.8	86	4,440.2	34,610	Japanese 99%, some Koreans, indigenous
Korea, Republic of (South)	99	48.9	50.9	82	1,203.6	29,110	Korean
Korea, People's Dem Republic (North)	121	23.3	26.2	60	—	—	Korean
China, People's Republic of	9,634	1,350.4	1,402.1	51	7,150.5	7,640	Han Chinese 92%, Zhuang, Mongolian, Tibetan, Uygur
China, Macao SAR	0	0.6	0.8	100	—	—	
Hong Kong (to China, 1997)	1	7.1	8	100	304.3	47,480	Chinese, Europeans
Mongolia	1,567	2.9	3.4	63	8.3	3,670	Mongolian 90%, Kazakh, Chinese, Russian
Taiwan	36	23.3	23.5	78	—	—	Taiwanese 84%, mainland Chinese 14%
East Asia Totals/Averages	**11,836**	**1,585**	**1,635**	**56**	**13,107**	**10,430**	

Source: *World Population Data Sheet 2012*, Population Reference Bureau; Microsoft Encarta 2005.

Table 4.2	**Populations of Major Urban Centers in East Asia**

TOP 10 MOST POPULATED URBAN CENTERS

City, Country	2012 (millions)	2025 (millions)
1. Tokyo, Japan	37.1	38.3
2. Shanghai, China	16.5	19.4
3. Beijing, China	13.1	14.5
4. Seoul, South Korea	12.8	14.0
5. Osaka-Kobe, Japan	11.3	11.4
6. Guangzhou, China	10.8	13.1
7. Shenzhen, China	9.6	12.2
8. Wuhan, China	8.6	10.3
9. Hong Kong, China	7.6	8.5
10. Tianjin, China	7.9	9.2

NUMBER OF MAJOR URBAN CENTERS IN EAST ASIA, BY POPULATION CATEGORY

Population Category	Number of Urban Centers
10.00 million and greater	6
5.00 – 9.99 million	13
2.50 – 4.99 million	28
1.00 – 2.49 million	59

Population Distribution and Density

East of a line drawn across China from northeast to southwest, just west of Harbin, Beijing, Lanzhou, and Chengdu, and extending across the Koreas, Taiwan, and Japan, population densities are over 10 per km² and exceed 100 in many areas. The densest areas are along the eastern and southern coasts of Japan, the western parts of the Koreas and Taiwan, north-central China, and the coasts and major river valleys of southern China. The western half of China and all of Mongolia are desert or mountain environments with densities of less than one person per km², except along major transportation routes (Figure 4.6).

Population distribution in China is marked not only by the contrasts between environmentally different areas, but also by the contrast in the proportions of growing urban populations and declining rural populations. Today, China is experiencing rapid urbanization of its people.

Japan is densely but unevenly populated. The main concentrations of people are on the small areas of lowland, and 70 percent of the total population lives in the Pacific coastlands from Tokyo to Osaka. This urbanized zone is often known as the "Tokaido **megalopolis**"—a series of almost continuous metropolitan centers with urban functions that exchange flows of people and goods with the surrounding areas and the rest of the world. As urban populations grew, the more remote and environmentally difficult areas in the mountains and along parts of the western coasts of Japan suffered depopulation. Such regions often have up to 25 percent of their population over 65 years old, and government grants are available to help people stay or settle there.

Mongolia has a dispersed population with falling natural increase and longer life expectancies now in the high 60s. Over 60 percent of Mongolia's population lives in urban places as the result of the pull of growing industrialization and the push

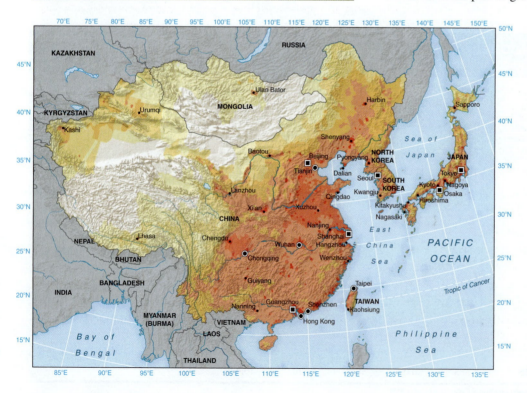

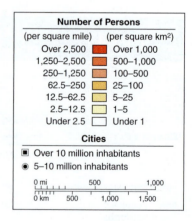

Number of Persons

(per square mile)	(per square km²)
Over 2,500	Over 1,000
1,250–2,500	500–1,000
250–1,250	100–500
62.5–250	25–100
12.5–62.5	5–25
2.5–12.5	1–5
Under 2.5	Under 1

Cities

■ Over 10 million inhabitants
● 5–10 million inhabitants

Figure 4.6 East Asia: **distribution of population.** Relate the main areas of high and low population density to physical features. Is this a satisfactory basis for explaining the differences?

of rural poverty. Mongolia's capital, Ulan Bator, houses nearly one-fourth of the country's population, including those living in large tented (yurt) areas (see Figure 4.7).

Population Growth

Changes in East Asia's population structure and distribution have important economic and social implications. China supports about a fifth of the world's population on only 7 percent of the world's arable land. As population policies in the 1950s and 1960s fluctuated between pro- and anti-growth, the country's population continued to rise, despite the soaring death rates and plummeting birth rates that characterized the period known as the "Great Leap Forward" when China experienced severe famine. Fertility reduction became a national priority in the 1970s. Later marriage, longer intervals between births, and fewer children were advocated, while modern contraception was made widely available.

The one-child policy was instituted to further decrease population growth. Growth rates fell from 2 percent per year in 1965 to 1.4 percent in the 1980s and to under 1 percent since 2001. From 1965 to 2012, total fertility fell from 6.4 to 1.5. Even this major achievement merely slowed the continued growth of China's huge population. More children survived early childhood despite fewer babies being born, as infant mortality fell from 200 per 1,000 live births in 1945 to 17 in 2012. The numbers of people living to old age also increased as life expectancy more than doubled from 35 to 75 years. Investments in education cut adult illiteracy from 80 to 19 percent.

Figure 4.7 **A yurt or ger settlement on the outskirts of Ulan Bator, the capital city of Mongolia.** Over 60 percent of Ulan Bator's population lives in these settlements. *Photo: © Olivier Cirendini/ Getty Images.*

Japan's total population increased from around 90 million in 1960 to 128 million in 2012. The combination of current low fertility (1.4) and population growth rates (natural increase of 0.2 percent) led to population stabilizing around the 2008 figure, but it is projected to fall to 120 million by 2025. Infant mortality at 2.3 per 1,000 live births is the world's lowest, and life expectancy at 83 years is tied with Hong Kong for the highest. As a result, Japan's population is aging. In 1970, 7 percent of the population was over age 65; by 2001, the proportion rose to 17 percent, and it is currently 24 percent. Meanwhile, smaller nuclear families averaging three persons per household in the 1990s became common. The aging of the population increases the needed pension provision, medical costs, and leisure provision for the retired, but it also adds to firms' wage costs as more employees reach senior positions. The Japanese are now worried that there will not be enough labor for their industries or funds to support the increasing numbers of elderly people.

South Korea's population of 49 million people in 2012 is likely to rise slowly to around 51 million by 2025 because of a low rate of population increase. Life expectancy rose from 47 years in 1955 to 81 years today. As in Japan, the population in older age groups will increase markedly in the next 20 years. North Korea had 23 million people in 2012, and this is predicted to rise to over 26 million in 2025. These two countries differ in that South Korea's modest rise is related to growing affluence, while North Korea's has a context of increasing poverty and lack of opportunity.

Ethnic Groups

The **Han Chinese** make up 92 percent of the Chinese population. The Han culture diffused by conquest and movements of people to administer new areas since the AD 200s and 300s. Most minority groups live along the northern borders with Russia and Mongolia and in western China. Central Asian ethnic groups dominate the north and northwest. Minorities such as the Mongols in the north, the Tibetans and Uighur in the west, and the more dispersed Manchu and Muslim Hui, each with their own language, are allowed limited jurisdiction over their affairs, particularly the expression of their culture in music and drama. Still, local resentment of the Chinese takeover of their lands causes the eruption from time to time of anti-Chinese demonstrations in the minority strongholds of Tibet and Xinjiang, but these are suppressed by the central government.

Most Japanese people stem from an ancient mixture of Asian continental and Pacific island stocks. However, three small minority groups remain important. On the northern island of Hokkaido the Ainu retain ethnic and language differences. On the largest island of Honshu, the Burakumin, numbering around 2 million people, were outcasts and allowed to work only in the lowest occupations until their emancipation in 1871. The third group consists of foreign nationals, three-fourths of whom are Koreans, moved forcibly to Japan in the early 1900s. Until the 1980s, they found it difficult to gain Japanese citizenship and are still subject to discrimination. Migration into Japan is restricted, and only 1 percent of the population is of foreign origin.

North Korea and South Korea—officially the People's Democratic Republic of Korea and the Republic of Korea, respectively—have a common ethnicity and had a similar history until the mid-1900s. After the Korean War (1950–1953) the two countries moved in different directions ideologically, politically, and economically. Both Japan and Korea are internally homogeneous in terms of language. Ethnically, the Mongol people dominate Mongolia, although it also has 10 percent who are Chinese, Russians, or Kazakhs.

Urbanization

The degree of urbanization in East Asian countries varies considerably from a low of slightly more than half in China to over 80 (86 and 82) percent in both Japan and South Korea (see Table 4.1). The region has some of the largest cities in the world, led by the massive Tokyo region in Japan with a population of 37 million (see Table 4.2). The growth of global and intraregional trade in East Asia led to the emergence of global city–regions as essential nodes linking this region to the rest of the world through business and political exchanges. See "Globally Connected Cities of East Asia," page 134.

Japan and the Koreas

While the region has always had large and important cities, rapid urbanization took place in the 1900s. In the short period from 1920 to the early 2000s, Japan changed from a largely rural country, in which industrial towns housed 25 percent of the population, to one in which post–World War II industrialization and the availability of jobs in manufacturing and service sectors in cities led to expanding urban areas. Today almost 90 percent of Japan's population lives in cities.

In equally urbanized South Korea, the high proportion living in towns is a sign of modernization. The South Korean urban population increased from 25 percent in 1950 to 50 percent in 1980 and is currently 82 percent. The main population and industrial facilities are in the northwest, centered on the capital Seoul, Inch'on, and Taejon and along the southern coast, including Pusan, Taegu, and Ulsan. Almost totally destroyed in the Korean War of the 1950s, Seoul was rapidly rebuilt.

North Korea's population is also concentrated around its western coasts, in the capital city, Pyongyang, and Namp'o. Countrywide, about 60 percent live in towns. Although possessing greater mineral resources than South Korea, particularly coal and metallic ores, North Korea has not developed a major manufacturing capability that draws more people to the towns.

Urban vs. Rural Population in China

Until the mid-1900s, most Chinese cities were market and administrative centers, often with walls that restricted expansion. The exceptions were some port cities, of which Shanghai became the largest, growing because of external trade contacts dating back to the 1800s. Today the largest cities in China (see Table 4.2) include the capital Beijing, Shanghai, Guangzhou, and

Shenzhen, all of which are treated as political provinces. Entirely urban Hong Kong became a special area within China after the end of its British colonial government in 1997. Shenyang, Wuhan, Chongqing, Chengdu, Xi'an, and Harbin each grew from around 1 million or fewer people in 1950 to multimillion-person cities. It is likely that further growth will take Beijing and Shanghai to over 20 million people, and Guangzhou and Shenzhen to around 15 million people each by AD 2025, when China will have over 100 cities with populations of more than 1 million people.

In China, policies toward urban growth fluctuated from the mid-1900s. Communist rule from 1949 alternately favored urbanization and movements back to the countryside. In the Mao Zedong era, the wish to encourage modernization through industrialization led to rapid urbanization from 1949 to 1960. In this period the urban population more than doubled, from 49 million to 109 million, representing an increase from 9 to 16 percent of the total population. Part of this urbanization resulted from industrialization around previously administrative cities such as Xi'an and Nanjing. Another part came from the development of new inland centers such as Baotou, the iron and steel center in Inner Mongolia. Public rhetoric, however, exhorted the Chinese to avoid the environmental dangers of urbanization.

From the 1960s to the mid-1970s, the rural-urban basis of the registration system (hukou), established in the 1950s, was used to prevent rural-born Chinese from becoming legal residents of cities. The Cultural Revolution was launched by Chairman Mao in 1966 to ensure that China would remain true to Maoist ideology and not slip back into traditional "bourgeois" ways. As part of the movement that involved mobilizing the country's youth, over 20 million young urbanites, as well as many others, were moved to rural areas to be "reeducated" in the virtues of rural life.

Recent Urban Migration in China

From 1978 to the early 2000s, economic reforms placed an emphasis on urban jobs, better incomes, and better access to services and led to a further rapid rise of urban population numbers. From the late 1970s onward, new urban-based industries recruited labor from the countryside on temporary contracts. The city of Shenzhen in the Pearl River Delta increased in size from 1.2 million in 1990 after it became China's first Special Economic Zone, with free-market policies. Its population was over 10 million in 2012. Many new urbanites do not bother to acquire temporary residence certificates. Over half of Shenzhen's population is composed of migrants who live in factory dormitories during the week and return to their homes on the weekend. Although the hukou system has been relaxed, large towns of now-settled migrants on the periphery of Shenzhen bear witness to the times when living in the city was restricted.

Although the debate over the relative merits of expanding the largest industrial cities or developing small and medium-sized towns continues, Chinese people are increasingly making the decision to move to large urban centers. Often the parents move to the cities for work, while the children are raised by grandparents in the rural areas.

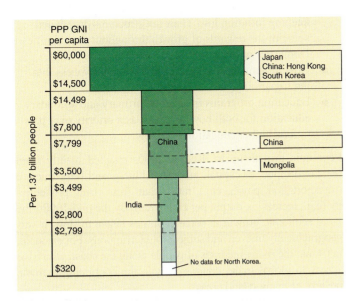

Figure 4.8 **East Asia: country average incomes compared.** The countries are listed in order of their GNI PPP per capita. Note the relative positions of Japan, China, and Mongolia. *Source: Data (for 2010) from* World Development Indicators, *World Bank; Population Reference Bureau.*

Global Economic Powers

Most countries of East Asia have a major involvement in the global economic system. They cluster toward the top of the world range of gross national income (Figure 4.8) and ownership of consumer goods (Figure 4.9). For U.S. traders, Japan is an established major competitor, while the People's Republic of China, South Korea, and Taiwan are considered emerging big markets. China in particular has become a major producer and consumer of goods. Trade patterns have also shifted; in 1985 the United States had a US$60 million trade surplus with China, which became a US$295 million trade deficit in 2010. North Korea and Mongolia, however, remain largely outside the world economy.

The recent economic growth of these countries was at first based on exporting products to American and European markets, but by the 1990s it was also bolstered by trade among countries within this region and Southeast Asia. Asian internal trade is now as great as external trade, which raises questions as to whether the region's economic future lies more with intra-regional trade or global trade. However, the gradual evolution of the Asia-Pacific Economic Cooperation (APEC) forum reflects the global trading links from East Asia to Southeast Asia, the South Pacific, and the Americas.

From Poverty and Defeat to Renewed Eminence

After World War II, poverty, political disruption, and cultural confusion were the rule in East Asia. In 1945:

- Defeated Japan was economically devastated and hated in the region for its wartime acts.
- The Koreas struggled with independence following decades of Japanese occupation and exploitation. Splitting Korea into North and South led to the 1950s Korean War, which destroyed much of both countries.

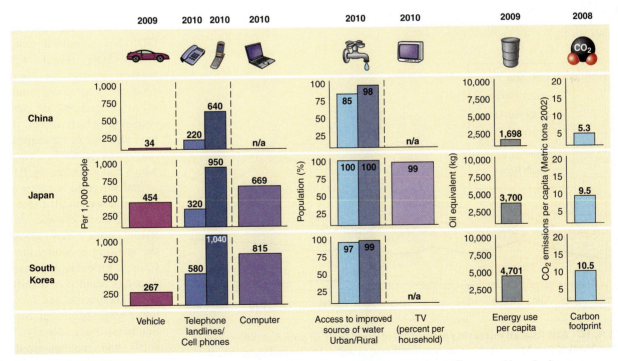

Figure 4.9 **East Asia: ownership of consumer goods, access to clean water, and use of energy.** How do these figures relate to economic growth? *Source: Data from* World Development Report, *World Bank 2012; Population Reference Bureau, World Population Datasheet, 2012.*

- In China, the warring Nationalist and Communist factions resumed their prewar conflict. When the Communists prevailed in 1949, the remnant of the Nationalist army fled to Taiwan.

Manufacturing provided the basis of modern Japanese growth. Modernization in Japan built on economic stability and government encouragement. New factories built in the coastal zone from Tokyo to Osaka made iron and steel, ships, automobiles, constructional engineering products, electronics, and textiles.

The newly industrializing South Korea, Hong Kong (before and after it became part of China in 1997), and Taiwan increased their incomes six- to sevenfold from 1980 to 2000. South Korea and Taiwan built economies that were based on exporting to the rest of the world. Over nearly 30 years from 1949 to 1976, the People's Republic of China under Mao Zedong built national cohesion, albeit at tremendous cost in terms of lives and policy changes. China experienced its most rapid economic growth after 1980, increasing its total GNI PPP from US$202 billion in 1980 to US$1,065 billion in 2000 and US$7,151 billion in 2012. However, neither North Korea with its declining economy nor Mongolia with its small economy experienced such economic expansion.

Weathering Economic Downturns

The East Asian economic growth in the later twentieth century was termed the "East Asian Miracle" in a 1993 World Bank report, which argued that distinctive Asian characteristics motivated and fueled economic growth:

- Successful countries had strong governments that managed stable business environments by such measures as keeping inflation low. Their tax structures distributed some of the rewards of growth throughout communities. Domestic savings increased.
- Governments encouraged export competitiveness that led to partial global integration.
- Education programs focused on primary and secondary education for both boys and girls as a priority over prestigious higher education for a few.

Economic growth up to the mid-1990s resulted mainly from increased inputs of capital, labor, machinery, and infrastructure (transportation, power, etc.), rather than from rising productivity (e.g., increasing output per employee). By the mid-1990s, the rapid economic growth in some countries began to slow. It was geographically uneven and subject to interruptions. For example, in China, the rising incomes of people along the coast, who have access to global trade, contrasted with the continuing extreme poverty of many inland areas.

In 2001, a World Bank report, "Re-Thinking the East Asian Miracle," exploded the assumption that following the "Asian Way" would produce continuing economic expansion. Geographic factors, beginning with the development of distinctive cultures in specific places through history and the interactions of people with the natural environment, bring a greater degree of diversity to this region than one simplistic prescription could change in a few years.

The International Monetary Fund made loans in exchange for an end to protectionist policies that kept Western foreign investments out of East Asia. South Korea recovered its economic impetus within a couple of years, but in that time it also suffered unemployment, corporation failures, and an increase in long-term poverty.

In recent years, the East Asian GDP has surpassed that of the United States (Figure 4.10). But weaker demand for East Asia's products and lower investment growth, functions of the global recession which began in 2008, have also slowed the regional economy. Even the rise in China's GDP, which had increased by 10 percent annually over the last three decades, slowed down in 2012 to 8 percent due to the economic downturn. Rising wages that accompanied the growing economy have made Chinese products more expensive. This has resulted in a shift in foreign investments and factories to lower-wage places in South and Southeast Asia. Nearly 700,000 small and medium-sized enterprises were shut down in China, leaving 5 million people unemployed, an interesting outcome of globalization.

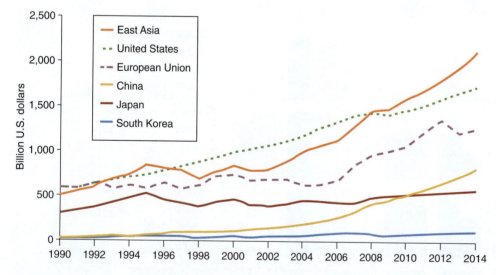

Figure 4.10 **East Asia: an increasing proportion of world GDP.** Growth and projections of GDP (1990–2014) of East Asia and major countries within it, as well as the United States and the European Union. Note how East Asia's GDP has surpassed that of the United States and that China has overtaken Japan and is likely to continue to gain ground. The GDPs of the United States, China, and Japan rank first, second, and third in the world. *Source: International Monetary Fund, World Economic Outlook Database.*

4.4 Geographic Diversity

East Asia (see Table 4.1) is dominated by two powerful countries, Japan and China. Japan's strength has been tested by the March 11, 2011, earthquake that devastated Tohoku, its northeastern region. China's growing dominance stems from its role in regional and global economies and its political-military strength. Mongolia is linked to China because of its location between China and Russia. Taiwan is also linked to China, which claims possession of the island. Although recent trends bring the two closer to reunification, many Taiwanese bitterly oppose such a move. As separate countries, North and South Korea have very different levels of global involvement but were united until 1950 and may reunite in the future.

Japan

Small and densely populated, Japan became the world's second-wealthiest country following a dramatic economic recovery after 1945. Japan's physical environments—from snowy Hokkaido in the north to subtropical Okinawa in the south and from the volcanic mountain spine to the coastal lowlands—combine with distinctive contrasts in urbanization, manufacturing emphases, and rural ways of life. Japan consists of four main islands: Hokkaido, Honshu, Shikoku, and Kyushu (Figure 4.11).

Ancient traditions place Japan's foundation in 660 BC. The principle of **Shinto** encapsulated Japanese traditional values in a national religion built on animism, ancient myths, and customs. It is a pantheistic religion of many gods and involves emperor worship and the concepts of Japanese superiority. In Shintoism, both living and nonliving objects possess spirits, with Mount Fuji and other mountains being given special reverence.

Japanese emperors resided at Kyoto, retiring from public life and delegating administration of the country to leading families of court nobles. From the AD 1100s, these feudal lords took over the imperial administration, and armed samurai warriors under a *shogun* (military leader) acted over the heads of the powerless court. When the last shogun resigned in 1867, the Meiji emperor regained his position as titular head of government. The royal capital moved to the shogun center of Edo, later called Tokyo ("eastern capital"). However, in the 1889 constitution, the emperor became a figurehead and political power was taken by leading figures in the parliament.

At the end of the 1800s, Japan's imperialist ambitions, coupled with its new military power, enabled the country to take over Taiwan and Korea, and to win a war against Russia that gave it Manchuria in northern China. During World War I, Japan advanced its economy and generated a trade surplus. However, its involvement with Axis powers during World War II and the dropping of atomic bombs on Hiroshima and Nagasaki in 1945 wiped out earlier economic gains. Even in 1960, Japan, despite 100 years of modernization, still had a gross national income per capita that was one-eighth that of the United States.

Agriculture

Japanese farms occupy small areas of arable land within the hilly and mountainous country, often intermingled with housing and industry which compete for the use of the land. This makes land prices higher than what farmers could afford without subsidies. Government support through subsidies maintains agriculture as a significant sector of the economy, although the country's agricultural population is waning. After 1970, only 12 percent of farmers worked full-time without off-farm jobs. The size of most farms remained stable, however, because land increased in value and many owners held onto their land as a family investment. In the early 2000s, Japanese farmers faced new shocks. Although Japan's agriculture remains highly protected from foreign competition, improved transportation links and refrigeration brought large quantities of Chinese tomatoes, eggplants, onions, and garlic bulbs to Japanese supermarkets. The 2011 earthquake and tsunami devastated Japan's northeastern coastal areas (Figure 4.12) and its agriculture and fisheries, while the radioactive contamination in the wake of the Fukushima Daiishi nuclear plant crisis has raised questions of food safety. Possible contamination also poses threats to the country's food production and food exports.

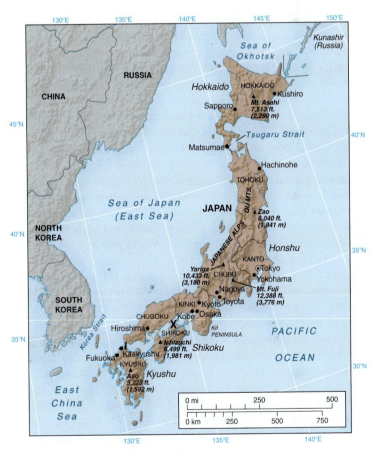

Figure 4.11 Japan: major islands, districts, and cities.
Assess from the map the significance of the mountainous spine and the long coastline for Japan's geography. The 'X' on the map shows where the Seto Great Bridge links the islands of Shikoku and Honshu.

Figure 4.12 Views of a section of the northeast coast of Japan before and after the tsunami of March 11, 2011. Note that many of the settlements along the coast have been destroyed. *Image: German Aerospace Center (DLR)/Rapid Eye/NASA.*

Industrial Development

After 1911, the Japanese government became more democratic, and political parties and labor unions were established. In the period leading up to World War II, industrial expansion was based on government-business links, cheap labor, large numbers of small businesses, and improved infrastructure, aided by cheap raw materials on the world market. Overseas, Japan became known for its cheap goods, often of mediocre quality. The 1930s economic depression—when other countries raised tariff charges against cheap Japanese imports—led to an internal pact between Japanese military leaders and the elite families, the *zaibatsu*, who owned the biggest industries. The military leaders took over political power and shifted industrial production to naval ships, air force planes, and other military equipment. This phase culminated in World War II.

In the following decades, Japan moved rapidly to the forefront of the global economy. Japan's export-led growth raised its gross national income one-hundredfold between 1960 and 2000 at current dollar values. From the 1970s, major Japanese industries had sufficient resources to invest in the United States, Europe, and Southeast Asian countries to manufacture goods for their domestic markets. Japan then invested in China. Japanese companies produced goods in China and Southeast Asia for their brands that sold worldwide. In the 1990s, however, the Japanese economy's growth slowed to 1.5 percent. Real estate values peaked in 1991, and the Tokyo Stock Exchange crashed in 1990–1992. Although it has grown, Japan's GDP is still just one-third of the U.S. GDP (see Figure 4.10).

Japanese industrial development is strongly localized, with heavy industry in coastal locations on reclaimed land because of the need for space and proximity to ocean transportation links to raw material suppliers and markets. Light industry is more widely distributed and intermixed with housing, with older firms retaining inner-city sites. Location in the inner city takes advantage of the availability of labor, linkages with other firms, and market outlets. Local governments encourage the establishment of light industrial estates on city margins, including science parks for high tech industries.

The Japanese government assisted and advised industry, with the **Ministry of Economy, Trade, and Industry** (METI; formerly the Ministry of International Trade and Industry) encouraging export sales through a worldwide network of market intelligence-gathering offices. Diversification required greater investment but raised the output of light industries such as those producing cameras and household appliances. After copying others' technologies and designs, Japanese firms became initiators. Japan's successful exports became known for their quality, reliability, and market-leading technology. In the 1980s, other countries built huge deficits of payments to Japan, causing the Japanese yen to double in value compared to other international currencies.

In the later 1900s, Japan's industrial economy shifted to producing goods such as autos and high tech goods for export markets. Japanese exports of manufactures and foreign investments led to the growth of huge multinational corporations such as automakers Honda, Toyota, Nissan, and Mitsubishi Motors, and electronics and electrical goods makers such as Sony, Sharp, Fujitsu, Panasonic, and Matsushita, who built worldwide brands manufactured in many countries. Japanese firms, especially those involved in electronics or automobile production, formed conglomerates known as *keiretsu*. These industrial groups were built around financial nuclei that usually included a major bank, and they produced giant trading and manufacturing companies. A slowing of Japan's economy in the 1990s resulted in its making investments in other countries. By the early 2000s, Japan was the world's second-largest economy after the United States (see Figure 4.10). Japanese multinational firms have been hard hit by the global economic downturn. Toyota, for example, announced its first loss in 70 years in 2008 and an annual net loss of over US$4 million in 2009. Ranked third on the Forbes 2000 list of the world's leading companies in 2009, it was not even on the top 20 leading companies in 2010. Recalls of nearly 20 million faulty vehicles since 2009 due to safety problems and agreement by the company to pay US$1.1 billion to settle a class-action suit added to Toyota's financial woes and cut into its sales, but the company regained its position as the world's largest car manufacturer by 2013. Recent increases in service occupations, from education and health care to retailing and tourism, have widened the range of employment and contributed to a diversified economy that is less dependent on the fluctuations of world markets than a manufacturing-dominated economy.

Japan's popular culture ranks high among the country's latest exports. In 2010, METI made the export of such popular culture items as Japanese music, video games, anime, comics (*manga*), films, and fashion a priority, hoping to add to the country's "soft power," which rests on making Japanese culture attractive to diverse groups of people.

New Challenges and Directions

Japan's Liberal Democratic Party (LDP), after ruling for several decades, was supplanted by the Democratic Party of Japan (DPJ) in 2009. Voted to power on a platform that assured comprehensive administrative reforms and no tax increases, the DPJ has not been able to fulfil its promises, while the opposition LDP is viewed with distrust in part due to perceived strong ties with the United States. Political problems were compounded by the devastation that occurred along 650 miles of Japan's northeast coastline during the 2011 earthquake and tsunami (see Figure 4.12). The natural disaster left 20,000 people missing or dead and had a direct economic cost of approximately US$210 billion. As Japan struggles to rebuild towns and cities, infrastructure, and its economy at costs estimated at hundreds of billions of dollars, it also has to reconsider its use of nuclear power and alternative sources of energy as most Japanese people are opposed to nuclear energy in the wake of the Fukushima Daiichi nuclear disaster.

Japan's population is expected to shrink by 30 percent from its current 128 million to 87 million by 2060, as the nation's aging accelerates and its birth rate remains low (see "Population Growth," p. 119). The country's average life expectancy dipped in 2011 after the March earthquake and tsunami, but an overall upward trend for life expectancy is likely to be seen in the years ahead. To deal with the labor force shortages and the care of the aged population, which will swell to 39 percent of the total population by 2060, Japanese companies are developing advanced robotics and automation systems that can take the place of human workers, as well as help take care of its rapidly aging population. In the meanwhile, Japan is gradually bringing in foreign labor, as the government has begun to relax restrictions on the employment of foreign workers who have skills that are in short supply in Japan, such as medical doctors and nurses.

The Koreas

Sharing a common ethnicity and culture, North Korea and South Korea (Figure 4.13a) also had a common history until the mid-1900s, but after that they moved in different directions. The countries occupy the hilly Korean peninsula that extends southward from the Chinese border along the Yalu River.

The area that is now North Korea was the center of the first Korean kingdom, around which the peninsula was unified in the AD 600s. After the Mongol invasion and retreat, Confucian principles of a hierarchical system of government and society were adopted. The Yi dynasty, founded in Seoul in AD 1392, ruled the kingdom of Korea until 1910, although it became subject to the Chinese in 1644. Close proximity to Japan attracted aggressive interest from that growing military power. In 1894–1895, as rebels opposed the Korean government, the

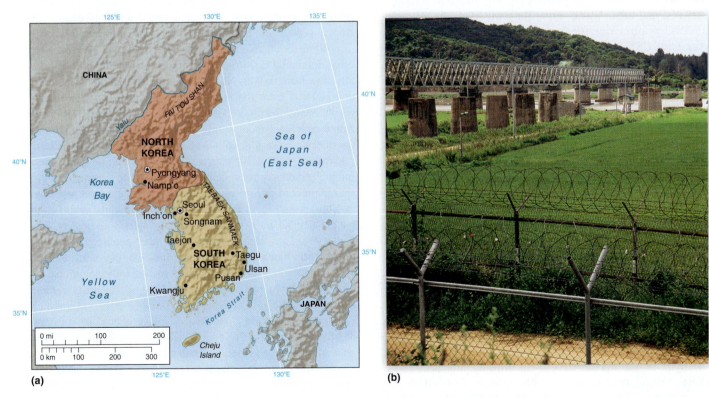

Figure 4.13 **The Koreas, North and South.** (a) After Japanese occupation in the early 1900s, the Koreas prefer the term "East Sea" to "Sea of Japan," but have not convinced others to change the name. (b) In this view of the Demilitarized Zone, a bridge crossing the frontier river has been destroyed and trees cut to maintain an open vista. *Photo: © Alasdair Drysdale.*

newly strengthened Japanese forces overran Korea, defeating both Chinese and Russian attempts to aid the Koreans. Japan annexed Korea in 1910.

At the end of World War II, the U.S. and Soviet armies defeated Japan and divided the "freed" North and South Korea at the 38th parallel. North Korean forces invaded South Korea in 1950. The United Nations sent forces under U.S. commanders to support South Korea, and, following destruction over much of the territory as warfare moved back and forth, the two sides agreed on an armistice in 1953. The agreement resulted in the formation of a Demilitarized Zone between the two countries (Figure 4.13b).

After World War II and the Korean War, South Korea established import substitution industries that developed manufacturing skills, substituted imports with locally produced goods, and hence reduced the need to pay hard currency for foreign goods. Its dictators forced the people to work for low wages while the economy improved. The transportation infrastructure built by the Japanese when they ruled Korea in the early 1900s, and rebuilt with external funds after the Korean War, helped develop the economy in its early stages of growth. South Korea built a series of major highways, and its public transport system is one of the world's best.

The devastating Korean War took place when three-fourths of the South Korean people still earned a living from farming. In postwar land reforms, the government took over large, inefficiently run estates and encouraged a new group of small landowners to apply imported fertilizers. Farm productivity rose, and South Korea became largely self-sufficient in food. The farming and rural emphasis gave way to manufacturing, services, and urban living. From the early 1960s to the early 1990s, few other countries' economies grew so fast. Exports rose from US$33 million in 1960 to US$419 billion in 2008. A country of subsistence farmers under largely feudal control was transformed in a generation to the world's main maker of large ships and memory chips, the fifth-largest automaker, and eleventh-largest world economy—with greater output than the whole of Africa south of the Sahara.

Contrasting Koreas

South Korea established iron and steel, shipbuilding, chemicals, automobile, and textile industries in the 1950s, and these continue to be important for its economy, although in the 1990s there was a swing toward high tech industry. The South Korean government supported large family-owned companies that developed into huge conglomerates known as chaebol, such as Hyundai, Daewoo, LG, and Samsung. These chaebol are similar to the Japanese keiretsu.

Although mostly noted for its economic growth, South Korea is increasingly making a regional contribution in popular culture that is also economic. The late 1990s saw the rise of "Hallyu," or "the Korean wave"—the growing popularity of all things Korean, from fashion and film to music and cuisine. As South Korea's economy plunged and GDP fell by 7 percent during the 1998 financial crisis in Asia, the government looked to hallyu as a tool of soft power, hoping to expand South Korea's

profile abroad along with demand for its cultural exports and tourism. As an international phenomenon, hallyu signals South Korea's growing influence across Asia. Korean television dramas are broadcast widely in China, Japan, and countries of Southeast Asia. Exports of Korean video games, television dramas and popular music (K-pop) have all doubled since 1999, while the value of cultural products exported since then has increased almost three times, to US$1.8 billion in 2008.

The typical rags-to-riches storylines of South Korean dramas that are family-friendly and emphasize Confucian values such as filial loyalty, honesty, commitment, respect for elders, and hard work appeal to Asians more than the western soap operas. The K-pop sensation, with its globally attractive dance beats and songs by good-looking youngsters, has developed a fan base not just in Asia but also in Europe, South and North America, and the Middle East. An example of the globalization of K-pop is South Korean singer-songwriter Psy's music video "Gangnam Style," which went viral and reached the top of the UK charts in 2012 and became the most popular video in YouTube history.

Since 2005 the South Korean government has given grants to organizations that introduce Korean culture overseas. Today, government and nongovernment agencies, commercial news media, and diasporic organizations seek to present Korea and its culture in a more positive light than they receive in foreign media. A survey conducted by the Chicago Council on Global Affairs in 2008 found that about 80 percent of respondents from China, Japan, and Vietnam, three of the largest markets for hallyu, hold South Korean culture in high esteem. Soft power is also connected to cultural tourism; South Korea has seen a spike of foreign tourists in recent years, many of whom cite Korean pop culture as the catalyst for visiting the country.

While South Korea became one of the newly emergent urban-industrial countries with support from the United States after the Korean War, North Korea remains poorly developed and isolated from the rest of the world under a Communist regime. After the loss of Soviet Union subsidies in 1991, North Korea depended increasingly on international aid to make up for internal economic collapse and famine. In 2000, the death rate rose 50 percent higher than in 1994, indicating widespread famine and disease, despite North Korea being the world's largest recipient of famine relief that year. The best food produced in North Korea feeds the army and party officials, while other people depend on donated food, often of lower quality. The country remains under suspicion by other countries of having nuclear and biochemical weapons. In 2006, the North Korean government announced that it had successfully completed its first nuclear weapons test, although some scientists remain skeptical of the test's success. In October 2012, after South Korea announced the United States' agreement to let it extend its missile range, North Korea stated that its missiles could now reach the United States, a claim that is also questioned.

By the late 1990s, there were signs of closer ties developing between the two Koreas. Presidential visits led to reunions of Korean families separated for 50 years. Talk of reunification of the two countries gathered some momentum, but both countries had reservations about such a move.

Recently there have been heightened tensions between the two countries. North Korea was held responsible for the 2010 sinking of a South Korean warship and for the deaths of 48 sailors onboard, as well as the shelling of Yeonpyeong island, allegedly in retaliation for artillery fired toward it by South Korea. Kim Jong-il, who ruled North Korea from 1994 until his death in 2011, was succeeded by his son Kim Jong-un. North Korea's new leader has set about modernizing Pyongyang, but whether he will usher in political and economic liberalization remains to be seen.

China, Mongolia, and Taiwan

The People's Republic of China (PRC) is the world's most populous country, with 1.35 billion people in 2012, and the third-largest by area (Figure 4.14). Hong Kong (now also commonly called Xianggang, the pinyin version of Mandarin Chinese) and Macau, previous colonies of the United Kingdom and Portugal respectively, were returned to PRC control in 1997 (Hong Kong) and at the end of 1999 (Macau). Mongolia (2.9 million) and Taiwan (2.3 million) are tiny by comparison but have distinctive roles in

Figure 4.14 **The People's Republic of China, Mongolia, and Taiwan: major geographic features, including towns and cities, major rivers, and provincial boundaries.** China has 23 provinces (capital letters), four cities with provincial status (Beijing, Shanghai, Tianjin, and Chongqing), and five autonomous regions that have large minority populations (Tibet, Inner Mongolia, Guangxi, Ningxia, and Xinjiang).

relation to China and the rest of the world. These two countries are discussed together after the section on China.

China: Early Civilizations and Kingdoms

By 2000 BC the first Chinese civilization developed in the Huang He (Yellow River) valley (Figure 4.15), based on wealth from agricultural surpluses and craft skills and controlling much of the lower Huang He and lower Chang Jiang (Yangtze River) basins.

The Zhou dynasty (1122–256 BC) used Iron Age technology, irrigation, and deeper plowing of the soil to support economic growth. During the political and social upheavals of this period, **Confucius** (Kong Fuzi), an administrator, called for better-trained and more able managers to focus on orderly conduct and proper relations in all social contexts. **Daoism**, the teaching of Laozi, disdained the Confucian system, preferring a return to local, village-based communities with little government interference. Daoism is based on the apparently opposing concepts of *yin* and *yang*: complementary, interdependent principles operating in space and time, emblems of the harmonious interplay or balance of all pairs of opposites in the universe (see Figure 4.2).

Around 220 BC, the Qin dynasty (spelled *Ch'in* in older forms of romanization, giving China its Western name), imposed strict laws to weld the separate feudal states into a centralized and culturally uniform empire with a standardized written script. Rulers sited the Qin capital near modern Xi'an. Southeast Asian Buddhism reached China in the early years of the first millennium AD, adding a spiritual dimension to materialistic Confucianism. Over the next three centuries, the Tang dynasty spread Chinese influence into the northeast of modern China, Korea, Japan, and northern Vietnam. Zen Buddhism, a form that centers on meditation and moments of insight in the course of mundane activities, arrived in Korea and Japan in the early 1200s.

Mongol Invasions

In the 1200s, the Mongols invaded China. Kublai Khan, the Mongol leader, established his capital in northern China at Beijing. He and his descendants ruled as Chinese emperors, driving the center of native Chinese intellectual and economic development southward. In the 1300s, growing Chinese resentment of the new rulers, disruptions, and taxes imposed on them, combined with crop failures and famines, opened the way for rebellion and the proclamation of the Ming dynasty in 1368.

Later Empires

Renewed Chinese expansion occurred under the Ming dynasty (1368–1644), during which the economy, literature, and education made advances. Chinese naval power under Admiral Cheng-Ho (1405–1433) ventured beyond India to the Persian Gulf, Red Sea, eastern Africa, and Madagascar. Although there was trade, the Chinese did not conquer new lands.

The succeeding Qing (Manchu) dynasty ruled from 1644 to 1911. By the mid-1700s, its rulers had expanded the Chinese realm to its farthest limits in the Northeast, Mongolia, Xinjiang (the Northwest), Tibet, Burma (Myanmar), and Taiwan.

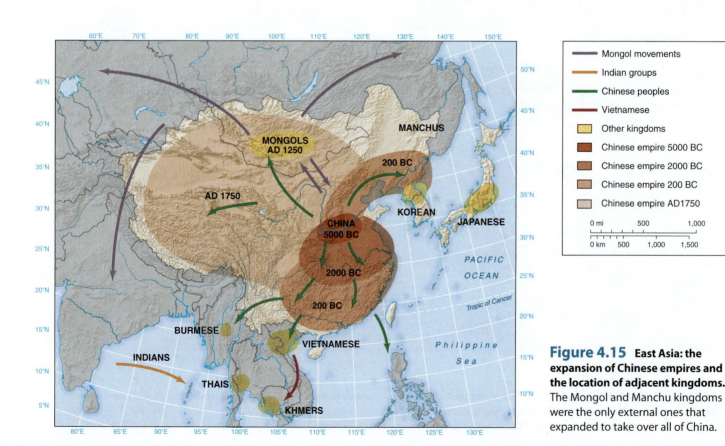

Figure 4.15 **East Asia: the expansion of Chinese empires and the location of adjacent kingdoms.** The Mongol and Manchu kingdoms were the only external ones that expanded to take over all of China.

In 1820, China was the world's wealthiest empire, producing nearly one-third of the world's GDP. But internal problems left China vulnerable to the imperialist forces of Japan, Russia, and the West. Following the country's defeat in the Opium Wars of the mid-1800s, China ceded Hong Kong to the British and was forced to open its ports to foreign trade and to allow foreigners to travel and live within its territory.

Disrupted Chinese Republic

In 1912 Sun Yat-sen and his Kuomintang Party (Guomindang, or the Chinese Nationalist Party) claimed overall political power. Attempts to reunify China under a common government in the 1920s and 1930s, however, were hampered by strife among local warlords, a Communist rebellion, and Japanese attacks. The main Japanese invasion of China began with a 1933 expansion out of the northeast territory that it had occupied earlier in the 1900s, followed by a general declaration of war in 1937. The Kuomintang and Chinese Communists united in resistance.

The Chinese Communist Party came into power in 1949 and established the People's Republic of China. The new government's first aim was to turn a backward, feudally divided agricultural country into a united, advanced, and centralized socialist industrial country. Mao Zedong, the leader and dictator of China from 1949 to his death in 1976, wished to make rapid and massive changes.

Collectivization and Communes

Modeling changes on the Soviet Union experience, China's Communist government prioritized large-scale industrialization and the collectivization of agriculture. Although little investment went into improving agriculture, rural social structures were revolutionized. Before 1950, most Chinese were landless tenants, working tiny parcels of land. The landlords or rich peasants who owned over 70 percent of the cultivated land made up fewer than 10 percent of the population.

In 1952, 300 million landless peasants received their own plots of land, houses, implements, and animals, free from debt. By early 1956, over 90 percent of the rural population joined cooperatives. In "advanced" cooperatives, the land became the property of the organization, apart from tiny garden plots for each family. A typical advanced cooperative included approximately 150 households farming a total of around 400 acres of land. This process, known as **collectivization**, was completed by late 1956.

The government also established communes. A **commune** combined agriculture, industry, trade, education, and the formation of a local militia. The commune leadership planned the agricultural year, other economic activities, and social functions. They guaranteed food, clothing, education, housing, and arrangements for weddings and funerals to members.

The Great Leap Forward and Cultural Revolution

The **Great Leap Forward** of the late 1950s aimed to increase China's rate of industrialization by producing basic industrial products, such as iron and steel. It attempted to integrate rural and urban areas by setting up small-scale production units in both. Throughout much of China, community efforts devoted to iron smelting, often in small backyard furnaces, caused a neglect of farm work and made worse the drought-based famine of 1960, in which 25 to 30 million people died. The plan was a disaster. Both industrial and agricultural production dropped, and China was forced to import wheat from Western countries. Stung by failures and misunderstandings, the Chinese leaders abandoned most Great Leap Forward policies.

After a few years of higher production, Mao Zedong attempted to change the whole basis of Chinese society and its administration, and the country once again turned inward. The **Cultural Revolution** of 1966 to 1976 diverted attention from increasing economic productivity to protests against unchanging local authorities and leadership cliques. Mao's "Little Red Book" summarized his wish for revolutionary actions to purge society. The Red Guards, composed mainly of young adults, carried out his tenets, destroying management structures at all levels of society and trashing cultural relics. Such fear-based tactics kept Mao in power until his death. His policies, however, swept aside many of the short-lived economic gains of the early 1960s in a massive persecution of the educated classes, technical experts, and former leaders.

China Joins the Wider World

After 1976, Mao's successor, Deng Xiaoping, brought in reforms that abolished most of the communes and engaged with the global economic system. His views, expressed by the saying, "It doesn't matter whether it's a black cat or a white cat, as long as it catches mice," enabled the adoption of more open policies in the economy. These policies attracted investment from outside China and generated the world's highest rate of economic growth from the mid-1980s to the early 2000s. Foreign investments led to the rapid economic growth of many coastal regions of China. The central Communist government, however, stayed in power, and its often oppressive measures restrained personal freedoms.

After Deng Xiaoping died in 1997, his successors developed the policies that brought economic growth and involvement in the global economic system. In 1998, further reforms enabled and encouraged families to buy their own housing, leading to a growth of speculative building. The acquisitions of Hong Kong and Macau in the late 1990s highlighted the process of reclaiming long-term Chinese territories and the end of foreign ownership of Chinese lands.

Through this period, many families' material expectations increased. In the 1970s, the desired items were a watch, a bike, and a sewing machine. Through the 1980s, they were TVs, washing machines, and refrigerators. In the 1990s, attention shifted to hi-fi equipment, microwave ovens, and air conditioning units, and by the 2000s focused on computers, cars, and apartments as people were encouraged to buy rather than save. China is asserting itself as a world financial leader through overseas aid and loans to countries in the Caribbean and Africa south of the Sahara. The Beijing Olympics were used to highlight China's rising economic and political power. The 2008 Olympics had a significant influence on Beijing's economic development, environment, and the growth of the country's advertising, television, Internet, mobile phone, clean energy, and sports sectors. Although the games made a profit of US$170 million and China won the most gold medals (51) of any country, they also called attention to China's lack of freedoms of religion, speech, and assembly and the environmental problems caused by its rapid industrialization and urbanization.

Farming and Rural Living in the 2000s

Chinese farming is both land and labor intensive as the country attempts to feed 21 percent of the world's population from 7 percent of the world's arable area. Output per unit of land is high. Most agricultural land is cropped two or three times per year, and competition is increasing between land-intensive crops—wheat, corn, soybeans, and cotton in the north and rice and sugarcane in the south—and labor-intensive crops that produce greater value per unit of land, such as vegetables and fruit.

The more profitable specialist crops, including industrial crops (cotton, soybeans), fruit, and vegetables, together with livestock products, threatened the output of some of the traditional staples, particularly wheat and rice. By encouraging grain growers to increase yields by using more imported potash fertilizer, China produced enough grain to put annual surpluses into storage.

In the 1980s, commune-based agriculture gave way to the **household responsibility system**, allowing families to decide which crops to grow. Small groups entered short-term contracts, leasing land to achieve production quotas set by the government. Farmers could sell any surplus crops or animal products in local markets (Figure 4.16).

While successful in terms of increasing production and raising incomes in many rural areas, the individualistic policy of household responsibility led to the neglect of some aspects of rural land use that communal working had encouraged, including the long-term processes of planting woodland and maintaining road and canal infrastructure. Since the early 2000s, other problems facing agriculture include long-term soil erosion and the expansion of urban-industrial areas that removed good land from production. As China transitions from a rural semi-subsistence to urban economy, its food consumption patterns will change, shifting demand from self-grown rice, wheat, and vegetables to more fish, meat, and processed foods.

Growth in Chinese Manufacturing

After 1978, the drive to greater efficiency and an open-door policy to encourage foreign investment became more important than self-sufficiency, and led to what has been termed the "Great Leap Outward." Today, China's manufacturing sector comprises a complex and uncertain intermarriage of local enterprise, foreign multinational capital, and public ownership. The country's industries grow rapidly in southern and coastal China, but those in the interior lag behind and need new plant and infrastructure. China's economy is now a mixture of state-controlled enterprises, military-industrial diversification, and

Figure 4.16 **China: rice paddy near Yinchang, Chang Jiang Valley.** A farmer uses a water buffalo to prepare land for planting rice. Nearby paddies already have rice plants. The older structure that housed both humans and animals has been augmented by a newer house for the farmer's family. Although Chinese farmers include the poorest people, many have experienced better incomes in the last 20 years. *Photo: © Michael Bradshaw.*

foreign multinational and local investment in manufactured goods for export.

From the 1980s, reduced central government subsidies and the "management responsibility system"—the industrial equivalent to the household responsibility system—placed more decision making in the hands of firms that were still government owned. By the early 2000s, 95 percent of industrial production sold at market prices. The central government, however, now focuses its own investment on only 1 percent of existing publicly owned enterprises, allowing others to be merged, taken over by workers, or go bankrupt.

In the late 1990s, bank crises due to overlending affected much of the rest of East Asia, slowing the rapid economic growth rates in the region's countries and reducing the markets for Chinese-made goods. Part of the problem faced by other East Asian countries was the expansion of Chinese output, causing oversupply to common markets.

One of the projects initiated by the Chinese government to support industry and transportation is the massive Three Gorges project on the Yangtze River. China has the world's greatest hydroelectricity potential, but even if it were all developed, it would supply just 6 percent of the country's needs. The building of the Three Gorges Dam from 1993 to July 2012 illustrates the environmental and social issues facing major hydroelectricity projects. The outcome should generate enough electricity to save the yearly burning of 45 million tons of polluting high-sulfur coal, and it will improve flood protection and navigation. The dam, however, created a lake 600 km (450 mi.) long below Chongqing, and 1.3 million people had to be resettled. Farmers on good valley land were moved to poorer uplands, and submerged industrial towns were rebuilt on higher land. But this project is Chinese government policy, and it was completed in spite of misgivings from environmentalists and some internal advisers.

Mongolian Isolation

Mongolia is a landlocked country of mountain ranges, the Gobi Desert, and semiarid grassy steppes akin to the northern steppes region of China. It has a small population and struggles to maintain its independence as a buffer state between China and Russia. The capital, Ulan Bator, is connected by rail to both China and Russia.

After being a Chinese province, Mongolia became independent in 1921 with Soviet Union backing, and a Communist government was installed in 1924. The Soviet Union modified Mongolia's East Asian character, replacing the Mongolian alphabet with Cyrillic script and reorganizing the education system. Mongolia became dependent on Soviet aid and trade. Russia's military forces were withdrawn only in 1992. In 1996, a coalition of democratic groups won an election to introduce reforms. After they lost Russian support, the Mongolian people strove to earn a living while attempting modernization in a landlocked, semiarid environment. In 2000, the former Communist Party was overwhelmingly reelected on the basis of policies that combined reform, social welfare, and public order.

Mongolia's Natural Resources

Mongolia has a strategic location in the heart of Asia but restricted economic potential. Many people still gain a living from herding livestock on the semiarid grasslands. Farmers responded to privatization and deregulation of meat prices in the early 1990s by massively increasing their herds. Cropland areas declined as agricultural cooperatives were split and land abandoned because there was no equipment to cultivate it.

Mongolia has minerals such as copper, gold, uranium, iron, coal, zinc, and molybdenum. Experts say that it has oil and rare earth minerals, too. The Republic of Mongolia is planning to leverage its natural resources to boost socioeconomic growth. Mining companies from around the globe are seeking to do business there, and the country hopes to raise US$25 billion in foreign investments by 2016. The mining boom led to GDP increasing by 7 percent in 2010. Mongolia calls itself a "wolf on the move," likening the growth of its economy to that of the Asian Tigers (Singapore, Taiwan, South Korea, and Hong Kong) in the 1980s. But its economy is likely to fluctuate with the demand for and price of minerals and raw materials such as cashmere. Although Mongolia is actively building economic ties with Russia, Australia, Canada, the United States, Turkey, Japan, and other countries, its main trading partner and the primary market for its mineral resources continues to be China.

Taiwan

Taiwan has 10 times the population on one-fourth the area of Mongolia. It is an island country with a mountain "backbone" along its eastern coast. At the end of the 1895 war, victorious Japan took Taiwan, along with Korea. After World War II, Taiwan returned to China. In 1949, Kuomintang forces escaping from China took control of Taiwan. The Kuomintang leaders brought their families to join several thousand military personnel who had been stationed on Taiwan to quell anti-Kuomintang riots. Taiwan prospered with high tech export industries and became a major investor in mainland Chinese industries.

Taiwan is densely populated. Originally settled by aborigines of Malay-Polynesian origin, the island's population today is nearly all of Chinese origin. The Taiwanese have demographic characteristics similar to those of other materially wealthy countries, including life expectancies of 75 years. Taipei is the capital and largest city of Taiwan, which has nearly three-fourths of its population living in urban areas as a result of the growing industrial and service-centered economy.

While Taiwan maintains its independence, the People's Republic of China insists it remains a part of the mainland country and intends to reincorporate it. The PRC attitude places Taiwan outside the normal niceties experienced by independent countries, such as membership in the United Nations. For 50 years after 1949, the PRC threatened to invade Taiwan and backed off only under U.S. military threats.

China's leadership wants to apply to Taiwan the "one country, two systems" formula adopted for the incorporation of Hong Kong. Around half the Taiwanese population already supports reunification with China based on a Hong Kong type of provision that would preserve Taiwan's prosperity and its role as a major investor in Chinese economic growth. Many Taiwanese, however, are watching events in Hong Kong before deciding whether to become fully part of China again. In the meantime, they continue to strengthen their military capabilities. Supporters of Taiwanese independence argue that it is different from Hong Kong: Taiwan is 170 km (100 mi.) offshore, has no built-in deadline date for return to China, and has a more mature democracy than Hong Kong and much of the rest of East Asia. The independence position has guarantees of Western support, especially from its strongest ally, the United States. But pressures grow on Taiwan as many of its best workers and managers move to the mainland, unemployment increases, and World Trade Organization members ask questions about the Chinese being able to use the Taiwanese ports and airports freely.

In the early 2000s, Taiwan became more enmeshed in mainland China's economy. China is Taiwan's third largest trading partner, following the United States and Japan. Nearly half of the investment in Shanghai's Pudong development project (see Figure 4.2) is Taiwanese. Although Taiwan's government officially limited Taiwanese investments in China, over US$60 billion followed indirect routes through Hong Kong, despite few return profits.

Taiwanese Economy

After 1949, the Kuomintang government encouraged rapid industrialization, rural change, and urbanization. The increasing rice surplus gave way to more diversified farm products, including sugar, tea, vegetables, and fruits. Expansion of urban-based industrial jobs and buildings, plus the mechanization of farming, reduced the farm population from 6 million in 1965 to fewer than 4 million in the 1990s, while an initial agricultural trade surplus gave way to increasing food imports.

Starting in the mid-1960s, export processing zones focused attention on manufacturing for export. By the 1980s, the emphasis switched to high technology developments. Private ownership of firms increased from 52 percent in 1960 to over 80 percent in the 1980s. The government retained control of banks, interest rates, and exchange rates to maintain a consistent financial environment. It cultivated a strong domestic manufacturing industry, at first based on transistor radios, followed in the 1970s by the assembly of Japanese electronic goods, and then promoted links with companies such as Philips (Netherlands) to build its own integrated circuit industry. Taiwan now leads the world in such technologies as integrated circuits, laptop computers, modems, and data communications.

By the 1990s, Taiwan had one of the largest world trade surpluses and was able to export capital, particularly to China and Thailand. As is the case with Japan and South Korea, direct investment in Chinese manufacturing contributed to many "Taiwanese" products. As Chinese costs undercut Taiwanese producers, Taiwan continues to invest in new technologies such as biotechnology and new integrated circuit designs. Such investment, however, demands so much capital that it poses major risks for the Taiwanese government such as overspending and eventually losing in competition with the Chinese. Taiwan's furniture and textile industries have already moved to mainland China, leaving increased unemployment on the island, and its electronics industries are under threat.

Still, Taiwan runs a large trade surplus, and its foreign reserves are the world's fourth largest, behind China, Japan, and Russia. Taiwanese President Ma Ying-jeou, who was elected in 2008, advocates closer ties with the mainland. Although Taiwan lost some 4,400 manufacturing jobs in 2012, this was compensated by over 77,000 new jobs in the service industry, most of them in hotels, restaurants, and stores. About 2.3 million tourists from the PRC were expected to visit Taiwan in 2012, giving its economy a boost. Closer links with the mainland bring greater economic opportunities to Taiwan, but also pose new challenges as the island becomes more economically dependent on China, while political differences remain unresolved.

4.5 Contemporary Geographic Issues

China: A New World Power

China has the potential to become a major world power alongside Japan. It is the world's most populous country, and between 1980 and the early 2000s, its "open door" policies were the basis for economic growth at the fastest sustained rate of any country. This economic growth multiplied the value of its output fivefold and raised 600 million people out of absolute poverty. Chinese factories now make many goods bought in Western stores, from electronics to clothing.

In 1991, around 100,000 private businesses employed 1.8 million people; by 2001, 24 million worked in 1.5 million large private businesses and around 30 million in smaller, individual concerns. In that decade, private industry grew from a few percent to 40 percent of total Chinese industrial output.

In 2010 China had the most foreign direct investment among the world's less materially wealthy countries, reaching a record high of US$105.7 billion that year. Companies from the United States, Europe, Japan, South Korea, and Taiwan moved much of their low-end assembly to China, although there is a rise in high-end manufacturing and some services as well.

To support such economic investment, China's huge supply of labor includes college graduates and computer engineers who are paid one-tenth as much as their Taiwanese or Japanese counterparts. However, Chinese entrepreneurs themselves face difficulties. Their government is slow to expand the list of companies allowed to go public with share offerings. Many entrepreneurs find it difficult to persuade Western financiers to support them in joint projects after intense scrutiny by external analysts and investors. Most investments in China come from Chinese in Hong Kong and Taiwan, and from other Chinese living abroad—but also increasingly from Japan.

Western and Asian multinational corporations trading in China include oil companies, electronics groups, auto manufacturers, and restaurant chains. For example, Kentucky Fried Chicken (Figure 4.17) opened a huge outlet in China next to Tiananmen Square in central Beijing in 1987. Despite the 1989 incident in this square, when Chinese army tanks attacked peaceful demonstrators, McDonald's followed in 1992, and both KFC and McDonald's continue to expand. The first Starbucks outlet

Figure 4.17 China: central Beijing. A KFC outlet alongside Chinese stores, together with a BMW and lots of bicycles: a mixture of global economy and local trends. *Photo: © Michael Bradshaw.*

in China—serving coffee in a tea-drinking country—opened in 1999, and by 2010, 376 shops held franchises in 13 Chinese cities, although concentrated mainly in Beijing and Shanghai. The most contested site for Starbucks was in a Beijing Forbidden City souvenir shop, with 70 percent of 60,000 people surveyed opposing it. But it continues to serve coffee, and Starbucks proposes to have 1,500 outlets in mainland China by 2015. Such symbols of an intruding global economy, however, generally excite little public antagonism. Many Chinese regard the expansion of multinational food franchises as a positive sign that they are involved in the global economy.

China joined the WTO in November 2001. The Chinese leaders expected WTO membership to boost exports and foreign investment, but it also forced its industries to become more competitive in world markets in both quality and price. WTO rules require China to cut import tariffs and allow foreign businesses to compete in its highly protected areas such as telecommunications, automaking, insurance, and banking industries. China has, however, been accused of flouting WTO rules on steel and other commodities. A dispute between China and the United States over a case of import duties levied by China on US-produced specialty steel was settled in 2012 with the WTO ruling in favor of the United States.

China's GDP growth has surpassed that of the Asian "miracle" economies. It is the world's top location for outsourced manufacturing. It has witnessed an exponential increase in its exports and trade. In 2011, China was the United States' second-biggest trading partner (after Canada), and the European Union's second-biggest trading partner (after the United States). China is Japan's biggest trading partner, although Japanese exports to China fell sharply in 2012 in the wake of political tensions between the countries.

Although growth in sectors like construction and manufacturing is slowing down, China's industrial growth has expanded into areas such as high-speed trains, pharmaceuticals, and nuclear power plants. The country is looking to become a global power in software and services as well. It is making inroads into outsourced global services such as processing medical claim forms, applications for loans and credit cards, and even grading examinations from the developed countries of the world.

Following a strategy of more regionally balanced development, new centers for back-office work are located away from the coasts—for example, in Dalian, a city in northeastern China, and in Xi'an, the capital of Shaanxi province. Many of Dalian's residents speak Japanese and Korean due to historical connections, giving it an advantage in performing back-office services for these two countries. The Xi'an High-Tech Industries Development Zone is one of China's largest technology parks. Currently housing 7,500 companies, it covers 35 km² and is expected to expand to 90 km².

China's disciplined and trained work force includes a large number of university graduates well educated in basic computing and mathematics, and the lower wages paid to Chinese engineers and programmers make the country highly competitive in the global arena for this work. Well-known information technology companies like IBM, Hewlett-Packard, Microsoft, Siemens, and Infineon all have been in China for several years

now. However, it may take China years to catch up with India, as the latter has the advantage of a large English-speaking, educated population and greater dominance in research and design.

China has an expanding influence in Africa, with bilateral trade reaching a record US\$166 billion in 2011. China has also helped in infrastructure development on the continent through the construction of roads, railways, and airports, but it is also a major supplier of conventional arms to African states. Critics say that these are means for it to acquire ownership of Africa's natural resources.

The countries surrounding China remain wary of its military and new economic strength. China is currently involved in territorial disputes with Japan, Vietnam, Taiwan, the Philippines, Malaysia, and Brunei.

Environmental Problems in China

In Maoist China, every effort was made to use nature to build the Chinese economy, largely without any recognition of the environmental disasters, deforestation, air and water pollution, and even famine that such policies would create. Several of these policies remained in place as China developed its industries and urbanization continued.

Water Shortage and Pollution

Two-thirds of China's 669 cities suffer from water shortages, more than 40 percent of its rivers are severely polluted, and over 300 million people in rural areas lack sufficient drinking water. These problems are compounded because China is one of 13 countries in the world with the most limited water resources. Additionally, water resources are distributed unevenly in China, with the 44 percent of its population that lives in the north having access to only about 15 percent of the water resources. Glacial retreat due to climate change is further reducing the water supply. China's per capita availability of renewable water resources is about 25 percent of the world average, while water consumption per unit of GDP is three times the world average due to industries that are intensive in their use of water, outdated technologies, low reuse rates, and waste. As urbanization and development continue, it is expected that the gap between the supply and demand for water will continue to grow. Water shortage is exacerbated by pollution of China's rivers and lakes due to urban sewage and industrial wastes. Harmful algae blooms have also increased over the last decade, and 80 percent of China's lakes suffer from eutrophication.

Air Pollution

China is the world's biggest emitter of carbon dioxide. The air of about 11 percent of China's cities is heavily polluted with particulate matter, sulfur, nitrogen dioxides, and ozone, and only 56 percent had air quality that met the standards of the State Environmental Protection Administration of China (SEPA). Much of the air pollution is related to industrial activity, the energy sector, urbanization, and traffic (Figure 4.18). The burning of bituminous coal, the main source of energy in China, releases smoke and toxic gases into the atmosphere. China is likely to remain dependent on coal for the next

Figure 4.18 Smog engulfs the city of Xi'an, a provincial capital in China. This air pollution is caused by heavy industry, massive construction projects, traffic, and the use of coal as a fuel. *Photo: © Justin Burner/Getty Images RF.*

several years. Acid rain resulting from air pollution is particularly prevalent in southern China. Approximately 49 percent of China's population is rural, and nearly all rural households are dependent on biomass and coal for household cooking and heating. Indoor air pollution from the burning of solid fuel in China is responsible for over 400,000 premature deaths each year.

As the number of motor vehicles in China increased from 6.2 million in 1990 to over 100 million in 2011, automobile exhaust has become another important contributor to rising air pollution. A study showed that high ozone levels in Shanghai were related to deaths caused by cardiovascular disease, especially in the cooler months. The Chinese government is trying to increase energy efficiency and institute measures that will prevent further deterioration of air quality. Limiting the use of automobiles on the streets of large cities like Beijing and Shanghai has been attempted, but with limited success.

Urbanization

Rapid urbanization has taken place since the economic development and socioeconomic changes of the 1980s. More than half (51 percent) of China's population now lives in cities. The consequent growth in the production of domestic waste in cities and its inadequate treatment have resulted in serious environmental problems in the air, soil, and water. China's urban population is predicted to reach nearly 1 billion by 2025 as people continue to move from rural to urban areas, further exacerbating the problem.

As increasing amounts of agricultural land are converted to urban use, China strives to achieve higher agricultural production through the use of fertilizers and pesticides, which have contaminated farmland and escalated water pollution. In fact, China has become the world's largest consumer of fertilizers and the second largest consumer of pesticides. In some areas of China, crops are irrigated with water which has been polluted with heavy metals such as cadmium, arsenic, chromium, and

lead due to upstream mining activities. For example, mercury mining in Guizhou Province in southwestern China has resulted in significant pollution of the local ecosystem. The impacts of environmental problems are significant and include damage to human health, social conflicts, and economic losses. Many premature deaths are traced to pollution, while the deterioration of the environment is an important cause of social unrest in Chinese society. Economic losses from environmental pollution and ecological damage were estimated to be between 7 and 20 percent of the annual GDP during the past 20 years.

Initiatives for Environmental Protection

The Chinese government has initiated many efforts to control its environmental problems. Since the early 1980s, a series of national plans, policies, and laws have been enacted. In 1983, China made environmental protection one of its basic national policies. In 1994, a strategy was laid out to achieve sustainable development. Two years later, the first five-year plan on environmental protection was developed. More recently, a balanced relationship between humans and nature was proposed as an overarching concept. At the same time, efforts have also been made in ecological restoration and soil and water conservation, control of desertification, flood control, climate-change mitigation, and conservation of biodiversity. In 2011 the government outlined a plan to expedite water conservation and achieve the sustainable use and management of water resources in a decade.

China is trying to change its mode of economic development and promote a sustainable economy. Progress has been made in many areas, including biodiversity conservation, the improvement of environmental policy, law, and funding mechanisms, and the implementation of international environmental treaties. In 2012 China pledged to spend US$372 billion on energy conservation projects and antipollution measures over the next four years. That same year it also cut a deal with the EU to reduce greenhouse gases through the development of Chinese emissions trading schemes. There are numerous instances of regional and local environmental rehabilitation successes. With the exception of Lhasa, Tibet, China's top five least polluted cities in 2011—Haikou in Hainan Province, Nanning in Guanxi Province, Fuzhou in Fujian Province and Funming in in Yunnan Province—were all in southern China.

But China still faces huge challenges in resolving its environmental problems at the national level. These challenges include the limited reserve of natural resources, the already tense human–nature relationship, the strong drive for economic development and urbanization, and the inadequacy of institutions and legislation to enforce environmental regulations.

Globally Connected Cities of East Asia

The growth of global and intraregional trade in East Asia led to the emergence of global cities as essential nodes linking this region to the rest of the world through business and political exchanges. Tokyo, Japan, is one of the world's top four global cities—with New York, London, and Paris—while Hong Kong is growing in importance. Seoul (South Korea), Osaka (Japan),

Taipei (Taiwan), and Beijing and Shanghai (China) are also major cities with significant employment growth in international business services.

Tokyo

Tokyo, the seat of the Japanese government, displaced Osaka as the country's commercial capital after World War II, when many Japanese companies moved their headquarters to the city. The growth of Tokyo paralleled the rise of Japan in the global economy. In the 1970s with the expansion of the Japanese economy and associated social change, Tokyo was transformed into the now familiar cityscape with skyscrapers, expensive shopping districts, hotels, and corporate headquarters. The Ginza area in Tokyo is Japan's most exclusive and expensive shopping and entertainment district. It has the most expensive real estate on earth; 1 square meter here costs US$10,000 (Figure 4.19). Today Tokyo is a world city that is a control point for international finance and transnational production and marketing. In 1960 only one multinational headquarters was located in Tokyo; by 1997 the city housed 18 of the top 100 corporate headquarters, including five of the largest.

In 1986, as well as being the center of government administration, Tokyo became the world's second financial center after New York and ahead of London. In the early 2000s, Japan was the world's third-largest economy with 8 of the 10 largest world banks, having built its world prominence on huge capital surpluses. Japanese banks have continued to increase their absolute and relative asset strength. Top-ranked global banking groups such as the Mitsubishi Tokyo Financial Group and the Mizuho Financial Group have their headquarters in the city. The Mitsubishi Financial Group merged with UFJ Holdings, another large Japanese bank, to form the world's largest privately owned bank with about US$1.6 trillion in assets. Headquartered in Tokyo, the megabank has rapidly increased its overseas presence.

Tokyo has the largest metropolitan economy in the world; its GDP of US$1.32 trillion would put it in seventh place among national economies of the world. It is the center for Japan's transportation, publishing, and media industries. Nearly 70 percent of Japan's anime production companies are located in Tokyo, and recently the first global animation center was opened in the city. It is also an important center of international advertising, housing five of the world's top 20 advertising agencies. However, critics claim that Tokyo continues to be a predominantly Japanese rather than a global city, lacking a cosmopolitan culture and truly global linkages. The city does not attract as much international labor (less than 2 percent of Tokyo's labor force is foreign) or foreign capital.

Seoul

Seoul has been the capital of Korea for over 600 years. It evolved from a monocentric city during Japanese occupation to a growing metropolis with multiple centers during the latter half of the 1990s (Figure 4.20). Seoul houses most of the country's industry and is connected via "technobelts" (systems to link industry and research via different modes of communication) to new centers of research and development.

Figure 4.19 **Japan: Ginza area in Tokyo.** The Sukiyabashi pedestrian crossing in Tokyo's ritzy shopping and entertainment district. *Photo: © Corbis RF.*

Figure 4.20 **Seoul, capital of South Korea.** The Chongno tower and other modern buildings dominate the center of the city, which was largely destroyed in the 1950s war against North Korea. The volume of road traffic reflects the country's economic growth. *Photo: © Jose Fuste Raga/Corbis.*

In 1996 South Korea abandoned its policy of state-led capitalism and switched to a neoliberal market-oriented economic model. Following the deregulation of capital outflows and tremendous increases in foreign direct investment, Korean banks, investment and insurance firms have joined the global financial market. As Seoul's economy globalized, its connectedness with the rest of Asia and the world increased. Seoul-based multinationals such as Samsung, Hyundai, Daewoo, and LG have led the way in making overseas investments, reaching beyond investment in China and Southeast Asia. They now have operations in Russia and the countries of Eastern Europe and Africa. Numerous foreign banks and financial organizations operate in the city. Seoul hosts 71 percent of Korea's overseas-based service industries, half the nation's international hotels and trading companies, and nearly all of its communication services. Practically all of the country's stock brokerages, foreign bank offices, offices of foreign media and broadcasting networks are in the city. Seoul has also become an important international airline hub. Increases in the number of air linkages to cities around the world and the number of international flights to and from Seoul speak to the city's increasing global connectedness.

Employment growth in international business services in Seoul has been significant over the last 20 years. In addition to flows of global capital, Seoul has also witnessed an influx of foreign migrant workers. The Industrial Training Program instituted in 1993 permits recruited foreign labor to work for specified lengths of time in low and unskilled jobs. Workers from developing countries of Asia, including China, Vietnam, Bangladesh, the Philippines, Nepal, and Indonesia, came to South Korea through the program, although many continue to stay as illegal workers.

The Seoul Metropolitan Government identifies the four future images of Seoul as a "historical, humane, cultural, and world city." As stated previously, Seoul exports Korean popular culture to East and Southeast Asia. The 1988 Summer Olympic Games held in the city and the 2002 World Cup soccer finals, co-hosted in Seoul and several Japanese cities, further enhanced Seoul's international profile and its global reach via television, webcasts, and other media.

Hong Kong

Hong Kong comprises an island (centered on Victoria) and mainland peninsula (Kowloon) that were ceded to the British in the mid-1800s. A more extensive area of mainland, the New Territories, was leased to the United Kingdom for 99 years in 1898. All returned to Chinese rule in 1997. The Basic Law, under which Hong Kong operates within China, guarantees retention of its free-market system for 50 years and treats the city as a complementary financial center to Shanghai. By 2007, rivalry continued between Hong Kong and Shanghai over becoming China's dominant financial sector. While Hong Kong had its own dollar currency, linked to the U.S. dollar, and gains in international markets, Shanghai's currency is the Chinese yuan, not easily exchanged with Hong Kong dollars, but giving it great advantages within China.

The world's busiest port, Hong Kong has an excellent natural harbor with deepwater access and extensive water frontages (Figure 4.21). Goods received from inland China are sent to worldwide destinations, while others collected from the rest of the world are sent into China. These are the features of an **entrepôt**. As the Chinese economy opened up and expanded after 1976, particularly in the special economic zone around Shenzhen, the value of Hong Kong's trade multiplied sevenfold. At the time of political transfer, Hong Kong was already deeply enmeshed with interior China in trade, investment, and personal contacts. As part of China, Hong Kong began with advantages over its internal and external rivals for dealings inside the growing country. New ports opening along the Zhu Jiang delta in the early 2000s relieved the huge pressure on Hong Kong's port facilities.

The population of Hong Kong, 7.6 million in 2012, is almost totally Chinese. As the main urbanized area became overcrowded from the 1950s, new towns spread into the New Territories. They now accommodate around one-third of the total population. The high population density causes half of government expenditure to go for roads and public transportation. In the late 1970s, electrification upgraded the railroad route inland to Guangzhou (Canton), and its use increased. Two cross-harbor tunnels and an underground transit system opened in the 1980s. A new Hong Kong International Airport was built offshore north of Lantau Island and linked by bridges to the mainland. It replaced the older airport in Victoria Harbor that could not be expanded on its limited site.

Hong Kong's growth as a manufacturing center took off in the 1970s, and by 1990, manufacturing provided one-third of all employment. The production of electronic goods and scientific equipment increased, but the textiles and clothing industries declined because of foreign competition and rising labor

Figure 4.21 **People's Republic of China: Hong Kong.** View from Victoria Peak on Hong Kong Island to Kowloon across the harbor. Hong Kong has some of the highest population densities in the world. *Photo: © Alasdair Drysdale.*

costs in Hong Kong. In the 1980s, manufacturing moved out of Hong Kong and expanded just over the border in southern China, with its cheap land and labor costs. The higher wages in factories and service industries force farming labor costs up because farms have to match the wage levels.

By 2000, the shift out of manufacturing as factories moved inland led to service industries accounting for 85 percent of Hong Kong's GDP and 80 percent of employment (from around 50 percent in 1980). Hong Kong is one of the world's major finance centers and the third-largest gold market, accounting for up to 15 percent of the world total. Hong Kong provides expertise in finance, trade, and public administration and has an important stock market. It is the regional headquarters for 800 foreign companies. Tourism became Hong Kong's second major industry. Many new hotels catered to this influx of people that added an extra 10 percent to the population at peak times. After Hong Kong's transfer to China in 1997, tourism continued at a high rate, reaching 42 million visitors in 2011. Today, Hong Kong is considered a city with a high degree of global connectivity. Its geographical position and history make it a prime location to service clients in the growing Chinese market. Hong Kong is poised to make history as a green and sustainable city. The city's agenda for cleaner air, higher energy efficiency, and modern waste management is pursued through legislation to raise the energy efficiency in all its buildings, tax incentives to promote green transport and encourage the use of electric vehicles, and the preservation of its rural areas. After Haikou, Hong Kong was China's least polluted city in 2011.

Beijing

Beijing, a city of a mere 1.4 million people in 1949 when it was reinstated as the political capital of China, grew to a city of over 15 million by 2010. An ancient settlement, over the course of nine centuries it has been governed by Manchu, Mongol, and Chinese rulers and was occupied by the Japanese. Beijing's walled Old City, 25 square miles in area, was built on a grid system with narrow lanes (hutong) lined by single-story courtyard homes (Figure 4.22). Some of these still stand, although most hutong disappeared as they were replaced by wide streets and high rises. As the city grew, suburban developments pushed it outward. Today Beijing occupies some 6,000 square miles, and the Old City is encircled by six concentric beltways or ring roads that facilitate transportation in the enormous metropolis. The city's population, which is still mostly concentrated in the core, continues to increase and is projected to exceed 20 million by 2025. About 40 percent of Beijing's population is composed of migrants.

In the 1980s after Deng Xiaoping's market reforms, Beijing became an industrial center with nearly 150 types of industries. It still has China's largest petrochemical industry and is the leading producer of rubber products, plastic, and refrigerators; it is second in pig iron and washing machines; third in power generators, wool cloth, cars, and color televisions; fourth in internal combustion engines; and fifth in sewing machines and beer. However, many of the older heavy industries were closed or moved to neighboring Hebei in preparation for the 2008 Olympics as officials attempted to reduce pollution. New specially

designated industrial parks here include Zhongguancun Science Park, Yongle Economic Development Zone, Beijing Economic-technological Development Area, and Tianzhu Airport Industrial Zone. The share of the service and high-tech sectors here is on the rise, and finance makes up the largest segment (14 percent in 2010) of Beijing's GDP. The city is second only to Tokyo in the number of Fortune Global 500 companies and has the most banks of any Chinese city. It also has the world's second busiest airport, the best universities in China, nearly 100 Starbucks, and a new subway system. Despite the government's pledge to improve air quality, Beijing continues to suffer from air pollution that can be traced to automobile traffic, industries in towns surrounding the city, and seasonal dust storms from the deserts to its north and northwest. In a 2011 World Health Organization study of the world's most polluted cities, Beijing was ranked the fifth most polluted city in China, and the world's 10th most polluted capital, ranking lower than Lagos in Nigeria.

Human Rights

Traditional Confucian concepts of rights emphasize the duty of the state to ensure stability and encourage economic prosperity rather than civil liberties for its citizens. Western countries have criticized China's record of human rights violations since the formation of the Communist state, pointing to the 25 to 30 million deaths that occurred during the Great Leap Forward (1958–1961) when a combination of natural and human-caused disasters resulted in a drastic reduction in food grain production and caused a famine. The Chinese government did not seek international aid at this time even as its people were dying. The

Figure 4.22 **Liangguochang Hutong, Dongcheng District, Beijing.** Hutongs are alleys formed by lines of traditional courtyard residences. Many Beijing hutongs have been demolished to make room for new buildings and roads. *Photo: © Fotosearch/Getty Images RF.*

lower numbers in the 50- to 54-year age group in the 2012 age-sex diagram (Figure 4.23) for China reflects the impact of this major population disaster.

North Korea has one of the worst records for violations of human rights. Dissidents and family members of suspected dissidents have disappeared. Offenses punishable by death include "treason against the Fatherland" and "treason against the people." The country has experienced famine several times, and millions of people are currently going through the worst hunger in a decade. Many are forced to scavenge for wild foods. The government fails to ask for adequate international aid, even going so far as refusing to accept food aid from the United States after March 2009.

Civil Liberties and Political Rights

The Chinese government handles the tensions associated with globalization in various, often contradictory, ways. While educated Chinese now have increased opportunities to shape careers, go abroad, or pursue research interests without party interference, the army and security forces continue to respond firmly to dissent and unrest. The justice system is poor and subject to political pressures that prevent magistrates from acting fairly: for most people outside political and business elites, trials and imprisonment remain arbitrary.

Amnesty International estimates that over 500,000 people are being detained in China's jails without charge or trial, and millions are unable to access the legal system. Human rights defenders face house arrest, harrassment, and even imprisonment. The Internet and other media are highly censored. Multiparty democracy is not yet mentioned as even a remote possibility. Writer and human rights activist Liu Xiaobo was awarded the 2010 Nobel Peace Prize while serving an 11-year prison sentence for opposing Communist single-party rule in China. He remains imprisoned, although many, including Mo Yan, a fellow citizen of China and the 2012 Nobel Prize winner for literature, have called for his release.

The PRC government argues that human rights should focus on "positive rights" such as adequate food, shelter, and clothing rather than on "negative rights" of freedom of speech, press, and assembly. "Positive" rights demand a more active role by the state, while "negative" rights suggest noninterference by the state. Editors of newspapers and magazines that stray from government directives or are critical of the People's Republic are fired or their publications banned. Opposition and organized protest against the government are not tolerated. The brutal military operation to crush a peaceful protest by students who occupied Tiananmen Square in 1989 for seven weeks demanding democratic reform was condemned by the international community. Incidents of torture, forced confessions, and forced labor are also reported. According to Amnesty International, thousands of people are put to death in China for a variety of crimes, including those of a nonviolent nature. China has the most executions of any country, accounting for almost 80 percent of all such deaths globally. While acknowledging deficiencies, China asserts that its human rights situation is improving.

The state in North Korea controls all media, and retaliation for dissent and noncompliance is brutal. Listening to unauthorized broadcasts and retaining or disseminating information can result in "correction" in labor camps from two to five years. Officials from the Ministry of Public Security regularly conduct inspections in private homes to ensure that citizens comply with state laws. Freedom of religion is also severely restricted. North Koreans sent to prison camps are often subjected to torture and inhuman treatment, including forced hard labor, inadequate food, beatings, lack of medical care, and unhygienic living conditions, as a result of which many prisoners fall ill and die in custody or soon after release.

Minority Rights

Several ethnic minority groups in China, Japan, and Korea reportedly receive poor treatment. Minorities, such as the Tibetans, Mongolians, and the Uygurs who live in rural western China, are politically and culturally suppressed for fear that they will push for independence (Figure 4.24). At the same time, significant numbers of Han Chinese are moved into these areas, arguably to help develop them, but with the result that

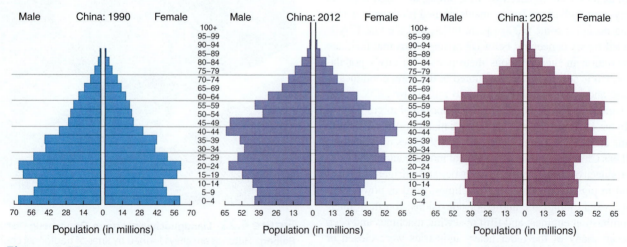

Figure 4.23 **China: age-sex diagrams for 1990, 2012, and 2025.** *Source: U.S. Census International Database: http://www.census.gov/population/international/data/idb/region.php.*

Figure 4.24 Urumqi, Western China. Ethnic Uygur men leave a mosque after a religious service in Urumqi in the largely Muslim Xinjiang Province. Note the mix of traditional and Western attire.
Photo: © John Ruwitch/Corbis.

local cultures and populations are diluted. Journalists and local activists report high levels of repression in the minority strongholds, including the confiscation and burning of local history texts, forced labor on government projects without pay, and suppression of information about human rights abuse, including arbitrary detention and torture. In Japan, social outcasts such as the Burakumin, native ethnic minorities like the Ainu and Okinawans, and foreign ethnic minorities like the Koreans still suffer from discrimination. They are often not considered for positions in good companies and are trapped in poorly paid jobs with little hope for advancement.

While religious freedom is guaranteed in China's 1982 constitution, members of religious groups such as the Falun Gong are arrested and imprisoned, sometimes for long periods without trial. All religious organizations must register with the government. Contrary to the official position on cultural freedom for minority groups, young people in Xinjiang are not allowed to attend mosques or receive religious instruction until they have completed nine years of compulsory schooling. Similar restrictions occur in Tibet, where the Buddhist community is repressed, as are Christians who worship outside state-sanctioned churches. The government of China restricts Internet access to what it considers sensitive subjects, such as websites with content that is religious, pertains to Tibetan and other independence movements, freedom of speech, democracy, and police brutality as well as indigenous and minority rights. Censored sites include Voice of America, BBC, Yahoo!, Google, Geocities, and Wikipedia.

Gender Equality

While most countries in East Asia have made significant strides in achieving gender equality, much remains to be achieved. The age-sex diagram for China shows that there are more women than men in the older age groups in the country but fewer in the younger groups (see Figure 4.23). This gender imbalance is attributed to a clash between the one-child policy (Table 4.3) and a traditional preference for sons in Chinese society, resulting in the abortion or even the killing of girl babies in order to save the family "space" for a boy child. The resulting disproportion between the sexes could pose major social problems for China. Parents and other family members often indulge every whim of single male children, leading to what is referred to as

Table 4.3	**Debate:** Population policies in China
Arguments for One-Child Policy	**Arguments against One-Child Policy**
The one-child policy results in fewer births that place claims on resources.	Family planning is most successful when parents act on their own initiative, perhaps guided by information.
It brings slower population increase and a more balanced population structure in the early stages.	It upsets the population balance between sexes (abortion and murder of baby girls) and age groups (in the later stages when there are small numbers of young people and increasing numbers of older people).
It allows better planning for urban expansion, housing, education, job availability, and transport and utility infrastructure.	It ignores other factors in population growth, including migration, the effect of natural disasters (as during the Great Leap Forward) and diseases, and the impact of political chaos (as in the Cultural Revolution).
It provides control by the central government and local officials in the interests of the whole population.	It often requires draconian methods of control that violate human rights.
It brings a rapid realization of the importance of family planning in the face of imminent overpopulation disaster. Other policies had failed.	It ignores cultural and economic factors that favor large families.
It is important to maintain this process.	The one-child family is at risk and fragile in placing all hopes on a single child: most families would consider two children ideal. Only children are often unduly pampered and obese.

the Little Emperor syndrome: a thoroughly spoiled child with a sense of entitlement. Demographers predict that by 2025, tens of millions of marriageable men will not be able to find wives. The shortage of women is already reflected in the increased trafficking of women and girls from inland China and the neighboring countries of Burma and North Korea for prostitution and marriage.

Mao Zedong in the 1950s said that "women hold up half the sky." Women's position in Communist China improved as their greater participation in economic and political arenas was encouraged. But as capitalism makes greater inroads into China, many of the gains are being lost. Women are increasingly working in manual, lower-paid jobs in agriculture and industry. Women workers in village and township enterprises routinely hold the low-level positions and are the first to be laid off. They also have to retire at age 50, five years before the retirement age for men.

Throughout Japan's modernization, its people maintained a traditional culture that encouraged attitudes based on loyalty to the emperor, family, and workplace. The main Japanese traditions, from Shintoism to Buddhism and Confucianism, agree on the subordinate status of women. Although women's status is changing with their greater education and involvement in the labor force, they still work for lower wages, and few attain management positions or lifetime employment status. In 1985 Japan passed its first legislation against workplace gender

discrimination. Yet, between 1985 and 2008, the proportion of full-time female employees fell from 68 percent to 47 percent, largely because women were moved into part-time and contract positions.

Currently a woman cannot ascend to the Chrysanthemum Throne and become the Empress of Japan. A bill presented to the parliament by Prime Minister Junichiro Koizumi in early 2006 sought to amend the Imperial House Law and allow women to succeed to the throne. Although progressive Japanese and the members of the prime minister's advisory council support female succession, traditionalists and nationalists have so far strongly opposed the idea.

In South Korea as well, traditional Confucian mores have resulted in an overall lower status for women. The country has one of the world's highest rates of sex-selective abortions and homicide of females. Gender disparity is still prevalent in access to education, job opportunities, and social status, although there are efforts toward greater equality by the government and NGOs. The country's female labor force participation has risen to over 42 percent, but like in Japan, many of these workers are in temporary and part-time positions, few reaching top-level and prestigious jobs. An interesting reversal of traditional gender roles can, however, still be found in the coastal villages of Cheju Island, where women dive for marine products such as oysters and seaweed and are considered the main breadwinners of the family.

Geography at Work

China's Landscapes and Global Change

Erle Ellis and his team of collaborators study the impacts of population growth, economic development, and the use of modern industrial technologies and fossil fuels on environment and agriculture in China (Figure 4.25). The country stands as one of the few places where agricultural systems that had sustained high yields continuously for centuries might still be available for study. By using a multiscale approach integrating regional data on land cover, terrain, and climate with higher-resolution satellite imagery, household surveys, interviews with village elders, and field measurements of soils and vegetation, the team measured ecological changes with unique precision at five rural sites selected across environmentally distinct regions of China.

Erle's early hypothesis, that the secret to sustaining long-term high agricultural productivity was through planting nitrogen-fixing legume "green manure" crops and by efficient nitrogen recycling, was disproved. However, he discovered that nitrogen was added to the soil via sediments harvested from canals and purchased oilcakes (residues from pressing oil from imported soybeans and rapeseed). Another important finding was that because of the ready availability of synthetic nitrogen, sediment was now left to fill in village canals and had increased soil nitrogen concentration over time, producing a dramatic net increase in nitrogen storage across the region's rural landscapes. Potentially, this nitrogen "sink effect" has global implications. As carbon

Figure 4.25 Dr. Erle Ellis and his team carry out a humidification study in China. *Photo: Courtesy Erle Ellis.*

storage is tightly linked to nitrogen storage in soils, this nitrogen sink can help explain the "missing sink" for atmospheric carbon dioxide that continues to complicate our understanding of anthropogenic climate change. Erle's research shows that fine-scale local changes in landscape management could have major regional and global consequences that are observable only by detailed site-based measurements.

Bangkok, Thailand. *A street in Bangkok, Thailand's capital and largest city. Despite the introduction of rapid transit systems and an elevated expressway network that serves over 1.5 million vehicles each day, Bangkok's traffic infrastructure has not kept pace with its rapid growth, resulting in frequent traffic jams.*
Photo: Elizabeth Chacko.

LEARNING OBJECTIVES

After reading the chapter, you should be able to:

- Identify Southeast Asia as a region of diverse cultures linked by maritime trade.

- Use a map to locate and describe the mainly tropical climate, landforms, distinctive ecosystems, and natural resources of the region.

- Describe the natural hazards and their effects on people in the region, and describe the human impacts on the environment.

- Trace the history and recent economic development of the region.

- Locate the Mainland and Insular Southeast Asian subregions on a map and explain how these subregions and the countries within them differ in terms of their cultures, population, politics, and economics.

- Describe the cultural, political, and economic linkages among Southeast Asian countries and their degree of global connectivity.

Southeast Asia

Figure 5.1 **Southeast Asia: physical features, country boundaries, and capital cities.** The region, which is situated almost entirely within the tropical latitudes, has the Indian Ocean to the west, the Pacific Ocean to the east, and a land boundary with China to the north.

Elevation (ft.)

below sea level 0 500 1,000 2,000 5,000 10,000

-10,000 -5,000 -500 0

0 mi 500 1,000 1,500

0 km 500 1,000 1,500 2,000 2,500

Lambert Azimuthal Equal-Area Projection

5.1 Diverse Region, Close Internal Ties

Southeast Asia lies between China to the north and Australia to the south, and between the Indian and Pacific oceans (Figure 5.1). It includes the continental countries of Vietnam, Laos, Cambodia, Thailand, and Myanmar, and the essentially island countries of Malaysia, Singapore, Indonesia, East Timor, Brunei, and the Philippines. The region has only one short land boundary (see Figure 5.1) with China (Chapter 4) and India (Chapter 6) in the north. The other boundaries are in the ocean, with stretches of water separating this region from others.

Southeast Asia has an increasingly important position in the globalizing world. The region's location is at a global crossroads where peoples and environments converge and the influence of empire and economic globalization are evident. Traders from China, India, and the Arab world exchanged goods in the region and influenced local human geographic patterns prior to the establishment of European colonial enterprise. Commerce and conquest brought a lasting imprint of Buddhism, Hinduism, and Islam to the region. Southeast Asia is a geographic transition zone from the Indian Ocean to the Pacific Ocean and from the Northern to the Southern Hemisphere. Varied beliefs and traditions intersect here with contemporary economic forces. An increasingly global view and a delicate balance of competition and cooperation mark this region. Proximity to water and coastal access dominate the region's physical situation. Many regional countries have hundreds of islands. The region has a history of population movements in and between the mainland and islands over millennia. The diversity of environments and peoples (Figure 5.2) resulted in many discrete local identities. The cultural diversity of Southeast Asia includes ancient Hindu and Buddhist peoples and the world's most populous Islamic country. European trading from the late AD 1400s led to many of the regions becoming European colonies, which later formed the basis of new independent countries from the late 1950s. New trade and political links join the region in one direction with East Asia and the Americas. It also looks west to the Indian Ocean and even European countries.

5.2 Distinctive Physical Geography

The natural environments of Southeast Asia range from expansive mainland areas to thousands of small islands, from high elevation conditions to vast stretches of humid tropical environments, and from volcanic islands and coral reefs that are forming now to some of the world's oldest landscapes.

Continents and Islands

Tectonic plate movements over the last 200 million years led to the huge continent of **Gondwanaland** breaking up to form parts of the region (see Figure 9.3). Huge earthquakes and volcanic activity along convergent plate margins accompanied

Figure 5.2 **Southeast Asia: diverse populations.** "Long-necked" Karen tribe children, Western Thailand. *Photo: © Ian Coles.*

the movements. Volcanic eruptions and earthquakes continue today, part of the Pacific Ocean "Ring of Fire." As Australia on the Indian plate (see Figure 1.5) drove northward toward Asia, it combined with the Philippines and Pacific plates pushing east to raise the Southeast Asian islands. In the west the Indian plate dived beneath the Eurasian plate and created the high mountain ranges of Myanmar and the volcanic line of islands extending from Sumatra and through Java. To the east, the plate movements formed the Philippines and brought other continental fragments into collision with a central group of islands that are mainly Indonesia today. Rising ocean levels at the end of the Ice Age formed islands and drowned bridges, leaving most of the South China Sea and the Java Sea as shallow continental shelf areas. The Indochina Peninsula and eastern Thailand (Khorat Plateau) are hilly areas formed from the erosion of ancient rocks brought to the surface by uplift.

Varied Climates and Distinctive Ecosystems

The majority of Southeast Asia is geographically situated in the tropical latitudes. Tropical rainy climates affect a zone up to 10 degrees of latitude from the equator (Figure 5.3), including Malaysia, Indonesia, East Timor, and the southern parts of the Philippines. Places receive 2,000 to 3,500 mm (80 to 140 in.) of rain per year, and their temperatures vary little, from 25° to 30°C (80°F plus). Local weather in the islands is dominated by daily wind reversals in convective circulations. During the day, air heated over the land rises, drawing in cooler, humid air from the ocean. Such sea breezes bring moisture that rises and condenses to form clouds over the land in the afternoon, often leading to thunder, lightning, and downpours of rain. Some islands have thunderstorms over 300 days a year. At night, air cools over the land and flows back to the oceans; the skies are clear.

Outside this zone, still within the tropics, the rains are seasonal. Total rainfall levels are generally lower over the Asian continent, the Philippines, and many Pacific islands. Temperatures range from around 30°C (80°F) in summer to 20°C

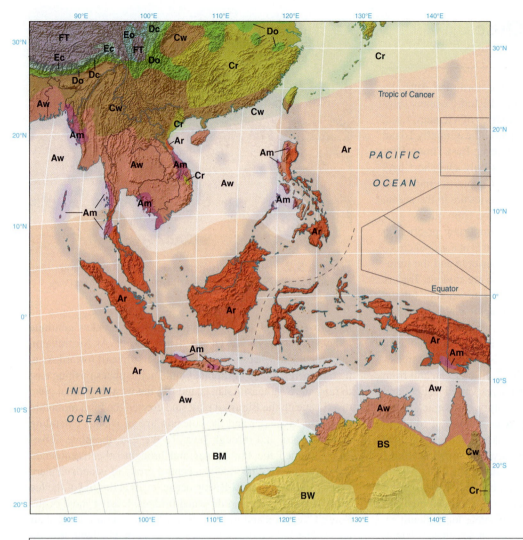

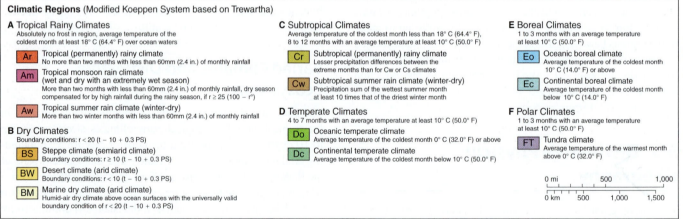

Figure 5.3 **Tropical Southeast Asia.** Ocean currents, trade winds, sea breezes, and elevation all modify the influence of Southeast Asia's geographic situation in the tropical latitudes.

Climatic Regions (Modified Koeppen System based on Trewartha)

A Tropical Rainy Climates
Absolutely no frost in region, average temperature of the coldest month at least 18° C (64.4° F) over ocean waters

Ar Tropical (permanently) rainy climate
No more than two months with less than 60mm (2.4 in.) of monthly rainfall

Am Tropical monsoon rain climate (wet and dry with an extremely wet season)
More than two months with less than 60mm (2.4 in.) of monthly rainfall, dry season compensated for by high rainfall during the rainy season, if r ≥ 25 (100 − r°)

Aw Tropical summer rain climate (winter-dry)
More than two winter months with less than 60mm (2.4 in.) of monthly rainfall

B Dry Climates
Boundary conditions: r < 20 (t − 10 + 0.3 PS)

BS Steppe climate (semiarid climate)
Boundary conditions: r ≥ 10 (t − 10 + 0.3 PS)

BW Desert climate (arid climate)
Boundary conditions: r < 10 (t − 10 + 0.3 PS)

BM Marine dry climate (arid climate)
Humid-air dry climate above ocean surfaces with the universally valid boundary condition of r < 20 (t − 10 + 0.3 PS)

C Subtropical Climates
Average temperature of the coldest month less than 18° C (64.4° F), 8 to 12 months with an average temperature at least 10° C (50.0° F)

Cr Subtropical (permanently) rainy climate
Lesser precipitation differences between the extreme months than for Cw or Cs climates

Cw Subtropical summer rain climate (winter-dry)
Precipitation sum of the wettest summer month at least 10 times that of the driest winter month

D Temperate Climates
4 to 7 months with an average temperature at least 10° C (50.0° F)

Do Oceanic temperate climate
Average temperature of the coldest month 0° C (32.0° F) or above

Dc Continental temperate climate
Average temperature of the coldest month below 10° C (50.0° F)

E Boreal Climates
1 to 3 months with an average temperature at least 10° C (50.0° F)

Eo Oceanic boreal climate
Average temperature of the coldest month 10° C (14.0° F) or above

Ec Continental boreal climate
Average temperature of the coldest month below 10° C (14.0° F)

F Polar Climates
1 to 3 months with an average temperature at least 10° C (50.0° F)

FT Tundra climate
Average temperature of the warmest month above 0° C (32.0° F)

0 mi   500   1,000
0 km   500   1,000   1,500

(70°F) in winter. The northern Philippines, Indochina's mainland, Thailand, and Myanmar experience a **monsoon tropical climatic environment** in which summer rains are brought by winds from the oceans. Some Pacific islands have little relief, and those farther east receive little rainfall. Hilly islands often have rainy and arid sides depending on the direction of the trade winds: from the northeast in the Northern Hemisphere and from the southeast in the Southern Hemisphere. Windward slopes are wet; leeward slopes are drier.

The tropical western Pacific Ocean is subject to destructive **typhoons** (hurricanes) that impact the Philippines, and sometimes, Vietnam. Countries bordering the Indian Ocean are affected by tropical storms called cyclones.

The region's tropical oceans have high water temperatures throughout the year, which fuel the formation of intense tropical weather systems. Recent climate research shows that weather in this region is related to the El Niño fluctuations off western South America. When El Niño brings rain to Peru, tropical Southeast Asia and the nearby islands are dry; for most of the time between these events, Peru is arid and rainfall in this region is plentiful. During the major 1997–1998 El Niño event, Indonesia suffered intense drought and forest fires while Peru had unusual deluges of rain.

The diversity of climates, relief, and degrees of long-term isolation from other land areas as the Gondwanaland supercontinent split apart produced a variety of natural vegetation and animal types in the region. In the mid-1800s, Alfred Russell Wallace, a botanist, mapped the "Wallace Line" between Sulawesi and Borneo (see Figure 5.1) as a major division within the region's biogeography. The main contrast is between the continental Asian species and the unique groups of species inhabiting the islands of eastern Indonesia. The tropical rain forest areas of equatorial Indonesia and Malaysia contain diverse species of trees and other vegetation. The monsoon forests, from Myanmar to Indochina and the Philippines, have somewhat fewer species; extensive areas are dominated by teak forest.

Natural Resources

The mineral resources of Southeast Asia include tin, iron, gold, and precious stones. Much of the tin is mined from river deposits after the tin-bearing veins in the rocks were eroded and the heavy mineral dropped and concentrated in river transport. Indonesia, Malaysia, and Vietnam are among the top global tin producers. Due to the position of tectonic plates, deposits of oil and natural gas occur in Indonesia, Malaysia, Brunei, and Myanmar. The U.S. Geological Survey estimates that there are probably reserves of 21.6 billion barrels of oil and 299 trillion cubic feet of natural gas in the region, particularly in basins in the South China Sea, Sulu Sea, and Gulf of Thailand. Indonesia has significant reserves of coal, tin, nickel, copper, gold, and bauxite and is rivalling Australia as the world's leading coal exporting country. Monsoon forests from Myanmar to Indochina and the Philippines and tropical rain forests of Indonesia

and Malaysia contain hardwood trees such as teak and mahogany. Forests here are being clear felled for their valuable timber, causing loss of animal habitats and imbalances in ecosystems.

Borneo is the largest of the many islands in this world region. It is mostly in Indonesia (where it is known as Kalimantan), but its northeastern coast is shared by Malaysia (Sabah and Sarawak) and tiny Brunei. A hilly-to-mountainous island, Borneo was once covered by tropical rain forest in which lived a rich and distinctive fauna and flora, together with tribes such as the Dyak and Punan, many of whom continue to live on forest resources. One of the world's last true wilderness areas is threatened by loggers and oil palm plantations. From 1985 to 2001, 56 percent of protected lowland forest was cut in Kalimantan. If the forest cover continues to be depleted at current rates, there will be hardly any left in 15 years' time (Figure 5.4). Cutting began outside the protected areas, but buffer areas and then the protected areas were also cut. Much of the logging was illegal.

The forest would be cut over six years and the land planted with oil palms. Much of this land is part of the Betung Kerihun National Park (see Figure 5.4), which at present can be reached only by traveling four days in a motorized canoe. It is the hilly core of Borneo, from which nearly all the main rivers radiate. Cutting the forest would not only destroy important ecosystems, but the smoke from burning forest debris would drift across Singapore and Malaysian cities. Upstream erosion and flooding and downstream sedimentation would follow. Moreover, oil palms grow much better on lowland areas than on such hilly terrain. In 2008 the Malaysian government said it would prohibit forest clearing for the establishment of oil palm plantations. But Sarawak continues to log forests for oil palm plantations, citing long-existing plans for agricultural extension.

The Indonesian agriculture minister backs this project, which might bring 500,000 new jobs and reduce the prosperity gap between the Malaysian and Indonesian parts of Borneo. Palm oil is the world's most consumed edible oil. World

1950

1985

2020 (projected)

Figure 5.4 **Logging in Borneo.** Much of the tropical rain forest on this island is being removed without replanting. The Malaysian states of Sabah and Sarawak were first affected, but Indonesian Kalimantan is likely to be deforested by 2020. Logging, often against government regulations, and the planting of palm oil plantations are responsible. *Source: Data from* The Times *(London) 2006.*

supply in 2012 was dominated by Indonesia (52 percent of world supply) and Malaysia (36 percent) with continued expansion expected. Palm oil is in demand for cooking and consumer products, and has a future in the production of biodiesel fuel. The latter could reduce Indonesian oil imports and produce major exports as world oil prices rise. The project is opposed by the World Wildlife Fund (WWF), which is lobbying for a Heart of Borneo protected area along the international border. The WWF quotes unheeded 1980s cautions over logging in Sumatra, where the rain forest has now nearly all vanished. Among the local people, some want to keep the forests and their ancestral way of life, but others look favorably to the prospect of palm oil jobs and the housing, education, health care, and modern amenities they could buy. This is a common modern debate affecting local lives, ecologies, and sustainable economies, which are impacted by global investors and supported by government policy.

Water resources are plentiful in the tropical rainy areas. On the Asian continental area, the major rivers are the Irrawady and Salween rivers in Myanmar; the Mekong River that rises in China and flows through Laos, Cambodia, Vietnam, and along the border of Thailand; and the Red River in northern Vietnam. The Mekong is one of the region's longest rivers, and its development potential includes hydroelectricity, agricultural irrigation, and transportation. Improved flood control would also aid economic development of the lands along the Mekong's length. In 1995, Cambodia, Laos, Myanmar, Thailand, Vietnam, and China's Yunnan province signed an agreement for the future integrated development of the Greater Mekong Sub-Region based on a plan submitted by the Asian Development Bank. However, the difficulties of securing cooperation among former enemies and potential economic competitors within the Mekong's watershed remain.

Natural Hazards

The region shares vulnerability to natural hazards of different types, highlighted by the December 2004 tsunami wave that struck coastal Sumatra (see "Geography at Work: The Tsunami of December 2004," p. 162), Thailand, Malaysia, and places around the Indian Ocean. This is an example of the volcanic eruptions and earthquakes caused by the active plate tectonic margins in the region. Volcanic eruptions occurred in the 1990s in the Philippines, and numerous deadly earthquakes struck Indonesia in the early 2000s. In Indonesia, the eruption of volcanoes Mount Merapi in 2010 and Mount Luzon in 2011 necessitated the evacuation of hundreds of thousands of people, and the Mount Merapi eruption caused 353 deaths. Threats from typhoons and cyclones and resulting floods affect coastal areas. Typhoon Bopha hit the Philippines' Mindanao island in December 2012, causing torrential rain and landslides. Over 600 people were killed and nearly 1,000 declared missing. In 2008 Cyclone Nargis hit Myanmar, causing catastrophic destruction estimated at US$1 billion and at least 138,000 fatalities. Intense tropical rainfalls cause flooding, the collapse of steep slopes, and landslides of mud, especially where forest cover is removed. In 2011, heavy monsoonal rain caused severe floods

in Thailand, Philippines, Cambodia, Myanmar, Vietnam, and Laos, killing over 2,800 people and bringing life in many areas to a standstill. Intense rain also caused widespread flooding in Myanmar in 2012. The alternating patterns of weather associated with El Niño led to forest fires, smoke, and widespread breathing difficulties in parts of Indonesia, particularly Borneo and Sumatra. Global climate change causes ice melt and rising ocean levels, threatening the future of low-lying coral atolls and coastal settlements.

The tropical environments harbor many infectious diseases, such as malaria, cholera, typhoid, and rabies. Improvements in medical services, better sanitation, and the elimination of mosquito breeding sites have lowered the impact of such hazards and increased life expectancy.

Human activities in natural environments create their own hazards. The overuse of fertilizers and pesticides pollute ground and stream waters. In the Vietnam War of the 1960s and 1970s, U.S. planes defoliated forests to expose Vietcong positions. The barren slopes created by these actions still have not recovered. Mining affects downstream water and air quality. Gold mining in Sulawesi, Indonesia, has resulted in mercury and arsenic pollution of water bodies, causing skin and nerve diseases among local populations. Expanding cities create their own environments, and more controls are needed to reduce the pollution from vehicle exhausts and factories.

5.3 Distinctive Human Geography

Cultural Diversity

The population of Southeast Asia is characterized by its multiethnic composition. Prehistoric migrants came to Southeast Asia from the north, filtered southward through the large river valleys, and diffused outward across the island archipelagos. The varied precolonial and colonial histories combined to create the diversity present in the contemporary human geography of the region. The current political boundaries of Southeast Asia are almost entirely the legacy of colonial territories.

Although there are signs that the countries of this region are moving closer to each other politically, culturally, and economically, there are still many differences at each geographic scale. There is diversity within and between individual countries in Southeast Asia. The people in the region come from Malay, Chinese, Indian, and European stock (Table 5.1). The "Eastern" religions of Hinduism and Buddhism mix with Islam and Christianity. The Philippines and East Timor are the only countries with dominant Roman Catholic allegiance. Political control varies from democratic to military dictatorship, including single-party countries and Communist governments.

Early People Movements

The Mon and Khmer people occupied present-day Cambodia from the north, and between the AD 800s and 1200s, it was the center of the Khmer Empire. Later, the Vietnamese, Lao, and Burmese peoples established other centers. Traders from

Table 5.1 — Southeast Asia: Data by country, area, population, urbanization, income (gross national income purchasing power parity), ethnic groups

Country	Land Area (km²)	Population (millions) Mid-2012 Total	Population (millions) 2025 Est. Total	% Urban 2012	GNI PPP 2010 Total (US$ billions)	GNI PPP 2010 Per Capita (US$)	Ethnic Groups (%)
Cambodia, Kingdom of	181,040	15	18.0	21	24.9	2,080	Khmer 90%, Vietnamese 5%
Lao, People's Democratic Republic of	236,800	6.5	7.9	27	15.3	2,460	Lao 50%, Thai 14%, Meo and Yao 13%
Myanmar, Union of (Burma)	676,580	54.6	61.7	31	93.5	1,950	Burman 68%, Shan 9%, Karen 7%, other tribes
Thailand, Kingdom of	513,120	69.9	72.9	34	565.8	8,190	Thai 75%, Chinese 14%, Malay 3%
Vietnam, Socialist Republic of	331,690	89	101.6	31	267	3,070	Vietnamese 90%, Chinese 2%, tribal groups
East Timor (Timor-Leste)	9,486	1.1	1.6	30	4	3,600	Timorese (Malayo-Polynesian) 97%, Chinese 2%, Portuguese 1%
Indonesia, Republic of	1,904,570	241	273.2	43	1,008.2	4,200	Javanese 45%, Sundanese 15%, Madurese, Malays
Malaysia, Federation of	329,750	29	34.8	63	403.9	14,220	Malay 59%, Chinese 26%, Indian 7%
Philippines, Republic of	300,000	96.2	117.8	63	370.7	3,980	Malay 95%, Chinese 2%
Singapore, Republic of	620	5.3	5.8	100	283.3	55,790	Chinese 76%, Malay 15%, Indian 7%
Region Totals/ Averages	**4,483,656**	**586.1**	**708.7**	**44**	**2460.7**	**9,582**	

Source: *World Population Data Sheet 2012*, Population Reference Bureau; *World Development Indicators*, The World Bank, 2012; Microsoft Encarta 2005.

India brought Hindu and Buddhist religions, leaving imprints of traditions and architectures such as the temples at Angkor Wat, Cambodia, and those of the island of Bali. The arrival of Islamic groups in the AD 1200s introduced new cultural elements. Muslim merchants controlled the trade in products such as spices. They established sultanates in Malacca (now Malaysia), the islands of modern Indonesia, and Brunei. Such flows of people widened the range of languages spoken in the region. Throughout the period, the region acted as a melting pot for people from China and India who lived alongside and mixed with local people. The imprint of Indian and Chinese traditions and architecture resulted in the Western label of "Indochina" for Cambodia, Laos, and Vietnam.

Colonization

From the late 1400s, a series of European traders and colonists came into this multiethnic and multicultural complex (Figure 5.5). European trading demands reoriented local economies, political colonization determined modern country boundaries, and their ways of life increasingly impacted the region's cultures. Colonial languages joined the established creole varieties as common means of communicating.

The Portuguese first traded in the Philippines and annexed Malacca in 1511. In the late 1500s, Spain forced them out of the Philippines, building their capital at Manila on Luzon and bringing in priests to convert the local people to Roman Catholicism. The Dutch took control of trade in what became known as the Dutch East Indies or Spice Islands (modern Indonesia). The British then contested trade with the Dutch until an 1804 agreement resulted in a separation of interests: the United Kingdom took Malaya, building a new port in Singapore in 1819, and the Dutch took the East Indies.

During the early 1800s and 1900s, the French occupied Indochina (modern Laos, Cambodia, and Vietnam), building roads and railways and encouraging manufacturing. In the late 1800s, the British colonized Burma as part of the British Indian Empire, developing rice growing in the Irrawady River delta, building a new port at Rangoon (modern Yangon), and bringing in workers from India. Only the Kingdom of Siam (modern Thailand) remained independent with its strong monarchy, although it also imported Western ways.

During the colonial phase, more Chinese people were brought in to labor in Dutch and British mines and plantations. Many stayed and became important in commercial activities. Some converted to Christianity, adding further cultural diversity.

Cultural complexities within countries are illustrated by Thailand. The people of the mountainous north and west are mainly Buddhist with similarities to Myanmar; minority groups like the Karen (see Figure 5.2) moved to Thailand after persecutions in Myanmar and China. In the northeast, long isolation resulted in distinct cultural identities, and people share languages, artistic traditions, and Buddhism with Lao groups. The central Chao Phraya River basin around Bangkok is the Thai

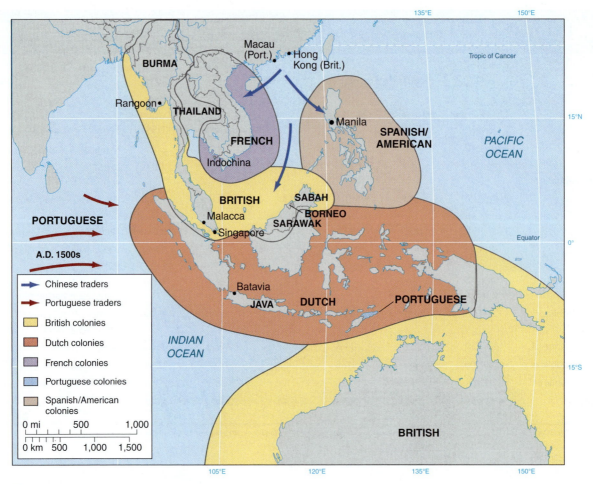

Figure 5.5 Southeast Asia: colonization. Americans, British, Dutch, French, and Portuguese assumed control of areas easily accessed by sea.

cultural heart, whose Buddhism is more akin to that of the Khmer people of Cambodia. Some Hindu Brahmins also live in this area and provide skills used in directing royal and official ceremonies and producing an annual calendar. Economic growth in this area attracted people from the poorer regions of the country. To the south and along the Cambodian border, Chinese settled from the 1800s as workers on sugarcane plantations, and they now work in timber mills and run small stores. Muslim Malay people increase in numbers toward the south of the narrow southwestern peninsula.

Population Distribution and Dynamics

The historic movement of people and their interactions with the natural environment resulted in variations in population distributions, densities, and attributes (Table 5.1). Dramatic contrasts in the regional population distribution are evident in this region (Figure 5.6). Population on the mainland portion of Southeast Asia is clustered in lowland river valleys and along coastal zones. Some of the islands of Southeast Asia exhibit very high population densities. The island of Java, in the country of Indonesia, has some of the most densely populated land in the region, as do many of the Philippine

islands. Rugged terrain and steep slopes complicate settlement on numerous regional islands, resulting in even higher population densities.

Within Southeast Asia, the highest densities are based on the historic development of wet-rice agriculture as in the fertile valleys and deltas of the Irrawady, Mekong, and Red rivers as well as modern urban expansion. Much of Borneo, northern Sumatra, and Irian Jaya, where forest has not been cleared until recently, have sparse populations.

In Indonesia the contrasts between highly populated Java and the less populated islands led to a transmigration program, in which over 6.5 million people were resettled from 1950 until the program ended in 2001. Despite its democratic rationale, many saw the program as political manipulation, moving people who were regarded as government supporters to outlying islands. Other problems stemmed from a poor selection process, which did not adequately consider age, skill levels, ethnicity, or appropriate farming systems. Furthermore, few unoccupied tracts of good land remained in the less settled islands. Nearly half the migrants failed to raise their own standard of living. Conflicts between local groups and settlers who cut the forest to plant oil palms or illegally produce timber have sometimes resulted in fatalities.

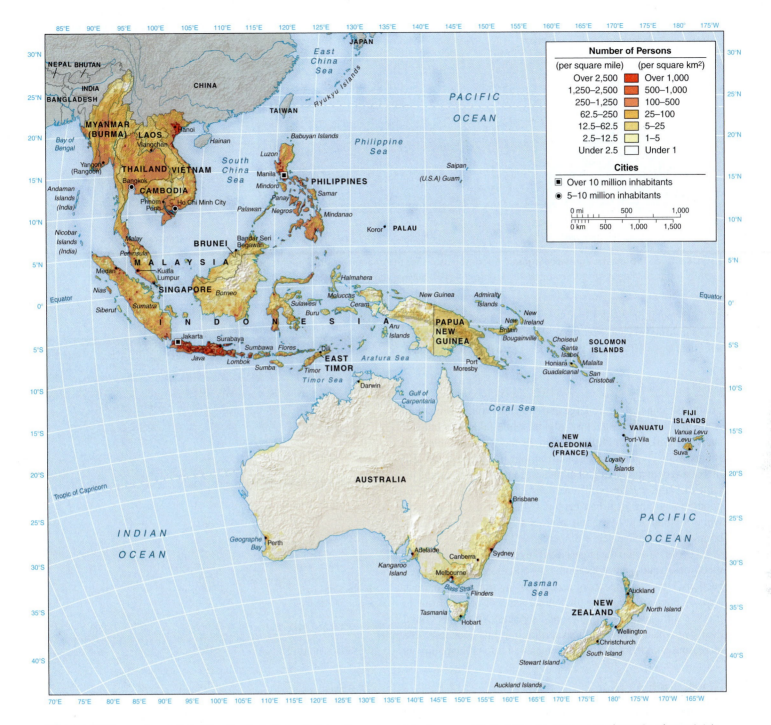

Figure 5.6 **Southeast Asia: distribution of population.** Steep mountain slopes and limited island territory complicate Southeast Asia's high population densities.

Population Growth

Like other aspects of this region, the contrasts of a few years ago are giving way to more similar conditions. The rapid population growth of Southeast Asia from 1965 (275 million) to 2012 (608 million) slowed as natural increase rates dropped from 2–3 percent to 1.2 percent. Average fertility rates in 1965 were around 6, but by 2005 they were all under 4 in all countries, and in 2012, closer to replacement fertility of 2.1. In Thailand, family planning policies reduced fertility to well under 2. The age-sex diagram for Indonesia (Figure 5.7) demonstrates the stability of

birth numbers and greater life expectancies as a result of medical progress. However, Cambodia and Laos had low rates of natural increase during the wars in the 1950s and 1970s but now have higher rates of more than 3.

HIV/AIDS increasingly affects health and well-being, particularly in Thailand and Myanmar. The disease is not confined to the communities of drug users, sex workers, and their clients. The large sex market, which is partly aimed at foreign tourists, may involve men who catch the disease and pass it on through sex to their wives. The ironic outcome affects Asian women, who are unlikely to have extramarital sex. From 2009 to 2011,

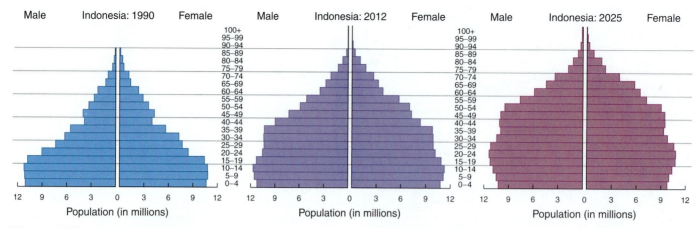

Figure 5.7 Indonesia: age-sex diagram. Suggest reasons for the changing demographic profiles shown by the three pyramids and identify the social issues raised. *Source: U.S. Census International Database: http://www.census.gov/population/international/data/idb/region.php.*

of the population aged 15 to 49 years, 0.8 percent had HIV/AIDS in Malaysia and Myanmar. Women in this age group had a higher rate of HIV/AIDS (0.7 percent) than men in Cambodia.

Rapid Urbanization

Southeast Asian cities are large and growing rapidly. Singapore is the area's major global city–region, while Jakarta in Indonesia and Manila in the Philippines are its megacities (Table 5.2). Bangkok in Thailand and Kuala Lumpur in Malaysia are also increasingly involved in global transactions. But most countries with the largest cities still have less than 50 percent of their population living in urban centers. Malaysia has 63 percent, while Vietnam and Myanmar are tied at 31 percent, and Cambodia and Laos have below 30 percent.

People flow into the largest cities in response to the availability of jobs, health care, education, transportation, and global connections. However, the rate of people moving from rural to urban areas is often too rapid for the authorities and private firms to provide sufficient housing, jobs, and services. Newcomers build their own shantytowns and expand the informal economy. The increasing number of motorized vehicles adds to urban pollution.

New Countries

The end of colonialism for countries in Southeast Asia came within 20 years of the end of World War II. At first the new countries attempted self-sufficient nationalism. However, subsequent regional and global trends drew the countries into producing export goods.

The republics of the Philippines and Indonesia gained independence in 1946 and 1950 following the growth of nationalist demands before World War II and the often horrific experience of Japanese occupation during the war. Burma gained independence in 1948 and was renamed Myanmar by its military leaders in 1989. In Malaya the British fought the local Communist guerrillas, but made Malaya independent in 1957. The Federation of Malaysia was established in 1963, joining Malaya, Singapore, Sarawak, and Sabah. Brunei did not join, and Singapore was expelled in 1965.

Table 5.2	Populations of Major Urban Centers in Southeast Asia	
TOP 10 MOST POPULATED URBAN CENTERS		
City, Country	**2012 (millions)**	**2025 (millions)**
1. Jakarta, Indonesia	12.1	15.4
2. Manila, Philippines	11.8	13.6
3. Ho Chi Minh City, Vietnam	7.3	9.2
4. Bangkok, Thailand	7.1	8.3
5. Hanoi, Vietnam	4.8	6.8
6. Singapore, Singapore	4.2	5.1
7. Yangon, Myanmar	4.2	5.9
8. Bandung, Indonesia	4.0	6.4
9. Surabaya, Indonesia	3.9	4.7
10. Kuala Lumpur, Malaysia	2.8	3.4
NUMBER OF MAJOR URBAN CENTERS IN SOUTHEAST ASIA, BY POPULATION CATEGORY		
Population Category		**Number of Urban Centers**
10.00 million and greater		2
5.00 – 9.99 million		2
2.50 – 4.99 million		7
1.00 – 2.49 million		12

In French Indochina the countries of Laos and Cambodia gained independence in the early 1950s. In Cambodia, the Communist Khmer Rouge ruled from 1975 to 1979, causing the death of 1.7 million people from disease, starvation, or execution. Many of the killers continued to live in Cambodia until 2001, when King Sihanouk put them on trial. Both Cambodia and Vietnam carried out genocidal eradication of hill tribes, linking political crusades to ethnic abuse.

Communist fighters took the northern part of Vietnam from the French in 1954. They eventually added the southern part of the country when U.S. forces left in 1975 after 30 years of civil war. Thailand, which had never been colonized, stemmed the flow of the Communist advance from the 1960s, attracting much financial aid from the United States.

Growing Economies

There is a wide range in per capita income figures for the countries of Southeast Asia. The countries at the top of the diagram in Figure 5.8 have the highest incomes and are the most involved in the global economic system. The range of material well-being is reflected in consumer goods ownership, access to piped water, and energy usage in different countries (Figure 5.9).

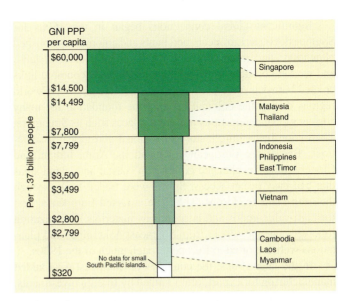

Figure 5.8 **Southeast Asia: country per capita incomes compared for 2010 data.** The data are GNI PPP. *Source: Data for 2010 from* World Development Indicators, *World Bank 2012; Population Reference Bureau.*

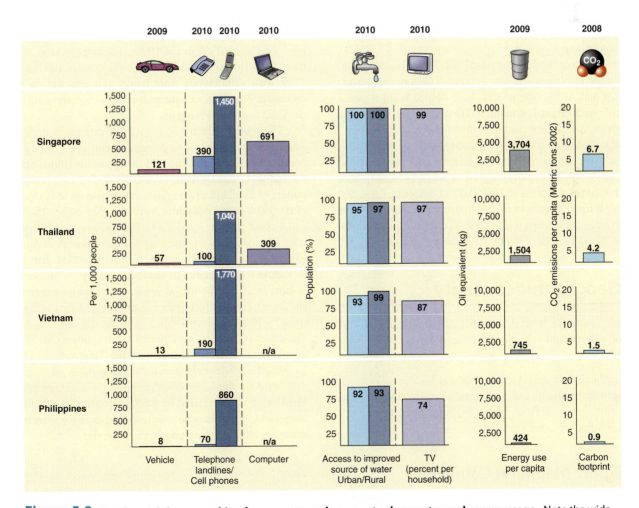

Figure 5.9 **Southeast Asia: ownership of consumer goods, access to clean water, and energy usage.** Note the wide range of affluence among these countries. *Source: Data from* World Development Indicators, *World Bank 2012.*

Significant global connections began in the 1600s for Southeast Asia as European merchants and colonists established plantations and mines utilizing a combination of local labor and workers brought in from China and India. Economic links created between some parts of Southeast Asia and other world regions during the colonial era laid the foundation for many current global connections present in places such as Singapore. The region has a long history of regional and global associations. Before the European colonizations, Arab, Indian, and Chinese traders exchanged goods with coastal and island areas that had easy ocean access. Spices were traded back through the Arab lands to Europe. The trade attracted Europeans, first for high-value goods and later for raw materials. Colonization followed. Tin and rubber from Southeast Asia mines and plantations were exported to Europe through much of the 1900s.

In the later 1900s, the countries of Southeast Asia added manufacturing for export, often funded by foreign investments, and developed major tourist industries. Thailand has a large electronics and electrical industry, producing hard disk drives, integrated circuits, and semiconductors. It is also known for vehicle assembly in the region. Malaysia and Indonesia have petroleum and natural gas, textiles, apparel, footwear, mining, cement, chemical fertilizers, plywood, rubber, and food industries. Countries with low labor costs such as Cambodia, Laos, and Myanmar attract clothing and footwear manufacturing units. Sportswear company Adidas recently moved its factories from China, where it pays workers monthly wages of US$500, to Myanmar, where workers are paid US$130 per month. Intraregional trade has increased, and economic organizations such as ASEAN and APEC (see p. 158) have helped build regional cooperation and regional economies. However, the emphasis on export goods makes most countries in the region competitors.

Southeast Asia is a region containing some of the world's wealthiest countries and others that are materially impoverished (Table 5.1). Per capita wealth is highest in Singapore and Brunei, each with incomes in excess of US$50,000. Cambodia, Laos, and Myanmar (Burma) remain in material poverty. The contributions of the service sector through tourism, health care, and business processing to the economies of Southeast Asian countries are rising.

5.4 Geographic Diversity

Two subregions arise from the interactions of natural environments and cultural and economic histories of the countries (Figure 5.10):

1. Mainland Southeast Asia: continental Myanmar (Burma), Thailand, Laos, Cambodia, and Vietnam.
2. Insular Southeast Asia: Malaysia, Singapore, Indonesia, the Philippines, and East Timor.

Subregion: Mainland Countries

Since the 1990s, all these countries have opened their economies to varied levels of external investment and associations, including joining ASEAN, although economic growth in Cambodia and Laos remains very slow. Myanmar, for example, paid Chinese contractors to build bridges, railroads, airports, and hotels and entered joint arrangements with European and Japanese investors to develop tourism and manufacturing for export. However, Myanmar and particularly the Wa tribe, with the tacit backing of the government, is a major world source of opium and amphetamine production. This strains relationships with neighboring Thailand.

Of the mainland countries, Myanmar was a military dictatorship from 1962 to 2011, while Laos and Vietnam have Communist governments coming to terms with a globalizing world. In 1989 Myanmar's military leadership adopted the new name to replace "Burma."

Thailand

Thailand (see Figure 5.10) stands out as the one country that did not become a colony, that retained a traditional monarchy despite a number of military coups, and that resisted the Communist advance after 1950. It also adopted an open attitude to the expanding global economy, developing export manufacturing based on foreign investment and a service industry that includes tourism and health care. Thailand's GNI PPP is twice the total of the other mainland countries in this world region, and per capita incomes are three times as much. However, Thailand's economy reflects both sides of becoming global.

The income from major vehicle manufacturers and oil refineries leveled off in the 1990s, leading investors to change their focus to land and buildings. This asset bubble burst in 1997 with withdrawals from banks and pressure on the baht currency. The Thai economy recovered slowly with external help and further investment.

The tourist industry is subject to fluctuations in demand as Asian visitor numbers fell after such events as the late 1990s economic crisis in the area, the 9/11 attacks in the United States, and the Indian Ocean tsunami of late 2004. Fighting between supporters of Thailand's two main political parties led to the closure of Bangkok's Suvarnabhumi Airport on November 25–26, 2008, and violent riots in April 2009. Both events and the two-week state of emergency imposed by the government in April negatively affected international tourism to Thailand. But the tourist industry is still vibrant, attracting nearly 16 million foreign visitors in 2010. The illegal sex and drug trades generate crime and health crises. The number of imprisoned drug offenders has risen by 10 times since 1990, swamping the courts and prisons. The drugs come across the border from southern Myanmar and merely take different routes when the Thai military intervene. *Ya baa*, the methamphetamine "crazy pill," is of less concern than heroin to the world community, but it is a major problem in Thailand. It was first used by workers needing to remain alert on long shifts, but it became a recreational drug for students. Drug addiction among young people is on the rise.

Vietnam

After the 1975 unification of North and South Vietnam, continued military spending backed forays into Cambodia through the 1980s and an on-and-off border war with China. Attempts

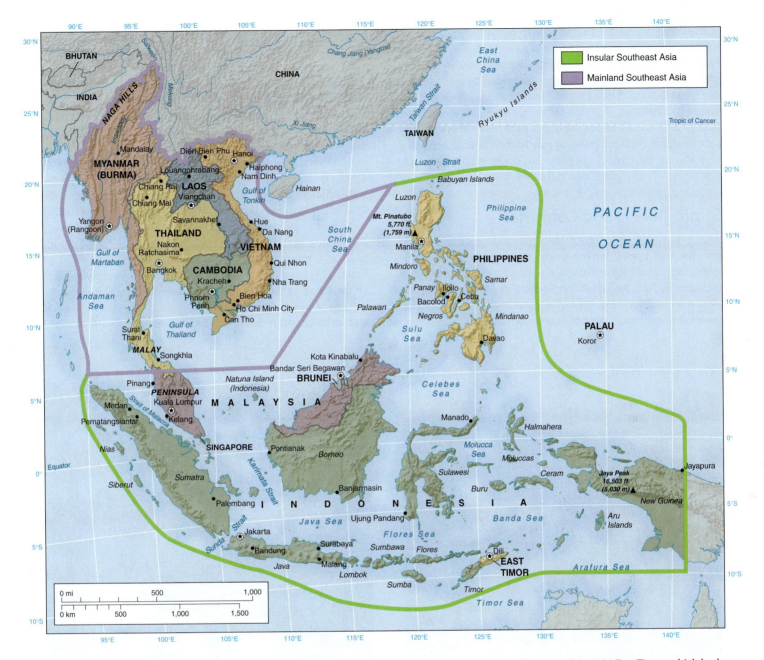

Figure 5.10 Southeast Asia: countries and major cities. How important are ocean routes in this region? In 1999 East Timor, which had been a Portuguese colony until 1974 and then occupied by Indonesia, gained its independence. Civil disturbances occur on other Indonesian islands.

to move economic output from agriculture to heavy urban industries were not successful, and in 1989 the backup finance from the Soviet Union ended. At this time, Vietnam's plight was made worse through economic blockades by the United States and ASEAN countries. Still a Communist country, in the 1990s Vietnam changed its economic approach to follow China's example of openness to global trade and investment. This led to greater interaction with its ASEAN neighbors.

The changes of outlook in Vietnam began with the 1985 law that allowed crops to be sold directly in private markets instead of through the communes and cooperatives selling to price-setting government ministries. After 1990, the Vietnamese dong currency was devalued and foreign investment allowed.

Within Vietnam, the southern area centered on Ho Chi Minh City found it easier to develop a marketplace economy than the northern area around Hanoi, which the Communist government controlled for a longer time. Japanese and South Korean corporations were the main investors in manufacturing industries until the United States normalized relations with Vietnam in 1995, the year Vietnam joined ASEAN.

A further Vietnamese development in the 1990s was its progress from a marginal coffee producer to one that now rivals Brazil. In late 2012 Vietnam became the world's largest exporter of coffee, which is grown in the central highlands. Farmers moved into the hilly area from the crowded lowlands as world coffee prices rose. When other world producers also boosted output, an oversupply crisis caused Vietnam to cut robusta coffee production and grow more higher-quality Arabica coffee trees. Vietnam's exports of rice are second only to those from Thailand, and it is the global leader in the production of cashewnuts and black pepper. However, agriculture's contribution to Vietnam's GDP has fallen in recent years. Manufacturing, information technology, and high-tech industries now form a significant and growing part of the national economy. It is also an increasingly popular investment destination, surpassed in Asia by only China and India.

Subregion: Island Countries

The main island countries of Malaysia, Indonesia, Singapore, and the Philippines have become the centers of economic growth in the subregion. Although nominally democracies, they experienced long periods of single-party and autocratic leadership, with intervening periods of internal conflict for the central leadership and loss of control in outlying areas.

Farming remains basic to the economies of many parts of these countries, particularly the cultivation of wet rice, or padi (Figure 5.11). The tropical rainfall is often augmented by irrigation to obtain maximum crops. The new rice technology that developed at the International Rice Research Institute of Los Banas in the Philippines in the 1960s led to the **Green Revolution**. Crop yields increased by combining high-yielding new varieties of rice with the use of fertilizers, pesticides, mechanization, and irrigation. Padi yields doubled from 1969 to 1989, allowing Indonesia and the Philippines to become almost self-sufficient at a time of rapid population increase. The main beneficiaries were owners of larger farms with good growing conditions, available labor, and capital to invest in the costly procedures. Smaller farms without access to capital gained little. The commercialization of farm production also had social impacts. Former communal land was sold to wealthy farmers or for urban expansion, creating many landless laborers who lost previous access to jobs at planting and harvest. Wage labor and mechanization destroyed traditional systems and resulted

(a)

(b)

Figure 5.11 **Southeast Asia: farming.** (a) Wet rice, or padi, cultivation in Bali. (b) A young farm worker in Thailand equipped to apply chemical sprays, including pesticides. *Photos: (a) © Corbis RF; (b) © David Zurick.*

in large numbers of disaffected people who supported alternative Communist and Islamist pressure groups.

Under colonization, European companies established commercial tree crops, grown on large plantations and sometimes by local farmers on smaller plots, including rubber, palm oil, spices, and pineapples. Most plantations were taken over by country governments after independence. Good plantation management produces year-round harvests and employment. Malaysia dominated world production of rubber, but failure to recruit enough cheap labor led to Indonesia and Thailand competing in this market. In Malaysia, less labor-intensive oil palms provided an alternative crop in great demand for Chinese cooking, since palm oil produces higher temperatures than other vegetable oils. Indonesian plantation crops include coffee, spices (cloves, nutmeg, cinnamon), tea, and rubber. In the Philippines, coconut products, sugarcane, tropical fruits, and coffee make up one-fifth of exports.

The exports of tropical hardwood timber from Malaysia and Indonesia are greater than those from the Brazilian Amazon and Africa combined. From the 1990s, prices for such timber rose by 50 percent as crop and mineral prices fell or remained the same. Wood industries grew in importance in Kalimantan (Indonesian Borneo) and Sumatra, but rapid cutting has had a negative environmental impact and caused social conflict. Due to logging and the conversion of forests to oil palm plantations in Sumatra, the habitat of orangutans is being rapidly depleted, threatening the future existence of these great apes. In response to resented Western objections to unmanaged cutting, Malaysia claims to have responsible sustainable logging, but it continues to export tropical logs and sawn timber, reducing forest cover by up to 2 percent annually. In 2010 Indonesia, known for its rapid deforestation, signed a letter of intent with Norway to place a two-year moratorium on new logging concessions in exchange for up to US$1 billion.

Mining adds to commodity exports. Malaysia has a history of tin mining, dominating world production for many years. Bangka and Billiton, Indonesian islands, also produce tin. Currently, Indonesia is the world's second-largest tin producer. Oil was an important mineral export of the region, with Indonesia being a major producer from its fields around Kalimantan and Sumatra. However, few new resources have been found. Indonesia stopped exporting oil in 2007 and is now an oil importer. Oil is also produced in the Malaysian territories of Sabah and Sarawak, together with neighboring Brunei. Indonesia is a major bauxite producer in the Riau islands and southwest Kalimantan. The bauxite is transported to the Kualatajay smelter near Sumatran coal mines to be processed into aluminum.

The main economic trend starting in the late 1900s was the shift to export manufactured goods and service industries. This shift intensified the income gaps between urban and rural areas. At independence, most countries had some factory-based industries and imposed tariffs and quotas to protect import-substitution industries producing for small local markets. Government licensing restrictions made local products expensive, and **crony capitalism** arose from bureaucrats and entrepreneurs working together. The combination of government-owned processing facilities for mining and agricultural products and small-scale, often Chinese-owned, textile and consumer goods firms remains important in the economies of these countries. Multinational corporations based in the United States, Japan, and South Korea invested in new factories from the 1970s, especially in Malaysia and Thailand, shifting their emphasis from labor-intensive to high tech industries. Much of this export-oriented industrialization is labeled **ersatz capitalism**, since it relies on imported investment, management, machinery, and skills rather than local enterprise and savings.

The importance of government employment, and the growth of health, education, office, retail, and wholesale sectors, resulted in a huge expansion of service sectors and middle incomes in these countries. Tourist industries attracted increasing numbers of foreign visitors and foreign currencies. Thailand, Malaysia, Singapore, Indonesia, and the Philippines together went from 7 million tourist visitors in 1983 to 20 million in 1990, and over 120 million in 2011. Their modern, cosmopolitan cities, precolonial and colonial historic buildings, tropical weather, and spectacular beaches and upland areas with resort facilities are visited by people from Asia, Australia, North America, and Europe.

Of the three small countries, Singapore, with only 5.3 million people, is a world leader in business services, and Brunei enjoys prosperity based on oil wealth. East Timor at last gained its independence after decades of colonial and Indonesian domination, but it continues to suffer internal conflict.

Indonesia

The Republic of Indonesia is the largest country in this world region in both area—when the land and ocean are combined—and population (see Table 5.1). The 5 million km^2 territory includes 1.9 million km^2 of land. The country extends over 5,000 km (over 3,000 mi.) from east to west and 1,760 km (1,100 mi.) from north to south. It incorporates around 15,000 islands, of which 6,000 are inhabited. Indonesia has an increasingly strategic position astride major sea lanes. It is marked by extreme internal diversity that is difficult for any government to control. The five main islands (see Figure 5.10) are Kalimantan, Sumatra, Irian Jaya, Sulawesi, and Java. Kalimantan (Borneo) is shared with Brunei and Malaysia; Irian Jaya shares New Guinea with Papua New Guinea, and Timor is shared with East Timor.

Indonesia is the world's fourth most populous country and the largest with a Muslim majority, with 88 percent declaring to be Muslim in the 2010 census. Archaeological remains and ancient buildings testify to a varied history. In the period from the AD 600s to 1300s, Hindu and Buddhist kingdoms formed on Sumatra and Java. Spice-trading Arabs brought Islam, which became the main religion, replacing collapsed Hindu and Buddhist kingdoms. In the 1500s Portuguese traders found many small, vulnerable states in what they termed the East Indies. By the 1600s, the Dutch took over as the most powerful Europeans in this group of islands, relying on indirect rule to impose their will on, and extract income from, traditional local elites. The Dutch established large plantations and forced cultivation on Java, but brought in more liberal conditions in the 1900s. In 1942 Japanese armies captured Java to obtain local oil. They released General Sukarno, who acted as a puppet but also planned for independence, which Indonesia declared in 1945. The Dutch tried to reoccupy their colony, but international pressure caused them to acknowledge Indonesian independence in 1949.

Two generals governed the country for nearly 50 years. Sukarno (1949–1966) made a point of bringing the varied island groups under political control, but achieved little economic growth. This was the priority of General Suharto's government (1976–1997), although, like his predecessor, his long-term government ended with economic crisis and popular uprisings. In the Suharto years, annual per capita income rose from US$70 to US$1,142 and social services were the most extensive in Southeast Asia. However, in the mid-1990s the lavish lifestyles of Indonesia's political elite, centered on the Suharto family, drained the country's resources. The growing middle class found upward mobility capped by corrupt officials. As foreign debt doubled, the 1997 Asian economic crisis brought an end to Suharto's government. Subsequent elected governments strove to put matters right, but faced a reduction in oil income, conflict in several islands, and natural disasters. Events such as the 2002 Bali terrorist bombing, the 2004 tsunami, forest fires, atmospheric pollution, flooding, earthquakes, and volcanic eruptions had social and economic consequences. The island conflicts included independence movements in Aceh in north Sumatra and East Timor, fights between resettled and local groups on Kalimantan, and Muslim-against-Christian conflicts in Sulawesi.

The complexity of its island populations makes it difficult to assess the total Indonesian population. The island of Java has the world's highest population density, contrasting with parts of Kalimantan that have extremely low numbers of people. In the west most people are Malaysian, but in the east Melanesian characteristics prevail, and elsewhere mixtures and origins are complex. Many Javanese, Sundanese, and Batak peoples speak of themselves as nations rather than Indonesians. In Borneo the Dyak and Punan groups have totally different lifestyles. The Indonesian language, taught in schools, is closely related to Malay and began as a **lingua franca** to enable communication among more than 700 ethnic groups. The Pribumi people are considered natives of Indonesia, and there are frequent conflicts between them and non-Pribumi people, especially those of Chinese origin.

Jakarta is the capital and largest city, with over 12 million people in the urban area in 2012. It was renamed from the Dutch Batavia by the Japanese to emphasize local wishes. It is a crowded and cosmopolitan city, with the local Betawi people and a large Chinese community. Transportation by road and rail is overloaded in rush hours. Traditional bicycle rickshaws were banned from major roads in 1971 and finally dispatched in 1991. Jakarta is a center of universities, tourist attractions, and new shopping malls. Like many huge cities in developing countries, it is subject to overcrowding, especially during the working week because of poor transportation, and the wet season brings flooding from clogged sewage systems and waterways.

A producer of oil and natural gas, Indonesia was the only Asian member of OPEC outside Southwest Asia, but suspended its OPEC membership in 2009. Indonesia's oil production fell significantly in recent years due to aging oil fields and lack of investment in new equipment. Today, Indonesia exports some crude oil and imports oil products. In the 2000s the long-term importance of tin, bauxite, and silver production was largely replaced by copper, gold, and coal. External investment increased coal production manyfold. In 2011 Indonesia produced 325 million tons of coal, making it the world's sixth-largest producer.

From the late 1980s, Indonesia encouraged foreign investment to produce clothing, machinery, transportation equipment, and footwear for export. Following the severe impacts caused by the late 1990s economic crisis and a drought that forced Indonesia to import rice, new investment stopped. By 2006 measures to improve these failings resulted in resumed economic growth, although rising world oil prices created further stress. In 2011, the Indonesian government announced a new Masterplan to encourage increased investment, especially in infrastructure projects across the country.

Singapore

Singapore is envied by many other countries. It is the smallest country in Asia—an urbanized small island (Figure 5.12a) without natural resources—but is by far the most materially wealthy in GNI PPP per capita in Southeast Asia (see Table 5.1). It is a global city–region equal to any but the world's three largest, a multicultural and multinational center of international business services rivaling Hong Kong, Frankfurt, Sydney, and Chicago. However, not every Singaporean gains the benefits of its major world role. Even in such a wealthy country, there are many poor—materially and socially.

Singapore ("Lion City") had a long history as a minor trading and fishing port before Sir Thomas Stanford Raffles of the British East India Company made an 1819 treaty with its owner, the Sultan of Johore (Malaya), to develop it as a trading point and settlement. Singapore's growth was rapid, and it became a United Kingdom crown colony in 1867. Its strategic location on developing shipping routes between Europe and China made it an entrepôt port—where goods from surrounding areas were transhipped to larger ships and goods from farther away were distributed in smaller boats (see Figure 5.12a). During this period, Chinese and Indian people migrated to Singapore, with the former group now 75 percent of the total population.

After World War II occupation by the Japanese from 1942 to 1945, Singapore became self-governing in 1959. In 1962 it entered the fully independent Federation of Malaysia with Malaya, Sabah, and Sarawak. However, irreconcilable differences over ethnic policy led to the federation expelling Singapore in 1965, when it gained official sovereignty. At first, the newly independent Singapore struggled to deal with unemployment, housing shortages, and a lack of natural resources. The government took strong actions to reduce unemployment, improve living standards, build large-scale public housing, increase national defenses, and remove racial tensions. Most former rural land is now covered with housing projects and regional commercial centers to reduce overcrowding in the old central area on the south coast. Entry to the central area is controlled by electronic road pricing.

The Singapore government redirected its economy from port-centered activities to industry. The economy depends heavily on exports, particularly in consumer electronics, information technology products, pharmaceuticals, and on a growing service sector that includes banking, medical services, tourism, higher education, and even gambling (the city–state opened its first casino in 2010). As a transportation and communications hub, Singapore became the regional headquarters of multinational corporations. The government remains involved in the business and social life of the country. It is one of the cleanest, safest, and most orderly urban areas in the world. However, its actions attract criticism, including the harsh punishment of seemingly minor crimes. It restricts length of stay for low-skilled and domestic workers, while encouraging well-educated businesspeople to settle. For some years, Singapore even encouraged intelligent and financially successful people to have more babies than the "less gifted."

Singapore has a multicultural population composed of people from many groups that speak different languages and have allegiance to different religions.

Difficulties still exist in relationships with Malaysia as both countries compete for major port customers. Singapore depends on Malaysia to supply freshwater: catching the tropical downpours on Singapore Island supplies half of its needs, but self-sufficiency is promised by extending reservoirs and building desalination plants.

Singapore positions itself as Southeast Asia's financial and high tech center (Figure 5.12b). Singapore's main thrust is to attract investments as an international business center. Its main trading partners are Malaysia, Indonesia, China, Japan, Hong Kong, and the EU and the United States.

East Timor

East Timor (Timor-Leste in Portuguese, Timor Lerosa'e in Tetum), occupies around half of the island of Timor (see Figure 5.10) and is the newest of the countries in Southeast Asia. Portuguese colonial occupation of Timor in 1702 followed centuries of passage and occupation of the mountainous island

(a)

(b)

Figure 5.12 **Singapore.** (a) The initial port and city center is where the name is printed. Changi Airport is one of the world's busiest. Most of the island is now built up with housing and commercial development, leaving a small area for nature reserves and gardens, but also keeping a number of historic buildings and the Science Center. What do the features shown on this map say about the tiny country of Singapore? (b) A part of the Central Business District with skyscrapers, located near the mouth of the Singapore River. It is one of the most densely populated areas of the city-state. *Photo: (b) © Corbis RF.*

successively by Australoid, Melanesian, and Malayan peoples, resulting in linguistic diversity. The largest group are the Tetum, concentrated around the capital city, Dili. Chinese and Indian trading networks from the 1300s valued the local sandalwood, slaves, honey, and wax. Pressure from the surrounding Dutch-controlled islands caused the western part of Timor to be ceded in 1859. The former Portuguese colony is 90 percent Roman Catholic.

After the World War II battles against Japanese forces, in which over 50,000 Timorese were killed, Portugal resumed control but abandoned its colony in 1975, and independence was declared on November 28. Nine days later Indonesia invaded before international recognition could occur, alleging that East Timor could have a Communist government. This overcame the initial Australian and U.S. objections to the invasion. Timorese guerrilla forces fought the Indonesian military, who punished the civilian population. It is estimated that up to 250,000 Timorese were killed out of the 1975 population of 600,000. In 1999 a UN-sponsored agreement among Portugal, Indonesia, and the United States led to a referendum and a vote for full independence. Violent clashes instigated by Indonesian forces and local militia were ended by an Australian-led peacekeeping force, although the militias continued to attack from their bases in West Timor.

The 1999 conflicts led to the destruction of East Timor's weak economic structure, and 260,000 people fled westward. UN efforts resulted in the reconstruction of urban areas and rural infrastructure. By the time of full independence and UN membership in 2002, around 50,000 of those who fled had returned. Much remains to be done, and East Timor has the lowest per capita income of any world country. Its economic future will be helped by the development of oil resources in the Timor Gap between it and Australia. In 2005, after some years of bickering over previously undefined boundaries between the two countries, an agreement was reached and joint development of the oil resources planned.

5.5 Contemporary Geographic Issues

ASEAN and APEC

Two organizations originating in this region reflect the shifts from inward-looking countries toward greater regional and global involvements (Figure 5.13). Formed in 1967, the Association of Southeast Asian Nations (ASEAN) claims to be a political, economic, and cultural organization of Southeast Asian countries, although it is widening its associations with neighboring countries. The Asia-Pacific Economic Cooperation (APEC) forum was formed on Australian initiative in 1989 and looks to wider international involvements.

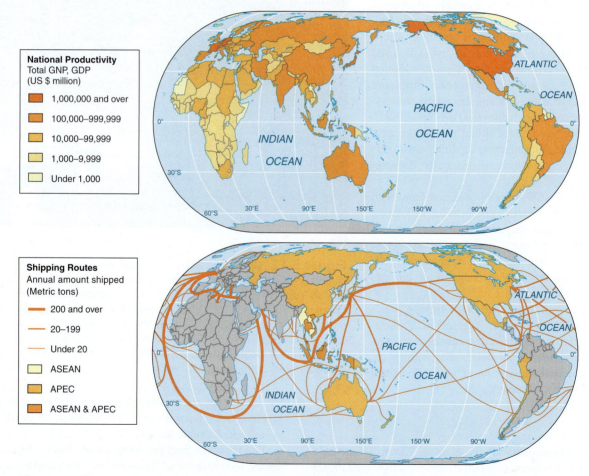

Figure 5.13 ASEAN and APEC: the countries of this world region in a global context. Compare the incomes of member countries with each other and those outside the membership, and comment on the location of major shipping routes.

Association of Southeast Asian Nations

The 1967 founding purpose of ASEAN and its members—Thailand, Indonesia, Malaysia, Singapore, and the Philippines—was to provide nonprovocative solidarity against Communist expansion in Vietnam and internal insurgency. The Bali Summit in 1976 proposed economic cooperation, which did not gain much support until 1991, when Thailand proposed a regional free trade area. Brunei joined ASEAN in 1984 and Vietnam, Laos, Myanmar, and Cambodia in the late 1990s. East Timor formally applied for membership in 2011. In 2010, ASEAN countries had nearly 600 million people and total GDP PPP of US$2,860 billion, compared to the EU (498 million people and US$14,794 billion).

ASEAN members are a politically, culturally, and economically diverse group of countries, including governments ranging from democracies to autocracies. The meetings of ASEAN leaders consider the economic and cultural development of member countries. From a pattern of meetings every three years from 1992, annual occasions took over from 2001, and meetings have occurred twice a year since 2008 with a rotating list of summit hosts. Regular features of these three-day events are meetings with the three East Asian Dialogue Partners (ASEAN+3: China, Japan, South Korea) and two other regional Dialogue Partners (ASEAN-CER: Australia and New Zealand). In 2005 an East Asian Summit brought together ASEAN and six Dialogue Partners (as above plus India) and established an ASEAN-Russia Summit.

ASEAN has regular meetings with other countries through the ASEAN Regional Forum, an informal group of 25 countries concerned about Asia-Pacific security issues that first met in 1994. They include Australia, New Zealand, Papua New Guinea, and East Timor, India, Pakistan, China, Mongolia, Japan, North and South Korea, Russia, the United States, Canada, and the European Union. In 2009 a free trade agreement was signed by the ASEAN regional block of 10 countries, New Zealand, and Australia. It is estimated that the agreement will increase agreegate GDP in these countries by tens of billions of dollars. The group is expected to establish a single market (similar to the European Union) in 2015. ASEAN's recent discussion of political and security issues, particularly those dealing with the South China Sea, is a departure from its customary focus on regional economic integration.

Asia-Pacific Economic Cooperation

APEC is composed of a group of countries around the Pacific Rim which meet to improve political ties and to promote economic integration and more open trade. It holds annual meetings by rotation in the member countries. Attending leaders dress in the national costume of the host country.

APEC arose partly from Australia's concerns that it would be isolated from relations with other countries after it had lost many of its European markets and received an initial lack of response from ASEAN. The first meeting in Canberra, Australia, attracted 12 countries. In 1993, U.S. President Clinton invited economic leaders to discuss promoting prosperity through cooperation in order to advance global trade talks. APEC established headquarters in Singapore.

Early APEC goals focused on free trade and investment, including the reduction of tariffs below 5 percent. In the early 2000s security and terrorism became concerns after confrontations with demonstrators outside the summit venues and the 9/11 events.

APEC membership increased from the original 12 (Australia, Brunei, Canada, Indonesia, Japan, South Korea, Malaysia, New Zealand, Philippines, Singapore, Thailand, United States) to include China, Hong Kong, and Chinese Taipei (Taiwan) in 1991, Mexico and Papua New Guinea in 1993, Chile in 1994, and Peru, Russia, and Vietnam in 1998.

In April 2006, Ambassador Tran Tong Tuan of Thailand, the executive director of the APEC Secretariat, noted changes in the nature and role of the organization. He stated that the rise of China in a cooperative and peaceful context required that a balance be achieved among the United States, China, Japan, and possibly India as a basis for prosperity in the Asia-Pacific. Human security is threatened by terrorism, epidemics, and natural disasters that are by their nature trans-boundary events that no single country can control. APEC has a major role in the Asia-Pacific region because of its wider-than-security basis and its ability to pioneer needed developments. At the APEC meeting held in Hawaii in 2011, members pledged to create a "seamless regional economy," to promote green growth, and to promote regulatory practices that facilitate trade and investment. Trade in the APEC region outperformed that in other parts of the world in 2012 despite the global economic uncertainty.

Conflicts: South China Sea

Ocean space is very important in this region, illustrated by territorial claims in the South China Sea by several countries surrounding it (Figure 5.14). The South China Sea is the world's largest sea body after the five oceans and is the world's second most used sea lane, being a major route for tankers delivering oil to China and Japan. It is almost surrounded by the countries of China, Vietnam, Cambodia, Thailand, Malaysia, Singapore, Indonesia, the Philippines, and Taiwan. There are 200 small islands, mostly uninhabited, and seamounts that do not break the surface. The main groups are the Spratly and Paracel Islands, the largest of which, Taiping Island, is only 1.7 km (less than 1 mile) long and 3.8 m (13 ft.) above sea level at its highest point. The whole area is subject to the surrounding countries' claims, which increased in intensity following the discovery of oil and natural gas potential. The proven reserves of oil total 7.7 billion barrels and estimates predict 28 billion.

Names for the sea mirror the claims on it. The European name is followed by China, but many Vietnamese call it the Eastern Sea and Filipinos the Luzon Sea. Recently, the name *Southeast Asian Sea* has been proposed for the water body. The 1982 United Nations Law of the Sea allows each country

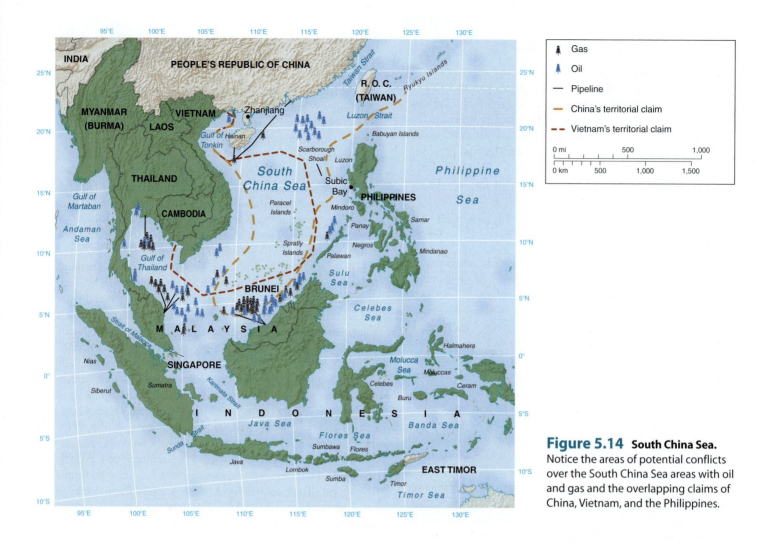

Figure 5.14 **South China Sea.** Notice the areas of potential conflicts over the South China Sea areas with oil and gas and the overlapping claims of China, Vietnam, and the Philippines.

to have an Exclusive Economic Zone of 200 nautical miles (371 km) beyond territorial waters. China, Vietnam, Brunei, Malaysia, the Philippines, and Taiwan all claim large portions. The People's Republic of China claims almost all of the sea and is building an aircraft carrier battle group to secure oil and natural gas lines. That country seized the Paracel Islands from Vietnam in 1974, and a clash between these countries in the Spratly Islands in 1988 led to over 70 Vietnamese deaths.

The ASEAN group does not want these territorial disputes to escalate. It established the Joint Development Authorities where claims overlap, in which joint projects develop areas and divide profits without settling the territorial claims. But there are sharp divisions within the organization on how to proceed with an ASEAN-China code of conduct for the South China Sea. President Benigno Aquino of the Philippines issued a proposal to transform the disputed area into a "Zone of Peace, Freedom, Friendship and Cooperation," but Laos and Cambodia did not send representatives to an ASEAN meeting in July, 2012 to consider this proposal, allegedly due to pressure from China.

Although tensions between China and the Philippines over ownership of the Scarborough Shoal (see Figure 5.14)

have eased, in June 2012, Vietnam reasserted its claims to the Spratly and Paracel Islands. As a Chinese state-owned company announced that it was opening nine blocks in the South China Sea to international bids for oil and gas exploration, thousands of Vietnamese protested against China's claims to the islands. Continued political tensions in the South China Sea could damage China's efforts to project its rise in Asia and the world as peaceful, and could hamper regional cooperation among ASEAN members (Figure 5.15).

Myanmar (Burma): Human Rights Violations and Economic Sanctions

From 1962 until 2011, Burma (Myanmar) was ruled by a military junta that suppressed almost all dissent and wielded absolute power despite international condemnation. The military government was accused of gross human rights abuses: the forcible relocation of civilians, the widespread use of forced labor, including children; excessive and even lethal force against peaceful protesters, torture by the army and police, the suppression and imprisonment of human rights activists, the lack of an independent judiciary, and restriction of Internet access through software-based censorship. Burmese citizens could be jailed for offenses such as owning a computer or a fax machine. The government is also accused of official discrimination against certain ethnic minority groups such as the Rohingya. Although many Rohingya have lived in Burma for generations, they are maligned as intruders who came from Bangladesh to steal scarce land. Clashes in 2012 between the Rohingya Muslims and Rakhine Buddhists led to thousands of houses being burned down and numerous injuries.

The EU, United States, and Canada imposed economic sanctions (restrictions upon international trade and finance that one country imposes on another for political reasons) on Burma. The country's military rulers were also sharply criticized by foreign governments and international aid agencies for delaying relief efforts following Cyclone Nargis, which struck in 2008 and killed more than 100,000 people and decimated the economy. Of the major economies, only China, India, and South Korea have invested in the country.

Political freedoms were lacking in Burma for decades (Table 5.3). Nobel Peace laureate Aung San Suu Kyi's National League for Democracy (NLD) party won a landslide victory in the 1990 multiparty elections, but was not allowed to govern. Placed under house arrest in 1989, Aung San Suu Kyi was detained for 15 years. Another general election was held in 2010 and could have been an important step in the transition from military rule to a civilian democracy. But allegations of fraud, continued domination of politics by the military, and widespread corruption remain. Still, the installation of a nominally civilian government under President Thein Sein in 2011 and the election to parliament of Aung San Suu Kyi in 2012 spell hope for democracy. Burma abolished direct media censorship in 2012, and more than 650 political detainees were freed, but many more remained imprisoned at this writing. Former Secretary of State Hillary Clinton said that the United

Figure 5.15 **South China Sea.** Chinese patrol boats sit docked at the pier in Sansha on an island in the disputed Paracel chain, which China now considers part of Hainan province. China has appointed military officers at a newly established garrison in the South China Sea in the country's latest step to bolster claims to disputed islands in the area. *Photo: © STR/AFP/Getty Images.*

States would consider easing sanctions if further progress was made toward political reform. The EU lifted all nonmilitary sanctions for a year in April 2012 and offered Burma more than US$100 million in development aid. Aung San Suu Kyi also called on the United States to ease sanctions and to make targeted investments in Burma, a possibility following President Barack Obama's visit to the country in November 2012.

Table 5.3	Debate: Economic sanctions in Myanmar
Advantages	**Drawbacks**
Economic sanctions are a better option than war to express aversion to country policies and provide incentives for change.	Black-marked countries such as Myanmar can circumvent embargoes by purchasing goods and finding investors elsewhere.
Totalitarian country governments rarely change their policies due to economic sanctions.	By damaging a nation's economy, we do harm to the people, not to those who are actually responsible for actions or policies that are flawed.
By forming an alliance and imposing trade embargoes, trading partners of the offending country can prevent the flow of items that could further the cause of conflict.	
Economic sanctions can undermine governmental legitimacy.	
Countries (such as the United States and Canada) that impose sanctions on Myanmar have an ideological benefit.	
Sanctions have rarely protected the rights and civil liberties of the people.	

The Tsunami of December 2004

Geographic knowledge and geographic tools were utilized immediately to assist in recovery efforts following the devastating Indian Ocean **tsunami** on December 26, 2004 (Figure 5.16). Centered just off western Sumatra, the submarine movements of ocean floor set off by earthquakes generated huge waves that hit local shores and within a few hours caused damage and loss of life around the Indian Ocean, including western Thailand, Sri Lanka, and even eastern Africa.

The analysis and computer manipulation of satellite images and aerial photos from planes and helicopters using geospatial tools such as geographic information systems (GIS) began within hours of the tsunami's huge waves hitting the coasts around the ocean. Before-and-after imagery made it possible to assess losses in the human landscape (houses, stores, hotels, roads, bridges, and port facilities). It also highlighted changes to the physical environment, such as significant alterations to coastal landforms, losses of mangrove vegetation, and creation of new inlets and islands as others were eliminated.

An understanding of the physical and human geography of Southeast Asia and other affected areas was essential in coordinating the efforts of aid agencies, regional governments, and the international community. Recovery efforts and subsequent continuing support required an in-depth geographic appreciation of such aspects as social relations and burial practices, political relationships, including local conflicts between government and insurgents, and transportation and communications networks available in the different areas affected.

Figure 5.16 **Indian Ocean tsunami: Koh Raya, Thailand.** People flee one of a devastating series of tsunami waves crashing on the shore near Phuket on Thailand's west coast. *Photo: © John Russell/ AFP/Getty Images.*

Kerala, India. A farmer races his paired oxen at a cattle race in the Palakkad district in Kerala, India. During this traditional race, which is held during the harvest season, two pairs of cattle are raced simultaneously on a 100-meter track on a flat and muddy field.
Photo: © Ranjithtravelogue.

LEARNING OBJECTIVES

After reading this chapter, you should be able to:

■ Recognize the physical boundaries and shared colonial history that define the South Asian world region.

■ Understand the physical forces that shape the region's climate, landforms, and natural resources, as well as the natural hazards and human-induced problems here.

■ Compare the population and urbanization characteristics of the different countries of the region, and describe the factors contributing to similarities and differences.

■ Locate the subregions of South Asia on a map, and explain the variations in economic growth experienced in different countries.

■ Describe the ethnic and religious diversity of the region, and explain why these diversities sometimes result in intrastate and interstate political tensions.

■ List the major resources and industries in the countries of South Asia, and describe the growth of different economic sectors and their global linkages.

South Asia

Figure 6.1 **South Asia: physical features, countries, and capitals.** The map shows how the Himalayas form the northern boundary of South Asia and are separated from the plateaus of southern India by the Indus, Ganges, and Brahmaputra river valleys.

6.1 Ancient Cultures and Modern Countries

South Asia, the smallest world region in terms of area, is the second most populous after East Asia. Most of the region consists of the Indian subcontinent enclosed by the Himalayan Mountains and reaching to the Indian Ocean (Figure 6.1). India, Pakistan, Bangladesh, Nepal, Bhutan, and Afghanistan are the countries located here. The region also includes the island countries of Sri Lanka and the Maldives.

South Asia bears the imprint of the rise and fall of civilizations and empires, the emergence and solidification of religions, and varied interactions between its people and the natural environment. Invading peoples, from the Aryans who entered the subcontinent in 1200 BC to the colonizing British who considered India the star in the crown of the British Empire, gave the region a shared history and perceived common culture but uneven political, social, and economic geographies.

Dominating the region with an area of over 2.6 million km² (1 million mi²) and a population of 1.3 billion, India is the world's seventh-largest country in area and second largest in population. Pakistan, although second in area in this region, is only a third as large as India. This Islamic country, which was once a part of British India, now has closer ties to its Muslim neighbors in Southwestern Asia. Bangladesh is also a Muslim country on India's eastern border and is almost completely surrounded by its larger neighbor. Comprising mostly the floodplains and delta of the mighty Ganges and Brahmaputra river systems, it is one of the most densely populated countries in the world. Nepal and Bhutan are landlocked Himalayan countries. The former is open to outside influences, while Bhutan limits its interactions with the global system. Sri Lanka, whose relatively high level of social development sets it apart from other South Asian countries, bears scars from the effects of prolonged civil war. The Maldives consist of over a thousand islands and depend on fishing and tourism.

6.2 Distinctive Physical Geography

The natural environments of South Asia are impacted by the Himalayan ramparts, the great plains of its major rivers, coastal influences, and the long dry seasons and uncertain rainfall that are the hallmark of the dominating monsoon climate. The mountains, plains, and hilly areas influence the distributions of settlement and economic activities. Furthermore, South Asia's large population places stresses on its densely settled and fragile natural environments.

Contrasting Landscapes

World's Highest Mountains and Deep Valleys

The Himalayan Mountains, forming the northern boundary of the region (see Figure 6.1), include Mount Everest (8,850 m; 29,035 ft.), the world's highest mountain, and nearly 50 other peaks that rise over 7,500 m (25,000 ft.). This mountain wall is the southern margin of the high and broad plateau of Tibet (China). The combination forms a wide barrier to both climatic influences and historic incursions of invading peoples.

The Himalayas resulted from the continental plate fragment that is peninsular India crashing into the Eurasian continental plate (see Figure 1.5b). The resulting collision thickened the crustal rocks to produce the world's highest mountains. The continuing thrust of peninsular India beneath Asia, signaled by earthquakes, raises the mountains by 6 cm (2.4 in.) per year. As they rise, however, glaciers and rivers cut into them and wear them down. At present, an approximate balance exists between the rate of rise and the rate at which these mountains are being eroded.

The headwater streams of the Indus and Ganges river systems breach the Himalayan wall in the far northwest. Deep valleys, carved by glaciers and rivers, provided lower routes, such as the Khyber Pass through the Hindu Kush range, that gave access to Afghanistan and Central Asia and formed historic points of entry for invading Aryan and Central Asian groups. At the northeastern end of the Himalayas, the ranges are lower as they turn southward into Myanmar, and they include breaks eroded by the Brahmaputra River system that afford Indian access to and from China.

Peninsular Hills and Plateaus

Peninsular India is a tilted continental plate fragment of very ancient rocks. The highest points along the western coast—the Western Ghats—rise to just over 2,500 m (8,000 ft.). Much of the peninsula consists of plateaus and hills sloping eastward to the Bay of Bengal—a gradient followed by many rivers. Along the east coast, a broken line of hills, the Eastern Ghats, rises in places to 1,500 m (5,000 ft.). Coastal plains of varied widths encircle the peninsula. They are at their widest where rivers such as the Krishna and Godavari create deltas on the eastern coast.

As the continental plate fragment forming the peninsula of South Asia broke away from the other southern continents over 150 million years ago, layers of volcanic lava poured out through cracks in Earth's crust, covering large areas of older land. The lava flows form the Deccan Plateau in the northwestern peninsula.

Major River Basins

Between the Himalayas and the peninsular plateaus, three major river systems—the Indus, Ganges, and Brahmaputra—cross a wide lowland zone formed by their deposits. The melting Himalayan snows and the monsoon rains combine in powerful flows in these rivers. The strength of flow can be gauged from the fact that the Ganges and Brahmaputra sweep debris 3,000 km (nearly 2,000 mi.) out to sea in the Bay of Bengal.

Rock particles and fragments worn from the Himalayas and carried by these rivers built deposits of **alluvium** up to 3,000 m (10,000 ft.) thick beneath the plains. The surface of the deposits is generally flat but marked by small-scale relief of a few or tens of meters where the rivers cut new channels in older material. Along the northern margins of this plain, large **alluvial fans** of gravel mark the junction of the steep mountains with the lowlands. The sudden lowering of gradient as they enter the lowlands causes the rivers to drop the coarse gravel and sand they carry along their steeper-gradient mountain valleys. In the drier areas of the northwest are sections of **badlands topography**, where occasional rainfall runoff cuts dense networks of steep-sided gullies into unvegetated alluvial deposits. The badlands were traditionally home to *dacoits* (robbers) who preyed on travelers. At their mouths the Ganges and Brahmaputra join to form the world's largest delta; low-lying, flood-prone land that occupies most of Bangladesh. These plains are densely populated and intensively farmed, using river water for irrigation. The Ganges is considered divine by Hindus, for whom it has great religious significance. Hindus believe that bathing in the Ganges cleanses them of sins, while dying on its banks helps attain salvation (Figure 6.2).

Monsoon Climates, Vegetation, and Soils

The monsoons dominate the climatic environment and life in much of South Asia (Figure 6.3). Monsoon winds bring heavy summer downpours of rain over much of the Indian subcontinent but little rain at other times of the year. Although the precise cause of the monsoons is still debated, these seasonal winds are affected by air pressure, the shifting position of high-velocity jet stream winds in the upper atmosphere, and heating and cooling of the Indian subcontinent (Figure 6.4). In the winter season of the dry monsoon, the Himalayan Mountains cut off South Asia from Central Asia and the freezing winds that blow from a high pressure area over the Tibetan Plateau. South

Asia remains warm and very dry as winds flow outward from high atmospheric pressure over the northwest. Only northern Sri Lanka and southeastern India receive rain at this season—from winds that blow out from the continent, over the Bay of Bengal, become moist by evaporation from the ocean surface, and turn toward the land.

As the land warms in early summer, temperatures in South Asia become very high (over 30°C, 86°F) and air rises, lowering atmospheric pressure. When the wet monsoon breaks in June or July, winds are drawn into the low atmospheric pressure area over the continent. Southwesterly winds from the Indian Ocean bring moisture, causing heavy rainfall on the western coastal mountains of the peninsula. The lift forced by humid air flowing up and over mountains (the orographic effect) adds to the amount of condensation and precipitation. Some of the world's largest annual rainfall totals, over 10,000 mm (400 in.) per year, are recorded in the Assam hills of northeastern India. In the peninsula, most of the monsoon rains fall on the Western Ghat mountains, leaving the eastern lands in a rain shadow, which receive lower summer rainfalls that vary sharply from year to year. Precipitation falls as snow at high elevations on the Himalayas, and the summer meltwaters add to river flows. Each year during the summer, a small number of tropical cyclones occur in the Bay of Bengal, bringing flooding and death to the Ganges-Brahmaputra Delta in Bangladesh. The concentrated rainfall and consequent flooding is also life giving, as it supplies the high water needs of crops like rice and jute and replenishes the rich alluvial soils of the river valleys and deltas.

The monsoon rains miss most of the northwestern parts of South Asia, including Pakistan and Afghanistan, which remain dry throughout the year, forming one of the world's major arid regions, the **Thar Desert**. The main source of water in this part of the region is the rivers, such as the Indus and its tributaries that are fed by the melting snows of the Himalayas.

Forests and Soils

Much of South Asia was originally forested, including the peninsula, northern river plains, and Himalayan foothills. After centuries of clearance by expanding populations for fuel and to increase the cultivated area, the teak forests of southern India and the forests on the Himalayan slopes are almost all that remain. These are being reduced further in size.

The most fertile soils occur in areas subject to annual flooding, on the lava plateaus or beneath the forests. Long use and the improper implementation of modern techniques have, however, reduced the soil quality. The drier area soils have been degraded by soil erosion, waterlogging, and salinization.

Natural Resources

The ancient rocks of the South Asian peninsula contain precious stones and mineral ores, including iron and radioactive thorium, while the newer rocks on top contain one of the world's largest coal reserves. Deposits of oil and natural gas occur in the thick sediments beneath the major valley areas and just offshore in both the Bay of Bengal and the Arabian Sea.

Figure 6.2 South Asia: the sacred river Ganges. The river at Varanasi draws Hindu pilgrims to bathe in its waters. The river is regarded as a goddess. It is supplied from Himalayan glacier meltwater and is prone to flooding. *Photo: © Alasdair Drysdale.*

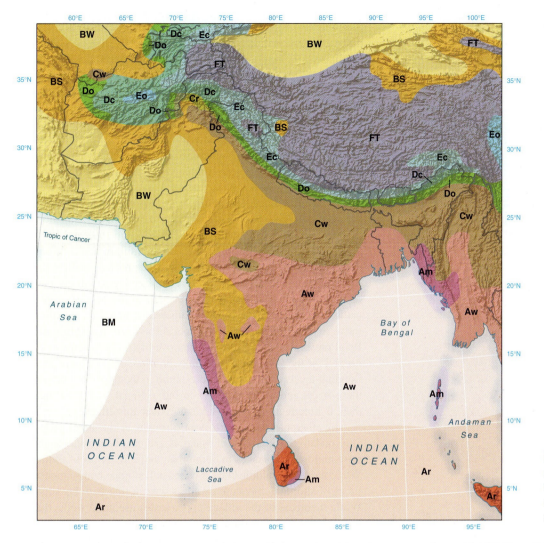

Figure 6.3 **South Asia: climates.** Climatic map, showing the extent of full monsoon conditions, lesser summer rains, arid areas, and highland climates.

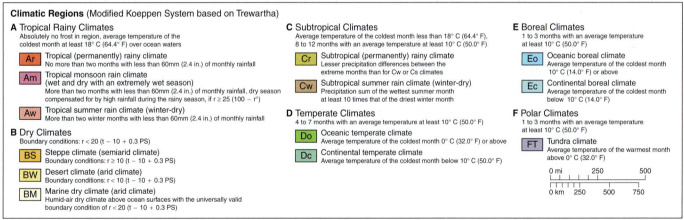

Climatic Regions (Modified Koeppen System based on Trewartha)

A Tropical Rainy Climates
Absolutely no frost in region, average temperature of the coldest month at least 18° C (64.4° F) over ocean waters

Ar Tropical (permanently) rainy climate
No more than two months with less than 60mm (2.4 in.) of monthly rainfall

Am Tropical monsoon rain climate (wet and dry with an extremely wet season)
More than two months with less than 60mm (2.4 in.) of monthly rainfall, dry season compensated for by high rainfall during the rainy season, if r ≥ 25 (100 − r°)

Aw Tropical summer rain climate (winter-dry)
More than two winter months with less than 60mm (2.4 in.) of monthly rainfall

B Dry Climates
Boundary conditions: r < 20 (t − 10 + 0.3 PS)

BS Steppe climate (semiarid climate)
Boundary conditions: r ≥ 10 (t − 10 + 0.3 PS)

BW Desert climate (arid climate)
Boundary conditions: r < 10 (t − 10 + 0.3 PS)

BM Marine dry climate (arid climate)
Humid-air dry climate above ocean surfaces with the universally valid boundary condition of r < 20 (t − 10 + 0.3 PS)

C Subtropical Climates
Average temperature of the coldest month less than 18° C (64.4° F), 8 to 12 months with an average temperature at least 10° C (50.0° F)

Cr Subtropical (permanently) rainy climate
Lesser precipitation differences between the extreme months than for Cw or Cs climates

Cw Subtropical summer rain climate (winter-dry)
Precipitation sum of the wettest summer month at least 10 times that of the driest winter month

D Temperate Climates
4 to 7 months with an average temperature at least 10° C (50.0° F)

Do Oceanic temperate climate
Average temperature of the coldest month 0° C (32.0° F) or above

Dc Continental temperate climate
Average temperature of the coldest month below 10° C (50.0° F)

E Boreal Climates
1 to 3 months with an average temperature at least 10° C (50.0° F)

Eo Oceanic boreal climate
Average temperature of the coldest month 10° C (14.0° F) or above

Ec Continental boreal climate
Average temperature of the coldest month below 10° C (14.0° F)

F Polar Climates
1 to 3 months with an average temperature at least 10° C (50.0° F)

FT Tundra climate
Average temperature of the warmest month above 0° C (32.0° F)

Water is a critical natural resource. Historically, the monsoon rains made it possible to support large populations at subsistence level. The Ganges, Brahmaputra, and Indus rivers brought economic life to the subcontinent's lowlands, especially after irrigation and water storage techniques were developed. The waters flowing northward from Sri Lanka's central hills also fostered a civilization based on irrigation. Water was instrumental in the Green Revolution. In the states of Punjab and Haryana, and parts of Uttar Pradesh, production rose and prosperity increased by irrigating more land, applying more fertilizer, and using more pesticides. Irrigated cropland across South Asia increased from 18 percent (132 million hectares; 326 million acres) in 1960 to 32 percent (174 million hectares; 430 million acres) in 1980, topped 35 percent in 2000 and reached 42 percent in 2011.

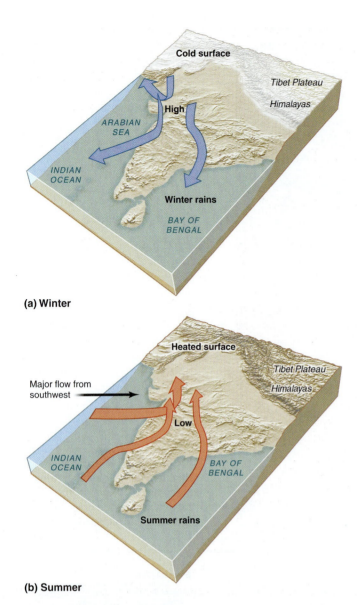

(a) Winter

(b) Summer

Figure 6.4 **South Asia: the monsoons of the Indian subcontinent.** (a) In winter, high atmospheric pressure in the northwest of the region leads to dry winds that blow outward. (b) In summer, air rises over the heated continent, drawing in very humid air from the Indian Ocean—the source of water for the monsoon rains.

Environmental Problems

People living in South Asia face a number of environmental hazards. Some are an integral part of the dynamic natural environment, but others are the outcomes of human clearance of land, population increase, and being a part of the global economic system.

In the dynamic natural environment, the clashing plates that produced the Himalayas continue to set off earthquakes, such as those at Latur, Maharashtra, in 1993 and Bhuj, Gujarat, in 2001. More frequent earthquakes along the base of the Himalayas are usually less devastating, although very large ones occurred in 1905 and 1934. The October 2005 earthquake in

Kashmir (Pakistan) measured 7.6 on the Richter scale. It killed at least 86,000 people and injured more than 69,000, as well as causing widespread destruction of buildings and roads. Earthquakes in 2008 and 2011 in Balochistan, Pakistan, were felt in the surrounding countries, but caused less destruction than the 2005 earthquake as they occurred in more sparsely populated areas. An earthquake in northeastern India and Nepal in 2011 caused damage to buildings and resulted in at least 16 deaths.

Flooding is a major problem in the lower Ganges and Brahmaputra valleys and their combined delta. Snowmelt in the Himalayas combined with heavy rainfall, the funneling effect of the Bay of Bengal, and storm surges from tropical cyclones, deforestation, low topography, and a very dense population make flooding a frequent disaster. Peak floods in 1998 and 2004 affected over 30 million people. Nearly 1 million houses were destroyed due to flooding in Pakistan's Sind province in 2011. In India and Bangladesh, floods in 2011 displaced tens of thousands of people from their homes and resulted in the loss of crops, cattle, and housing. It would be impossible, however, to either move so many people to unfloodable areas or construct effective flood prevention measures. But improved flood preparedness is needed for the vulnerable materially poor population. Increased flooding levels result from deforestation in the upper reaches of these rivers and their tributaries in India, Nepal, and Tibet, but there is little regional cooperation to reduce this effect.

In other parts of South Asia, drought is the main problem, and the availability of water is critical. Water shortages raise major social issues in both urban and rural areas of India. While the urban middle classes can install storage tanks against times of shortage, the slum dwellers wait in line with buckets. While members of the village upper castes use good wells, the untouchables must find their water elsewhere.

In the future, the Maldive Islands and coastal areas such as much of the Ganges-Brahmaputra Delta that are only a few feet above sea level face drowning by a rising ocean level as global warming proceeds. The high costs of raising dikes and building sea defenses make such measures unlikely. Coastal areas are also at greatest risk during natural disasters such as the Asian tsunami in December 2004. The South Asian countries most affected by that event were Sri Lanka, where over 31,000 people died, and India (primarily the state of Tamil Nadu), which reported over 10,000 deaths.

Human-Induced Environmental Problems: Air and Water Pollution

The huge rise in South Asia's population and the necessary growth in national economies leave landscapes of exploitation, degraded resources, and pollution.

Making the herbicides and pesticides used in the Green Revolution resulted in toxic concentrations of chemical factories. On December 3, 1984, at the Union Carbide pesticide plant in Bhopal, India, water leaked into a methyl isocyanate storage tank, triggering chemical reactions and a cloud of toxic gases, causing the world's worst industrial accident. At the time it killed 3,000 people immediately and another 12,000 subsequently in the adjacent slum areas (Figure 6.5). These official

Figure 6.5 **South Asia: pollution sources.** The Union Carbide chemical plant and nearby poor housing in Bhopal, India. *Photo: © Jagdish Agarwal/The Image Works.*

figures are thought to have been around one-third of the true numbers. Nearly 30 years after the event, the site, now owned by the state of Madhya Pradesh, remains highly contaminated. Both the state and Dow Chemical (the U.S. parent company of Union Carbide) deny responsibility for cleaning up the site. This was a well-publicized, but not isolated, instance of the multinational corporation re-siting "dirty" manufacturing facilities in India.

Increasing damage to the Taj Mahal because of air pollution led to the closure of metal workshops in the surrounding area to reduce the sulfur gases that damage building stones such as limestone and marble. However, it is only possible to take such expensive action on a local scale.

About 40 percent of the Indian population lives in the middle and lower Ganges River basin. The land is intensively cultivated, and there are many factories and urban areas. Although considered sacred and health-giving by Hindus, the Ganges suffers from pollution from various sources and poses health risks to those who use its water. For example, heavily polluted effluent from a concentration of leather tanning works enters the Ganges River close to some of the main Hindu ritual bathing sites. In recent years, an upsurge of trash

dumping in the Ganges, combined with funeral pyres along its banks, further worsened the water quality. The 1980s Ganga Action Plan aimed to clean up the river, but, although a few projects were implemented, lack of funding, continuing industrial malpractice, illegal dumping of wastes into the river, and illiteracy brought the program few successes. In 2011, the Indian government with support from the World Bank launched the National Ganga River Basin Project, another cleanup program.

6.3 Distinctive Human Geography

South Asia is a region of diverse peoples and cultures with its many languages and distinctive religious orientations. Home of one of the world's oldest civilizations (in Mohenjodaro, Harappa, and Kalibangan in the Indus Valley), the region has been impacted by the incursion and assimilation of different groups of people, the rise and fall of powerful local empires, and the effects of British colonial rule that lasted for over a century from the mid-1800s onward. Waves of people, including the Aryans and Mongols from Central Asia and Turks and other Muslims from Western Asia, entered through the northern mountain passes to access the rich plains. Later, traders including the Muslims (from AD 1100) came from the ocean, as did the colonizing Europeans (from 1500). Each group left its marks on the land, people, and culture (Figure 6.6). Among South Asian innovations that benefited the world are the decimal system and zero. Today, this region interacts with the rest of the world from a complex cultural base that fuses traditional values and modern ideas.

South Asia's total population of 1.7 billion in 2012 is predicted to rise to over 2 billion by 2025, compared to less than 1.7 billion in East Asia. The region has a high average population

Figure 6.6 **South Asia: historic buildings.** The entrance to the mausoleum of Akbar, the greatest of the Mughal emperors, near Agra, India. *Photo: © Bill Westermeyer.*

density of 186 persons per square kilometer and a high proportion (39 percent) of arable land. This means that additional growth will place severe pressures on the environment when the region has few current or prospective means of supporting its people at improved levels of life quality. India's population of 1.26 billion in 2012 is the world's second largest after China's, and its rate of increase is greater. Estimates for 2025 raise India's population to 1.46 billion. Growth continues because of the huge number of people of childbearing age (Figure 6.7) and increased life expectancy. In the demographic transition process, India remains in the population increase phase. The rapid increases of population in South Asia are caused by birth rates that remain higher than death rates. Birth rates are highest among the materially poorer groups, especially in rural areas, leading to significant changes in the social balance and political pressures from hitherto disadvantaged groups.

Bangladesh had a slightly higher population than Pakistan until 1980. By 2006, however, Pakistan's population exceeded that of Bangladesh (166 million compared to 147 million), and estimates for 2025 suggest a greater gap, with Pakistan having 230 million and Bangladesh 183 million people. The greater rate of population increase in Pakistan is the result of a continuing high total fertility rate of 3.6, compared to 2.3 in Bangladesh. Both have life expectancies in the mid- to late sixties.

Given its smaller area, Bangladesh has greater problems of feeding its growing population, an issue that stimulated its government to institute effective family planning programs in the 1970s that resulted in significant reductions in fertility.

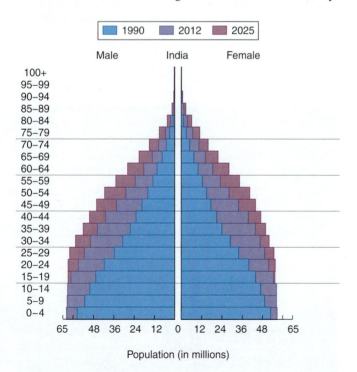

Figure 6.7 India: age-sex diagram. Even if fertility declines, the large numbers of young people will produce continuing population growth for many years to come when they reach childbearing age. *Source: U.S. Census International Database: http://www.census.gov/population/international/data/idb/region.php.*

Currently over 50 percent of married women use contraception, up from 8 percent in 1975. Despite a continuing subordinate role for women, new policies encourage their social and economic development.

Pakistan's need for an educated work force is undermined by the combination of influences from Islamic leaders, feudal landlords, and poor government that kept Pakistan's school population low until the late 1990s. Its enrollment rates in primary and secondary education were only 30 percent, compared to 70 percent in India and over 80 percent in Indonesia and China. In the mid-1990s, aid donors' demands led to new national policies for education, health, and population. Government funding shifted from elite university education toward universal primary education.

Of the countries on the margins of this world region, Afghanistan's population grows most rapidly at around 2.8 percent per year, followed by Pakistan with 2.1 percent. The Maldives' and Sri Lanka's populations increase at just over 1 percent. In most countries, total fertility rates are declining but remain over 6 in Afghanistan and over 3 in Pakistan. In Nepal, a strong family planning program, widely disseminated, reduced the fertilty rate to 2.6. Lower birth rates could be linked to the Nepalese female adult literacy rate, which was 3.7 percent in 1971 but rose to 35 percent by 2004. As life expectancies rise, the population increases rapidly and there are many young people.

Distribution and Density

As a relatively small world region that is home to the world's largest population, South Asia's population numbers and distribution reflect interactions between global and local influences. Apart from a few almost uninhabited areas, this region has high densities of population. The greatest concentrations are in the lowlands of the Indus, Ganges, and Brahmaputra rivers and around the southeast coasts of Sri Lanka (Figure 6.8).

Areas of very low population density include the desert areas of Pakistan and the western area of India, the mountainous areas of western and northern Pakistan, Afghanistan, the swampy Rann of Kutch that is mostly within India, and the Himalayan region in northernmost India, Nepal, and Bhutan. These areas hold little prospect for future population expansion. Such areas of very low population form breaks along the region's political borders and help to clearly define South Asia.

The rest of the region has moderate population densities, which often reflect the availability of resources that support rural employment. Over most of the region, improving water supplies for irrigation scarcely keep up with rising numbers of people, and so these areas of moderate densities offer few prospects for supporting more people.

Around 20 million ethnic Indians live abroad, with the largest groups in Nepal, South Africa, Malaysia, Sri Lanka, the Persian Gulf states, the United Kingdom, and the United States. The Indian diaspora has introduced elements of Indian cultures in these countries in the form of Indian food, Indian films, and even ethnic religions such as Hinduism and Sikhism through the erection of temples. In recent years, flows of professionals to the United Kingdom and United States for better-paid

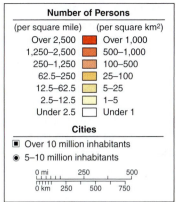

Figure 6.8 **South Asia: distribution of population.** Note the contrasts within South Asia: high population densities occur in the major river valleys and low population densities occur in high mountains and arid areas.

opportunities have created a significant "brain drain." Many Pakistanis, Bangladeshis, and Sri Lankans also live and work abroad, and there was a major period of migration to the United Kingdom from the 1960s when former colonial people were allowed a British passport. More recently, workers from these countries moved to jobs in the oil-producing countries of the Persian Gulf and to Southeast Asia. Their remittances to families in South Asia form important income contributions.

Cultural Diversity

Religion

The cultural diversity of South Asia includes numerous religions, languages, and dialects. Several ethnic religions (**Hinduism**, **Sikhism**, **Jainism**) and **Buddhism**, a global religion, originated in South Asia. Additionally, other global religions such as Islam and Christianity have a strong presence in the region. Hinduism, the religion of 80 percent of India's population (see Figure 1.18a), is believed to have crystallized in the Indus Valley in 1200 BC.

Most accurately described as "the religion of the people" and "a way of life," Hinduism is an ethnic religion into which people are born. Today, nearly all of the world's 800 million

Hindus live in India. In the Himalayan kingdom of Nepal, almost 90 percent of the population is Hindu. In Sri Lanka, Hindus are a substantial minority group (about 15 percent of the population, mostly Tamil). About 35 percent of Bhutan's population is composed of Hindus of Indian and Nepalese origin.

Caste is a Hindu concept that was solidified by invading Aryans by at least 1500 BC. The **caste order** is similar to ethnic or class divisions elsewhere. Aryans grouped society into four categories: the priests (*Brahmins*—the top group); warriors and rulers (*Kshatriyas*); and commoner merchants and artisans (*Vaishayas*). Those outside these divisions, mostly non-Aryans, were the lowest caste of menials and servants (*Shudras*). Those who did not belong to one of these groups were outcasts and labeled "untouchables." Caste membership became hereditary, based on intermarriage within the group; birth determined status in society. Caste permeates all South Asian society and may still determine where people live, work, and who they marry, especially in rural communities.

Buddhism and Jainism were founded in the Ganges River valley in the 500s BC in reaction to aspects of Hinduism such as the rigid caste system and hundreds of gods. Asoka, the most celebrated ruler of the Mauryan Empire that held sway over most of the Indian subcontinent from 321 to 181 BC, adopted Buddhism. He propagated it and sent Buddhist missionaries abroad to spread their faith. As a result, much of Sri Lanka was converted to Buddhism, which also spread eastward and northward into Southeast and East Asia. After Asoka's death, Buddhism declined and Brahman Hinduism regained its importance in India.

Sri Lanka has a Buddhist majority (70 percent), composed of the dominant Sinhalese group. Bhutan is a Lamaistic Buddhist country like its neighbor, Tibet. Nepal has a small proportion of Buddhists who have had a major impact on the local character of Hinduism. Afghanistan was once home to a fusion of Buddhist and Persian cultures evidenced in artifacts such as ancient giant Buddha statues in Bamiyan, which were destroyed by the Taliban government in 2001 despite international efforts to save them.

Jainism, which emerged in the 600s BC, is associated with its founder Mahaveera. Built around a code of nonviolence, it is designed to save followers from an endless cycle of birth and rebirth. This nonviolent code was later taken up by Mahatma Gandhi in his early-1900s campaign for India's independence from the British. Forbidden to farm (and kill worms and insects), Jains became an exclusive group. Many, especially in the Mumbai (Bombay) area, are wealthy traders. They make up a small proportion (less than 1 percent) of the Indian population today, but include over 3 million people.

Islam was brought to India by Arab traders in the AD 700s and spread by Sufi mystics and teachers in former Buddhist strongholds in northwestern and eastern India, particularly Punjab and Bengal. Muslim invasions from the 1200s onward and associated conversions under Muslim sultans and emperors, especially the **Mughal** (**Mogul**) dynasty, solidified Islam's position in India. Under the Mughal dynasty, established in the 1700s by the Persian Turk Babar, India experienced developments in such fields as architecture: the Taj Mahal and many examples of Muslim buildings built in the 1600s (see Figure 6.6) survive throughout the region. Muslims form an important religious minority in India with about 177 million people in 2010, who comprised 14.6 percent of the country's population, making India the country with the third-largest Muslim population in the world. Pakistan and Bangladesh are Muslim states whose pre-independence history is closely tied to that of India. Arab (Muslim) control of medieval Indian Ocean trade routes included the Maldive Islands, which stretch southward into the Indian Ocean. The islands became solidly Muslim by the AD 1100s and remain so. Afghanistan is 99 percent Muslim, of which 84 percent are Sunni and 15 percent Shiite.

Sikhism, which emerged in the early 1500s, combines aspects of Hinduism such as belief in reincarnation with strict monotheism, borrowed from Islam. Sikhism preaches universal toleration and is marked by a strict code of conduct; many temples have kitchens providing food for all. The Golden Temple at Amritsar is the holiest temple. Although small in number (approximately 20 million in India, composing 2 percent of the population), many Sikhs became wealthy from their bumper crops in Punjab. In 1980, extremist Sikhs in Indian Punjab occupied the Golden Temple in Amritsar and used it as a base for killing rival Sikhs until they were ousted in 1984 by a major military offensive. In revenge, they assassinated Prime Minister Indira Gandhi, after which Sikh areas of Delhi were attacked and burned by mobs.

Religion, with diverse expressions, remains important in South Asian life. Religious festivals, traditions, and rituals are celebrated with fervor; religious leaders, from the local Sai Baba of Shirdi to Sri Sri Ravi Shankar with his global Art of Living Foundation, have thousands of followers. However, religion is a tinderbox within South Asia, and minority religious groups remain vulnerable. Feelings of alienation have caused Christian groups in Nagaland State in northeastern India and the Sikhs in Punjab to lobby for a separate state or even an independent country (Sikh Khalistan, the land of the pure). In Sri Lanka, the dominant ethnic groups (Sinhalese, Tamil) have different religions (Buddhism, Hinduism) as well, adding to the tensions between them.

Languages and Dialects

The variety of languages in South Asia rivals its religious diversity. India alone has over 1,600 languages and dialects. Hindi, India's national language, is spoken and understood by about 40 percent of the population. It is merely one of 16 official languages recognized by Indian states and the one most used in movies and TV. Of India's official languages, the Indo-Aryan languages are prevalent in the north and Dravidian languages that originated in India predominate in the south. English remains the lingua franca, the common language used by the legal system, and is spoken by the educated elite. It is often used in modern Indian literature and university courses and is a common form of communication in commerce and national politics. Speakers of minority languages are at an increasing disadvantage as globalization affects more of India.

Bengali is the language of 98 percent of the people of Bangladesh, who originated from Indo-Aryan stock that mixed with local ethnic groups. The Biharis form a smaller group of non-Bengali Muslims who speak Urdu and migrated to East Pakistan after independence in 1947.

Urdu, a form of Hindi with Arabic script used by Muslim people before independence, is the official language of Pakistan. The Muhajirs, migrants from Hindu India to Pakistan in 1947, speak Urdu. Punjabi and Sindhi are also widely spoken in Pakistan. Many Pashtun groups live in the hills bordering Afghanistan. Rivalries among groups and the elitism of Urdu-speaking feudal landowners and military officers continue to dominate social and political life in Pakistan.

Fifty percent of Afghanistan's population speaks an Afghan form of Persian, which is the dominant language. Pashtuns dominate the south, while groups of Central Asian peoples, such as the Tajiks and Uzbeks, inhabit the north. Each of these ethnic groups has its own language and dialects (Figure 6.9). In Sri Lanka, the Sinhalese speak Sinhala, a language that belongs to the Indo-European language family, while Tamil, a Dravidian language, is the mother tongue of the Tamils.

Global Linkages and the Impact of Colonialism

Historically, the material wealth of South Asia was built on gem and metal deposits and also on external trading links to African, Arab, and Southeast Asian lands. Although the region has had

Ethnolinguistic Groups

- Baluch
- Pashtun
- Hazara
- Nuristani
- Ismaeliens
- Turkmen
- Uzbek
- Tajik
- Kyrgyz
- Other

Cities

- ■ Over 1 million inhabitants
- ● Over 100,000 inhabitants
- ○ Over 50,000 inhabitants
- —— Roads

0 mi 100 200
0 km 100 200 300

Figure 6.9 Afghanistan. The complexity of ethnic groups and their links across the political borders is a major factor in the difficulties of governing Afghanistan. The small number of border crossings in this mountainous country link some of the groups.

links with European countries for centuries, its seaborne trade was controlled by the Arabs since the AD 800s. The great wealth of this region became a magnet for European adventurers in the mid-1400s searching for an alternative route to get Indian products and riches to Europe that bypassed the Arab middlemen. In 1498, the Portuguese explorer Vasco da Gama reached India after sailing around southernmost Africa. The Europeans who arrived by sea found few defenses. During the 1500s, the Portuguese and Dutch established trading stations, such as Goa, around the coasts of South Asia. The first British trading post followed in 1612. During the 1600s, the Dutch forced the **(British) East India Company** out of the East Indies, and the British switched to India, ousting their Portuguese and Dutch rivals from most trading centers. The British East India Company took the Portuguese port of Bombay ("good bay," now Mumbai), built a new port at Madras (Chennai), and began trading with the most populous area of Bengal around modern Calcutta (Kolkata), which became the company's main center. At the beginning of the 1700s, India's GDP rivaled China's as the largest in the world.

The British Empire

By the early 1700s, political chaos caused by internal factions splintered South Asia into small and large kingdoms, ruled by Muslim or Hindu princes. Foreigners took control of the increasing overseas trade in cotton, cloth, rice, and opium. Over the next 100 years, the British East India Company, backed by the British Indian army that employed Indians as foot soldiers (sepoys), increased its hold on South Asia, taking a particularly strong position in Bengal and the peninsula. The company's area of influence was extended southward into Ceylon (now Sri Lanka) in 1798 and westward into Punjab in the mid-1800s.

A major mutiny of its Indian sepoy troops in 1857, which many Indians see as "The First Independence War," led the British government to take full political control of South Asia. It abolished the British East India Company and established the **British Indian Empire**, often referred to as the British Raj (*raj* means rule or government). It included almost all of the subcontinent. However, 40 percent, or almost one-fourth of the total population, remained under 600 "independent" princely family governments, who governed their territories in harmony with British policies (Figure 6.10).

The British saw their role as "civilizing" India through Western education, new technology, public works, and a new system of law. However, benefits went both ways. British interests redirected India's farms to produce raw materials for British industries. The region entered the expanding global economy of the later 1800s, when the export of Indian cotton compensated for negative British trade balances. Furthermore, the British Indian army became a tool of attempted imperial expansion into Afghanistan and Burma in the 1880s and into Tibet in 1903–1904. As British colonies, Ceylon and the Maldives were also incorporated into the world economic system that focused on the demands of the colonizing country and other wealthier markets rather than on local needs.

British imperial rule had massive effects on the geography of the peninsula. Its mercantile economy focused on primary products to be exported to Europe, defined the resources that could be developed, determined what was produced, altered patterns of land control, selected areas for development and cities for growth, and controlled external trade. British engineers irrigated land in the Indus and Upper Ganges river basins to produce cotton for export to its textile mills while the Indian textile industry was suppressed in favor of British cotton goods. They built railroads from the main ports to move troops and exports. Former communal land was reallocated to larger and smaller landowners, forming interest groups that were expected, in return, to support the colonial administration. The cities of Calcutta, Bombay, Madras, and Karachi grew faster than others after the British East India Company designated them as foci where lines of overseas trade and internal communications met.

The British colonizers also set up schools to educate the native population and teach them English. As a result, English became the lingua franca and unifying language among the traders, lawyers, and elite families. India has the second-largest population of English speakers (about 125 million) after the United States. Fluency in this global language has helped South Asian participation in global arenas of trade, business,

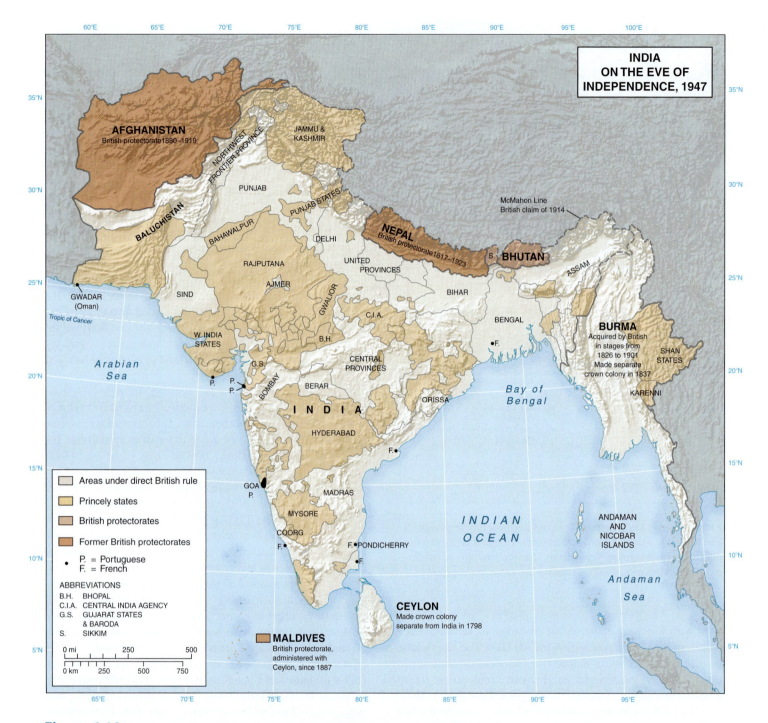

Figure 6.10 **South Asia: the British Indian Empire just before independence in 1947.** The princely states had working arrangements with Britain but were not ruled as part of the Indian Empire. At independence they agreed to be part of the new countries of India and Pakistan and subject to their governments. Kashmir is still disputed between India and Pakistan.

education, and a wide variety of professions. Within India, an English-speaking, college-educated work force is the basis of a low-cost service industry in back-office work such as call centers and other information technology-enabled services.

Although the countries of South Asia have made considerable progress in economic and social arenas and are increasingly involved in the global economy, they remain among the world's poorest (Figure 6.11). Low possession of consumer goods, access to clean water, and energy usage (Figure 6.12)

show that even the most advanced countries here have a long way to go toward development.

Rural and Urban Contrasts

South Asia is a study in contrasts. Extremes of abject poverty coexist with tremendous wealth and modern luxuries; traditional village life in many rural areas contrasts with cities and their high-rise office blocks and prestige apartments, interna-

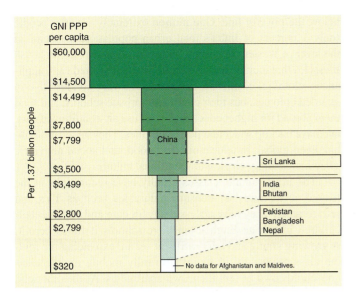

Figure 6.11 **South Asia: country average incomes compared.** The countries are listed in order of their GNI PPP per capita for 2010. *Source:* World Development Indicators, *World Bank; Population Reference Bureau.*

tional hotels, modern conveniences, and global influences. Even though the region has some of the largest cities in the world, South Asia's population remains largely rural (Figure 6.13). Rural areas still have large proportions of the semisubsistence and low-paid farming lifestyles that are linked to long-term poverty, and they lag behind urban areas in education and health care provision. By 2025, it is projected that India's rural population will be down to 55 percent of the total.

Rural Poverty

Laborers have little chance of escaping a lifetime of poverty, and debts are inherited by children. Cash incomes are small. Often unexpected obligations, such as sickness, the death of plowing oxen, or a demand for a dowry on the marriage of a daughter, deplete savings. In India, a manual laborer pays up to four times his annual wages for a daughter's dowry. Money lenders help with loans, but inability to repay leads to mortgaged and often lost land, while members of the family may be taken into bondage. Countermovements exist to combat rural poverty. Cooperative peasant holdings exist in places such as the Indian state of West Bengal. The most encouraging are the women-run cooperatives and small rotating loan projects such as the Grameen Bank, originally established in Bangladesh but now found throughout the region. In 2005 the Indian government moved to extend a guarantee of employment to all rural areas, with jobs mainly in public work projects at minimum wage levels.

Not all farmers are poor. In India and Pakistan the Green Revolution led to the emergence of wealthy farmers who enlarged their holdings and invested in commercial production, particularly in Punjab, Haryana, and Uttar Pradesh. They employ labor from the poorer states and campaign for even larger government subsidies for seeds, pesticides, fertilizers, and irrigation water. Agribusiness replaces semisubsistence farming, bringing higher living standards, mainly for farmer-owners.

While wealth largely determines social status in South Asia's cities, caste discrimination is most obvious in rural areas, where one's status is known to all. Outcasts were traditionally referred to as "untouchables" and later given the official designation of "scheduled castes." Today, *dalit* (meaning "ground

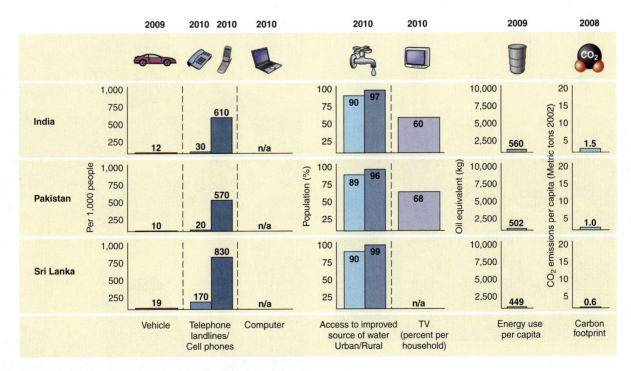

Figure 6.12 **South Asia: ownership of consumer goods, access to piped water, and energy usage.** How do these figures compare with other Asian countries? *Source: Data from* World Development Indicators, *World Bank 2012; Population Reference Bureau World Data Sheet, 2012.*

Figure 6.13 **India: farming.** Farmers in the north India plains share a new tractor for fieldwork. *Photo: © David Zurick.*

down" or "oppressed") is more commonly used and preferred. Dalits are still barred from using the same wells as upper-caste Hindus, dalit homes are still burned, and caste-based political control remains in many rural areas.

After Indian independence in 1947, positive discrimination for the newly designated scheduled castes gave them access to government employment, rural land ownership, and reserved places in public colleges. Jagjivan Ram, a dalit, became India's Deputy Prime Minister in 1979.

Urban Differences

In the cities the contrasts are equally obvious. As of 2011, India had 125,500 millionaires (in U.S. dollars), who make up 0.01 percent of the country's population, while 30 percent lives

below the poverty line. One million extremely wealthy people (only 1 percent of India's total urban population) consume at the level of Western countries, and spend their money on lavish lifestyles that include luxury vacations and imported cars, appliances, clothing, cosmetics, and food. The very wealthy live in guarded colonies, bus their children to private schools, and send them abroad for higher education. They use air-conditioned cars, and have backup water and power supplies. They are cosmopolitan, even global, in education, livelihood, and social activities. They increasingly break with traditional culture.

The urban middle classes usually have secure jobs in the formal sector and buy consumer durables such as TVs and refrigerators. They have access to health care and educate their children in private English-language schools if funds permit. A growing young and fairly affluent set are expanding the middle class, especially in India. Upwardly mobile middle class households own or rent their own apartments. At the other end of the middle class housing spectrum, many live in overcrowded tenements. Life is a struggle against impoverishment.

The urban poor, on the other hand, struggle for existence in shantytown slums (*bustees*) or on the sidewalks (Figure 6.14). Many slum dwellers are rural migrants with no title to housing or rights to establish businesses. They work in the unorganized and unprotected informal economy where child labor is common. Children work for their living in tea stalls, industrial workshops, construction sites, scavenging, or begging (Figure 6.15).

Figure 6.15 **India: informal sector.** A young girl carries bricks on her head at a construction site. Why is child labor still prevalent in South Asia? *Photo: © Phillipe Lissac/Godong/Corbis.*

Figure 6.14 **South Asia: shantytown.** Slum in Mumbai with dwellings made of sacking, plastic, wooden boards, and metal sheeting. Compare these structures with the apartment blocks in the distance. *Photo: © Scott Eells/Bloomberg/Getty Images.*

Few can afford to obtain an education, and there are few schools in the slums. There is a widening gulf between the marginalized urban poor and the more affluent, middle class families.

South Asia has a large and growing manufacturing sector ranging from small-scale craft industries (Figure 6.16) to large-scale steelworks and modern high tech facilities. Small-scale industries in India employ 140 million people and produce 35 percent of all manufactured goods. An example of such an industry is the leather works of Kanpur, which produces "quality handmade goods" such as shoes and purses for national and global markets in unsafe work environments, using poorly paid and often, child labor. In contrast, quality specialty steel is

Figure 6.16 **South Asia: major geographic features, countries, and cities.** India dominates the region by its size and centrality; Bangladesh and Pakistan are two Muslim countries created as one at independence but since broken apart. Afghanistan, Nepal, and Bhutan form the Northern Mountain Rim, while the Maldives and Sri Lanka are the Southern Island Rim. This is a region of huge and growing cities. In Pakistan A = Waziristan, B = Tribal Area, and C = Northwest Frontier.

produced using the most modern techniques in Jamshedpur, at prices undercut by only the best South Korean mills.

6.4 Geographic Diversity

A new diversity within South Asia emerged in separate countries following the end of the British Raj (see Figure 6.16 and Table 6.1). The three groupings recognized here are:

- The Republic of India, the world's second-largest country in total population, has a dominant presence within the region.
- Bangladesh and Pakistan are two Muslim countries, each with over 100 million people, resulting from the breakup of the original Pakistan in 1971.
- The Mountain and Island Rim, comprising Afghanistan, Nepal, and Bhutan in the mountainous north and Sri Lanka and the Maldives in the island south, includes a group of smaller countries on the margins of South Asia.

India

India has the world's second-largest population and dominates South Asia in land area, resources, and economic activity. In 2012, the Republic of India's population of 1.260 billion (see Table 6.1)

was second only to China. As an increasingly significant major power, India is a product of its history and despite limited resources to underwrite change, the country has an expansive vision of its future.

India became an independent country in 1947 as a republic with a federal union government. The former princely states and British Indian provinces were admitted as states within India's federation (see Figure 6.10). There are 28 states and seven union territories, each with its own elected assembly. State boundaries relate to the distribution of languages, reflecting the huge range of languages spoken in the country. The number of states increased after independence, mainly as the result of the original states not coping with the interests of ethnic groups.

India is the world's largest democracy. Every adult has a vote. Women have a stipulated minimum number of seats in local government. The union (federal) government has responsibilities for defense, foreign affairs, and currency. It raises most of the taxes and makes grants to the states. In the 1990s, states increased their functions, particularly in the delivery of social programs, including health and education.

Agriculture

The country's economic profile is moving from a farm base to factories and offices. Between 1965 and 2008, farming decreased its contribution to GDP from 44 to 17 percent without causing famines. Due to the adoption of Green Revolution technologies that

Table 6.1	**South Asia:** Data by country, area, population, urbanization, income (gross national income purchasing power parity), ethnic groups						
Country	**Land Area (km²)**	**Population (millions)** Mid-2012 Total	2025 Est. Total	**% Urban** 2012	**GNI PPP 2010** Total (US$ billions)	**2010** Per Capita (US$)	**Ethnic Groups (%)**
INDIA							
India, Republic of	3,287	1,259.7	1,458.2	31	4,159.7	3,400	Indo-Aryan 72%, Dravidian 25%
BANGLADESH AND PAKISTAN							
Bangladesh, People's Republic of	144	1,52.9	183.2	25	269.7	1,810	Bengali 98%
Pakistan, Islamic Republic of	796	180.4	229.6	35	484.4	2,790	Punjabi, Simdhi, Pashtun, Muhajir (Indian origin)
MOUNTAIN AND ISLAND RIM							
Afghanistan, Islamic State of	652	33.4	47.6	23	36.5	1,060	Pashtun 38%, Tajik 25%, Hazara 19%
Bhutan, Kingdom of	47	0.7	1.3	35	—	4,990	Bhhutia 50%, Nepalese 35%, Sharchops 10%
Nepal, Kingdom of	147	30..9	35.9	17	36.2	1,210	Nawar, Indian, Tibetan, Gurungi, Sherpa
Sri Lanka, Dem. Socialist Republic of	66	21.2	21.9	15	104.6	5,010	Sinhalese 74%, Tamil 18%, Moor (Arab) 7%
Maldives, Republic of	0.3	0.3	0.4	35	—	8,110	Sinhalese, Dravidian, Arab, African
Region Totals/Averages	**5,139**	**1,679.5**	**1,978.1**	**30**	**5091.1**		

Source: *World Population Data Sheet 2012*, Population Reference Bureau; Microsoft Encarta 2005.

use high-yielding crop varieties, chemical fertilizers, pesticides, and irrigation, food grain production in India almost quadrupled from 50 million tons in the 1950s to over 235 million tons today. In 2002, India became the world's second-largest exporter of rice in good monsoon years (after Thailand) and second-largest producer of wheat (after China), positions it still holds. However, the benefits of the new technologies are geographically and socially uneven. The farmers with larger land holdings in the states of Punjab and Haryana benefited the most. Among the drawbacks of the Green Revolution is the need to buy new hybrid seeds for each crop cycle and use increasing amounts of fertilizer and pesticides to maintain the high yields. Large areas of the hilly peninsula that have inadequate irrigation grow coarse grains such as millet and sorghum that withstand dry conditions.

The vagaries of the monsoon still determine the harvest: good years enable India to export food; bad years require imports. Despite adequate food to feed its population, hunger and malnutrition are still prevalent. Between 1990 and 2005, the Indian economy doubled in size, but the number of hungry rose by 65 million. In 2007, in poor and rural parts of India as much as 42 percent of children under the age of five suffered from malnutrition, a rate that was higher than in Africa south of the Sahara. Important commercial crops are cotton (India is the world's second-largest cotton producer), tea grown on plantations in the hills of northeastern India and high ranges of Tamil Nadu and Kerala in southern India, and jute, a commercial fiber cultivated on the western margins of the Ganges-Brahmaputra Delta.

Manufacturing

India has a large and growing manufacturing sector ranging from small-scale craft industries to large-scale steelworks and modern high tech facilities. The distribution of manufacturing in India reflects historic craft specialties, raw material locations (mines, farms), the British legacy of transportation hubs, ports, and administrative centers, varied encouragements offered by the union and state governments since independence, and entrepreneurial skills.

The most mineral-rich part of India is the Chota Nagpur Plateau in the northeastern part of the peninsula (see Figure 6.1), where plentiful iron ore and coal deposits formed the basis of local steelmaking. India's rocks also contain other metal ores such as bauxite, copper, gold, and manganese. As part of its self-sufficiency policies after independence, the Indian government developed a large-scale, heavy industrial sector of steel, chemical, and aluminum works, many built in conjunction with Western corporations such as Union Carbide and ICI.

The Indian government followed economic policies of self-sufficiency and import substitution in the decades soon after independence. In the 1990s, it adopted new policies that involved India more in the global economy by raising production and exports and being more open to foreign investment. Major multinationals such as General Motors, Ford, Chrysler, Peugeot, Volvo, Hyundai, Mitsubishi, Daewoo, and Volkswagen have begun producing cars and trucks for the local market. Indian company Tata, which makes the Nano, the cheapest car

in the world, is now set to produce it for a global market. The Indian pharmaceutical industry is challenging world markets.

Between 1965 and 2008 industry rose and then stablized at 29 percent, while services rose from 34 to 53 percent of the GDP. Textiles and garments (mostly of cotton, but also blends with synthetic fibers and jute) comprise over one-fourth of India's exports, mostly to the United States and European Union countries. Further economic modernization, however, depends on India's ability to supply the infrastructure of power, telecommunications, and road transportation that will attract and keep productive businesses. Unfortunately, power shortages still affect many areas. India experienced one of the world's biggest blackouts on July 31, 2012, affecting 600 million people, particularly in the northern and eastern states. The failure of the northern, eastern, and northeastern power grids toppled power supply in 20 out of India's 28 states for six straight hours. India remains deficient in transportation infrastructure. Its railroad system has a major role, but it needs updating to compete with the gradual improvement of roads and the efficient new private airlines.

Human Development

Over the years since independence, India achieved self-sufficiency in food, literacy rates doubled, life expectancy rose from 33 to 65 years, infant mortality fell (from 165 per 1,000 live births in 1960 to 47 in 2012), income poverty was reduced, and overall income substantially grew. Forecasts indicate that India's population will have a growing working-age group in contrast to most other countries.

Despite these improvements, based on the UN's Multidimensional Poverty Index (MPI), in 2010, there were 421 million MPI poor people in eight Indian states (Bihar, Chhattisgarh, Jharkhand, Madhya Pradesh, Orissa, Rajasthan, Uttar Pradesh, and West Bengal) compared to 410 million in the 26 poorest African countries combined. India's 2011 census showed that overall literacy had increased to 74 percent, but rural literacy was 69 percent and female literacy in rural areas only 59 percent. Progress is slow, although India's total GNI PPP is ranked fourth in the world (China is second).

On the political front, in the 1990s the world's largest democracy became increasingly caste-based and regional. The Congress Party, a dominant force since independence, gave way to local groups in the states, and some countrywide political groupings began to have greater roles. In 1998, the Bharatiya Janata Party (BJP) became the new government, with a strong Hindu nationalist mandate. But in 2004, the Congress Party regained control of the union government. India's problems of poverty, unemployment, weak demand, and low levels of investment continue to be major challenges.

Bangladesh and Pakistan

Bangladesh and Pakistan were partitioned from British India at independence in 1947 on the basis of their Muslim majorities, to form a single country (Pakistan) divided into East Pakistan and West Pakistan. Perceptions that resources from East Pakistan were being siphoned off to benefit West Pakistan resulted in the East Pakistanis revolting for a separate state. In 1971, after a brutal civil war, Bangladesh (formerly East Pakistan) was established as a separate country with India's military help. War-ravaged Bangladesh built itself anew with huge quantities of aid from the world's materially wealthier countries. Since independence, its internal politics have been marked by military takeovers, coups, and assassinations. In the 1990s, a settled democracy brought major positive changes. Over 90 percent of the population of both countries is Muslim. Bangladesh was established as a secular republic, unlike Pakistan whose 1973 constitution is founded on Islamic law. However, since 1977, Bangladesh's constitution has described it as an Islamic country.

In Pakistan, General Pervez Musharraf seized power in 1999 and dismissed the democratically elected Pakistan parliament. Although the Pakistan Supreme Court backed this move, it insisted that an election be held within three years: the general was reelected in 2002, but in 2004 he agreed to step down as head of the military while he remained in office until 2007. Following the assassination of opposition leader and former Prime Minister Benazir Bhutto, President Musharraf resigned. Asif Ali Zardari, Benazir's husband, was elected president in September 2008 and is still Pakistan's head of state.

From 1980 to 2012, Pakistan's literacy rate more than doubled, although it was still only 55 percent; electrification of rural households rose from 16 to 61 percent; and television reached 75 percent of city dwellers and 50 percent of the rural population. In late 2001, as a result of its support for the U.S. antiterrorist coalition, sanctions against Pakistan (imposed after its 1998 nuclear tests) were dropped, and it was given increased access to European markets as a reward for opposing terrorists and the drug trade. The United States has pledged development and military aid to Pakistan in exchange for help in combating armed Islamist groups such as the Taliban and al-Qaeda, which have made inroads into Pakistan's Swat district. However, U.S. forces discovered and killed Osama Bin Laden, the al-Qaeda leader and alleged mastermind behind the 9/11 attacks, in Abbotabad, Pakistan, where he had been hiding since 2004, without the help of Pakistani authorities.

Agriculture

Agriculture is the mainstay of the economy of both Pakistan and Bangladesh. Pakistan is a country of arid lowlands and high mountains, whereas Bangladesh is mostly low-lying apart from the small hilly eastern region inland of Chittagong, and is well watered by rain and rivers. Water is scarce in Pakistan, but flooding is a major problem in Bangladesh. Nearly half of the Bangladeshi labor force is engaged in farming, producing 19 percent of GDP in 2010. Bangladesh grows over half of the jute that enters world trade (Figure 6.17), as well as rice, tea, and sugarcane. The floodplains and deltaic lands that support most of Bangladesh's agriculture are low-lying and vulnerable to ocean surges during tropical cyclones. The slow expansion of agricultural production in Bangladesh at 2 to 3 percent per year coupled with population growth means that food security is still an elusive goal.

Figure 6.17 **South Asia: jute harvest near Tangail, Bangladesh.** The fiber is stripped from the plant stalk using the plentiful water of the Ganges River. What is jute used for?
Photo: © James P. Blair/National Geographic Image Collection.

Cotton is Pakistan's chief commercial crop. Wheat, rice, and sugar cane are grown as irrigated food crops. Some farmers are beginning to look to more profitable labor-intensive crops, such as vegetables and fruit. Much of Pakistan's best farmland is owned by a few wealthy landlords. The rich landowners' linkages with politicians gains them access to jobs in the government bureaucracy and the ability to veto measures that might decrease their dominant position in Pakistani society. Land is treated as a source of political power rather than a resource to enhance productivity.

Manufacturing

Bangladesh's limited manufacturing sector was traditionally based on the processing of agricultural products such as jute (see Figure 6.17), rice, tea, and sugarcane. The jute industry and other agro-industry units require restructuring, but the Bangladesh government faces the opposition of trade unions in trying to privatize the industry.

In the 1990s, following the adoption of structural adjustment policies that reduced trade restrictions and encouraged the external financing of garment factories, Bangladesh doubled its exports. The garment industry, with retailing customers in Europe, the United States, and East Asia, has become a major export earner. Enterprise zones at Chittagong and Dhaka (Dacca) are further attempts to expand industrial income and employment. Low wages for apparel workers in Bangladesh (9 to 20 U.S. cents per hour compared to 20 to 30 cents in Pakistan and over $8 in the United States) as well as the use of child labor has drawn criticism. During the 1996 elections, political strikes of half the 1 million poorly paid garment workers led to economic reforms that returned people to work and the economy to growth. Privatization continues, but severe infrastructure bottlenecks (roads, power stations, communications) slow the economy's expansion. Losses by state-owned enterprises continue to be a heavy burden for the limited national budget to bear.

Pakistan has a longer experience of manufacturing than Bangladesh. In exports, primary products decreased from 48 to 10 percent of the total between 1975 and the early 2000s, while manufactured goods rose from 52 to 89 percent. Textile manufacture, especially of cotton goods, dominates industries that include food processing, chemicals, and car assembly. During the 1990s, growth in output and productivity in manufacturing increased jobs, even in rural areas, at a rate that equaled the rate of new entrants to the labor force. However, such gains did not benefit the poorest groups of people. The new government elected in 1997 gave signs that it would help make producers more internationally competitive. It also wished to tackle problems that kept so many of its people poor and socially deprived. A military government replaced the elected government before these aims were met.

Both Bangladesh and Pakistan have small reserves of oil and natural gas, although neither exploited them fully until the 1990s. Pakistan's important chemical industry is built on deposits of gypsum, rock salt, and soda ash. Fishing in the coastal areas of both countries is poorly organized.

Service industries employ a growing proportion of the labor force in Pakistan (36 percent of women and 18 percent of men) and are expanding in Bangladesh (30 percent of men, 12 percent of women), and both countries produce around half of their GDP from such industries. Health, education, and financial services are growth areas. Neither country attracts many tourists.

Human Development

The Human Poverty Index is under 50 percent for both countries. Pakistan's GDP total is almost twice that of Bangladesh, but both have adverse trade balances and depend on aid donations. In Bangladesh, the problems stem from overpopulation (very high densities with a low resource and infrastructure base). Fewer than half the population has access to sanitation or legitimate electricity supplies and only 14 percent to garbage disposal.

In Pakistan, the inequitable distribution of wealth still leaves large numbers very poor, although the average per capita income is higher than that of Bangladesh. Pakistan lagged behind other low-income countries in social conditions into the mid-2000s, with low levels of literacy among women, high infant mortality (68 per 1,000 live births, compared to the average of 45 in all less-developed countries), and continuing high total fertility. Following the failure of earlier efforts, plans in the 1990s included a social action program that claimed major advances in the first three years, but the program suffered from slow bureaucratic responses, a lack of community participation, and changing political fortunes. The Pakistan government has increased social sector spending with a view to improving income poverty, education, and health.

Both countries lack many basic provisions that might encourage more manufacturing industry to locate there. Power supplies are subject to shortages, although Pakistan hopes to double its power provision with the aid of private investment. In the 1990s, it completed major hydroelectricity projects in the Himalayas, while in 1997 a thermal power station was completed with foreign capital and expertise at Hub near Karachi.

Mountain and Island Rim

The countries of the northern mountains—Afghanistan, Nepal, and Bhutan—have environments that isolated them from global connections and fostered internal strife through tribal rivalries. Events in these mountain countries are now influenced not only by what happens in the rest of South Asia, but also by interactions with neighboring countries in Central Asia, Iran, Russia, and China. The island countries of Sri Lanka and the Maldives have easier ocean connections with the global economic system but experience other problems that restrict their fuller development.

In 2008, Afghanistan, Nepal, and Sri Lanka each had 20 million to 33 million people. Bhutan and the Maldives had tiny populations of fewer than 1 million people between them. The mountain countries are among the world's poorest, while the Maldives and Sri Lanka have better living conditions. Nearly all these countries suffered internal conflict since the 1990s.

Afghanistan

In the late 1800s the British and the Russians delineated the borders of the state of Afghanistan that lay between their empires. The northern border was seen by the British as a buffer to Russian expansion. The country brought together groups fragmented by language, creed, geography, and historic cultural traditions. The Pashtuns formed the majority in the southern parts, but Tajik and Uzbek people dominated the north (see Figure 6.9). King Abdur Rahman (1880–1901), backed by British money and weapons, tried to create the foundations of a centralized Afghanistan. This and later attempts to build and unify the country were often ruthless, resented by the intensely independent groups, and generally unsuccessful.

During the early decades of the 1900s, internal feuding and external influences held back attempts to modernize Afghanistan. Tensions with Pakistan emerged over the northwest frontier lands that Pakistan swiftly annexed after its own independence in 1947. Afghanistan then linked more closely to the Soviet Union until 1979, when the latter invaded and occupied it. During the period of Soviet occupation, local warlords supplied with weapons by the United States through Pakistan, fought the Soviet armies and established their own military units. After the Soviet Union withdrew in 1992, the warlords and their armies fought each other for control of the country. Most government systems broke down. Education became irregular, while public health and health care were weakened.

In the late 1990s, the Taliban group conquered virtually all of Afghanistan, although fiercely opposed by northern tribes who resisted the imposition of hard-line Sunni Islamic tenets. From the west, Iran supported Shia groups, including Hezbollah, but they failed to overthrow the Taliban. The rest of the country lived under the repressive Taliban order, which expelled most international aid agencies. Strong opposition to the Taliban among the Afghani people resulted in their accepting and using U.S. military intervention and military aid to depose the

Taliban government. A new government was established and the first "democratic" elections held in 2004 (though marked by local coercion).

Nomadic herding of sheep and the cultivation of wheat, rice, and other cereals are important to Afghanistan's agricultural base. In a politically motivated act in 1998, the Taliban destroyed the fertile central farming region of Afghanistan, the Shardi Plain near Kabul. They cut fruit and nut trees, burned villages and wheat fields, blew up irrigation channels, and left crops to wither. Kabul's industrial district had 200 factories in 1992; the number was reduced to 40 under the Taliban, and only six still operated in late 2001. The 1992 fighting destroyed most of the Jangalok steelworks outside Kabul, and the machinery was taken to Pakistan. Even the army factory that made cannons, guns, and ammunition is largely abandoned.

Opium poppies became an important crop for Afghanistan's beleaguered farmers. In the 1990s Afghanistan supplied 70 percent of the world's heroin as the country's other farm products dwindled through war and as Turkey, Iran, and Pakistan enforced strict drug control laws. Afghanistan's isolation, disrupted government, lack of alternative commercial products, and warlord control of the drug trade make it difficult to stop the production of opium and heroin and their entry into illegal drug trafficking routes. Although Afghanistan was threatened with the withdrawal of aid because of its increasing production of these drugs, payments continued for fear of angering the warlords. Planting continues, with the farmers becoming indebted to the opium traders, who take their land and even their women.

Afghanistan faces a difficult and uncertain future in the light of its history of factional strife, oppressive rulers, devastated economy, and poor relationships with surrounding countries. Money from the sale of opium continues to fund local groups and undermines overall political stability. The Afghan International Office for Migration encourages the return of qualified Afghans who fled the Russian- or Taliban-dominated country, but it has difficulties in attracting professionals who settled comfortably in the United States, Europe, or Australia, where they have higher salaries and prospects of asylum or citizenship.

Nepal and Bhutan

From the 1850s, the Hindu kingdom of Nepal had a close relationship with Britain, and many Nepalese served in the British Indian army. Following British attacks in the 1800s, Nepal remained independent under pro-British rulers who hired out Gurkha soldiers (from different Nepalese ethnic groups) to the British. Nepal became a democracy in the 1980s. Despite this political orientation, alternative political parties were banned until 1990 while left-wing (Maoist) guerrillas caused major disruption in rural areas and attacked remote army and police barracks. In 2001, Nepal's monarchy was almost all murdered by the heir to the throne in a family argument. In 2005, the embattled king suspended all constitutional freedoms. In May 2006, following weeks of street protests and strikes, a cease-fire was declared between Nepal's monarchy and the Maoist rebels, and the king agreed to cede power to a multiparty alliance. The Nepalese Constituent Assembly abolished the monarchy in 2008.

In Bhutan, the main group of people, the Bhote, drove out Indian peoples in the AD 800s and built impressive fortified Buddhist monasteries in the 1500s. The isolated kingdom of Bhutan was annexed by Britain in 1826 and given autonomy in 1907. Today, it remains a tiny buffer state between India and China. It has few cross-border contacts, preferring to maintain its own culture and to limit Western tourists and influences. Bhutan's official policy emphasizes "Gross National Happiness," placing cultural, spiritual, and environmental well-being and the happiness of its citizens ahead of material wealth.

Sri Lanka

Sri Lanka's (formerly Ceylon's) proximity to the Indian peninsula and its openness to ocean-borne trade and conquest attracted competing groups. Indo-Aryan peoples with Buddhist beliefs from northern India developed a civilization based on irrigated rice agriculture in the north-center of the island that lasted from around 100 BC to the AD 1200s. During the later part of this period, Tamils—Dravidian peoples from southern India—established kingdoms in northern Sri Lanka around Jaffna. As Tamils expanded their influence, the majority Sinhalese peoples drifted southward, abandoning the irrigation systems and relying on rainfed agriculture. They grew spices such as cinnamon. Arab traders settled in the country and controlled the overseas spice trade.

The Portuguese who were early traders and colonists in Ceylon left a strong legacy of Roman Catholic missions. In 1795, the British took Ceylon, which remained a crown colony until its 1948 independence. British companies set up plantations for growing tea, rubber, and coconuts, instead of the traditional spices (Figure 6.18). The emphasis on export crops, however, was accompanied by the neglect of traditional agriculture, leading to a decline in the rice crop and a need to import half the island's food needs.

After independence, Sri Lanka's government by the dominant Sinhalese people instituted reforms that were disliked by the Tamil minority, leading to a civil war. From a time when Sinhalese and Tamils lived side-by-side in Colombo, the capital, there were greater separations of people, over 60,000 deaths in the fighting and urban terrorism, and the emigration of many Tamils to India, Europe, and Canada.

Sri Lanka has tea, rubber, and coconut commercial plantations. These provide 35 percent of the country's exports, down from over 90 percent in 1950 as a result of diversification in the economy since independence. Sri Lanka's adoption of new rice varieties in the late 1960s converted the major food deficit—which arose out of the colonial focus on export crops—into a virtual self-sufficiency in rice today. Sri Lanka is an important source of gemstones and also has graphite reserves. Local agricultural processing, together with the making of carpets and footwear, are the most common industries. Much manufacturing is in small-scale factories or cottage industries based on the dispersal of production processes in homes. Government policies since independence resulted in a diversified manufacturing base in Sri Lanka, which also produces steel, textiles, tires, and electrical equipment. By the 1990s, however, expansion placed pressure on existing transportation and telecommunications

Figure 6.18 **Sri Lanka: tea plantation.** How are the tea bushes arranged on the hills? *Photo: © Alasdair Drysdale.*

facilities, while power shortages reduced output and exports of garments and other textiles. In early 2004, tourism accounted for 10 percent of Sri Lanka's GDP, but the industry is still reeling from the devastating effects of the 2004 Indian Ocean tsunami.

Maldives

Fishing and tourism are the mainstays of the economy of the Maldives. Here, tuna accounts for 60 percent of exports. The islands also have commercial coconut plantations. A popular tourist destination, the Maldives saw a dramatic increase from a few thousand visitors in 1980 to nearly half a million by the early 2000s. Almost 80 percent of the economy of the 1,190 islands in the Maldives depends on tourism. The tourist industry employs over 25 percent of the working population, up from only 4 percent of the total labor force in 1985. The limits of expansion are, however, being reached, signaled by coastal pollution, groundwater contamination, and the need to bring in labor to cope with the increasing demand. Rising sea levels due to global climate change are a growing concern for this island nation, which seeks to find other areas where its population can be relocated.

6.5 Contemporary Geographic Issues

Population Growth and Patterns

By 1961, India's population was 440 million, and Indian government estimates of future population growth caused it to institute policies to slow the population increase. Efforts to spread the use of birth control methods gained acceptance slowly among a largely illiterate people. The threat of forced imposition of male sterilization was a major political issue in the 1977 elections that ousted Indira Gandhi's government until 1980. Greater effectiveness of family planning in the 1980s and 1990s built on rising levels of literacy, increased urbanization, and improvements in the status of women. Hindu women of higher castes use family planning more than those among the poorer castes and poorer, less-educated Muslims. The new urban middle class provides signs of the Indian population stabilizing around fertility rates of two to three children. More women adopt family planning, including sterilization, once they have given birth to a boy, and this is the most successful aspect of the program. Many families in rural areas, however, continue to want large families to provide extra manual labor and care of parents in later life. Moreover, many of the 120 million Muslims living in India disapprove of birth control.

The Indian age-sex diagram records fewer females than males (see Figure 6.7). The 2001 Indian census of population showed that India has a sex ratio of 107 males per 100 females, slightly lower than in 1991 but higher than that of its neighbors. Within India, sharp differences exist across the states, from 95 males per 100 females in Kerala to 116 in Haryana. Furthermore, in the under-7 age population, the sex ratio increased from 105.8 in 1991 to 107.8 in 2001, with the sharpest rises in the more prosperous states. It has been suggested that there are millions of missing females in India because of female infanticide, excused by some as saving the girl from a lifetime of suffering. In 1997, the *Times of India* estimated that mothers or village midwives kill 16 million girl babies a year. In 1994, after discovering that most aborted fetuses were female, the Indian government outlawed tests that could determine the sex of a fetus. Laws now prohibit doctors from revealing the sex of an unborn baby, although they are flouted among wealthier families. The 2011 census of India showed, however, that sex ratios in northern India have improved.

The escalation of the practice of dowry in all sections of Indian society is causing a further devaluation of girls and women. For example, a man with a steady job in Tamil Nadu State expects a dowry of US$1,000, 100 grams of gold, household goods, and a car. Wives may be badly treated if their dowries are not up to expectations, and there are cases where the wife has been murdered by being burned alive so the husband can remarry for a better dowry. It is also a wifely duty to have sons; families with more daughters than sons are often implicated in dowry murders. Families with daughters start saving at birth but still require loans.

From the 1990s, HIV/AIDS emerged as a major health and development threat. An estimated 1 percent of India's adult population (or at least 5 million—but probably many more) is affected by the disease—the highest number for a country after South Africa. Worst affected are the southern Indian states (apart from Kerala) and Manipur in the northeast next to the Myanmar boundary. Together, these states account for 80 percent of the reported cases of HIV/AIDS in the country. However, rates in some of the northern Indian states that do not have checks may be just as high. The cities of Mumbai, Chennai, and Delhi also have a high prevalence of the disease. HIV/AIDS is spreading fast in India's rural areas, where health care and knowledge of the disease are poor.

Alarmed by the predicted effects of an unchecked epidemic on its population and economic development, the government launched an HIV/AIDS prevention campaign. Its mission is to educate the Indian people on the means of transmission of the virus and offer support for prevention measures. Private organizations and NGOs are also involved in these efforts. Despite being the country with the third-largest number of people with HIV/AIDS (after South Africa and Nigeria), India has a low rate of the disease. There has been a 50 percent decrease in new HIV cases in India over the last 10 years. It was estimated that in 2007 the infected population was less than 2.3 million, considerably less than the projected 5.5 million. Reactions to those with HIV/AIDS typify Indian social situations. While vulnerable groups such as prostitutes and truck drivers are particularly liable to be infected, the low status of women in families increases the risk of transmission, and those infected by husbands are cast out by their in-laws. Many HIV-positive men lose their jobs. Indian pharmaceutical companies Cipla and Ranbaxy produce generic antiretroviral drugs for the Indian and developing country markets, selling them at a fraction of what they usually cost in developed countries.

Urbanization

Five of the world's 20 largest cities are found in predominantly rural South Asia. The metropolitan areas of Mumbai, Delhi, Kolkata, Dhaka, and Karachi each house over 10 million people (Table 6.2). The largest cities in all countries are immense and growing. Attracted by new jobs in factories and offices, better access to education and health services, and opportunities in the informal sector of the urban economy, migrants from rural areas and small towns continue to pour into cities.

Urban Populations

Indian cities continue to house under 30 percent of the country's growing population. However, the total population of urban areas is projected to rise from 250 million people in the mid-1990s to over 600 million by 2025. From 1980 to 2008, the number of Indian cities with over a million people jumped from 12 to 37. By 2015, Mumbai (Bombay), Delhi, and Kolkata (Calcutta) will be three of the world's 10 largest cities (Table 6.2). Several Indian cities, including Mumbai and Delhi, have increasing global corporate services such as accounting, advertising, banking, and law, while Bengaluru and Hyderabad have global connections through their high tech industries.

Like India, Bangladesh and Pakistan remain largely rural countries. Pakistan's urban population was 35 percent in 2008, while Bangladesh's towns and cities account for 24 percent of the country's population. The major cities in both countries struggle to cope with the arrivals of rural people and the shantytowns and social tensions that this influx creates.

Karachi is the largest city in Pakistan and its major port and commercial center. Violence in the 1990s among rival ethnic Muslim groups here was made worse by the glaring contrasts

Table 6.2	Populations of Major Urban Centers in South Asia

TOP 10 MOST POPULATED URBAN CENTERS		
City, Country	2012 (millions)	2025 (millions)
1. Mumbai, India	20.7	26.4
2. Delhi–New Delhi, India	18.9	23.5
3. Kolkata, India	15.6	18.9
4. Karachi, Pakistan	14.2	19.1
5. Dhaka, Bangladesh	14.0	22.0
6. Chennai, India	8.7	10.8
7. Bengaluru, India	8.7	11.4
8. Hyderabad, India	7.9	10.1
9. Lahore, Pakistan	7.7	10.5
10. Ahmedabad, India	6.5	7.7

NUMBER OF MAJOR URBAN CENTERS IN SOUTH ASIA, BY POPULATION CATEGORY	
Population Category	Number of Urban Centers
10.00 million and greater	5
5.00 – 9.99 million	7
2.50 – 4.99 million	10
1.00 – 2.49 million	38

between rich and poor, the large numbers of unemployed teenagers, and the availability of guns following years of warfare in Afghanistan. Other major Pakistani cities are inland, where Lahore is the major center of Muslim culture and Faisalabad is the cotton industry center. Islamabad, the new capital, is close to the Kashmir border, while Peshawar is the commercial center of the far north at the eastern end of the Khyber Pass into Afghanistan.

Bangladesh has fewer large cities. Dhaka is the capital, Chittagong the main port, and Khulna is a growing industrial center. The pressure of people on scarce farming land and lack of farm jobs for adults forced migration into cities, where jobs demand educational qualifications that most migrants do not have. From the 1980s, the shortage of jobs in Bangladesh resulted in many laborers seeking employment outside the country, such as in the Persian Gulf countries and Malaysia.

In the mountainous countries, the main concentrations of people are in lower areas that are accessible to major transportation routes. Kabul (Afghanistan), Colombo (Sri Lanka), and Kathmandu, capital of Nepal, are the largest cities of the subregion and contain extensive shantytowns. The Himalayan states have most of their people near the Indian border. Sri Lanka's population is focused on the rainier southwestern coastlands around Colombo. In 2008 Nepal was only 17 percent urbanized, while the other mountain and island countries were over 20 percent.

Changing Urban Landscapes

The urban landscapes of South Asia reflect the waves of cultural influences that washed across the region. Many ordinary buildings were constructed of materials such as wood that do not have a long life, but most towns have more permanent religious, royal, or military buildings that have lasted for centuries. Hindu temples are often very ornate (Figure 6.19). Muslim mosques and tomb gardens also provide distinctive landscape elements (see Figure 6.6).

Older precolonial sections of towns left a heritage of high housing densities with poor transportation access along winding alleys. Shops and artisan workshops encroach on walkways. Few buildings are more than two stories tall, and often there is no clear distinction of residential, commercial, and industrial functions. People of similar caste or trade live and work together in localities.

Sections added to towns during the British rule from the 1700s to the mid-1900s included a central market, administrative offices, and frequently a clock tower. New residential areas were built on European plans with wide streets and open places separating business and residential functions. Britain built and developed major port cities such as Kolkata, Mumbai, Chennai, Karachi, and Colombo initially as a means of establishing trading security. The ports were then joined to inland centers by railroads that remain the major mode of transportation available to most Indians. Some smaller cities, such as Bhopal, have no planned infrastructure and, even in major cities, funds for maintaining infrastructure are scarce. Shantytowns are common.

Modern cities have major industrial areas, including steel-works and aluminum plants, chemical factories, textile mills and garment-assembly factories, vehicle makers, and electrical goods production. Concentrations of factories are most common around the metropolises of Mumbai, Delhi, Kolkata, and Chennai and in urban areas located close to raw material sources.

Up to the mid-1990s, the wealthy lived in the city centers and the poorer people toward the outskirts, but many of the wealthy have moved to new suburbs on the fringes of cities. Slums and shantytowns spring up to accommodate the continuous flow of poor urban migrants. Some slums have brick or block dwellings with windows, doors, and corrugated roofs; others are shacks made of any available material, often at risk from fire. The poorest ragpicker settlements adjoin and merge into refuse heaps. Where slums are recognized by municipalities, they have piped water, legal electricity connections, and shared sanitation. Many are not. There is no street lighting, cleaning, or repairs. Municipal policy is generally to control and raid the slums rather than protect them. Water is supplied by tankers and public standpipes, and there are few toilets apart from open spaces. Yet homes are kept clean. Disease (diarrhea, tuberculosis) is endemic or occurs in rampant outbreaks of

Figure 6.19 **South Asia: contrasts.** An Internet café is close to a Hindu temple in Bengalaru (Bangalore) India. *Photo: © David H. Wells/Corbis.*

cholera, hepatitis, and typhoid. HIV/AIDS is a major scourge, as are drug dealing and drinking illicit alcohol.

The contrasts between the wealthy and the poor are clearly evident in cities such as Mumbai, India's primary commercial and industrial center. The city accounts for over one-fourth of India's manufacturing output and is home to over half of India's top 100 companies. Mumbai handles one-fourth of India's trade in its port and 70 percent of its stock exchange transactions. It is the headquarters of nearly all of India's commercial banks and has factories making a wide range of products from textiles to pharmaceuticals. India's US$1.5 billion film industry is the largest in the world, making over 1,000 movies a year, more than twice Hollywood's output. Mumbai is the center of India's movie industry (often called "Bollywood"), with half of all films made in India being filmed here. Real estate prices in Mumbai rival those of New York City, making it difficult for all but the wealthy to own their own homes.

Dharavi, allegedly the largest slum in Asia, is also found in Mumbai (see Figure 6.14). Poor housing, very high densities, and limited access to water, sanitation, and other infrastructure facilities characterize the slum. However, over 100,000 people produce US$500 million in goods each year from small businesses such as bakeries, metal workshops, recycling, tanneries, and potteries in Dharavi. Families who profited and moved out continued the business connections. The Mumbai government hoped to turn the area into a showpiece development by 2010, but it has yet to be completed. The economic recession and meddling by powerful local political parties have been blamed for the delays.

Pollution due to improper and inadequate disposal of industrial effluents and domestic and industrial wastes greatly detract from the quality of life in South Asian cities. For example, Delhi generates nearly 4,000 tons of trash each day but clears only 2,500 tons, leaving the rest on the streets. Of 1,800 tons of sewage, only two-thirds are collected.

Air pollution in cities like Delhi and Kolkata are at their worst during the winter months when toxic gases, hydrocarbons, and particulate matter are trapped near the surface by cold air. Contributors to poor air quality are vehicular traffic, coal and kerosene used as cooking fuels, coal-fired power plants, and the poor condition of city streets. According to the WHO, more than 11,000 people die each year in Kolkata due to air pollution-related illnesses, and at least half the city's children have excessive amounts of lead in their blood. The city struggles to improve its air quality. Delhi, which in the mid-1990s had the most air pollution of any Indian city, has succeeded in dramatically improving its air quality by mandating a switch to cleaner fuels like liquid petroleum gas and compressed natural gas for city buses.

Liberalization and the Rise of High Tech Cities in India

Consequent to greater openness to foreign investments and internal private enterprise following structural adjustment and liberalization in the 1990s, India emerged as one of the world's fastest-growing economies. In 2005, a report by America's National Intelligence Council likened the emergence of India and China in the early twenty-first century to the rise of Germany in the nineteenth and the United States in the twentieth, with "impacts potentially as dramatic."

An important component of this economic revolution is the growth of high tech industries in India. The rise of this sector was assisted by a large youthful population, many with excellent scientific and technical training, a work force that speaks English, and legal systems that offered good copyright and contract law protection. The total output of India's modern high tech industry rose from US$2 billion in 1995 to over US$50 billion in 2007. India is moving from established services such as chip design, back-office work such as call centers (Figure 6.20), medical transcription, and information technology (IT) consulting to high-end ones such as research for banks and other financial agencies and research in drug development and engineering. High tech industries are found not only in large, established metropolises such as Mumbai and Delhi, but also in smaller, growing regional centers such as Bengaluru and Hyderabad in southern India.

Well into the 1970s, Bengaluru (Bangalore) was known as the "Garden City" for its open areas, parks, tree-lined streets, and single-story residences. The city's economy was largely based on public sector industries such as munitions and aerospace and some private engineering and textile firms even in the 1980s. In the early 1980s Texas Instruments set up a factory here, but it was with the opening up of the Indian economy in the 1990s that Bengaluru began to attract numerous multinational corporations. The main new industries here are electronics, computer engineering, software and services, telecommunications, aeronautics, and machine tools. Multinational corporations in the city include 3M, AT&T, Digital, Ericsson, Hewlett-Packard, IBM, and Motorola. Bengaluru's attraction as a center for computer software and hardware development was complemented by a growing pool of skilled labor from its various engineering colleges and from migration to the city from other parts of India. Information technology and IT-enabled services accounted for over 60,000 jobs in the Bengaluru area by the late 1990s. More recently, biotechnology industries have joined the IT sector firms.

As its population and industries burgeoned, Bengaluru expanded in area. Technology parks such as Electronic City (Figure 6.21) and the Whitefield International Tech Park that house many electronics, telecommunications, and computer software and services firms emerged on the outskirts of the city where land was available. These planned industrial parks have facilities that are similar to those found in their counterparts in Western countries. Electronic City has over 100 firms including Motorola, Siemens, ITI, and Indian software companies Infosys and Wipro, set on 330 landscaped acres. IT companies seeking more accessible city locations even set up offices in buildings zoned for residential use, in violation of city codes. Over 1,000 IT companies with an investment of US$1.3 billion are based in Bengaluru, which is now considered India's Silicon Valley and IT capital.

New townships and residential areas (some of which are gated communities) have developed to house the city's growing IT worker population. International construction companies are vying with each other to develop new housing in the area. A 25 km-long IT corridor, anchored by Electronic City and International Tech Park at either end, is designed to cater to a million people by the year 2021. The corridor will include business

Figure 6.20 **India: call center.** A call center in Bengaluru (Bangalore). This service industry is a major part of Indian development of IT outsourcing. Jobs in this area are set to rise to over 1 million and sales were US$14 billion by 2007. *Photo: © Sherwin Crasto/Reuters/Corbis.*

Figure 6.21 India: modern technology. The Infosys Technology Campus, Electronics City, Bengaluru. Infosys employs over 14,000 staff and is India's top software exporter with clients like Bank of America and Citigroup. *Photo: © Reuters/Corbis.*

parks, commercial centers, six townships, two universities, hospitals and polyclinics, and two golf courses.

While Bengaluru's status as India's premier high tech city is widely acknowledged, other urban centers such as Hyderabad, Pune, and Noida are also developing their IT sectors. As in Bengaluru, new technology parks and industrial campuses such as Hitech City are planned and developed in Hyderabad, a fast-growing IT hub. A progressive state government promoted the establishment of new centers of technical, scientific, and business education, provided the infrastructural support to help new firms and institutions of higher education, and initiated the development of infrastructure such as highways and a new international airport, transforming the former capital of a princely state into a modern city. Companies such as IBM Global, Google Online, Mindtree Consulting, and iSoft all have operations in Hyderabad.

Ethnic Conflicts: Intra-State and Inter-State

All South Asian countries have sizeable ethnic and other social groups whose antagonism toward each other spills over into conflicts. Local rivalries often escalate into conflicts that affect wider areas. Tensions between Hindus and Muslims that led to the partition of British India into India and Pakistan continue. In secular India these came to the fore when in 1992, militant Hindus pulled down a Muslim mosque at Ayodhya, a site that is thought by some to be the major historic center of Hinduism and the birthplace of the god Rama. In 2002, Hindu-Muslim tensions erupted in violence in Gujarat as Hindu mobs burned Muslim areas after 58 Hindu pilgrims were killed in a train fire. These and other events led to the rise of *Hindutva* (Hindu View) as a political movement that is fertile ground for right-wing Hindu chauvinism and anti-Muslim and anti-Christian sentiments. The Rashtriya Swayamsevak Sangh (RSS, National Volunteers Association), founded in 1925 with an agenda for national unity on the basis of Hindu supremacy, and its political arm the Bharatiya Janata Party (BJP) have built a strong base in India.

In Nepal, after eight years of insurgency, Maoist guerrillas control the rural areas and threaten to take over the government in Kathmandu. Similar Maoist groups known as Naxalites operate within India and, although they talk to state governments as in Andhra Pradesh, they refuse to disarm. Bangladesh is becoming involved with the spread of Islamic extremism linked to attacks in northeastern India. A plan to dredge the Sethusamudram Ship Canal through the shallow Palk Strait between Sri Lanka and India is causing fresh tensions following India's decision to go ahead despite Sri Lankan opposition. This canal would save over a day's sailing for ships between east and west Indian ports, but would harm local fisherman, Columbo's transshipment trade, and local environments.

South Asia may be turning into a "**shatter belt**" of conflict and disruption linked to the troubled areas of southwestern Asia (from Iran, through Iraq to Israel-Palestine) through the Pakistan-Afghanistan linkage. The interregional conflicts generated by such events hold back the type of cooperation and regional trading agreements that have become common in other world regions. The South Asian Association for Regional Cooperation (SAARC) set up in 1985 generated little economic interaction apart from some academic studies of the potential. Moreover, external and internal rivalries raise defense expenditures and affect development. Additionally, the region has many of the ingredients for nationalism and ethnic conflict. The geographic concentration of different ethnic groups, strong feelings of cultural pride, and perceptions of being excluded and discriminated against politically, culturally, and economically by a dominant group all form a potent mix that can lead to nationalistic fervor and ethnic strife.

Civil War in Sri Lanka

Sri Lanka was a politically stable country with a growing economy after independence and until the 1980s when northern minority Tamil people began to fight the dominant Sinhalese for their rights and a defined Tamil territory, Tamil Eelam (*eelam* means "homeland"). The Tamils comprise two distinct groups—the "Sri Lankan Tamils" who have lived on the island since before the Christian era, and "Indian Tamils" who are descendants of plantation laborers brought in the 1800s and early 1900s by the British colonizers. Sri Lankan Tamils are concentrated in the northern Jaffna area and the districts along the eastern coast, while the Indian Tamils are found mostly in the tea plantation areas of the Central Province (Figure 6.22).

Problems between the majority Sinhalese and the minority Tamils arose after independence due to government policies that allegedly favored the Sinhalese. Attempts were made to repatriate the Indian Tamils, many of whom were denied citizenship in post-independence Sri Lanka. The Sinhala Only Act of 1956 made Sinhala the only official language of the island nation, restricted many government jobs to Sinhala speakers, and changed university admission policies in favor of the dominant group, causing the Tamils to lose educational opportunities as well.

An attack on the military by the Liberation Tigers of Tamil Eelam (LTTE) in 1983 and retributive anti-Tamil rioting in Colombo and other Sinhalese-dominated areas, in which up to 2,000 Tamils lost their lives, can be considered the beginning

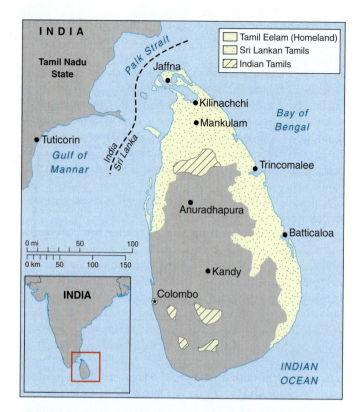

Figure 6.22 **Sri Lanka: basis of civil war.** The Tamil population lives mainly in the north and northeast of the island.

of the civil war. Peace talks between the Tamils and the government in 1985 failed. In 1987, although the Sri Lankan government succeeded in pushing rebels to the northern city of Jaffna, the LTTE and other Tamil guerilla groups set off several bombs in Colombo, causing hundreds of deaths. This landmark year also saw the signing of the Indo-Sri Lanka Peace Accord under which the Sri Lankan government conceded to some Tamil demands, including greater power to the provinces and making Tamil an official language. The Indian Peace-Keeping Force (IPKF) was established to oversee the cease-fire and help maintain order in the Tamil-dominant north and east, but met with resistance from both Sinhalese and rebel Tamils. Hostility toward the IPKF and its brutality culminated in the 1991 assassination of former Indian Prime Minister Rajiv Gandhi, under whose leadership the IPKF was established.

Following the IPKF withdrawal in 1989, the LTTE recaptured significant areas of the north. Both sides intensified their attacks in a struggle for control of territory and power. Government forces attacked civilian buildings such as Tamil temples, churches, and schools that were used as safe havens by refugees fleeing war-torn areas, causing many Tamil casualties. The LTTE stepped up its attacks in public places and on important political figures. A Tamil suicide bomber killed Sri Lankan President Ranasinghe Premadasa in 1993. In 1996, the Central Bank in Colombo was bombed, and in successive years so were the Sri Lankan World Trade Center and the Temple of the Tooth, a holy Buddhist shrine in Kandy. In 1999, the LTTE attempted to kill Chandrika Kumaratunga, Sri Lanka's president at that time.

However, the 1990s also saw greater attempts to bring peace to the island through domestic and international efforts. The LTTE, which had been banned as a terrorist organization by governments of India, the United Kingdom, and the United States, announced a unilateral cease-fire in December 2000 in a show of willingness to explore strategies to end the war while safeguarding Tamil rights. But they attacked Bandaranaike International Airport the following year, destroying planes and wreaking havoc on the country's tourist trade.

In 2001 a fragile cease-fire was brokered by the government of Norway. Since then, the island has been in an uneasy state of truce, complicated by accusations of covert operations by both the LTTE and the government. In 2005, newly elected President Mahinda Rajapakse promised to pursue peace talks with the Tamil rebels. However, in his election campaign he had called for renegotiation of the cease-fire and a tougher stance against the LTTE. By the end of 2005, the Tamil rebel group had renewed its attacks on the Sri Lankan army and navy, while a parliamentarian linked to the Tigers was assassinated during a Christmas mass.

An estimated 65,000 lives, both Tamil and Sinhalese, have been lost in 26 years of fighting. In May 2009, the LTTE was defeated and its top leaders, including LTTE chief V. Prabhakaran, were killed in a Sri Lankan army offensive. Thousands of civilians were killed and tens of thousands injured in the 2009 military attacks. Nearly 300,000 Tamil civilians who fled the war zone were placed in detention camps. An exile faction of the LTTE has declared that a new "transnational government" will continue the political struggle for a separate state of Tamil Eelam. Although the transitional government is not recognized, many of the 700,000 Tamils who left Sri Lanka in the aftermath of the 1983 riots are fervent Eelamists and may continue to support the formation of a separate Tamil state.

The Indo-Pakistan Dispute over Kashmir

The events of 9/11 and the subsequent attacks on Afghanistan brought to light many links of agreement and conflict across sets of countries. One of these was the long-lasting dispute between India and Pakistan over the Kashmir Province at the northern end of their common border (Figure 6.23). By early 2002, India and Pakistan were sending military forces to the border, and war looked possible. As in its relationships with the former Taliban government in Afghanistan, however, the war on terrorism caused the Pakistan government to hold back and even repudiate groups of Islamic radicals it had supported and trained for many years. The situation along the Indian–Pakistani border remained tense—as it had since the partition adopted at independence in 1947.

Some history of the Kashmir situation explains present events. In the mid-1800s, the British sold the rulership of the Muslim state of Kashmir to a Hindu maharajah for around US$50 million at today's prices. At independence, Kashmir opted to be part of India. Although the maharajah tried to avoid the need to join India or Pakistan, an invasion by Pakistani tribesmen forced his hand, and he chose India in return for military help. Indian Prime Minister Jawaharlal Nehru, in his eagerness to get popular ratification for the accession, brought Pakistan's 1947 invasion to the notice of the United Nations and

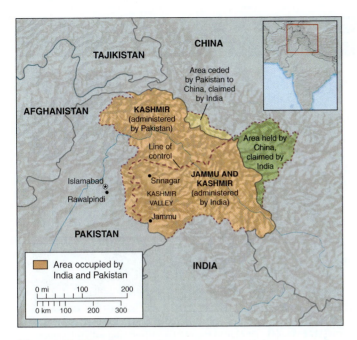

Figure 6.23 Kashmir. This map shows a complex geography. A line of control separates Indian and Pakistani areas of present occupation. India also has border disputes with China.

urged Pakistan forces to withdraw until a referendum allowed Kashmiris to choose between India and Pakistan. After the 1948 cease-fire, Pakistan held onto the one-third it had conquered, much of it close to the city of Rawalpindi and later the new capital, Islamabad. The Pakistanis never withdrew, and Nehru never put the matter to a vote. The maharajah initially handed over control of defense, foreign affairs, and communications to India, and Kashmir had its own prime minister until 1965.

Over the years, India progressively integrated Kashmir, moving many of its Hindu population to Delhi and setting them up with craft industry employment making rugs, woolen clothes, and decorated plates and selling them in tourist outlets. In 1964, India extended to Kashmir the right of the federal government to dismiss state governments. Each decision, however, seemed to thwart the popular will of Kashmiris. India and Pakistan fought over Kashmir in 1965 and 1966, when India accused Pakistan of backing insurgents.

War broke out again in 1971, when Bangladesh (East Pakistan) became independent, backed by India. Kashmir was another source of fighting in that war, after which the 1972 Simla Agreement divided Kashmir into the Pakistan state of Kashmir and the Indian state of Jammu and Kashmir. This agreement established today's line of control.

Any goodwill generated by the first genuinely free Jammu-Kashmir elections in 1977 that brought back Sheikh Abdullah collapsed with the ousting of his son from power in 1984. Antigovernment Muslim groups contested the 1987 state elections, but Delhi annulled some of their successes. From 1989, separatist violence that India alleges is backed by Pakistan was matched by Indian military repression.

By the late 1980s, Pakistan looked friendlier to Kashmiris than India despite having a very different culture based on Islamic

rules and military dictatorships. Indians believe their democratic constitution can provide liberation for Kashmiris. In fact, most Kashmiris see themselves as prisoners of either country and are weary of continuous fighting. They prefer independence.

It is difficult to see who speaks for Kashmiris. The state government based in Srinagar is very unpopular. Some Kashmiris look to the United States and its allies as potential saviors that can push India into a settlement. But India and Pakistan are the real arbiters. India shifted from inaction to attempts to talk with Pakistan or separatists. The current Pakistani crackdown on terrorists causes Indian cynicism about Pakistan's promises when banned groups such as Lashkar-e-Taiba continue to act after merely changing their names. India, however, keeps a security force of 400,000 men in Kashmir and hopes the crackdown will moderate the militancy. There is speculation that Pakistan will return support to more moderate Kashmiri groups such as Hizbul Mujahideen.

Any election or referendum faces major problems. If separatists who deeply distrust India participate, they must declare themselves Indian citizens to qualify to vote. Without the separatists, however, Kashmiri Muslims will mock the result as worthless. The Indian government rejects proposals for foreign election observers or outside mediation as questioning its authority. This attitude continues a high-handed and superior approach based on the original partition decision, Indian democratic institutions, and resistance to terrorist attacks but avoids discussion of how India alienated the Kashmiri people.

Partition of Kashmir between India and Pakistan with a degree of autonomy for both parts of Kashmir might then occur. Pakistan conceded much after 9/11, withdrawing support for the Taliban in Afghanistan and shutting down a potential war against India after terrorists attacked the Indian parliament in Delhi.

Former Pakistani President Pervez Musharraf was a military dictator who gained international disapproval when he dismissed the previous democratic (but corrupt) government in 1997, ordered a nuclear bomb test in 1998, and oversaw training camps for Islamic extremists. Under pressure from the United States, he ended ties with the Afghanistan Taliban and banned radical Islamist groups. However, such groups continue to exist. On November 26, 2008, members of the Pakistan-based Lashkar-e-Taiba, considered a terrorist group by many countries including the United States, embarked on a series of shootings and bomb attacks across Mumbai, India's financial capital. Over three days the terrorists killed at least 173 people and wounded over 300.

Both India and Pakistan took advantage of U.S. policies arising from the war on terrorism from late 2001, but the Kashmir question continues to threaten the uneasy peace. In the early years of the global war on terrorism, the United States supported Pakistan monetarily and politically to try and eliminate al-Qaeda, and continues to do so to restore stability to the volatile Afghanistan–Pakistan region. India criticized Pakistan for allowing anti-India terrorist organizations to operate in that country and for its slow response in bringing the perpetrators of the 2008 Mumbai attacks to justice. The U.S. government said that it would work closely with both countries on regional security and to combat terrorism.

Battling Infectious Diseases

Dr. Mohammed Ali (Figure 6.24) uses his skills in geographic information systems (GIS) to study the spatiotemporal patterns of infectious diseases such as cholera, dengue fever, and acute lower respiratory infection in developing countries. People's environments and their health behaviors vary in space. The population groups to which individuals belong and the neighborhoods in which they live influence their lifestyle, health, and health-seeking behaviors. Epidemiologists and medical geographers study environmental risk factors at different geographical/ecological scales to understand health problems.

A native of Bangladesh, Dr. Mohammed Ali works at the International Vaccine Institute (IVI) in Seoul, South Korea. He and his colleagues at the IVI have conducted vaccine evaluation studies in research sites in China, Vietnam, Pakistan, and India. They created household GIS databases to investigate the ecological factors that influence vaccine coverage and to map geographic variations in vaccine use. The IVI team uses GIS, satellite remote sensing, and spatial modeling techniques in research that is grounded in geographic theories of human-environment interaction. The team modeled the emergence and fluctuation of cholera in central Vietnam and in Matlab, Bangladesh. By integrating spatial data sets on sea surface temperature, rainfall, vegetation, land use, climatic variables such as monthly temperature and monthly rainfall, and sociodemographic data such

Figure 6.24 Dr. Mohammed Ali (at left) working with colleagues in Guanxi Province, China. *Photo: Courtesy Dr. Mohammed Ali.*

as population distribution and socioeconomic status, they examine the effects of individual and ecological factors on the emergence of disease. They believe that associations between these variables and cholera incidence in Vietnam and Bangladesh can be used to predict future epidemics in other parts of the world.

Marrakech, Morocco. Women in traditional attire walk along a street in the medina quarter of the fortified old city. The medina in North African cities typically has narrow streets where pedestrians, donkeys, and mules are more commonly seen than motorized transportation.
Photo: Elizabeth Chacko.

Northern Africa and Southwestern Asia

LEARNING OBJECTIVES

After reading the chapter, you should be able to:

- Distinguish this world region in terms of similar natural environments, dominance of Islam, and global significance of its energy resources.

- Describe the physical features of the region and explain how these are linked to the area's natural resources and environmental problems.

- Identify the region's spatial patterns of language and religion, and compare population and urbanization characteristics among the different countries.

- Explain the region's political geographies from the birth of major civilizations and empires to colonial control and independence, linking this to the region's modern Pan-Arabism and Islamism.

- Contrast the region's oil wealth with water scarcity, and give examples of how the geography of these resources contributes to political tensions within and outside the region.

- Locate the main subregions and countries of Northern Africa and Southwestern Asia on a map, and describe how these can be distinguished in terms of their physical, cultural, economic, and political geographies.

- Explain Israel's unique place in the world region and understand the current conflicts with its territories and neighboring countries.

- Identify the major human rights issues, from the role of women to ethnic conflict in Sudan and the situation of the Kurds.

- Describe the complexities involved in rebuilding a democratic Iraq and the underlying causes and results of the Arab Spring.

Figure 7.1 **Northern Africa and Southwestern Asia: physical features, country boundaries, and capital cities.** Northern Africa is bounded to the north by the Mediterranean Sea and to the south by the almost empty Sahara Desert area, except in the east, where the Nile River joins Sudan to Egypt. Southwestern Asia is almost surrounded by seas, except in the narrow Sinai Peninsula at the northern end of the Red Sea and the northern mountain boundaries between Turkey, Iran, the Caucasus countries, Turkmenistan, Afghanistan, and Pakistan.

7.1 Global Center of Importance

This world region incorporates North Africa to include much of the Sahara and what Westerners commonly called the "Middle East" (Figures 7.1 and 7.2). The term *Middle East* is an ethnocentric, specifically a Eurocentric, term that describes lands by their location relative to Europe. The term originated from an early 1900s French view of the "Near East," incorporating modern Turkey, Syria, Lebanon, and Palestine, and the British concept of the "Middle East" that included Egypt, Arabia, and the Persian Gulf area. The "Far East" (Pakistan farther east)

indicated greater distance from Europe. Rather than apply the Eurocentric term of *Middle East*, the more global and culturally sensitive term *Southwestern Asia* is used in this book. In this chapter, the boundaries of Northern Africa and Southwestern Asia are primarily coastlines with a line drawn in the south through the Sahara along country boundaries. Mountain ranges mark the northern and eastern limits of Turkey and Iran. The region has a largely arid natural environment, making fresh water a political issue. Nevertheless, despite extensive deserts, more productive mountain and river plain environments are significant as well.

(a)

(b)

(c)

(d)

Figure 7.2 **Northern Africa and Southwestern Asia: diverse landscapes and peoples.** (a) McDonald's in central Cairo, Egypt, is one example of global connections. (b) Hassan Mosque in Casablanca, Morocco, was built for the sixtieth birthday of former Moroccan king Hassan II in 1993. It is the second largest religious monument in the world after Mecca. It has space for 25,000 worshippers inside and another 80,000 outside. The 210-meter minaret is the tallest in the world. (c) Weekly animal market at Sinaw, Oman. Although some camels are sold, most of the animals traded are goats. Oman is an oil-producing country, but many Omanis still depend on agriculture and pastoralism for their livelihoods. (d) Village of Bilad Sayt in the Hajar Mountains of Oman. Although all of Oman's cities and towns are connected by good roads, this and many other villages can still only be reached by dirt roads. Note that the village is built on a bluff, as good farmland is scarce.
Photos: (a) © Joseph P. Dymond; (b) © Heike Alberts; (c) © Alasdair Drysdale; (d) © Alasdair Drysdale.

Today, Northern Africa and Southwestern Asia is one of the most influential as well as one of the most politically contentious world regions. Prominence began early in history with some of the world's first urban civilizations (Mesopotamia and Egypt). The three major monotheistic religions of Judaism, Christianity, and Islam all originated here before spreading worldwide, and the region remains important to all three religions today. Europeans took control of most of the region in the 1920s and created new countries that reshaped identities and relationships. Hostilities toward Western influences grew and intensified after 1948 with the founding of the country of Israel as a homeland for Jews. The boundaries that the Europeans drew also are at the root of the struggle for the Kurdish and Sudanese peoples, contribute to the politics of oil and water resources, and have left the material wealth generated by oil revenues unevenly distributed between and within countries. Broader Western involvement since the 1950s has been a catalyst for Islamic fundamentalism, the Islamic Republic of Iran, and al-Qaeda, and has led to the wars in Iraq.

Global geopolitics entwined with local issues and led to the rise of many dictatorships and authoritarian governments that stifled the development of democratic movements and institutions. By late 2010, discontent erupted into widespread demonstrations for democracy, especially by the region's youth, who have grown tremendously in number but feel disenfranchised. By early 2011, the leaders of Tunisia and Egypt quickly succumbed to pressure and resigned, while the leaders of Libya, Syria, Yemen, Bahrain, and others fought to hold onto power. These local conflicts created by global geopolitics continued to be reshaped by global geopolitics as various countries of the international community and the region pressured governments to change, especially in Libya and Syria. The outside pressures have consisted of sanctions and military supplies, even military assistance in the case of Libya where the government finally fell by October 2011. At the writing of this edition, the Syrian government still was resisting international efforts as conflict in Syria became increasingly bloody.

7.2 Distinctive Physical Geography

The natural environments of Northern Africa and Southwestern Asia help to explain the region's geographic distinctiveness and internal diversity. Hot, dry plains exist alongside snowcapped mountain ranges and well-watered river valleys.

Clashing Plates

High mountain ranges and extensive plateaus contrast with wide river valleys (see Figure 7.1) and stand out where vegetation is sparse. It is more common to see bare rock than grassy or wooded slopes. Such landform differences are related to geologic activity along tectonic plate boundaries of the African, Arabian, and Eurasian plates (see Figure 1.5). Such rigid sections of Earth's crust underlie the extensive plateaus of North-

ern Africa and Arabia. The ancient rocks often are covered by flat layers of sedimentary rocks, such as the limestones and sandstones forming the plateaus on either side of the Nile River valley.

Collisions of the African and Arabian plates with the Eurasian plate formed the highest mountains: the Atlas in North Africa and the Taurus, Zagros, and Elburz ranges in Turkey and Iran. The Atlas Mountains rise to nearly 4,000 m (12,500 ft.) and the Elburz to over 5,500 m (17,000 ft.). Both are parts of the young folded mountain ranges that extend from North Africa and Southern and Alpine Europe, through Southwestern Asia, to the Himalayas in South Asia. Rocks, which were forming in the seas that filled the basins between the plates, were caught up as the collision closed the seas and raised them as parts of the mountains. Volcanic eruptions along the geologically active belt injected mineralized fluids that solidified as veins in the rocks. Earthquakes are common and threaten populations. Compression folded and lifted up these rocks, and large quantities of eroded rock were deposited in downwarped sections of the crust, where oil and natural gas could be trapped. The Persian Gulf area became the environment for the world's largest concentration of these fuels.

The Red Sea is in a rift (the Great Rift Valley of Africa) where two tectonic plates are pulling apart, causing the Red Sea to widen. At its northern end, the rift splits in two around the Sinai Peninsula. The eastern branch forms the valley into which the Jordan River flows from the uplands around the Sea of Galilee toward the Dead Sea, some 400 m (1,312 ft.) below sea level. The western branch falls short of entering the Mediterranean Sea but was extended to that sea by the construction of the Suez Canal, which opened in 1869.

Dry Climates and Desert Vegetation

Dry climates, namely **steppe climate** (semiarid climate) and **desert climate** (arid climate), are those in which evaporation rates are greater than precipitation. They dominate virtually the whole region (Figure 7.3). The region boasts the world's highest recorded shade temperature (58°C, or 136°F, at Al'Aziziyah, Libya), but the lack of cloud cover makes nights cool or cold, and freezes are possible in winter.

Most places have some rain, but it falls irregularly and with increasing uncertainty as aridity increases. The rainy winters of the coasts of Northern Africa and the eastern Mediterranean lands are caused by the seasonal southward shift of midlatitude frontal weather systems bringing Atlantic moisture. The mountains of Turkey and Iran force air to rise, often adding lift to the frontal zones, and precipitating rain and snow. Winter snowfall provides a source of meltwater that feeds the Tigris-Euphrates River system in spring. In southern Sudan, summer rains come from the northward movement of the equatorial rainy belt.

The arid climatic environments support only drought-resistant desert plants that partially cover the ground. Large areas of desert have little vegetation and are gravel-strewn, rocky, or sand-covered: sand seas cover one-fourth of the Sahara, which has a mostly rocky or gravel-covered surface. The vegetation cover thickens and becomes denser on uplands, where there are

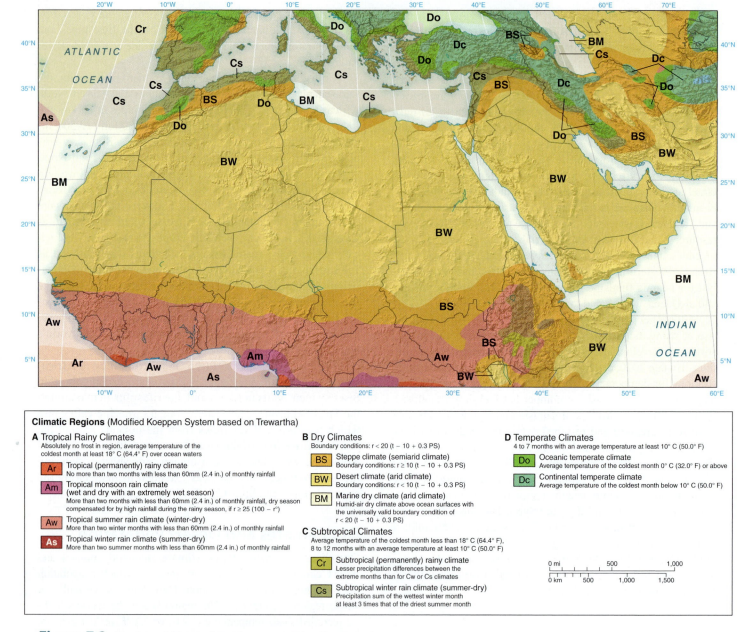

Climatic Regions (Modified Koeppen System based on Trewartha)

A Tropical Rainy Climates
Absolutely no frost in region, average temperature of the coldest month at least 18° C (64.4° F) over ocean waters

Ar Tropical (permanently) rainy climate
No more than two months with less than 60mm (2.4 in.) of monthly rainfall

Am Tropical monsoon rain climate
(wet and dry with an extremely wet season)
More than two months with less than 60mm (2.4 in.) of monthly rainfall, dry season compensated for by high rainfall during the rainy season, if r ≥ 25 (100 − r°)

Aw Tropical summer rain climate (winter-dry)
More than two winter months with less than 60mm (2.4 in.) of monthly rainfall

As Tropical winter rain climate (summer-dry)
More than two summer months with less than 60mm (2.4 in.) of monthly rainfall

B Dry Climates
Boundary conditions: r < 20 (t − 10 + 0.3 PS)

BS Steppe climate (semiarid climate)
Boundary conditions: r ≥ 10 (t − 10 + 0.3 PS)

BW Desert climate (arid climate)
Boundary conditions: r < 10 (t − 10 + 0.3 PS)

BM Marine dry climate (arid climate)
Humid-air dry climate above ocean surfaces with the universally valid boundary condition of r < 20 (t − 10 + 0.3 PS)

C Subtropical Climates
Average temperature of the coldest month less than 18° C (64.4° F), 8 to 12 months with an average temperature at least 10° C (50.0° F)

Cr Subtropical (permanently) rainy climate
Lesser precipitation differences between the extreme months than for Cw or Cs climates

Cs Subtropical winter rain climate (summer-dry)
Precipitation sum of the wettest winter month at least 3 times that of the driest summer month

D Temperate Climates
4 to 7 months with an average temperature at least 10° C (50.0° F)

Do Oceanic temperate climate
Average temperature of the coldest month 0° C (32.0° F) or above

Dc Continental temperate climate
Average temperature of the coldest month below 10° C (50.0° F)

Figure 7.3 **Northern Africa and Southwestern Asia: climate regions.**

higher precipitation totals and lower rates of evaporation. In the uplands of Northern Africa, Lebanon, western Syria, Turkey, and Iran, grassland and woodland vegetation increase with altitude as temperatures and evaporation levels fall and make the low-to-moderate precipitation more effective.

Soils are poor and undeveloped through most of the region: the best occur in the rainy coastal areas and along the valley floors of rivers where annual floods deposit fertile alluvium. Soil erosion resulting from plowing for grain and cultivation on steep slopes is a major problem in most of the region.

Climate change left its mark on the region and affected the history of human settlement. The present aridity began some 5,000 years ago, forcing many people into the watered valleys of the Nile and Tigris and Euphrates rivers, thus concentrating populations. The desert margins continue to fluctuate. They retreat in

series of wetter years and advance in drier years—or as human actions remove vegetation and lower the groundwater levels.

Natural Resources: Water

The arid climate makes rivers unusual in Northern Africa and Southwestern Asia. Rivers that exist often receive their water from surrounding mountains or from rainy areas outside the region. For example, the Tigris-Euphrates River system is fed by snowmelt on the high mountain ranges of Turkey and Iran. The Nile River system has two major headwater branches that begin in the rainy equatorial area around Lake Victoria (White Nile) and the seasonal rainy area of the Ethiopian Highlands (Blue Nile). The two major Nile River branches join near Khartoum (Figure 7.4). The White Nile River, flowing from Lake Victoria on the equator,

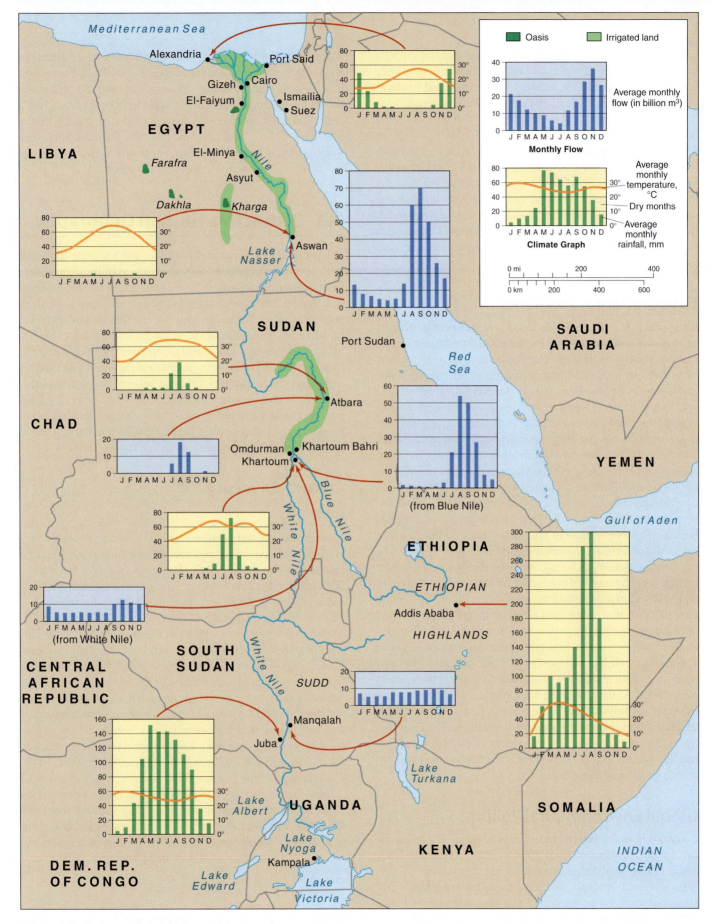

Figure 7.4 **The Nile River valley: rainfall and river flow compared.** The relatively small areas of irrigated land are shown in green. The climate graphs (yellow background) and the river flow graphs (blue background) show how the main flow comes from rainfall in the tropical south, and that the arid area through northern Sudan and Egypt receives little rain. The White Nile flow is moderate, but consistent; the Blue Nile annual flood dominates flows downstream of the confluence.

has tributaries that are fed by rains through the year and supply a fairly constant flow of water. The Blue Nile River flow is more important because it is less subject to evaporation and produces the annual September Nile River flood in Sudan and Egypt.

Israel, Gaza, and the West Bank occupy a small area of land between the Mediterranean coast and the Jordan River valley. The terrain includes a coastal plain and mostly hilly land with lower areas around Lake Tiberias (Sea of Galilee) and along the Jordan River to the Dead Sea. Although Israel receives winter rain, the total rainfall is low, and summer drought brings water shortages. Great efforts made it possible to supply water to dry areas and to manage the environment efficiently. Despite such careful management, internal groundwater sources are now fully used.

Environmental Problems

Ecosystems in arid environments are particularly fragile, as they are easily destroyed by human activities and do not recover quickly. As this region's population rapidly grows, new industries and cities are built in sensitive arid environments, and higher living standards raise the demand for water. Irrigation farming in arid areas requires good management to prevent the high rates of evaporation from drawing so much salt to the surface soil that crop productivity is reduced or ended. This process is known as **salinization**. Under careful management, maintaining good drainage allows water to flush the salts downward. Adding too much water waterlogs the soil and concentrates salts at the surface. In ancient irrigation schemes, 60 percent of the land in the Tigris-Euphrates River lowlands became unusable because of poor management. Modern usage led to loss of farmland for similar reasons.

The oil industry is a major polluter throughout the world, and the concentrations of production and distribution centers in the Persian Gulf area pollute the atmosphere and waters around oil wells, surface seepages, and where unwanted gases are flared off. Leakages pollute the waters near ocean terminals. Particles in the atmosphere cause fogs that worsen respiratory ailments. The Persian Gulf had lost most of its plant and animal life as a result of pollution since oil production first began there in the early 1900s. At the end of the Gulf War in 1991 the retreating Iraqis set oil wells on fire, adding carbon gases and particles to the atmosphere, but the particles fell on desert sands. They also released a huge oil slick, damaging plant and animal life, but it had less impact than it would have if the Persian Gulf had not already been heavily polluted.

Global Environmental Politics

Countries in Southwestern Asia increased their CO_2 emissions by almost 152 percent from 1990 to 2008. Over this same period, the United States' emissions increased about 15 percent and Europe's emissions declined. Southwestern Asia's CO_2 emissions were surpassed during this period only by China's 192 percent increase. This was the period in which the Kyoto Protocol (1997) sought to have countries on its Annex I list reduce their 1990 greenhouse gas emissions by 5 percent. Though many

countries in Northern Africa and Southwestern Asia ratified the Kyoto Protocol, they were not on the Annex I list and have done little to curb their emissions. When measuring CO_2 emissions per person, the top emitters in the world have been Qatar, United Arab Emirates, Bahrain, and Kuwait, followed by the United States and Canada, and then Saudi Arabia. Though China is the world's largest total CO_2 emitter, its emissions per person are low compared to many Southwest Asian countries.

7.3 Distinctive Human Geography

This world region is culturally diverse, and a historic center for artistic, scientific, and technological innovations. Along with having differing languages, the region is the hearth for Judaism, Christianity, and Islam, related to one another through worshipping the God of Abraham. Muslim empires flourished in the Middle Ages before this world region became a focus of geopolitical strategies by major world powers in the 1800s and 1900s. Its location at the junction of other world regions meant that this region acted as a source of diffusion outward and as an inward-directed focus of external influences. The discovery of oil has provided wealth and influence to some, but has also become a geopolitical issue. Growing populations live mainly in urban areas, which stresses water resources in this arid region.

Cultural Diversity

Languages

Northern Africa and Southwestern Asia is linguistically very diverse, with numerous languages among four major language families spoken (Figure 7.5). The most common are Arabic, Berber, Farsi (Persian), Turkish, Kurdish, and Hebrew. Arabic is spoken by just under 50 percent of the people in the region. It is also the preferred language in the Qu'ran and Muslim prayers. "Arabs," who were once defined as living in the Arabian Peninsula, now include all those using Arabic as their first language. Berber is linguistically connected to Arabic but older and is found primarily in Morocco and Algeria, especially in the Atlas Mountains and the Sahara Desert where it best resisted the spread of Arabic (Figure 7.6). Farsi (Persian) is the official language of Iran and is linked to the Shiite Islamic beliefs of Iran's leaders and majority population. Kurdish speakers straddle Turkey, Syria, Iraq, and Iran. Hebrew is the official language of Israel. The language of the Hebrew Bible and religious services is an archaic form, superseded in everyday usage by modern Israeli Hebrew. The movement devoted to creating a Jewish state from the 1800s, known as Zionism, revived the use of Hebrew, modernizing it for secular usage to provide a common religious and secular element among Jews from many countries.

Religions

Religion and related traditions contribute greatly to peoples' identities in Northern Africa and Southwestern Asia. The early animist religions had many gods linked to natural phe-

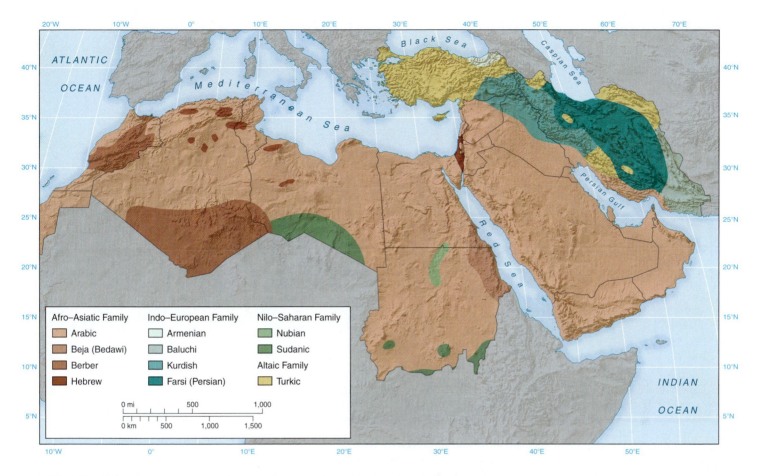

Figure 7.5 Language families and languages of Northern Africa and Southwestern Asia.

Afro–Asiatic Family
- Arabic
- Beja (Bedawi)
- Berber
- Hebrew

Indo–European Family
- Armenian
- Baluchi
- Kurdish
- Farsi (Persian)

Nilo–Saharan Family
- Nubian
- Sudanic

Altaic Family
- Turkic

nomena. Some human emperors were treated as gods. After approximately 1000 BC, dominance by religions with many gods gave way to religions based on a single god: **monotheism**. Judaism was the first monotheistic religion and was followed by Christianity and then Islam. These three monotheistic religions share common religious figures and religious texts. All three diffused from their hearths in Southwestern Asia to Europe, Africa, and Asia (Figure 7.7).

Judaism is a religion whose adherents worship Yahweh, seen as the only God, creator, and lawgiver. It began in the area now known as Israel and the Palestinian territories some 2000 years BC, where Abraham and his descendants settled after moving from Mesopotamia. Jewish beliefs focus on the historic role of family based on the line from Abraham, persecution beginning with slavery in Egypt, redemption as Moses led the people out of Egypt, and their occupation of the promised land. Yahweh intervened to support and punish through times of trouble and deportation, promising a messiah to save Jews from domination by others. In AD 70, the Roman army destroyed the holy city of Jerusalem and dispersed Jews through Southwestern Asia, Northern Africa, and Europe to form a major **diaspora** (dispersed community).

Christianity stemmed from new interpretations of the beliefs and teachings of Judaism in the early years AD, focusing on the teachings of Jesus of Nazareth. Christians began as Jews who believed that Jesus of Nazareth was the messiah, which

Figure 7.6 **Berbers in Morocco.** These two Berber men guide tourists into the Sahara Desert close to Erfoud, Morocco. *Photo: © Heike Alberts.*

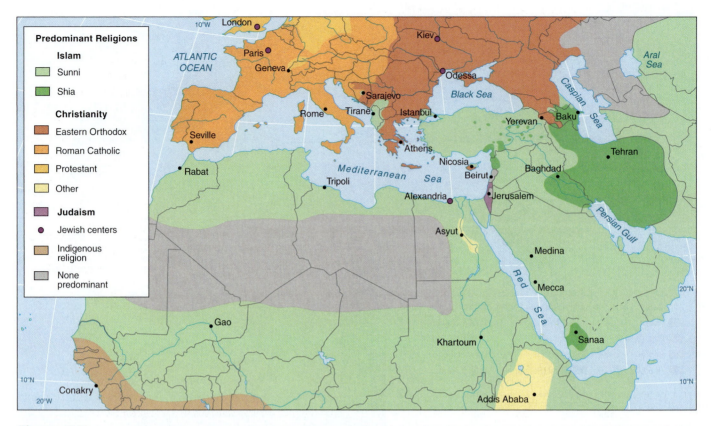

Figure 7.7 **Major religions in Northern Africa and Southwestern Asia.** Beginning in Mecca and Medina, Muslim Arabs conquered the area shown in green after Muhammad emerged as a new religious leader around AD 600. Islam also spread along trade routes. Christianity became the religion of the Roman Empire in the AD 300s, spreading through Europe. Jews were dispersed at the fall of Jerusalem in AD 70, and the modern country of Israel was not formed until 1948.

translates to "Christ" in Greek. Christian churches spread across Southwestern Asia and into Africa and Europe. Controversies over how much Jesus was man and how much God resulted in divisions among the churches of this region. A later division occurred between the eastern (Orthodox) and western (Catholic) groups of churches in Europe. From AD 395 to 1453, the eastern church was centered in Constantinople (modern Istanbul, Turkey). The largest Christian sect in this region today is the Coptic Church, with a pope who resides in Alexandria, Egypt.

Muhammad founded Islam in Arabia, notably in Mecca and Medina, during the early AD 600s. **Islam** means "submission to the will of Allah (God)," and the followers of the religion, **Muslims**, are "those who submit to Allah." Muhammad carried on many Jewish and Christian beliefs, such as monotheism. Seen as the prophet of Allah, Muhammad is believed to succeed earlier prophets such as Moses, David, and Jesus. The **Qu'ran** (holy book) is believed by Muslims to be the word of Allah revealed to Muhammad.

After Muhammad's death in 632, Arabs spread Islam rapidly westward to Northern Africa and Spain as well as eastward into central Asia, uniting the Arab peoples and creating a series of empires that went on to convert Persians, Turks, and people in India through conquest and trade. It led to a Muslim Golden Age of artistic and scientific achievements alongside further military expansions from the 800s to the 1100s. New forms of art and architecture developed, and Muslim mosques

remain dominant features of town landscapes, often doubling as centers of religious and secular activities (Figure 7.8).

Early divisions within Islam continue to be significant today. **Sunni Muslims**, or Sunnites, are the majority Muslim group and base their way of life on the Qu'ran, supplemented by local traditions. Political power was at first given to a succession of leaders, or caliphs, descended from historic Muslim leaders. Many Sunni Muslims are moderates in maintaining traditional Islamic practices. Exceptions include the Wahhabi sect that dominates Saudi Arabia, and the Taliban faction that ruled Afghanistan from 1996 to 2001.

A minority of Muslims also follow the Qu'ran but dispute the caliph leadership succession accepted by Sunnis. They look to descendants of Ali, the fourth caliph of Islam and a cousin and son-in-law of Muhammad. His son, Hussein, was defeated and killed by the Sunni caliph of Damascus in 680. Supporters see Ali as the only *imam*—authoritative interpreter of the Qu'ran—and are known as *Shi'at 'Ali* ("partisans of Ali"). For most of their history, these **Shia Muslims** (Shiites) were known for commemorating and lamenting Hussein's death by flagellation and withdrawal from the world. Today, Shiites comprise large proportions of the population in a few countries, including 90 percent of the Iranian population and a majority in Iraq. Until recent times, they shunned politics and looked to clerics, who were supported by alms, for leadership. The Iranian Revolution in 1979 marked a new era of Shiite political activism as Shiites took control of Iran's

Figure 7.8 **Islam and architecture.** Mosques are a common feature of urban landscapes in Muslim countries. Older (1300s) and more recent (1800s) mosques tower over Cairo, Egypt. *Photo: © Joseph P. Dymond.*

government. The religious government of Iran promotes Islamic isolationism from Western ways and supports extremist groups such as Hamas (Palestine) and Hezbollah (Iran, Syria, Lebanon).

Population

The population of Northern Africa and Southwestern Asia tends to be concentrated near the Mediterranean and Red seas, the Persian Gulf, and in the fertile valleys of the Nile and the Tigris-Euphrates rivers (Figure 7.9). Moderate densities are also found in the rest of Turkey and western Iran. Much of the rest of the region is almost empty of people.

With high birth rates and low death rates, the region's population will continue to grow rapidly over the next 20 years, with the total population increasing by almost 30 percent (Table 7.1). As the age-sex diagram in Figure 7.10a (Egypt) indicates, the large numbers of young people will produce continuing high rates of population growth as they reach childbearing age. In 2012, several countries had high total fertility rates, with the highest being Yemen (5.2), Iraq (4.6), the Palestinian Territories (4.4), Sudan (4.2), and Jordan (3.8). In contrast, the total fertility rates were below 2.1 (the rate necessary to maintain populations at their current sizes) in UAE (1.8), Bahrain, Lebanon, Iran (each 1.9), and Turkey (2.0).

In recent years, many of the region's countries have implemented policies to slow birth rates. For example, Tunisia set a minimum age for marriage and instituted a successful family planning program soon after it obtained independence. Morocco and Algeria implemented family planning and maternal and child services. Egypt's support for family planning resulted in a drop of that country's total fertility rate from over 5 to 2.9 between the 1970s and 2012.

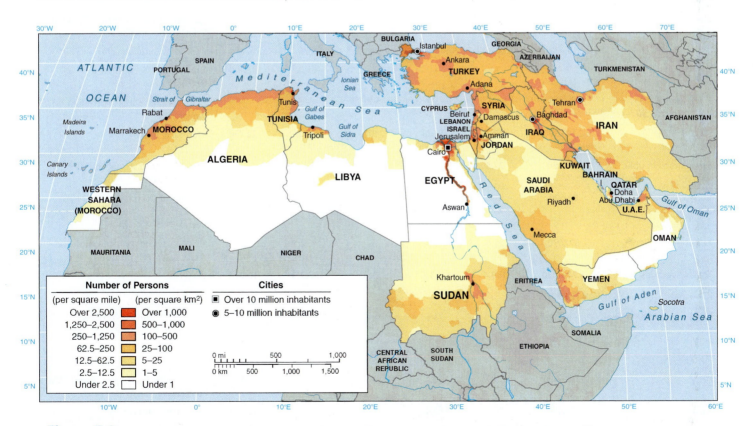

Figure 7.9 **Northern Africa and Southwestern Asia: population distribution.** Note where the highest and lowest densities of population occur.

Table 7.1 — Northern Africa and Southwestern Asia: Data by country, area, population, urbanization, income (gross national income purchasing power parity), ethnic groups

Country	Land Area (km²)	Population (millions) Mid-2012 Total	Population (millions) 2025 Est. Total	% Urban 2012	GNI PPP 2010 Total (US$ billions)	2010 Per Capita (US$)	Ethnic Groups (%)
NORTH AFRICA							
Algeria, Democratic and Popular Republic of	2,381,740	37.4	42.0	72	287.2	8,100	Arab 83%, Berber 16%
Socialist People's Libyan Arab Jamahirya	1,759,540	6.5	7.5	78	105.7	16,880	Arab-Berber 97%
Morocco, Kingdom of	446,550	32.6	36.9	58	149.3	4,600	Arab-Berber 99%
Tunisia, Republic of	163,610	10.8	12.1	66	95.6	9,060	Arab-Berber 98%
Totals/Averages	**4,751,440**	**87.3**	**98.5**	**66**	**637.8**	**7,306**	
NILE RIVER VALLEY							
Egypt, Arab Republic of	1,001,450	82.3	102.0	43	491.3	6,060	Egyptian-Bedouin-Berber 99%
Sudan, Republic of	1,861,484	33.5	46.8	41	88.6	2,030	Sudanese Arab 70%, Fur, Beja, Nuba, Fallata
Totals/Averages	**3,507,260**	**115.8**	**148.8**	**42**	**479.9**	**4,144**	
ARAB SOUTHWEST ASIA							
Bahrain, State of	680	1.3	1.6	100	26.0	24,710	Bahraini 63%, Asian 13%, other Arab 10%, Iran 8%
Iraq, Republic of	438,320	33.7	48.9	67	108.1	3,370	Arab 75–80%, Kurd 15–20%, Assyrian, Turkmen
Jordan, Hashemite Kingdom of	89,210	6.3	8.6	83	35.1	5.800	Arab 98%, Circassian, Armenian
Kuwait, State of	17,820	2.9	3.7	98	—	—	Kuwaiti 45%, other Arab 35%, Indian, Pakistani 9%, Iranian 4%
Lebanese Republic	10,400	4.3	4.8	87	59.5	14,090	Arab 93%, Armenian 5%
Oman, Sultanate of	212,460	3.1	4.0	73	68.3	25,190	Omani Arab 75%, Indian, Pakistani 21%
Qatar, State of	11,000	1.9	2.2	100	—	—	Arab 40%, Pakistani 18%, Indian 18%, Iranian 10%
Saudi Arabia, Kingdom of	2,149,690	28.7	36.2	81	609.8	22,750	Arab 82%, Yemeni 13%
Syrian Arab Republic	185,180	22.5	26.5	54	104.6	5,120	Arab 90%, Kurd, Armenian, Turkmen and others 10%
United Arab Emirates	83,600	8.1	9.9	83	351.0	50,580	UAE Arab 19%, other Arab 23%, South Asians 50%, other expatriates 8%
Yemen, Republic of	527,970	25.6	36.7	29	60.1	2,500	Mainly Arab, with African-Arab and South Asian
Totals/Averages	**3,726,330**	**138.4**	**183.1**	**65**	**1,422.5**	**10,278**	
ISRAEL AND THE PALESTINIAN TERRITORIES							
Israel, West Bank, Gaza	21,060	7.9	9.4	92	210.8	27,660	Jewish 82% (born in Israel 62%; white 26%, African 7%, Asian 5%), non-Jew (mainly Arab) 18%
TURKEY AND IRAN							
Iran	1,648,000	78.9	90.5	69	840.0	11,490	Persian 60%, Azerbaijani, Turkic 25%, Kurd 7%
Turkey	779,450	74.9	85.4	77	1,129.9	15,530	Turkish 80%, Kurd 17%
Totals/Averages	**2,427,450**	**153.8**	**175.9**	**73**	**1,969.9**	**12,808**	
Region Totals/ Averages	**14,433,540**	**495.3**	**603.5**	**62**	**4,720.9**	**9,531**	

Source: *World Population Data Sheet 2006*, Population Reference Bureau; Microsoft Encarta 2005.

Variations in immigration make Israel's population growth irregular. The annual growth rate fell from 1965 to 1980 but rose again in the mid-1990s as a result of immigration. In the early 1990s, over 1 million Russian Jews moved to Israel. The 2012 resident Israeli population of 7.9 million included 24 percent non-Jews, most of whom were Arabs. The fluctuating number of immigrants and the higher proportion of older people means that Israel's population structure differs from that of the region's Arab countries (Figure 7.10b).

Immigration also plays a significant role in many Arab Southwest Asian countries, particularly the six Gulf Cooperation Countries (Bahrain, Kuwait, Oman, Qatar, Saudi Arabia, and UAE). The booming oil business requires more labor than many of the oil-producing countries possess. Imported labor comes from other parts of the region (e.g., Jordan, Lebanon, and Yemen) and from South and East Asia (e.g., India, Pakistan, Bangladesh). In 2009, foreign workers accounted for 30 percent of the total populations of Oman and Saudi Arabia and 80 percent of the total populations of Qatar and UAE. They comprised 95 percent of the total work force in the UAE.

Most of the region's countries have life expectancies around 70 years. Israel has the highest with almost 82 years. Life expectancy in Yemen and Sudan remains lower, at 65 and 60 years respectively. In Iraq, it fell for a time below 60 after years of war and deprivation.

Urban Patterns

Arid environments, coupled with the growth in oil industry and government employment, make most countries in the region very highly urbanized (Table 7.2). For example, between 1950 and 2012, Kuwait and Qatar went from 50 percent to over 98 percent urban, while Saudi Arabia went from 10 to 81 percent urban. The rate of urban expansion means that many cities are dominated by new buildings (Figure 7.11). However, the expanding demand for housing by poorer people was often more than governments could meet. Shantytowns are features of cities across the subregion, known as *bidonvilles* in Casablanca, Morocco, and as *gourbivilles* in Tunis, Tunisia.

The Fertile Crescent, which runs along the Tigris-Euphrates rivers, along the eastern shore of the Mediterranean, and up the Nile River valley, is the location of some of the world's oldest and most historically significant cities. These older cities that were central to agricultural and trading economies survive as enclaves within today's expanded cities. Some are much changed by the clearing of crowded buildings that made way for new highways. High densities of homes, commercial premises, and public buildings inside city walls marked the traditional small towns of the region and their central **medinas** (Figure 7.12). Medinas, named after the sacred Muslim city in Saudi Arabia, are historic sectors of cities,

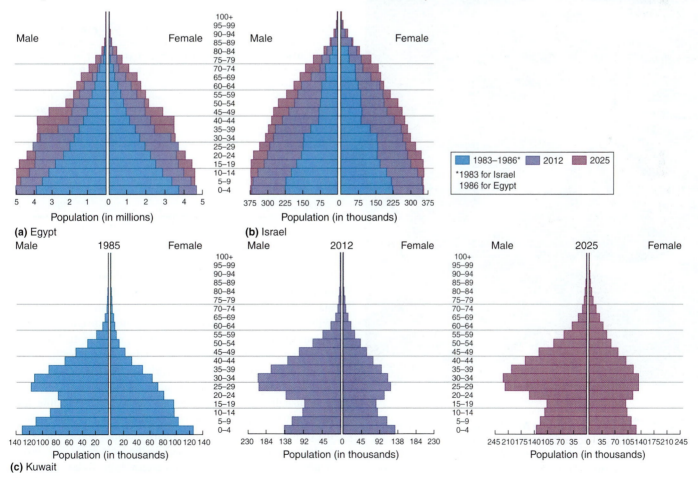

Figure 7.10 **Age-sex diagrams.** (a) Egypt. (b) Israel. (c) Kuwait. Account for the different patterns of change. *Source: U.S. Census International Database: http://www.census.gov/population/international/data/idb/region.php.*

(a)

(b)

Figure 7.11 **Southwestern Asia: Dubai.** (a) New buildings along Sheikh Zayed Road. (b) Oil wealth is funding such massive projects as these palm-shaped artificial islands in the Persian Gulf, built from 130 million cubic yards of rock and sand to create 74 miles of waterfront and new houses, hotels, and a marine park.
Photos: (a) © Alasdair Drysdale; (b) Photo courtesy Space Imaging Middle East.

Table 7.2	Populations of Major Urban Centers in Northern Africa and Southwestern Asia

TOP 10 MOST POPULATED URBAN CENTERS		
City, Country	**2012 (millions)**	**2025 (millions)**
1. Cairo, Egypt	12.8	15.6
2. Istanbul, Turkey	11.7	13.4
3. Tehran, Iran	7.9	9.8
4. Baghdad, Iraq	6.2	8.1
5. Riyadh, Saudi Arabia	5.0	6.3
6. Khartoum, Sudan	4.9	7.9
7. Alexandria, Egypt	4.5	5.7
8. Ankara, Turkey	4.2	4.8
9. Tel Aviv, Israel	3.4	3.8
10. Jeddah, Saudi Arabia	3.4	4.2

NUMBER OF MAJOR URBAN CENTERS IN NORTHERN AFRICA AND SOUTHWESTERN ASIA, BY POPULATION CATEGORY	
Population Category	**Number of Urban Centers**
10.00 million and greater	2
5.00 – 9.99 million	3
2.50 – 4.99 million	9
1.00 – 2.49 million	18

valued for their distinctive structure and social fabric. Their labyrinthine alleys, *souks* (commercial areas) (Figure 7.13), and artisan shops relate them to the past and attract tourists. Within these sectors existed a rigid pattern of land use. Prestigious craft workers, such as religious artisans, had premises close to a central mosque, castle, and square; those in lowlier occupations, such as leather workers, located on the edges. The residential population tended to concentrate in "quarters," usually based on religion so that separate Muslim, Jewish, and Christian quarters were common. The World Heritage list of historic cities contains many medinas across the whole region, from Fez and Marrakech in the west to the tall, brown mudhouses of Yemen in Southwestern Asia. Today, city walls remain only in cities where tourist interests are economically important, such as Fez and Marrakech (Morocco).

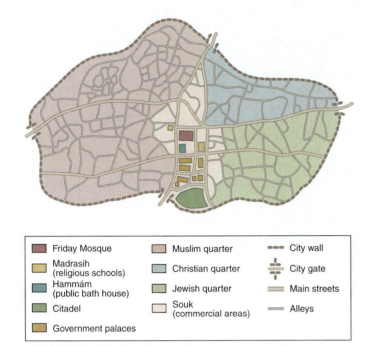

Friday Mosque

Madrasih
(religious schools)

Hammám
(public bath house)

Citadel

Government palaces

Muslim quarter

Christian quarter

Jewish quarter

Souk
(commercial areas)

City wall

City gate

Main streets

Alleys

Figure 7.12 **The typical layout of a medina.** *Source: From Iranian Cities by Masoud Kherabadi (University of Texas Press, 2000). Copyright Masoud Kherabadi. Reprinted by permission of the author.*

Just outside the old centers are newer high-rise apartments, hotels, and offices. Beyond are extensive suburban areas of housing, factories, and shopping facilities. Many older town areas were abandoned to poorer people when merchants and businesspeople moved to more spacious homes and commercial premises in the new suburbs. The factories and supermarkets changed patterns of working and social movements. Segregation by income and social group is common within the cities, especially in oil-rich Arab Southwest Asia where Arabs often live apart from immigrant workers.

Historically, Fez, Cairo, and Constantinople (Istanbul) were three important global cities and ranked as the third-, fourth-, and fifth-largest cities in the world as recently as the AD 1200s. Though having fallen in global significance, Istanbul (Turkey) and Cairo (Egypt) are the two largest cities in the region today (Figure 7.14). Abu Dhabi, Dubai (both in UAE), Riyadh (Saudi Arabia), and Tel Aviv–Jaffa (Israel) are considerably smaller in size but have many global connections. On the other hand, the isolation of Iran from much of the global economy leaves Tehran with few global city features though it is the third-largest city in the region.

Evolving Political Geographies

Empires to Countries

The Tigris-Euphrates River valley of Mesopotamia (modern Iraq) and the Nile River valley formed two of the world's early cultural hearths. From these places, many early human achievements diffused to the surrounding continents. Even in ancient times of slow and limited transportation, this region acted as a hub with constant movements of people to and from Northern Africa, Europe, and China. Over the centuries, a series of empires originating from both inside and outside the region dominated this part of the world. Some of the most noteworthy are the Assyrian, Babylonian, Egyptian, Persian, Roman, Byzantine, Mongol, Umayyad, and Abbasid empires. The last great empire of the region was that of the Ottoman Turks, which began in 1299 and lasted until 1923.

The Ottomans flourished in the 1500s and 1600s but turned inward-looking after the 1700s and did little to encourage

Figure 7.13 **In the medina.** The souk of Marrakech, Morocco, Northern Africa. *Photo: © Elizabeth Chacko.*

Figure 7.14 **Nile River valley: Cairo, Egypt.** Cairo is a modernizing city, the political and economic center of Egypt. *Photo: © Corbis RF.*

modernization or involvement in the expanding global economic system. By then, European empires were encroaching on the region and eventually took control in many places, including France (Tunisia, Morocco, Algeria), Italy (Libya), and Britain (Egypt, Sudan, and much of the southern Arabian Peninsula and Gulf coasts after the 1869 opening of the Suez Canal). Russia took land in the north. Some of the more isolated tribes in Arabia became largely independent under their own kings or sultans. During World War I, the British and French fueled Arab nationalism to expel the Turks from the areas that became Palestine, Syria, Lebanon, Jordan, Iraq, and the Arabian Peninsula.

The Arab Southwest Asian countries gained independence from their European colonial overlords in different ways. After the defeat of the Ottoman Empire in World War I, the League of Nations made Syria and Lebanon French protectorates and placed much of the rest of the area under British protection. The French encouraged republican governments. The British guaranteed the survival of the kingdoms they established, including Jordan (still a monarchy today), Iraq (a monarchy until 1958), and the small emirates along the Persian Gulf under traditional local rulers, or emirs. In the 1970s, seven emirs joined to form the United Arab Emirates (UAE). A homeland for the Jews was also established and eventually became the modern state of Israel in 1948.

Pan-Arabism

In the mid- to late-1900s, post-independence movements, generally opposed to Western economic colonialism, sought to unite Arab peoples on the basis of (mostly secular) nationalism into a single country with increased world influence. None succeeded in uniting the Arab world, though their efforts are noteworthy.

The **Arab League** was created in 1945 to encourage the united opposition of Arab countries to the establishment of Israel. Its seven founding members were the only independent Arab countries at the time, but its membership increased to 22 as more gained independence. Members of the Arab League eventually included the **Palestine Liberation Organization** (PLO), a political organization providing an umbrella for many smaller groups that demand a country for Palestinians, the people living in the lands used to create the state of Israel. After Egypt's 1979 accord with Israel, the Arab League expelled Egypt and transferred Arab League headquarters from Cairo to Tunis. From 1958 to 1961, Egypt and Syria joined as the United Arab Republic under the leadership of Colonel Jamal Nasser, with the intention of persuading other countries to commit their futures to a single **Pan-Arab country**. It did not attract others, and soon broke up.

Disunity, military defeats, and tensions over the Israel-Palestine issue and among countries with different resource bases weakened the Arab League and its ability to foster unity among Arab countries in the 1980s and 1990s. A major blow to the Arab League came during the Gulf War of 1990–1991, when one Arab country (Iraq) invaded another (Kuwait) and was defeated by a coalition of other Arab countries backed by the United States and other Western countries. Many Arabs felt betrayed by both Saddam Hussein's invasion of another Arab country and their own need to rely on outside help. The Arab League also failed to reach a consensus over the 2003 Iraq War. These conflicts showed that the interests of individual countries remained more significant than an overriding Pan-Arabism.

The Arab Spring that began in late 2010 as Arab citizens rose in protest against their oppressive governments gave the Arab League a new opportunity to be effective. After Colonel Muammar al Qadhafi in Libya violently cracked down on Libya's citizens, the Arab League suspended Libya's membership and worked to oust Qadhafi. After Qadhafi's death, it reinstated Libya's membership. In late 2011, the League suspended Syria's membership for similar reasons as it worked to stop Syria's leaders from ruthlessly repressing its protesters.

Islamism

In the 1970s and 1980s, as individual countries preferred their independence to a Pan-Arab identity, religious affiliation became more significant. In 1970, foreign ministers of Muslim countries set up the Organization of the Islamic Conference (OIC), which changed its name to the **Organization of the Islamic Cooperation (OIC)** in 2011. It now has 57 members, including such countries as Pakistan, Indonesia, and Nigeria. However, it is more successful in advancing individual member countries' interests than in defining and pursuing a common agenda to rival Western-dominated globalization. One of OIC's most important affiliates is the Islamic Development Bank, which is dedicated to economic development among OIC members.

In the 1970s, Islamic political groups also came to the fore, basing their ideology on the Qu'ran, interpretations of *jihad* as holy war, and references to past Islamic triumphs. These ideas contradicted many long-held Islamic beliefs and practices that were peaceful in nature. However, persecuted under nationalistic Arabism and secular governments, Islamic political groups called for the reestablishment of a single Islamic country and rejected the traditional view that involvement in politics should not be a concern of good Islamic practice. They hated the fragmentation of Islamic lands into separate countries, in which the religious establishment was subservient to those educated in Westernized ways.

Political Islam was fueled by the 1967 and 1973 Arab-Israeli wars. Similarly, the Israeli invasion of Lebanon in 1982 during the Lebanese civil war led to the formation of a group known as Hezbollah ("Party of God"). Meant as armed resistance to Israel, it has been funded by Iran and Syria. Similarly, Hamas ("Islamic Resistance Movement") was formed in 1987 during an uprising (*intifada*) against the Israeli occupation of Palestinian territories. As an offshoot of the Egyptian Muslim Brotherhood, its goal has been to wrestle control of the Palestinian territories from Israel and establish an Islamic state.

Political Islamists also have pressed to have Islamic law (*sharia*) adopted in countries where Muslims live. For example, not long after the Iranian Revolution in 1979, which brought Ayatollah Khomeini and Islamic fundamentalists to power, Islamic law became official in Iran in 1983. Saudi Arabia also has come to increasingly rely on Islamic law. It is noteworthy that Iran's religious traditions stem from the Shia form of Islam while Saudi Arabia's traditions are based a Sunni form of Islam rooted in Salafism (sometimes inaccurately called Wahhabism). These differing fundamentalist traditions help to explain the great rivalry that has developed between Saudi Arabia and Iran, illustrated by Saudi Arabia's support for Iraq in the Iran-Iraq War (1980–1988).

The 1979 invasion of Afghanistan by the former Soviet Union generated a jihad financed by the Gulf and Western countries. This unified Islamists around the world as international brigades from Egypt, Algeria, the Arabian Peninsula countries, Pakistan, and Southeast Asia worked together in guerrilla warfare as part of the Islamist armed struggle. In 1989, the Soviet Army withdrew from Afghanistan, causing Islamists to declare a great victory and gain immense self-confidence. However, subsequent failures to impose military solutions in Algeria and Egypt were major blows for Islamists. In general, most Muslim countries, including Libya, Syria, and many small Gulf countries, suppressed such ideas and imprisoned religious activists.

The Gulf War of 1990–1991 also angered many Islamists because Saudi Arabia and other Gulf kingdoms allied with the United States and Western countries to drive Iraqi forces out of Kuwait. In the process, Western troops were stationed in Saudi Arabia, the country of some of Islam's greatest holy sites. A radical Islamic fringe turned against the Arab kingdoms with rulers who repressed them and their international networks. Their hostility toward the United States also increased.

Meanwhile, Islamist failures caused most countries that had supported guerrillas to withdraw from extremist actions. Even in Iran, a democratic movement is building, supported by many young people who prefer peaceful coexistence with other countries rather than the radical, isolationist position that Iran's leaders have created. At the same time, extreme Islamists are frustrated at their lack of political success and have expanded their terrorist activities. Following their actions of September 11, 2001, organizations like al-Qaeda and individuals like Osama bin Laden gained great international attention. However, many Muslim countries held memorial services for the victims of 9/11, including Iran in its capital, Tehran. Thus, despite their actions on 9/11 and their support of suicide bombings in Israel, Islamic extremists have had little success in advancing their goals.

At the same time, Muslims in many countries have become increasingly frustrated with their economic situations and their lack of political rights. The Arab Spring that began in December 2010 toppled a number of governments and threatened others. Ironically, political Islam grew out of these kinds of frustrations, but Islamists did not plan and organize these revolutions. Yet, groups like the Muslim Brotherhood have come to the forefront as a leading alternative to fallen regimes in countries like Egypt.

Global Economics

The discovery of oil and the founding of new countries after World War I brought the region into world prominence again. The 1973 war with Israel triggered massive increases in the

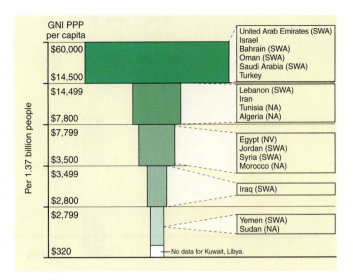

Figure 7.15 **Northern Africa and Southwestern Asia: country incomes compared.** The countries are listed in the order of their GNI PPP per capita. No data for Kuwait and Libya. *Source:* 2012 World Development Indicators, *World Bank 2012.*

price of oil, and the oil-producing countries in this region have built huge financial surpluses. Subsequently, many of the region's countries have become wealthier than other formerly poor countries in the world (Figure 7.15). Ownership of consumer goods (Figure 7.16) reflects a range of material wealth across the region.

Oil and Natural Gas Resources

The huge oil and gas production from Northern Africa and Southwestern Asia accounts for a large percentage of the world total. Despite the addition of new producers in Latin America, Asia, and Africa, this world region accounts for almost 37 percent of world total oil production and 53 percent of world total oil reserves (Figures 7.17 and 7.18). Saudi Arabia alone has 13 and 16 percent respectively of these world totals. For natural gas, this world region produces almost 20 percent of world total production and contains 42 percent of world total reserves.

Not all the countries of Northern Africa and Southwestern Asia, however, are major oil producers. Morocco, Turkey, Israel, and Jordan produce no oil. Tunisia, Sudan, Syria, and Yemen produce and export modest amounts. Countries that import oil face the burden of purchasing oil, whatever its price, and many go into debt during times of high prices.

Organization of Petroleum Exporting Countries

In the early to mid-twentieth century, international oil companies kept oil prices low for consumers in the world's wealthiest countries by paying little to the producing countries. In 1960, the producers around the Persian Gulf, together with Venezuela, formed the **Organization of Petroleum Exporting Countries** (OPEC). Today, eight of the 12 countries comprising OPEC's membership are in this world region. OPEC's main purpose is

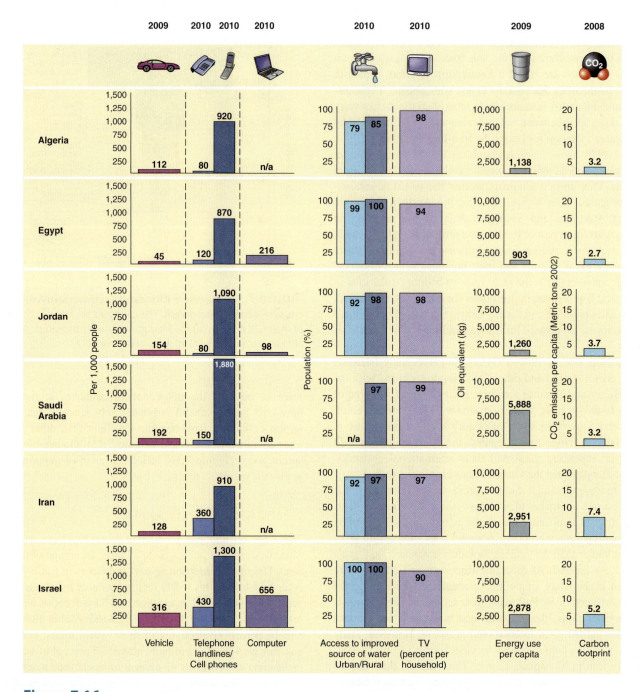

Figure 7.16 **Northern Africa and Southwestern Asia: ownership of consumer goods, access to good water, and energy usage.** Compare these levels with those of other countries in Africa. *Source: 2012 World Development Indicators, World Bank 2012.*

to work as a **cartel**—an organization that coordinates the interests of producing countries by regulating oil prices.

OPEC's oil embargo on the West during the Arab-Israeli War in 1973 was successful and allowed OPEC to control world oil distribution for a few years. As prices rose fourfold, revenues of the oil-producing countries boomed in the mid- to late-1970s. However, higher oil prices caused economic recession in the wealthier Western countries. Western oil companies opened new oilfields outside the OPEC area, including those in the North Sea (Europe), that had previously been too expensive to develop.

Beginning in the 1990s, Russia exported increasing quantities of oil and natural gas as its main source of currency. A world oil glut brought very low market prices and financial problems to the Arab producers. By the late 1990s, the OPEC oil producers and major industrial countries saw that their interests lay in a moderately high, but stable, oil price. War, other crises, and increased demand from growing economies like China pushed oil prices to record highs in the 2000s. Though these prices subsided during the world recession of 2007–2009, the growth of developing economies has continued to push prices high again.

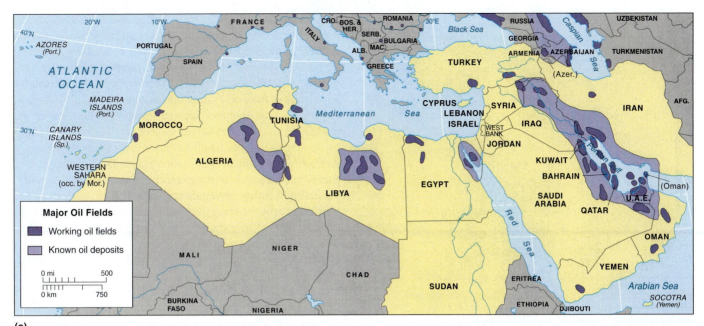

(a)

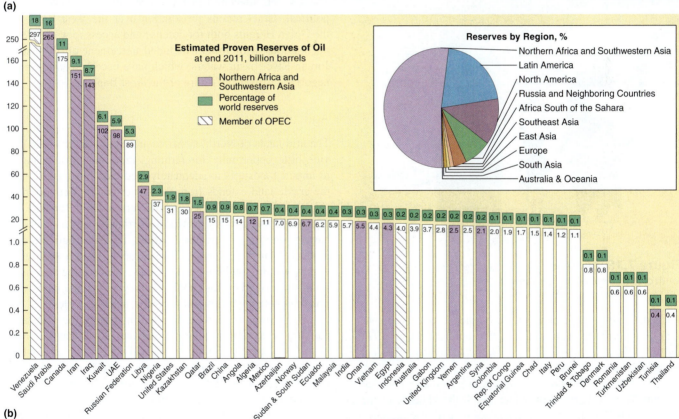

(b)

Figure 7.17 **Oil resources in Northern Africa and Southwestern Asia.** (a) Map of major oil fields. (b) World oil resources at the end of 2011, showing the continuing significance of this region. The Persian Gulf share, 67.8 percent of world reserves, is a little greater than 10 years ago, and reserves in this region are being used more slowly. Note which countries in the region have major oil reserves and which have little or no oil. *Sources: (a) The Economist; (b) British Petroleum.*

Water Politics

Much attention is given to this region's oil, but the issue of water is crucial in this arid region of the world. Like oil, however, water is not evenly distributed. Eighty percent of the region's fresh water is found in the Nile and Tigris-Euphrates river basins. The Jordan River, though much smaller, is crucial to countries that depend on it. However, each of these rivers flows through more than one country (see Figures 7.1 and 7.4), requiring cooperation among governments, and creating or exacerbating political conflict when agreements cannot be

7.3 Distinctive Human Geography **209**

Figure 7.18 **Oil in Iraq.** The Shaiba oil refinery, 20 kilometers south of Basra. *Photo: © AFP/Getty Images.*

reached. Conflict may increase as fresh water scarcity is becoming more of an issue as populations grow rapidly. Between 1975 and 2001, the amount of fresh water available to each individual dropped by more than half.

In 1959, the Egyptian and Sudanese governments signed the **Nile Waters Agreement**, which led to the building of the Aswan High Dam to store sufficient water and generate electricity. On completion of the dam in 1970, Lake Nasser behind the dam stored three times Egypt's annual water usage. Sudan receives only 13 percent of the annual flow. Along with disputes they may continue to have with each other, Egypt and Sudan also now contest the use of the water with those in the upper Nile River watershed, such as Ethiopia, Uganda, and Tanzania.

The Turkish government constructed a series of dams in southeastern Turkey on the upper reaches of the Tigris and Euphrates rivers as a part of a greater plan to irrigate lands to increase agricultural production (Figure 7.19). Because these

irrigation projects divert water from Iraq and Syria downstream, tensions have increased between Turkey and these two countries.

To provide fresh water, the region's wealthier countries have invested heavily in **desalination plants**, which make seawater usable. Northern Africa and Southwestern Asia account for almost half of the world's desalination capacity. Saudi Arabia and the United Arab Emirates account for about 30 percent of the world's capacity by themselves and contain the world's 10 largest desalination plants. Saudi Arabia receives about 70 percent of its fresh water from its desalination plants. Desalination is very energy intensive and consumes much of the region's oil supplies. Increasingly, many governments are planning nuclear power plants to supply the energy needed for desalination. Despite high costs, continued investment is increasing desalination capacity by about 10 percent per year.

Israel relies on external sources for 25 percent of its water. Brackish (slightly saline) water is used for some farming but can damage the soils if too much is used over many seasons. Urban water supply relies increasingly on coastal desalination plants. The droughts affecting Israel, Gaza, and the West Bank are very political. Israel retains jurisdiction over the West Bank and the Golan Heights both for defense and access to water. Peace negotiations with Jordan included agreement on the use of Jordan River basin waters, which have become central in the negotiations over the future of the West Bank.

Agriculture

The arid lands of Northern Africa and Southwestern Asia provide little opportunity for farming, requiring most countries to import foodstuffs to adequately feed their citizens. Most of the arable land extends from the shores of the Mediterranean (Figure 7.20) and Red seas, and the Persian Gulf. The Nile and Tigris-Euphrates rivers provide for irrigation. Interior deserts have given rise to nomadic herding. The most agriculturally productive areas are on the fringes of the region such as in Turkey and the southern Sudan. Indeed, the building of dams in southeastern Turkey has enabled Turkey to increase its production of cotton, soybeans, grains, fruits, and vegetables.

Huge sums have been invested in making the arid lands more agriculturally productive, but the payoff has not been much. For example, Saudi Arabia will end its domestic wheat production in 2016 because its water supply is being depleted by its attempt to grow wheat. So, many of the oil-rich Arab states are finding it worthwhile to acquire agricultural land in other regions. For example, Qatar has leased 100,000 acres in Kenya in return for financing a new port, and Saudi Arabia is investing US$125 million in a pineapple plantation and fruit processing plant in Zambia. Overall, Arab governments are looking at land in countries such as Australia, Brazil, Ethiopia, Kazakhstan, the Philippines, South Africa, Sudan, Turkey, Ukraine, and Vietnam. The Arab Organization for Agricultural Development (AOAD) expects to invest US$15 billion in agriculture abroad from 2011 to 2016 as part of the Arab Emergency Program for Food Security to reduce food shortages in Arab countries. Though Arab countries are obtaining much-needed food and

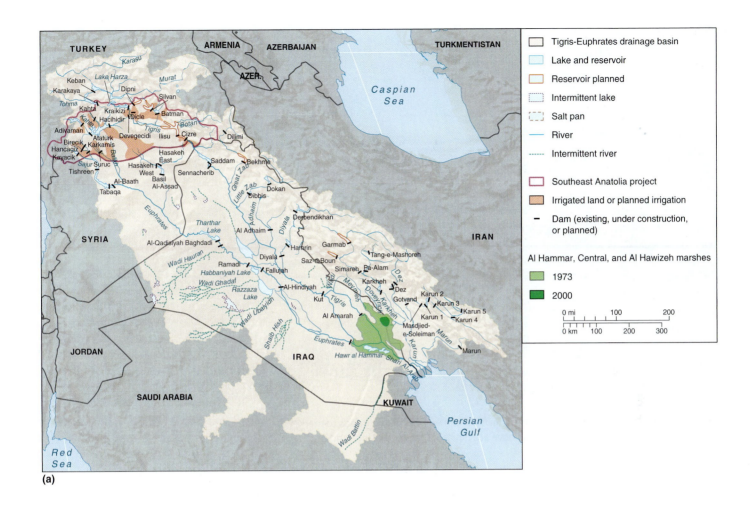

Legend:

- ☐ Tigris-Euphrates drainage basin
- ☐ Lake and reservoir
- ☐ Reservoir planned
- ☐ Intermittent lake
- ☐ Salt pan
- — River
- ---- Intermittent river
- ☐ Southeast Anatolia project
- ☐ Irrigated land or planned irrigation
- – Dam (existing, under construction, or planned)

Al Hammar, Central, and Al Hawizeh marshes

- ☐ 1973
- ☐ 2000

(a)

(b)

Figure 7.19 Tigris-Euphrates drainage basin. (a) Turkey's southeast Anatolia region has more than twenty dams to divert water from the Tigris and Euphrates rivers for irrigation projects, depriving Syria and Iraq of water. Dam building in Iran also has reduced the flow of water into wetlands of the lower Tigris River. In the early 1990s, Saddam Hussein's regime launched a program to drain marshes, partly to drill for oil and partly to eradicate the Marsh Arabs who opposed him. In southern Iraq, the Central Marshes and Al Hammar Marshes have completely dried up, and the Al Hawizeh Marsh is a fraction of its former size. (b) The Ataturk Dam on the Upper Euphrates River as it neared completion in 1992. *Source: (a) Data from UNEP; (b) © Ed Kashi/Corbis.*

Figure 7.20 **North Africa: rural landscape.** Bedouin women taking goods to market near Foudouk al Aouerab, Tunisia. The cultivated valley behind them and the bare hillsides above are typical of Northern Africa and much of the wider region. *Photo: © Kess van der Berg/Photo Researchers, Inc.*

building infrastructure in the host countries, it is a controversial practice because it is often done in countries that do not produce enough food for themselves, most notably in African countries.

7.4 Geographic Diversity

Northern Africa and Southwestern Asia is divided into five subregions: North Africa, Nile River valley, Arab Southwest Asia, Israel and the Palestinian Territories, and Iran and Turkey (Figure 7.21).

North Africa

The four countries of North Africa are Algeria, Libya, Morocco (with Western Sahara), and Tunisia (Figure 7.22), each with differing population sizes (see Table 7.1). Over 80 percent of Algeria's and Libya's territories are desert, but Morocco and Tunisia do not extend so far into the arid Saharan environment. The northern parts of Morocco, Algeria, and Tunisia are dominated by the Atlas Mountains, and that area is known as the Maghreb. It includes high ranges (Mount Toubkal in Morocco is 4,165 m, or 13,665 ft., in altitude), broken by internal plateaus and river valleys. The harsh, largely arid, and often mountainous natural environments of the North African countries restrict agriculture and most human settlement to a small percentage of the territory along the northern coasts and in the immediate mountain-

ous hinterland (Figure 7.23). Problems of water supply affect all these countries.

Arabs are dominant, but the Berber people help to make this subregion unique. European colonization, which began in the early 1800s, explains why much of the educated middle classes in the Maghreb countries speak French and those in Libya speak Italian. After World War II, nationalist groups fought for and obtained independence—in 1956 in Morocco and Tunisia, and in 1962 in Algeria. Morocco is politically stable under its moderate king, Muhammed VI, who allowed a new constitution in mid-2011 that now gives some new powers to parliament and the prime minister. Tunisia had 30 years of one-party rule when Islamic extremists were repressed and women's rights were established.

Algeria has been ruled by democratically elected governments with socialist policies based on central planning. The first-round election success of the fundamentalist Islamic Salvation Front (FIS) in the 1992 elections led to an army takeover of the government and civil war with the dispossessed Islamic militants. Terrorist activities and army repression through the 1990s devastated Algeria's economy and people, with over 100,000 deaths and the army's destruction of the FIS as a political party. The war essentially ended in 2002 after most of the remnants of the main guerilla groups disbanded, surrendered, or accepted amnesty. Elections held in 1999, 2004, and 2009 all were won by Abdelaziz Bouteflika, who was backed by the military.

Libya was a mainly desert area of little economic or political outside interest until Italy occupied it in 1911. After World

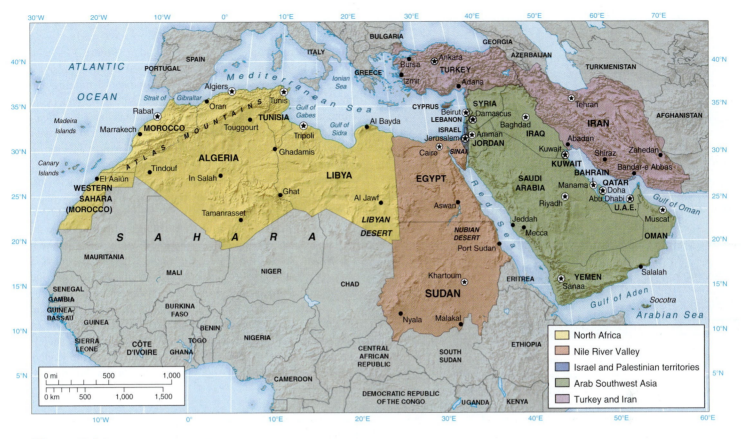

Figure 7.21 Northern Africa and Southwestern Asia: subregions and aspects of geography.

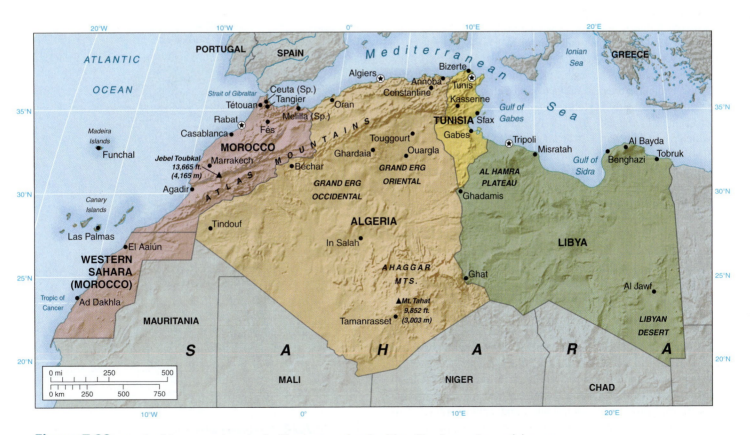

Figure 7.22 North Africa: countries, physical features, and main cities. "Ergs" are major sand dune areas.

Figure 7.23 **North Africa: mountainous environment.** The town of Moulay Idriss (north of Fez), named after a Moroccan saint who is descendant of the Prophet Muhammad, is located in a dry, mountainous environment where orchards are planted among the desert vegetation. *Photo: © Heike Alberts.*

War II, Italy was replaced by a British-French protectorate until Libyan independence was achieved in 1952. Colonel Muammar al Qadhafi seized power in 1969 and ran the country as a mixed military, socialist, and Islamic republic based on oil wealth. Libya was politically shunned by most countries in the 1980s and 1990s because its government supported international terrorists. The Arab Spring that began in neighboring Tunisia quickly spread to Libya by mid-February 2011. Qadhafi responded violently, which resulted in civil war and the death of Qadhafi in October 2011. A Transitional National Council, which formed in March 2011, then assumed power.

Many of the subregion's economic connections are northward to Spain, France, and Italy. Algeria, Morocco, and Tunisia, in particular, retain close ties with France and have strong connections to markets in Europe for selling products, buying goods, and sending emigrant labor. Morocco exports citrus fruits, vegetables such as tomatoes and potatoes, and cut flowers for European markets. It also exports cork from the bark of oak trees and fish: squid (for export to Japan) and tuna. Morocco, Algeria, and Tunisia also export phosphate for fertilizer. Almost half of Morocco's population is still dependent on agriculture. Morocco's manufacturing sector is substantially craft industries. Algeria also has light industries, including the manufacture of electrical components. Libya makes steel and aluminum, and Tunisia has a small steel industry. Algeria and Libya also export oil and gas. In Morocco and Tunisia, tourism is a major source of income, based on their sunshine, coastal locations, historic and cultural sites, and stable political environments. Most tourists come from northern Europe, with its cool and rainy climate.

Nile River Valley

The Nile River connects Egypt and Sudan (Figure 7.24) and provides them with water that has sustained a human presence in the dry eastern Sahara since the early days of farming and civilization (see Figure 7.4). However, the Nile River waters are finite, and pressures from growing populations make it difficult for Egypt and Sudan to continue to modernize and diversify their economies. Neither produces enough export income to buy the needed imported goods for its people.

Egypt and Sudan are similar in some ways, but their geographic positions, political environments, and products make them different, too. Egypt is by far the largest Arab country in population (82.3 million in 2012). Though Muslim and Arab, Egyptians constitute a discrete subgroup within the larger Arab population spread throughout the region. Egypt also plays a strong role in international relations. It retains control over the major global choke point of the Suez Canal and the Sinai Peninsula, which is the land route from Africa into Southwestern Asia.

In 1952, Egypt became fully independent after centuries of Ottoman and then British domination. It became a socialist state focused on its own internal needs. Under the leadership of President Gamel Abdul Nasser, Egypt developed more

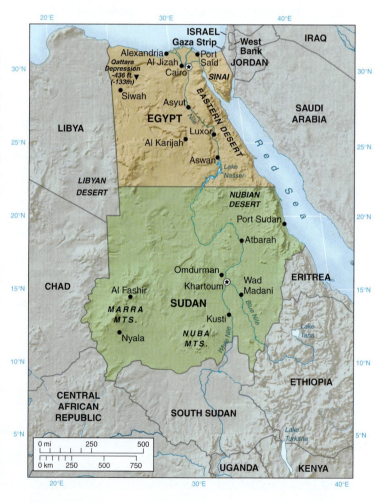

Figure 7.24 **Nile River valley: countries, physical features, and main cities.**

rapidly than the rest of Africa in the 1960s and 1970s. President Mohamed Hosni Mubarak ruled Egypt from 1981 until 2011, when the Arab Spring swept him out of power. The Muslim Brotherhood then assumed power in the elections that followed.

From 1956 to 2011, Sudan had twice the area of Egypt, but only half the population. It was very diverse with over 600 ethnic groups and 400 languages spoken (see Figure 7.5). Muslim Arabs dominate in the north and black Africans practicing traditional animistic religions or Christianity dominate in the south. Overall, the southern part of the country had characteristics similar to those of its neighbors Chad and Ethiopia, and it resembles other African countries more than those of Northern Africa.

In 1956, Sudan gained its independence from the United Kingdom despite Egyptian claims. The tensions resulting from the period of Egyptian occupation of Sudan and fears of future Egyptian expansion remain in the minds of many Sudanese and affect negotiations over the use of Nile River water. Internally, the Arabic Muslim peoples have controlled the government in a series of military regimes that have tried to impose Arabic and Islamic culture and traditions upon all the peoples of the country. Consequently, a civil war erupted from 1972 to 1982. A ruling military junta composed of a mixture of the military and Islamists took control in 1989. Sporadic violence continued but

began to die down in 1999 when Sudan's government moved toward multiparty politics internally and better international relations. Peace talks in the early 2000s led to the signing of several accords. In 2005, the final North/South Comprehensive Peace Agreement (CPA) granted the southern rebels autonomy for six years. After this, a referendum on independence took place in January 2011 and received overwhelming support. South Sudan became independent on July 9, 2011. A separate conflict erupted in the western province of Darfur in 2003 (see "Human Rights" section, p. 222).

Arab Southwest Asia

Arab Southwest Asia is the heart of the Arab and Islamic worlds. It comprises the Arabian Peninsula and the Fertile Crescent that includes the Tigris-Euphrates River basin and the Lebanon coast (Figure 7.25). Despite small populations, the countries of this subregion play a major part in world affairs because of their oil wealth and involvement in the Arab-Israeli peace process.

The countries of Arab Southwest Asia are distinguished by forms of government, emphases within Islam, and natural resources of oil and water. The differences in natural resources and economic management produce a wide range of economic status— from countries that remain materially poor to those that rival the world's wealthiest (see Table 7.1).

The governments of countries in this subregion mostly move slowly, if at all, toward democracy. Many of the oil-rich countries remain dominated by ruling families or dictators; Saudi Arabia is the only country named after a family, the Sauds, who still rule with the title of king and support a huge range of related princes. Other Gulf countries have sheikhs, emirs, and sultans with similar roles, and Jordan has a king. Syria has an authoritarian military-dominated regime. Until the fall of Saddam Hussein, Iraq had an authoritarian, military-dominated regime but is now moving toward democracy. In Lebanon, a greater degree of democracy now exists after a destructive civil war in the 1980s and 1990s between the Muslim and Christian groups spurred on by Islamic extremists, although Syria still exercises a major influence.

Arab Southwest Asia has major economic differences between countries with high oil

Figure 7.25 **Arab Southwest Asia and Israel: countries, physical features, and main cities.** The Fertile Crescent stretches from the Mediterranean coastlands through Syria and Iraq to the Persian Gulf. Israel declared Jerusalem its capital in 1950, but most countries of the world do not recognize this declaration and have their embassies in Tel Aviv.

revenues and those that produce little or no oil. The richer countries tend to border the Persian Gulf and sit on huge oil reserves (see Figure 7.17). The Saudi Arabian economy produces about three times the total income of any other country in the sub-region. These countries have small total populations and rely on immigrant labor. Wealth is not distributed widely or evenly among their populations. Menial, low-wage jobs and much of the commerce, especially in retailing, are left to Indians and other Asian immigrants. Lebanon, Jordan, and especially Yemen are much less developed economically. Iraq and Syria are the exceptions to this rich-poor duality, since until the mid-1980s they both had oil revenues, significant water resources and agriculture, and large labor forces.

In oil-rich countries, income benefits the ruling elites and is used partly to generate industrialization, intensify agricultural output, provide more and better roads, airports, health services, and education, and increase living standards. Up to one-third of revenues in some countries go to purchasing military hardware. The oil income makes it possible to shift economies toward sustainable development based on diversification. A **diversified economy** is one in which manufactured goods are more important than primary products, and a variety of manufactured products is joined by a growing service sector. For example, Bahrain and UAE now have aluminum smelters and Oman a copper smelter. Qatar and Saudi Arabia produce steel. Several countries also manufacture construction materials, including cement, to provide the materials for upgrading roads and building new housing and factories. Countries developed food and consumer goods industries. Services also became important in many of these countries where banking and government employment are increasing rapidly.

Oil-producing countries have reaped huge profits from high oil prices in recent years, as much as US$740 billion in 2012 with 60 percent of it from this world region. These profits, known as petrodollars, are invested around the world. Much is invested in real estate and corporations like Procter & Gamble, Hewlett-Packard, PepsiCo, Time-Warner, and Disney. The Abu Dhabi Investment authority invested US$7.5 billion in Citigroup after that corporation lost billions of dollars in risky home mortgages to Americans. Such large sums of money have helped to fuel the global economy, for example, keeping interest rates on loans and credit cards low. It also has allowed Americans to spend lavishly and go deep into debt.

Arab Southwest Asia is also rich in history and culture, and is the location of some of the world's earliest known towns. The historical sites and hot, sunny climate attract many tourists. The rock-hewn city of Petra in Jordan is one such example. Every year, millions travel—mostly as religious pilgrims—to Saudi Arabia to visit the Islamic holy sites in Mecca and Medina (Figure 7.26). Mecca is the birthplace of Muhammad, Islam's founder, and Medina became Muhammad's power base after he was expelled from Mecca.

Israel and the Palestinian Territories

Israel stands out in Southwestern Asia. It is a unique example of a country created by the United Nations for a particular ethnic or

Figure 7.26 **Arab Southwest Asia: pilgrimage.** Mecca, Saudi Arabia, the Al-Haram mosque, the main pilgrimage site for Muslims, who seek to visit it at least once in their lifetimes. *Photo: © Nabeel Turner/Stone/Getty Images.*

religious group, despite opposition from those living in and around it (see Figure 7.25). The Palestinian territories of the West Bank and Gaza are Israeli-occupied territories following conquests in the 1967 war, although the United Nations ruled that they should be returned to Syria, Egypt, and Jordan. Territorial disputes, terrorism, and refugee groups have major local geographic impacts and global implications. This subregion is marked by a dual society of different opportunities for Jew and non-Jew.

Israel's population is composed of Jews, Israeli Arabs (which include Druze and Bedouin), and Palestinian Arabs (see Table 7.1). The Arabs and some of the Jews trace their family existence in the country to before Israel's independence. The Sephardic Jews from southern Europe form the majority, but the Ashkenazi Jews of Central and Eastern European origin play important roles in politics. Other Jews are of Asiatic origin or arrived from Russia in the 1990s.

Israeli Arabs generally accept the Israeli state and rarely conflict with Israeli Jews. On the other hand, Palestinian Arabs conflict greatly with Israelis because they reject the state of Israel and would prefer to have their own Palestinian state (see "Israelis versus Palestinians" section, p. 218). Palestinian Arabs are most often simply called Palestinians, referring to the name of the land before the Israeli state was carved out of it in 1948. Interestingly, it was common to also call Jews Palestinians before Israel was created. Palestinian Arabs compose 90 percent of the population in the West Bank and Gaza and resent not having full governmental powers.

Israel's economy places its per capita income in the top quintile (20 percent). Ownership of consumer goods is high (see Figure 7.16). Its economy is diversified. Agriculture, which uses intensive reclamation and irrigation farming methods, produces fruits, vegetables, and flowers for export to Europe but now constitutes only 2.5 percent of total GDP (Figure 7.27).

Figure 7.27 **Israel: agriculture.** Vineyards and date palms.
Photo: © Corbis RF.

Manufacturing accounts for approximately 30 percent of Israel's economy. Polished diamonds are one of its largest exports, followed by chemicals and electronic, medical, and scientific equipment. Beginning in the 1990s, Israel became a major center of and leader in high technology development in manufacturing areas such as telecommunications, electronic printing, diagnostic imaging systems for medicine, and data communications.

Just under one-third of its exports go to the United States, with much of the remainder going to European countries, Hong Kong, India, and China. Israel has high trade deficits and owes much of its external debt to the United States. It is also by far the largest recipient of U.S. foreign aid, receiving US$3.1 billion in economic and military aid in 2013 and a total of US$115 billion in over the last 60 years.

Israel gains over 66 percent of its GDP from the services sector. Education and health care rank very high. The financial sector grows increasingly as Israel's economy opens to privatization and foreign investment. Tourism is a major industry that attracted over 2.8 million visitors in 2010 but often is disrupted by conflict and war. When possible, many visitors also travel to such places as Jerusalem, Petra (Jordan), and the pyramids (Egypt).

While Israel has a growing economy that places it ahead of its neighbors in development and lifestyles, the Palestinian areas of Gaza and the West Bank continue to have poorer conditions for human development. Palestinians accuse the Israelis of paying unequal attention to the needs of Palestinians in these territories compared to Israelis—a form of apartheid. Israelis accuse Palestinians of harboring terrorists. Though billions of dollars in international aid was given to Palestinians in the late 1990s, the intense and very destructive conflict between Palestinian terrorists and the Israeli army in the 2000s and 2010s has destroyed much of the infrastructure for Palestinians in both Gaza and the West Bank, in turn ruining the economy and leading to a social crisis. In Gaza, for example, 50 percent of the work force is unemployed and 40 percent of its population lives in refugee camps. Many families depend on emergency food supplies.

Turkey and Iran

Turkey and Iran occupy the northern and eastern margins of this region (Figure 7.28). They are largely mountainous countries lying along fault lines that make them subject to frequent and

Figure 7.28 **Turkey and Iran: countries, physical features, and main cities.**

devastating earthquakes. Their mountains receive precipitation, much of which falls as winter snow, which feeds rivers and provides fresh water for urban areas and farming.

The two countries are influential in Southwestern Asia and the wider world, sharing economic leadership of the region with Saudi Arabia, Egypt, and Israel. They have crucial strategic positions between the southern boundary of Russia and Neighboring Countries (see Chapter 3) and the Persian Gulf oilfields. Together their populations comprise nearly one-third of the total population of the entire region (see Table 7.1). Both countries have Kurdish minorities who desire independence, a problem they share with Iraq and Syria. Governments feel threatened by the Kurds, especially the Turkish government (see "Human Rights" section, p. 222).

Some differences distinguish Iran and Turkey. For example, Iran stands out in the region as a country where Shiite Islam dominates. Iranians (Persians) are a distinct people from Turks, and both groups are unrelated to Arabic peoples who dominate this world region. Separate histories of empire building and differing forms of government and economies add to both countries' distinctiveness.

In the early 1900s, Iran was ruled by shahs who kept the country largely under military control but allowed for the adoption of Western education and the development of a wide range of economic activities. Though Iran's economy grew, the last shah was repressive, and resentment toward him increased. In 1979, nationalist religious leaders, led by the Ayatollah Khomeini, seized political power. The shah fled to the United States, and the U.S. government, which had long supported the shah, refused to turn him over to Iran to stand trial. Consequently, anti-American feelings grew in Iran, culminating with the taking of hostages at the American embassy in Tehran. At the same time, Iran shifted to Islamic religious leadership and isolationism. The new and current constitution of the Islamic republic in Iran depends on a Shiite interpretation of Islamic government. Clergy are expected to establish a just social system and implement Islamic laws.

Turkey's political system is very different from that of Iran. After the disastrous defeat of the Ottoman Empire in World War I, Turkey became a nationalist and secular republic, putting the country before religion in questions of government. From the 1920s until 1936, the new leader, Mustafa Kemal Ataturk, ruled with a single-party government. Turkey gradually modernized and became increasingly involved in the global economy. Neutral during World War II, Turkey became a member of NATO (see Chapter 2) during the Cold War.

In recent times, both Iran and Turkey face tensions between liberalizing Western influences and pressures from Islamist political groups. Many young Iranians dislike the strict Islamic rules imposed by a fundamentalist few on a moderate majority. In contrast, Turkey faces continuing challenges from Islamic political groups to its long-term secular state principles. For example, many citizens do not like the ban on the religious practice of wearing headscarves.

Iran and Turkey also have different types of economic development. With 9 percent of world oil reserves (see Figure 7.17), Iran has the potential of earning great export income like Saudi Arabia. However, the takeover of the country by religious

leaders, U.S. sanctions, and the war with Iraq in the 1980s has inhibited economic development. The Iranian government's decision to invest in nuclear power plants in the 2000s and 2010s was condemned by the United States, which fears that nuclear technology employed to generate electricity also can be easily used to develop nuclear weapons.

Turkey has little oil but invested heavily in the development of its water resources for more agricultural output and hydroelectricity. Industrial expansion increased from 1950 and especially in the 1970s and 1980s. Turkey's real income rose steadily to exceed that of Iran until the late 1990s when Turkey's government implemented new policies that put Turkey's economy into recession. Nevertheless, Turkey has a more developed services sector than Iran. Government employment is very important in the economy. International tourism grew and annually brings in over a billion dollars. In 2010, Turkey was the seventh most popular tourist destination in the world, attracting 27 million visitors (up from 1 million in 1980). The development of tourist resorts along its sunny coasts and the availability of historic, often religious, sites made Turkey a major venue for Europeans. Turkey is the only country in Northern Africa and Southwestern Asia (outside of Israel) with such a diversified economy.

7.5 Contemporary Geographic Issues

Israelis versus Palestinians

Perhaps the best known world conflict today is between the Jews living in Israel and the Arab Palestinians living in the occupied Palestinian Territories of the West Bank and Gaza. Although the United States plays a significant role in trying to reconcile the two sides, the Arab countries see U.S. positions and policies as unfairly favoring Israel. The conflict's origins go back in history, before Israel was established as a new country in 1948. The continuing resistance of Palestinians to the heavily armed Israelis suggests an irreconcilable conflict, at least one that will not likely be resolved until an independent homeland is created for the Palestinian Arabs.

The land in question is a small part of what is known as the Fertile Crescent that cradled and connected the early civilizations of Mesopotamia and Lower (northern) Egypt. The hilly coastal lands provided a home for the ancestors of the Israeli nation (Hebrews). Roman armies subdued their lands, which were then governed as the Roman province of Palestine. Eventually, the Romans tired of Hebrew rebellions, sacked Jerusalem, and dispersed most of the people in AD 70 to create the Jewish diaspora. Long after the fall of the Roman Empire, Arabs inhabiting the area converted to Islam in the AD 600s, as Muslim armies spread the Islamic faith. Jerusalem became the third most holy site for Muslim pilgrimages after Mecca and Medina.

The Turkish Ottoman Empire governed "Palestine" from medieval times to the early 1900s. In the 1800s, the idea of a separate country for Jews arose out of Zionism, a movement that began in

Europe and Russia that called for the creation of a separate Jewish nation-state. Anti-Semitism in Europe and Russia, including the imprisonment and murder of Jews, caused waves of Jewish settlers, numbering 60,000 from 1880 to 1914, to migrate to Palestine. The settlers taught the modern Hebrew language in schools and established socialist institutions such as labor organizations and farming in **kibbutzim** (singular: kibbutz). In the kibbutzim, land is communally owned and decisions are made collectively. Settlers planned for mass migration into a new, independent country as a safe haven from a persecuting world.

During World War I, the Ottoman Empire sided with Germany and against the United Kingdom and France. To combat the Ottoman Empire, the British and French fanned nationalistic feelings among the Arabs of the Ottoman Empire. The British also put forth the Balfour Declaration (1917), which called for "the establishment in Palestine of a National Home for the Jewish people." After World War I in 1923, the Ottoman Empire was dismantled. Much of Southwestern Asia was granted to the United Kingdom and France as protectorates, including Palestine as the Jewish National Homeland.

British authorities tried to address Arab discontent over the creation of a Jewish homeland in their midst by restricting the number of Jews migrating to the territory. However, numbers had been building since the late 1800s. Then Nazi persecution in Europe in the 1930s caused another 350,000 Jews to move to Palestine by 1940, despite prohibition of such movement under the British mandate (Figure 7.29a). After World War II and the genocide of the Nazi Holocaust, a million more European Jews joined them. Violence among Jews, Arabs, and British forces increased. Pressure mounted for the creation of an independent Jewish state, while Arabs resisted turning over lands to such a state. The United Kingdom could no longer control its protectorate and handed jurisdiction over to the United Nations, which in 1947 proposed a Partition Plan that would have created separate lands for Jews and Arabs (Figure 7.29b). Though the Arabs rejected the plan, the British withdrew their forces in 1948. Jews took matters into their own hands and declared the establishment of a new independent state which they called Israel. Egypt, Syria, Jordan, Palestine, and Iraq declared war, but the new state of Israel repulsed attacks and also expanded the country's original territory by the end of the war in 1949. During the conflict, some 600,000 to 700,000 Palestinian Arabs, approximately 80 percent of the Palestinian Arab population, became refugees. By 2010, almost 4.9 million Palestinian Arabs were living in neighboring countries (Table 7.3).

Arab countries refused to accept the existence of independent Israel and fought it unsuccessfully again in 1956, 1967, and 1973. During the 1967 war, Israel extended its territory southward into the Gaza Strip and Egypt's Sinai Peninsula, eastward across the West Bank area of Jordan, and northward to the Golan Heights of Syria (Figure 7.29c, d). Despite UN resolutions, Israel refused to give up the occupied lands, apart from Sinai, for security reasons. Israel wanted surrounding countries to accept the existence of Israel and to renounce military

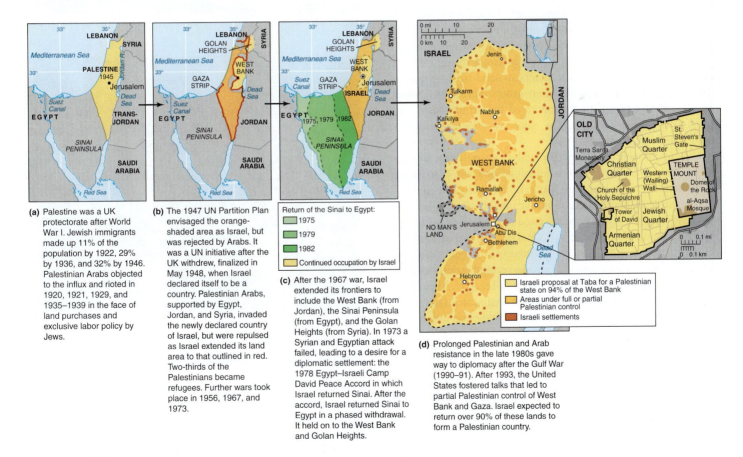

(a) Palestine was a UK protectorate after World War I. Jewish immigrants made up 11% of the population by 1922, 29% by 1936, and 32% by 1946. Palestinian Arabs objected to the influx and rioted in 1920, 1921, 1929, and 1935–1939 in the face of land purchases and exclusive labor policy by Jews.

(b) The 1947 UN Partition Plan envisaged the orange-shaded area as Israel, but was rejected by Arabs. It was a UN initiative after the UK withdrew, finalized in May 1948, when Israel declared itself to be a country. Palestinian Arabs, supported by Egypt, Jordan, and Syria, invaded the newly declared country of Israel, but were repulsed as Israel extended its land area to that outlined in red. Two-thirds of the Palestinians became refugees. Further wars took place in 1956, 1967, and 1973.

Return of the Sinai to Egypt:
- 1975
- 1979
- 1982
- Continued occupation by Israel

(c) After the 1967 war, Israel extended its frontiers to include the West Bank (from Jordan), the Sinai Peninsula (from Egypt), and the Golan Heights (from Syria). In 1973 a Syrian and Egyptian attack failed, leading to a desire for a diplomatic settlement: the 1978 Egypt–Israeli Camp David Peace Accord in which Israel returned Sinai. After the accord, Israel returned Sinai to Egypt in a phased withdrawal. It held on to the West Bank and Golan Heights.

- Israeli proposal at Taba for a Palestinian state on 94% of the West Bank
- Areas under full or partial Palestinian control
- Israeli settlements

(d) Prolonged Palestinian and Arab resistance in the late 1980s gave way to diplomacy after the Gulf War (1990–91). After 1993, the United States fostered talks that led to partial Palestinian control of West Bank and Gaza. Israel expected to return over 90% of these lands to form a Palestinian country.

Figure 7.29 History of modern Israel: a timeline of changing maps (a) to (d).

Table 7.3	Locations of Palestinian Refugees			
Country	Registered Refugees	% of Total Registered Refugees	% of Refugees Inside Official Camps	Number of Official Camps
Lebanon	455,373	9.1	50	12
Jordan	1,999,466	40.3	17.5	10
Syria	495,970	10	30	9
Gaza	1,167,361	23.5	44	8
West Bank	848,494	17.1	24	19
Total	4,966,664	100		58

Source: UNRWA [United Nation Relief and Works Agency for Palestine Refugees in The Near East] Statistics 2010: Selected Indicators. Programme Coordination & Support Unit-PCSU. November 2011.

activity against it. Syria would not acknowledge Israel's existence, so Israel held onto the Golan Heights. Unsuccessful in open warfare, Arab groups, both secular and religious, turned to terrorist methods from the 1970s.

In the 1970s and 1990s, U.S. presidents made attempts to reconcile differences. After the 1967 war, Israel returned the Sinai Peninsula to Egypt but retained control of the Israeli-occupied Palestinian territories of the West Bank and Gaza. The Jordanian and Egyptian governments gave up their rights to the West Bank and Gaza lands (which were previously parts of their territories) as a basis for creating a new Palestinian country by negotiation.

In 1994, following the Oslo Accord between Israel and the Palestine Liberation Organization (PLO), a Palestinian Authority was established and given limited jurisdiction and autonomy in Gaza and the West Bank. Israeli settlements in these areas (Figure 7.29d) were maintained, however, and their populations expanded. In 1996, progress toward the development of a Palestinian country halted as Israel resisted further devolvement of power and transfers of land to Palestinians. Terrorist activity by a Palestinian group known as **Hamas** increased. The Israeli military entered Palestinian territories to punish people for atrocities, such as suicide bombings, committed against Israelis.

A major source of tension and a serious problem for Israeli-Palestinian negotiations relates to how the Israelis use Jerusalem, Gaza, and the West Bank. Both Israelis and Palestinians claim Jerusalem as their rightful capital for their respective countries (Figure 7.30). Israel asserted its rights over Jerusalem by proclaiming the city to be its capital in 1950 and building the Gnesset (parliament) in the city's suburbs, though the United States and most other countries do not recognize Jerusalem as Israeli's capital and maintain embassies in Tel Aviv. After Israel took East Jerusalem in the 1967 war, it confiscated land in the occupied area and made Palestinians second-class citizens by denying them property rights.

Another problem stems from the Israeli government's decision to build Jewish settlements in Palestinian areas. The resulting violence has led to increased geographical separation as the Israeli government attempts to provide security to Jewish settlers. Because Israelis are in the position of power, they dictate the course of separation. To protect their settlements in the West Bank, they have been constructing a wall that runs through the territory (Figure 7.31). The wall prevents

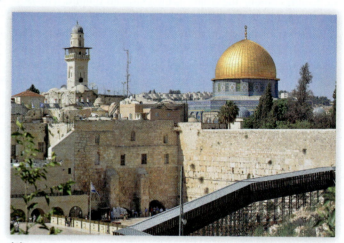

(a)

(b)

Figure 7.30 **Jerusalem: aspects of conflict.** (a) Dome of the Rock and the Western Wall. The Western Wall (sometimes referred to as the "Wailing Wall") is the holiest site for Jews because it is all that remains of Herod's Temple complex destroyed in AD 70 by the Romans. The gold dome of the Dome of the Rock, the holiest Islamic site in Jerusalem, is just 150 meters away on the Temple Mount (Haren al-Sharif) complex. (b) Israeli soldiers relaxing in the Old City. *Photos: (a) © Elizabeth Chacko; (b) © Michael Bradshaw.*

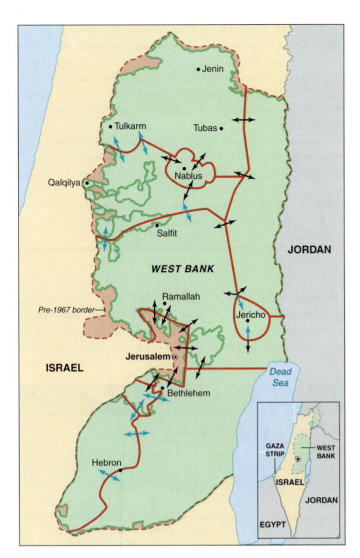

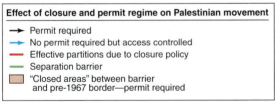

Effect of closure and permit regime on Palestinian movement

→ Permit required
→ No permit required but access controlled
— Effective partitions due to closure policy
— Separation barrier
▭ "Closed areas" between barrier and pre-1967 border—permit required

(a)

(b)

Figure 7.31 **Israeli barriers in the West Bank.** Since 2002, a wall separating Israeli and Palestinian settlements has been under construction. Other partitions hinder the movement of Palestinians. (a) Map noting the location of the wall. (b) Photo of the wall near Jerusalem. *Source: (a) Data from* The Economist, *October 4, 2007; (b) © Elizabeth Chacko.*

Palestinians from easily entering Israeli settlements, but the land for the wall comes from Palestinians who are then displaced. Moreover, while the wall is constructed to give Israelis easy access to the rest of Israel, many Palestinians find themselves cut off from their communities and workplaces. Palestinians are mainly restricted to low-wage jobs and denied access to higher-earning professions. Palestinians who wish to become doctors, for example, have to train abroad. The wall has resulted in much international criticism and has even been rejected by Israel's highest court, causing the Israeli government to modify its placement in many locations, though the government is forging ahead with its construction in the hopes that it will provide security for Jewish settlers. In 2005, the Israeli government decided to remove Jewish settlers from the Gaza Strip. Many Israelis supported the move in the hopes that it would foster peace. Many thought it too costly to provide security for 9,000 settlers occupying more than 25 percent of the Gaza Strip, which was inhabited by 1.5 million Palestinians mostly hostile to the settlers.

In July and August 2006, an armed conflict erupted between Israel and **Hezbollah**, a Shia Islamic militia and political organization in Lebanon supported by Iran and Syria. It began when Hezbollah launched missiles into Israel, captured two Israeli soldiers, and killed three others. Israel responded with deadly force. When it was over, approximately 400 Hezbollah fighters and 119 Israeli soldiers were killed. More than 1,191 Lebanese civilians and 43 Israeli civilians also died, and many more soldiers/fighters and civilians were injured on both sides. Nearly 1 million Lebanese and 300,000 Israelis were forced to flee their homes. Lebanon's infrastructure was largely destroyed, totaling US$7 to 15 billion, and Israel suffered US$1.6 to 3.0 billion in damage. Both sides claimed victory, but neither achieved many of its goals, and losses clearly overshadowed gains.

From the end of December 2008 to mid-January 2009, intense fighting commenced in Gaza between Hamas and the Israeli military, with both claiming that the other had started it. By the time of the cease-fire, 10 Israeli soldiers and three civilians died. More than 1,300 Palestinians died, more than half civilians. Many more were wounded, tens of thousands were left homeless, and more than US$2 billion in structural damage occurred.

While Israeli Jews and Palestinian Arabs contest each others' rights to their claimed homelands, it should be remembered that neither group is homogeneous or speaks with one voice. Both groups are very diverse and contain members with wide ranges of opinions. Each side has those who are conciliatory and willing to give up land for peace, and each side has hardliners, perhaps one in five inhabitants, unwilling to make any compromises or grant any concessions. The views in Table 7.4 are those that might be expressed by an Israeli Jew and Palestinian Arab.

Human Rights

Basic human rights for all citizens are a concern in some countries of Northern Africa and Southwestern Asia. For example, police abuses are widespread, individuals have few resources to defend themselves when accused of crimes, and penalties often are severe with the death penalty applied frequently. Turkey is the only country that has abolished the death penalty, which it did in 2002. Also, countries like Algeria, Lebanon, and Morocco are the only ones that have maintained moratoriums on executions for as much as 20 years. Otherwise, this world region accounted for 558 of the 676 reported executions worldwide in 2011 (which excludes thousands of unreported executions in China), according to Amnesty International. After China, Iran ranked second in the world with 360 executions, followed by Saudi Arabia (82), Iraq (68), the United States (43), and Yemen (41).

A wider variety of human rights abuses are found in the region. The role of women in society is a major concern, and so are specific armed conflicts within individual countries. The ones that have created the greatest number of deaths and refugees include the Israeli-Palestinian conflict (see previous section), the civil war in Sudan, and repression involving the Kurds, who are found in Turkey, Syria, Iraq, and Iran.

Women in Society

Gender inequalities are a continuing issue in Muslim countries. Male attitudes and the law work against equal opportunities for women. As with other aspects of Muslim culture, local interpretations and outcomes vary. Considerable debate exists in

Table 7.4	Debate: Israelis versus Palestinians
A Hard-Line Jewish View	**A Hard-Line Arab (Palestinian) View**
We were here first and have a longer history of occupying this land: it is ours, as claimed by Abraham and Joshua.	Our ancestors were here from the time of Abraham, whom we also recognize as a father of our people.
Our religion started first in this area. Our holy sites include Hebron (where Abraham is buried) and Jerusalem, both in, or partly in, the West Bank.	Jerusalem and Hebron are sacred to Muslims.
The United Nations agreed to the partition of Palestine and recognizes Israel with its Jewish majority.	The decision resulted from a combination of weak Arab support and strong U.S. and British pressure. We were ignored, although we made up most of the population here in 1948. Israel has taken large areas of land from us that were not part of the UN plan.
We have returned some territories we took in the 1967 war but hold onto the rest as a matter of national security and survival.	After the 1967 war, the United Nations ordered Israel to hand back the occupied territories, but it has not done so more than 40 years later.
Arabs deny our right to exist and want to wipe us off the map; we are exercising our right to defend ourselves. When joined, the surrounding Arab countries outnumber us, so we have to ensure strong security.	We were forced out of our land, we lived in crowded camps without amenities, and we now live in poverty. The Jews close checkpoints with no notice and interrupt our lives. We cannot argue with them because they are supported by the United States, and we now hate that country as well.
We regard them all as possible terrorists who blow up our restaurants and nightclubs, kill our athletes, assassinate our leaders, and drive suicide bombs into our neighborhoods.	They refuse to recognize our presence and nationality, suppressing our language, religion, and culture. Most of us want to live peaceful lives, but they treat us all as spies and criminals, abusing our human rights.
Jerusalem is our real capital city and is central to the Jewish faith. Some Jews want to remove the mosque on the holy mount.	Jerusalem is sacred to us, and any moves to destroy the mosque would be a declaration of all-out war that would unite Muslims.
No strong Palestinian nationality was expressed here before 1948. This area of land was merely a British protectorate carved out of the former Ottoman Empire, and most people knew of the intention to create a land for Jews. The present Palestinians are Arabs who should have been taken in by existing Arab countries. They have invented Palestinian nationalism as part of a plot to eliminate Israel.	We want our own lands and independence from Israel.
They are poor workers and earn only low wages. We go out of our way to employ them, but it would be better to employ only Jews.	They are well fed and materially wealthy. If we want to study for the qualifications that would earn us better jobs, we cannot do so in our country and have to go elsewhere. A doctor friend of mine, who works in a Jerusalem hospital, had to go to Greece to qualify.
The Palestinian Arabs have not repaid all the help we have given to them, raising their well-being above that of other Arabs in this region.	We can do nothing that is legal to improve our lot, so it is not surprising that some of us take to the gun and bomb.

Muslim countries over what the Qu'ran says on these matters. Saudi Arabia boycotted the 1995 Fourth World Conference on Women in Beijing on the grounds that it was anti-Islamic. In Iran, it is a criminal offense for a woman not to wear a headscarf. And yet, Muslim-dominated countries such as Turkey, Pakistan, Bangladesh, and Indonesia have elected women leaders in recent years. While Saudi Arabia excludes women from all public activity, Tunisia promoted women's rights after its 1957 independence.

New technology is changing relationships. For example, Saudi Arabia did not allow women to travel alone but recently removed this restriction for women over 52 years old and allowed younger women to travel alone with a "yellow slip" provided by their male guardians. Now men receive electronic notifications when their women and children enter and leave the country. *The Economist* reported that a woman tweeted the Saudi government stating that it also should electronically notify women when their husbands have chosen second, third, and fourth wives.

Sudan

Not long after Sudan became an independent country in 1960, internal conflict ensued, with the country experiencing numerous civil wars that have extended well into the 2000s. Much of the conflict has been between the Arab Muslim north and the black Christian and animistic south. The warfare created famine and misery in this drought-prone country. As many as 2 million people died. Many others were displaced and left homeless. Arab Muslims have controlled the government in recent times. However, years of war weakened the country's economy and made it difficult for the government to assert effective control over the country. The Sudanese government relied on undisciplined troops and paramilitary groups to fight those who opposed it. Consequently, human rights abuses during campaigns were widespread. Northern soldiers taking southern captives as slaves are just one example. The long history of the north inflicting misery on the south caused people in the south to want to be permanently divided from the north, which occurred when South Sudan became independent on July 9, 2011.

In the early 2000s, armed conflict erupted between the government and non-Arabic tribes of western Sudan. To put down the rebellion, the government armed local Arab tribes who were steadily moving into the western province known as Darfur (meaning "land of the Fur people"), a source of anger for the Fur and a partial cause of the rebellion. These armed individuals earned the name *Janjaweed*, an Arabic word meaning "devil on horseback with a gun." The name reflects the fact that these nonprofessional troops, though acting on behalf of the government, terrorized the local population, burning villages and raping women in what the international community labeled as genocide. The international community condemned the Sudanese government, and 22,000 international peacekeepers were sent to the region. However, the Sudanese government denied that it had supported the Janjaweed, despite having deputized many Janjaweed as police with the assignment to restore order in the western region. This policy encouraged human rights violations. By 2009, 2.7 million had been displaced, having fled to neighboring Chad. An estimated 300,000 had died, 20 percent of them from direct assault and the rest from diseases such as diarrhea, pneumonia, and malaria that they contracted after being displaced without adequate food, water, shelter, and medicine. After a sharp decline in deaths at the end of 2009, the UN no longer considered the conflict to be a war. In 2010, a ceasefire was achieved.

Kurds

Many Kurds have long desired their own nation-state called Kurdistan, and they have struggled and been persecuted for trying to establish their own state (see Figure 7.5). In 1984, the Kurds of southeastern Turkey launched a military campaign to free the Kurdish areas of Turkey. The Turkish military responded with force, leading to many human rights abuses. After the capture of a Kurdish leader who was then sentenced to life in prison, the Kurdish fighters changed their military struggle to a political one. In the early 2000s, the Turkish government, desiring membership in the European Union, succumbed to pressure from the European Union to grant more rights to its citizens. Kurds now hold political office and are allowed to speak Kurdish in schools.

While the situation of the Kurds in Turkey is much better today, life for Kurds in Iraq is far less certain. In the early years of Saddam Hussein's reign, Kurds enjoyed a number of rights. However, after a number of Iraqi Kurds joined with the Iranians in the Iran-Iraq war in the 1980s, Saddam Hussein persecuted Kurds, even dropping lethal mustard gas on them. After the Gulf War in 1991, the United States and its allies set up a no-fly zone in northern Iraq to prevent the Iraqi air force from bombing Kurdish areas. The act allowed the Kurds of northern Iraq to effectively set up their own governmental structures. Following the ouster of Saddam Hussein in 2003, the United States encouraged Kurds to participate in a new federally organized Iraq.

Kurds in Iran have suffered less persecution in recent years and have not engaged in armed conflict. The last armed struggle led by Iranian Kurds occurred right after the Iranian Revolution in 1979. Though Iran's Kurds have engaged in fewer armed struggles than Kurds elsewhere, they once succeeded in creating the only modern Kurdish nation-state, the State Republic of Kurdistan, with its capital in Mahabad. This nation-state was short-lived—it was both founded and destroyed in 1946.

Kurdish rights have been increasingly violated in Syria since 1963 when the Baath Party took power. The Baath Party is similar to the party by the same name in neighboring Iraq that governed under the leadership of Saddam Hussein. The Syrian Arabization program deprived Kurds of Syrian citizenship, ownership of their land, and since 1992, the right of children to use their Kurdish names when registering for school. As of 2013, false imprisonment, torture, and mass arrests of Kurds had been common for decades in Syria.

Iraq

Iraq occupies much of the land that was part of ancient Mesopotamia (meaning "land between the rivers," namely the Tigris and Euphrates). Until the end of World War I, it was within the Ottoman Empire and did not exist as a distinct political unit. In 1916, British military forces began occupying the area. Following the war, the three Ottoman provinces of Baghdad, Mosul, and Basra were joined together to create modern-day Iraq, which in turn became a British mandate.

As a European creation, Iraq has little internal homogeneity. The north is occupied by Sunni Kurds, the north-central portion by Sunni Arabs, and the southeast by Shia Arabs (Figure 7.32 and see Figures 7.5 and 7.7). All three groups have more in common with peoples in neighboring countries than with one another. Ethnic and religious diversity does not necessarily undermine the functioning and well-being of countries. However, well-functioning countries usually were created by their own people and have established a governing tradition that the country's citizens respect.

When Iraq first became a British mandate, the British tried to create a governing tradition by installing a king from the Hashemite clan, which traces its roots back to the prophet Muhammad. Iraq seemed stable when it was granted independence in 1932, but the Hashemite king was overthrown in 1958. Iraq then was governed by a series of military coups until 1968, when the Arab Socialist Baath Party took power. The Baath Party combined Arab nationalism with socialism and Pan-Arabism. A member of the Baath Party, Saddam Hussein, became president in 1979.

Secular in outlook, Hussein disliked fundamentalist Islam, which in turn made him distrust the Shiites of southeastern Iraq. A Pan-Arabist, he was suspicious of the Kurds in northern Iraq. He saw Iraq's ethnic and religious diversity as a threat to Iraq and subsequently believed that he must maintain unity through force.

Saddam Hussein's support came from the Sunnis in north-central Iraq, primarily in an area known as the Sunni Triangle, marked by Baghdad in the southeast, Ramadi in the southwest, and Tikrit (Hussein's hometown) in the north. As Hussein's reign progressed, he built an oppressive police state to promote unity. Kurds and Shiites suffered greatly.

In 1979, an Islamic revolution brought Shia fundamentalists to power in neighboring Iran under the leadership of Ayatollah Khomeini. Saddam Hussein feared that Iran's Shia Muslims would incite Iraq's Shiite population. In 1980, Hussein's military invaded Iran, allegedly to settle a border dispute. Initially successful, the invasion stalled and turned into a long war of attrition that claimed more than a million casualties before ending in 1988.

Relations between Iraq and Kuwait were poor, too. The Iraqi government never recognized Kuwait's sovereignty, believing for historical reasons that Kuwait belonged to Iraq. Hussein accused Kuwait of slanting drilling underneath the border between Iraq and Kuwait to illegally obtaining Iraq's oil, and he used this as a justification to invade Kuwait in August 1990. Iraqi forces quickly conquered Kuwait. However, UN economic sanctions followed, and a coalition of military forces largely comprised of American troops drove Iraqi forces out of Kuwait and southern Iraq in early 1991 in the Gulf War (also known as Operation Desert Storm).

Iraq lost the Gulf War, but its military remained formidable, and Hussein was seen by the international community as a continued threat. The United Nations continued its economic sanctions with the intent of forcing Hussein to disarm. The United States and the United Kingdom established no-fly zones (see Figure 7.31) over northern and southern Iraq that forbade the Iraqi air force from launching campaigns against the Kurds and Shiites. Hussein was uncooperative and campaigned to have the sanctions removed.

U.S. President George W. Bush accused Hussein of pursuing a nuclear weapons program and of not cooperating with UN weapons inspectors. In 2003, an American-led military force invaded Iraq and quickly toppled Saddam Hussein's regime. The United States and its allies set up a provisional government that later became more permanent. By the end of 2011, U.S. forces withdrew from Iraq. Though Iraqis were freed from Hussein's rule, the history of favoritism and oppression within

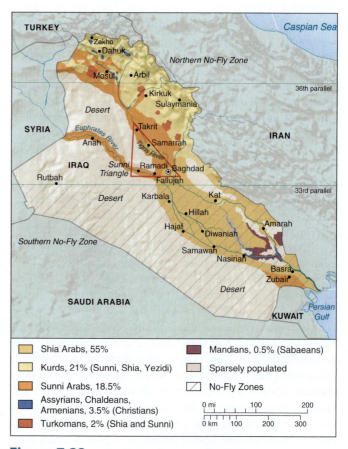

Figure 7.32 Iraq: ethnoreligious diversity.

Iraq left deep divisions within Iraqi society. For Iraq to become a peaceful, stable democracy, the Iraqi government will have to overcome these divisions.

The Arab Spring

The peoples and cultures of Northern Africa and Southwestern Asia are very old but many of the countries, their political boundaries, and their governments are not. Many of them came into being after World War I, while others were created later. Many of the existing countries had their boundaries altered. Global geopolitics and economics played a major role in creating this region's political map. The United Kingdom and France were major decision makers with Italy, the United States, and even Russia (the Soviet Union) exerting their influences at times in some parts of the region. This led to the rise of a number of dictatorships and authoritarian governments that stifled the development of democratic institutions. Many of these governments invested in educational systems but hampered economic development. As noted previously, population growth rates have been high in the region. Consequently, by 2010, many countries had large populations of young people who were well educated but unemployed and angry at the lack of both economic opportunities and political freedoms. Anger grew until it erupted in widespread demonstrations calling for reform in what quickly became labeled as the Arab Spring.

Tensions first exploded in Tunisia with the death of Muhammad Bouazizi, an unemployed university graduate who could not find a job. He tried to make a living by erecting a booth along a street to sell vegetables. Police confiscated his booth because Bouazizi did not have the proper permits (or probably did not pay the appropriate bribe). On December 17, 2010, he went to the town square, poured gasoline on himself, and lit himself on fire to commit suicide. Young people began to march in the streets to express their outrage at his plight. The police responded harshly, with more than 30 people dying by early January. Police actions provoked a revolution. On the evening of January 14, 2011, after 40,000 protesters gathered at the Interior Ministry in Tunis, President Ben Ali and his immediate family fled to Saudi Arabia. Some called it the Jasmine Revolution. It certainly inspired frustrated young people across the region to believe that youthful protest could topple dictatorial governments.

After the government fell in Tunisia, people took to the streets in Egypt, Lebanon, Morocco, Oman, Syria, and Yemen (Figure 7.33). Egyptian President Hosni Mubarak resigned in February. Within days, protests began in Libya. However, Libya's leader, Colonel Muammar al Qadhafi, responded violently, igniting a civil war. The United States and NATO declared a no-fly zone over Libya and supported the rebels with air strikes against forces loyal to Qadhafi. By the latter part of October, rebels captured and killed Qadhafi, effectively ending the civil war.

The mass demonstrations calling for the ouster of Syria's President Bashar al-Assad and his ruling Baath Party turned into the Syrian Civil War in April 2011 when al-Assad ordered his troops to open fire on protesters. By the writing of this book in early 2013, Syria had experienced great bloodshed and turmoil, almost 60,000 deaths and 650,000 refugees. Al-Assad's army was slowly losing the war, but the end result was hardly clear.

Demonstrations and some violence have occurred in the oil-rich Arab countries but on a far lesser scale. For example, the governments of Saudi Arabia and Bahrain have invested heavily in their economies to raise the standards of living of their citizens. This has prevented economic misery and blunted unrest, but the lack of political freedom may yet become an issue.

In countries like Egypt where revolution was successful, it remains to be seen whether democracy will develop. The previous dictatorships did not foster democratic practices, and Western governments' support of repressive regimes has turned many away from Western models to Islamism. For example, the Muslim Brotherhood won the election in Egypt and spoke of democracy but quickly implemented many nondemocratic policies.

Figure 7.33 **Tahrir Square, Cairo, Egypt.** In February 2011, anti-government protesters gather to listen to a speech by President Hosni Mubarak, who says that he will relinquish some powers but later resigns after continued protests. It is part of what some call the Jasmine Revolution or the Arab Spring. *Photo: © Chris Hondros/Getty Images.*

Geography in the War Zone

If you ask any U.S. service member returning from Afghanistan or Iraq what they did during the war, you may be surprised to find out how much time many spent talking and drinking tea with the locals or sitting in long meetings helping them to establish government institutions or build schools, soccer fields, or local infrastructure. The military's focus on counterinsurgency (COIN) means that in addition to building and maintaining the fighting skills traditionally associated with the military, soldiers, marines, airmen, and sailors are learning languages such as Arabic, Pashto, and Dari and spending significant amounts of time learning about the local customs, habits, and traditions of the areas where they are deployed. This demands a level of geographic literacy that goes far beyond memorization of random facts; it requires knowing how people interact with their immediate environments and one another. Helping them along the way are social scientists, including geographers such as Robert Kerr (Figure 7.34).

As a senior social scientist on Human Terrain Team (HTT) IZ03, Robert lived and worked with soldiers from the Army's 3rd Brigade Combat Team of the 4th Infantry Division and the 1st Brigade Combat Team of the 1st Cavalry Division in the northeast Baghdad neighborhoods of Sadr City, Adhamiyah, and Istiqlaal from the spring of 2008 to the summer of 2009. These areas are some of the most diverse neighborhoods of Baghdad, with Sunni and Shia Muslims as well as Kurds and Christians making up the mosaic of local cultures. In order for the Army units to effectively engage with the population to help them become self-sufficient in the post–Saddam era, it was essential that they develop a deep understanding of the cultural and physical patterns of these urban landscapes, and Robert's job was basically to help them become geographers.

In one example, local Iraqi law enforcement officials and Coalition Force leaders were puzzled by a sudden outbreak of crime in a northeast Baghdad neighborhood that was known for its peace and stability, even during the heaviest fighting early in the war. Robert, along with his teammates on HTT IZ03, set out to discover why the change had occurred so suddenly. Through many conversations with local residents, complicated geographical networks resulting from sectarian violence were discovered. The neighborhood in question was a mixed Sunni-Shia area that had escaped much of the sectarian violence that plagued Baghdad in 2006 and had become a refuge for members of both sects fleeing their homes in more violent neighborhoods. Seeing an opportunity to make money, most of the long-standing residents of the neighborhood divided their homes into rental units. However, this upset the local social structure in that local customs largely prevented

Figure 7.34 Geographer Robert Kerr in the Sadr City District Council Building in Iraq, July 2008. *Photo: Courtesy Robert M. Kerr, PhD.*

unrelated people from living together under the same roof. As more and more of the old residents saw an increasing number of unknown people in their neighborhood (and sometimes in their own houses), residents reported, they spent much less time out in public watching out for one another (as people in close-knit neighborhoods around the world usually do) and more time behind their own walls. The result of these processes was a sense of insecurity and instability that was far higher than anything they had ever felt. This led to a general decline in the sense of community among the old residents of the neighborhood, and as some of the disenfranchised refugees from other areas turned to crime to make a living, an environment of fear and lawlessness took hold.

When HTT IZ03 uncovered the geographical networks among the new refugees, neighborhood attitudes, and the refugees' places of origin, the Army unit working in the area was able to come up with creative ways to rebuild a sense of community among the neighborhood's old residents. This included organized community events designed to get people comfortable with coming back out onto the streets, as well as coordinated efforts with the local Iraqi government officials to help the refugees begin to move back and reintegrate into their old neighborhoods. By the fall of 2009 most of the homes in the neighborhood had been returned to single-family dwellings, many refugees were able to return to their old neighborhoods since sectarian violence had died down, and crime rates were reduced dramatically. This is just one way that the military uses geography to prevent insurgency.

Eastern Africa. *Africa is known worldwide for its diversity of wildlife, shown here by this lion crouching in the grass in the vast savannas of Eastern Africa.* Photo: © Joseph P. Dymond.

LEARNING OBJECTIVES

After reading this chapter, you should be able to:

- Identify the major physical features, ecosystems, and natural and mineral resources of the region.

- Connect the region's environmental problems to the uneven distribution of both populations and natural resources.

- Identify the major health concerns of the region, particularly the widespread advent of HIV/AIDS, and the subregions that contain the majority of this disease.

- Discuss the migration patterns of this region in both a historical and present-day context, relating these patterns to areas of civil and environmental conflict.

- Understand the European influence on the region of Sub-Saharan Africa and the effects this colonization had upon its people, especially in terms of political boundaries and oppressive government strategies such as apartheid.

- Identify the subregions of Sub-Saharan Africa on a map, and explain how these can be distinguished by their physical, cultural, and economic geographies.

- Locate the newest country of Sub-Saharan Africa, its reasons for independence, and current issues facing both this country and the surrounding region.

Sub-Saharan Africa

Figure 8.1 **Sub-Saharan Africa: physical features, country boundaries, and capital cities.** The region is bounded by the Atlantic and Indian oceans, and by the Sahara in the north.

8.1 The Challenge

The African continent south of the Sahara Desert is considered by anthropologists to be the cradle of humanity, from which people spread across the world (Figure 8.1). The natural environments of the region are endowed with huge mineral resources, major rivers and water resources, extensive areas where farming is possible, and the tourism resources of wildlife and landscape (Figure 8.2).

However, Sub-Saharan Africa is the poorest realm in the world today, and the majority of its 47 countries are affected by internal and/or external conflicts. Civil wars are a common threat throughout the region, along with many other geopolitical and environmental issues (Figure 8.3). Many countries have political leaders who siphon off scarce funds into their own pockets. The region's natural environmental issues include **desertification**, the process of overgrazing a grassland and converting that region into desert over time, deforestation, poor soils, and diseases affecting humans and their animals.

In this chapter we will examine the challenges of a world region where so many current human development trends are concerning, especially in terms of high birth rates, infant mortality, low life expectancy, and widespread disease. This realm has struggled to get on its feet after years of colonial rule by European countries that superimposed political boundaries on its land, followed by widespread independence of individual countries in the 1960s. However, new attention is being directed at the major task of helping African countries to improve their well-being. We start by assessing the physical landscapes of the region, transitioning to a view of human forces acting on the land over time to produce today's geographic diversity. We continue with a subregional examination of the current issues facing each location, interwoven into the overall cultural complexity of the Sub-Saharan Africa region.

8.2 Defining Sub-Saharan Africa

This world region is mostly defined physically by its steep escarpments along the coastlines, and a series of plateaus and valleys in the continental interior. Culturally, this region is bound by climate, European settlement history, disease, and tribal religions. The northern boundary is the Sahara Desert, which separates the dry, oil-rich Arabic world from the regions to the south. This is a diffuse boundary in that it is a transition zone and does not have a clear political dividing point.

Figure 8.2 **African safari in the TSAVO National Park, Kenya.** This photo, made on February 10, 2011, shows an elephant standing under a tree in TSAVO West National Park, some 350 kilometers southeast of Nairobi. A slowdown in the increase of Kenya's elephant numbers is raising fears among conservationists that hard-fought gains in saving the animals may be reversed amid growing demand for ivory. *Photo: © AFP/Getty Images.*

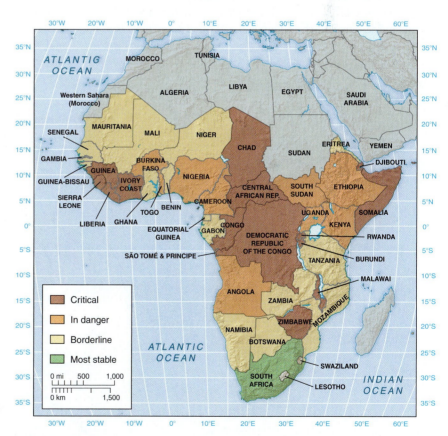

Figure 8.3 **Failed countries in Sub-Saharan Africa, 2012.** These are determined by indicators of instability such as high population growth, refugees, uneven development, economic decline, poor security, human rights issues, and external interference. Over 50 percent of the countries in Sub-Saharan Africa are in the Critical or In Danger categories. *Source: Data from* Foreign Policy, *2012.*

8.3 Distinctive Physical Geography

The natural environments of Sub-Saharan Africa provided stages for changing human activities over time, offered resources on which to base a set of choices, and imposed some restrictions at each stage of human development. These environments remain significant factors in today's geography.

Plateaus and River Valleys

The African landscape is dominated by interspersed valleys and plateaus (see Figure 8.1), which are underlain by some of Earth's oldest rock. The plateau edges rise steeply from the oceans in many places, and the interior is marked by steps from one level to another—steps that result in waterfalls or rapids on the major rivers. In eastern Africa, the plateau surface is actively being pulled apart by tectonic forces, causing scars in the land referred to as **rift valleys**. These often contain deep lakes along with very productive soils. Volcanic peaks such as Mount Kilimanjaro, the highest point in Africa, rise above the plateaus and form this region's main mountains.

The ancient rocks of African plateaus contain a wealth of mineral resources such as diamonds, coal, and petroleum. These were one basis for European interest and local economic development starting in the mid-1800s. However, the exploitation of several large mineral deposits awaits either peaceful conditions in their parent countries or the building of transportation links. Coastal and offshore oil deposits occur in some of the newest rocks and delta deposits. From Nigeria to Angola, coastal oil and natural gas resources are making the continent a major world source of these minerals. In South Africa, layers of sedimentary rocks contain vast coal deposits, of which the country is the world's sixth-largest producer.

Major Rivers

The largest river valleys include those of the Niger, Nile, Congo, and Zambezi. Each provides a major source of water and a transportation system, even though they all have channels containing rapids and/or waterfalls that interrupt navigation. For example, the largest river, the Congo, has a series of rapids from source to mouth. It enters the ocean through narrows with more rapids, preventing ocean-going ships from sailing upstream. Short lengths of road and rail are needed to transfer goods around the rapids. The next largest river, the Niger, has a large marshy delta at its mouth, and its northern loop flows through dry lands where the river remains shallow. In the south, Zambezi River navigation is also interrupted, including by Victoria Falls, the largest waterfall series in the world.

Unlike Europe, Africa does not have deep ocean harbors, and therefore there are few natural port sites. Access from the coast to the interior of the continent relied on rivers until

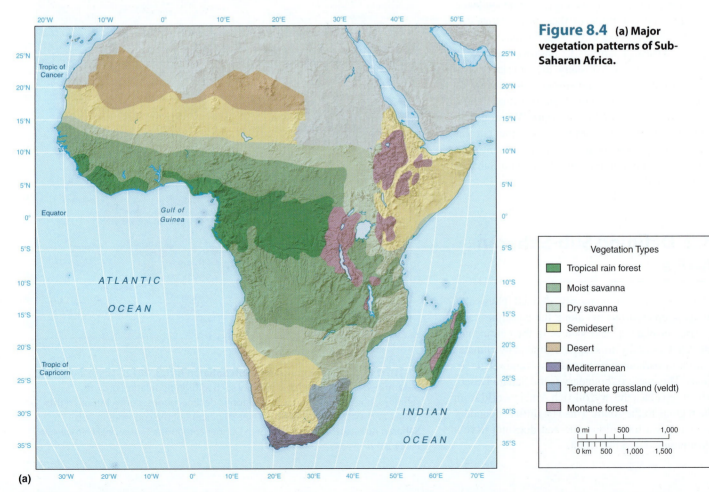

Figure 8.4 **(a) Major vegetation patterns of Sub-Saharan Africa.**

Vegetation Types
- Tropical rain forest
- Moist savanna
- Dry savanna
- Semidesert
- Desert
- Mediterranean
- Temperate grassland (veldt)
- Montane forest

(a)

modern transportation modes were introduced. Even to this day in the 21st century there are only a few railroads or permanent paved roads, mostly built to link interior mines and cities to ports. Air transportation alleviates some of these mobility difficulties, though it depends on the accessibility of airports and landing strips.

Tropical Climates

Almost all of Sub-Saharan Africa lies within the tropical and grassland climates (Figure 8.4). These range from tropical climates that are wet all year on or near the equator, to drier and more sparsely vegetated climates on either side of the equator. In the equatorial climates, temperatures average at least 30°C (80°F) year-round, and annual rainfalls average 3,500 mm (140 in.). In the drier climates, summer temperatures exceed 30°C (80°F) and winter temperatures reach 25°C (70°F), but rainfalls can average under 100 mm (4 in.) per year, qualifying them to be considered desert climates. Between these extremes the temperatures remain high all year, although rainfall is concentrated mainly in the summer season, totaling 500 mm (20 in.). The extensive plateau uplands provide cooler though drier conditions than the tropical latitudes suggest, featuring dry winters throughout these higher regions.

Tropical climates allow the production of a wider range of crops than in the world's mid-latitude regions. Examples include rubber, cocoa, coffee, cotton, and tropical fruits. They also attract tourists looking for sun and big animals on the world-famous safaris in Eastern Africa. However, such tropical environments harbor many diseases that restrict the productive ability of people, crops, and cattle. Malaria still affects millions of people, while the tsetse fly prevents cattle raising over large

Figure 8.4 **(b) Climate map of Sub-Saharan Africa.**

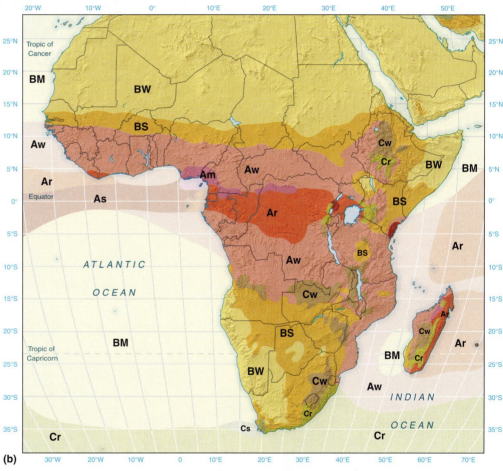

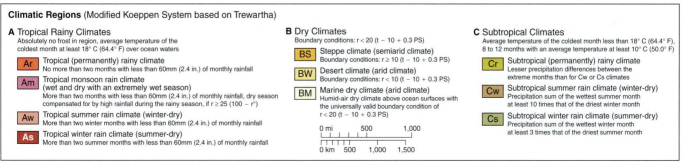

Climatic Regions (Modified Koeppen System based on Trewartha)

A Tropical Rainy Climates
Absolutely no frost in region, average temperature of the coldest month at least 18° C (64.4° F) over ocean waters

Ar Tropical (permanently) rainy climate
No more than two months with less than 60mm (2.4 in.) of monthly rainfall

Am Tropical monsoon rain climate
(wet and dry with an extremely wet season)
More than two months with less than 60mm (2.4 in.) of monthly rainfall, dry season compensated for by high rainfall during the rainy season, if r ≥ 25 (100 − r°)

Aw Tropical summer rain climate (winter-dry)
More than two winter months with less than 60mm (2.4 in.) of monthly rainfall

As Tropical winter rain climate (summer-dry)
More than two summer months with less than 60mm (2.4 in.) of monthly rainfall

B Dry Climates
Boundary conditions: r < 20 (t − 10 + 0.3 PS)

BS Steppe climate (semiarid climate)
Boundary conditions: r ≥ 10 (t − 10 + 0.3 PS)

BW Desert climate (arid climate)
Boundary conditions: r < 10 (t − 10 + 0.3 PS)

BM Marine dry climate (arid climate)
Humid-air dry climate above ocean surfaces with the universally valid boundary condition of r < 20 (t − 10 + 0.3 PS)

C Subtropical Climates
Average temperature of the coldest month less than 18° C (64.4° F), 8 to 12 months with an average temperature at least 10° C (50.0° F)

Cr Subtropical (permanently) rainy climate
Lesser precipitation differences between the extreme months than for Cw or Cs climates

Cw Subtropical summer rain climate (winter-dry)
Precipitation sum of the wettest summer month at least 10 times that of the driest winter month

Cs Subtropical winter rain climate (summer-dry)
Precipitation sum of the wettest winter month at least 3 times that of the driest summer month

areas. Poor people and poor countries do not attract the wider application of medical and veterinary treatments.

On the southern edge of the continent, warm, temperate climates with cooler winters provide some of the most pleasant living conditions within this region. Ocean winds help to moderate the warm summer temperatures that average 25°C (70°F), and subsequent mild winter temperatures of around 10°C (50°F). Rains come primarily in the summer with totals of 500 to 700 mm (20 to 30 in.).

Forests, Savannas, Deserts, and Their Soils

The natural vegetation regions of Sub-Saharan Africa (see Figure 8.4) correspond closely to the climatic regions, though they are becoming increasingly modified by human actions. The dense **tropical rain forest** of the equatorial rainy climate regions is marked by a huge variety of broadleaf, evergreen tree, and plant species. Though the great variety of plants provides a stock of biodiversity that many see as vital to the future of human existence on Earth, the multitude of insects and microbes in these biomes create diseases that affect humans, animals, and plants. Much of the rain forest has been cut for timber or to make way for commercial farming. Slash-and-burn forest clearance for short-term crop production is naturally replenished by a less-rich

Figure 8.5 **The physical landscapes around the city of Cape Town, South Africa.** Note the steep escarpment that borders the northern edge of the city (right side of the photograph), a barrier to interior exploration by Europeans for centuries. *Photo: © Martin Harvey/Digital Vision/Getty Images RF.*

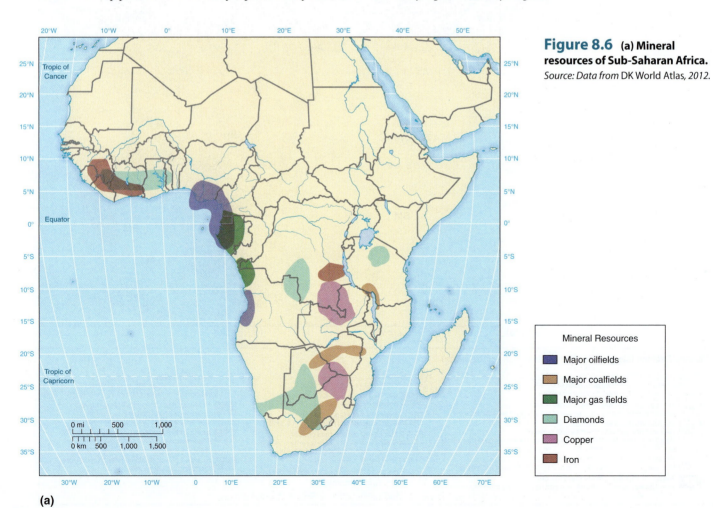

Figure 8.6 **(a) Mineral resources of Sub-Saharan Africa.**
Source: Data from DK World Atlas, 2012.

Mineral Resources

- Major oilfields
- Major coalfields
- Major gas fields
- Diamonds
- Copper
- Iron

(a)

secondary forest as soil quality deteriorates rapidly, and erosion is increased when the trees are harvested. Figure 8.5 shows Cape Town, South Africa, in which one of the first permanent European settlements was established along the coastline and the site of subsequent extensive deforestation.

The **savanna** grasslands (see Figure 8.17b) of the seasonally rainy areas have variable tree cover that depends on the length of the dry season. These are composed of grasses that encourage greater numbers of large herbivore animals such as elephants, giraffes, zebras, and antelopes (many of which provide meat for local people), and their carnivore predators such as lions. Today, viewing these animals in a safari adventure is a major focus of the tourist industry. Soils beneath the grassland are often workable, and many contain good proportions of plant nutrients that support agriculture. However, once exposed following cultivation or overgrazing, soils containing high

quantities of clay and iron minerals become cemented into red **laterite** that forms an almost impenetrable layer—for example, its main use is for blocks in house building. Vast quantities of mineral deposits such as diamonds, oil, coal, and gold are found throughout the region (Figure 8.6a).

The grasslands and seasonal rains lead into arid **deserts** north and south of the savanna regions. The margins of these deserts move back and forth as a result of climate change and also human destruction. The **Sahel** zone along the southern margin of the Sahara is a grassland transition zone that suffers periodic droughts made worse by overgrazing and firewood cutting, due especially to the unnatural political boundaries superimposed on this region by European colonists. The **desertification** of the Sahara and the dry margins in Southern Africa cause famines and destitution in countries that have few resources at their disposal. Drought has been a substantial environmental issue in this

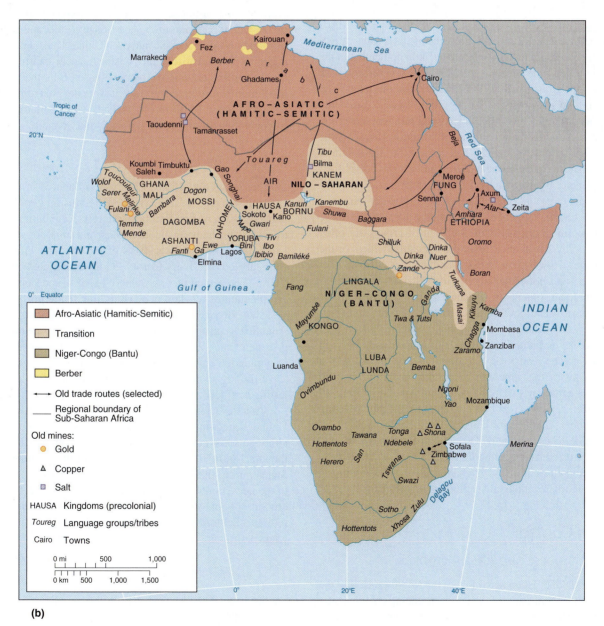

(b)

Figure 8.6 **(b) Historical language families and cultural groups of Sub-Saharan Africa.**

region: Lake Chad in the Sahel has shrunk by almost 90 percent in the last 50 years due to both drought and the siphoning of water for human-based irrigation of crops.

8.4 Distinctive Human Geography

Homo sapiens first emerged in this region about 15,000 years ago in East-Central Africa, from which modern humans pushed outward to populate the rest of the world. Indigenous societies developed characteristic cultures over many thousands of years before European and Arabic interventions resulted in a diverse set of geographies today.

African Ethnic Diversity

For thousands of years after the early dispersal of *Homo sapiens* to other world regions, population changes within Africa were mainly a matter of internal movements, as a wide variety of African peoples mixed and migrated. Cattle herders from the upper Nile River valley such as the Masai lived in tension with farming Bantu groups such as the Kikuyu, who migrated from Central and Western Africa. Bantu groups also reduced the populations of pygmy groups in the tropical rain forest and displaced the San and Khoikhoi hunters in Southern Africa. By the early 1800s, Bantu peoples who had moved into the southern African kingdoms of the Swazis, Zulus, and Zhosas established local residences, fighting each other for territory.

Members of groups (called "tribes" by European colonists) linked by kinship, language, and territory formed the basic social and political units (Figure 8.6b). Such groups are often distinguished by appearance, as in the tall cattle herders or the short pygmies. However, individuals and families changed allegiance as their economic and social circumstances changed.

Common constituents of African culture emerged from the long history of social development. Human, natural, and spiritual forces were contained in religious **animism**, which sees human success or failure as dependent on gods and spirits in rivers, tree groves, and rock outcrops. Land is traditionally a communal possession, inheritance, and responsibility in a still largely rural economy that is based on cultivating, herding, or hunting. Each person is part of a continuing life chain, revering ancestral spirits and the family unit. Large families are a blessing and childlessness a tragedy. The chief of a group is respected as a strong, trusted leader acting for the common welfare. These cultural foundations have been challenged by new religions, colonization, urbanization, and technological changes.

Shifting Control

Muslim Influences and African Empires

Starting in AD 650, Islam spread from the Middle East into the northern and eastern parts of Africa. Introducing camels and a new religion, these people traded via routes from the Mediterranean across the Sahara to the middle and eastern Niger River valley, centrally from Tripoli to the Lake Chad area, and into the Nile River valley in the east. As Islam moved into Africa, local practices such as polygamy and the use of the African drum were infused, creating **hybrid religions** in the region.

Trade in salt, gold, ivory, and slaves drove the advanced Western African empires of Ghana (AD 700–1240), Mali (1050–1500), and Songhai (1350–1600). Timbuktu, on the northernmost bend of the Niger River, began as a market for crops and cattle, and traded salt and gold by AD 700. Universities were established at Timbuktu and Djenne, employing scholars from Greece, Egypt, and the Middle East, and boasting libraries stocked with imported books. When the Mali emperor, Mansa Musa, undertook his Muslim pilgrimage to Mecca in 1324, he demonstrated his wealth with 500 porters each carrying a golden staff. Later Islamic expansion led to warriors and zealots settling grazing lands along the southern Sahara margins.

In Eastern Africa, Muslim traders established African ports such as Zanzibar to exchange goods with other ports around the Indian Ocean. Commodities such as ivory, gold, and slaves were exported to Arabia, Persia, and even China. Traders used the language of Swahili to conduct business, incorporating elements of African, Indian, and Arabic languages. Ethiopia resisted the Muslim invasions, and subsequent wars devastated much of the land. As a result, Muslims established coastal settlements around Ethiopia in modern Eritrea, Djibouti, and Somalia.

European Traders

Starting in the mid-1500s, European influences in the western and southern parts of Africa rivaled the Muslim intrusions in the north and east. At first European interests were limited to coastal trading by exchanging alcohol, guns, and sugar for slaves, gold, ivory, and palm products. Parts of the Western African coasts were labeled the "Ivory Coast," "Gold Coast," and even "Slave Coast." The demand for slave labor in North and Central American plantations resulted in a **triangular trade pattern** (Figure 8.7) that forced over 12 million enslaved Africans across the Atlantic Ocean and brought great wealth to European ship owners, merchants, and port cities. In Africa, the substantial loss of population and acquisition of guns and luxury goods by the slave-trading aristocracies inside the continent resulted in long-lasting conflicts and underdevelopment.

Antislavery movements beginning in the late 1700s led the United Kingdom to abolish Atlantic slave shipments in 1808, although some countries did not officially abolish slavery until the late 1800s. Humanitarian efforts in Europe and the United States during the 1800s led to the return of some freed slaves to the new country of Liberia, and the ports of Freetown (modern Sierra Leone) and Libreville (modern Gabon). However, the returnees seldom integrated with local populations, fueling later ethnic conflicts.

The Colonial Period

By the late 1800s, European explorations of the African interior, the opening of the Suez Canal (1869), and the demands of European manufacturers for raw materials led to changes. The

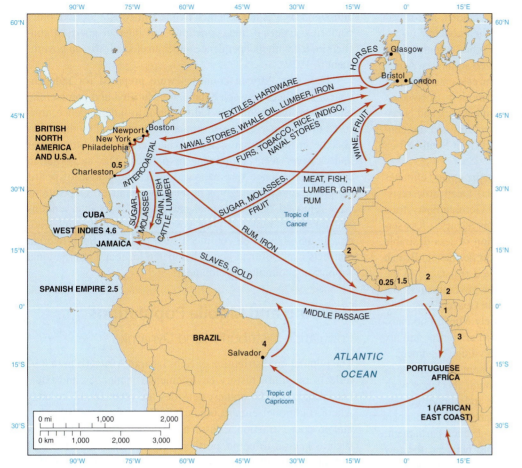

they called Hottentots. These people were often enslaved and forced into the desert margins. When the United Kingdom bought into this region in 1814, British colonists clashed with the Boers (Dutch farmers), forcing the latter to undertake the "Great Trek" to areas where they could be autonomous and use their Dutch-derived Afrikaans language. These Boer movements, combined with Zulu aggression, pushed the Nde-bele group northward into Shona lands.

The conflict between interests reached a high pitch when dia-monds and gold were discovered in the Boer Transvaal and claimed by British venturers such as Cecil Rhodes. Following British incite-ments, the Boers declared war in 1899. The British army soon took the major towns, but they failed to conquer the rural areas. In 1910, the four colonies—Cape, Natal, Orange Free State, and Transvaal, joined to form the Union of South Africa, an independent dominion within the British Empire. However, the rights

Figure 8.7 **Atlantic Ocean: slave trade.** The 1700s Atlantic economy traded colonial commodities with the home countries and brought slaves from Africa to the Americas. Numbers in millions are estimates of slave movements from Africa. The British ended most of this trade in the early 1800s. The country boundaries are those of today. *Source: Data from Thomas 1997.*

United Kingdom, France, Belgium, Germany, Spain, Italy, and Portugal divided these lands at the **Berlin Conference** in 1884. It is worth noting that no African nations were present at the conference that determined its new political boundaries. The extent of European colonization reached its peak by 1914 (Figure 8.8), with France and Britain controlling the majority of the land. Assets included plantation crops grown in the tropical rain forest, such as cocoa and palm oil, close to the coast and ports. Mines in the interior were connected to the ports by European-built railroads. Most colonists—the administrators, plantation and mine owners, and missionaries—stayed for short periods in Africa before returning home.

In some areas Europeans set up new homes, investing their lives and long-term prospects in the new lands. South Africa, with its warm, temperate climate, received the largest numbers of European settlers beginning with the Dutch, who built the port of Cape Town in the mid-1600s to supply ships trading with the Dutch East Indies (present-day Southeast Asia). Farmers in the immediate surroundings provided the port and ships with meat and crops. They took land from the indigenous Khoikhoi, whom

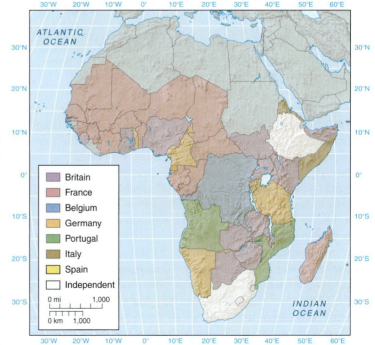

Figure 8.8 **European colonization in Sub-Saharan Africa by 1914.** The extent of colonies controlled by Europeans after being established at the 1884 Berlin Conference, and just before World War I.

of black African natives were again not considered, and it was over 80 years before they gained control of their country.

British commercial farmers settled in other colonies from Botswana to Kenya. After Germany colonized southwest Africa (modern Namibia) in 1884, local chiefs to the east opted for British rule in the colonies of Bechuanaland (modern Botswana), North and South Rhodesia (Zambia and Zimbabwe), and Nyasaland (Malawi). Cecil Rhodes's mining ventures and the construction of railroad connections northward from South Africa also supported such developments.

In Eastern Africa, the British settlers in the upland climates of British East Africa (Kenya and Uganda) faced little resistance. After World War I, the assignment of German East Africa (Mozambique) to British protectorate status attracted less settlement. In Western Africa, the combination of less desirable climate and local resistance resulted in fewer permanent British settlers. The British met considerable armed resistance in the early 1900s when they tried to combine the government of the coastal crop growers and plantations in Gold Coast (Ghana) and Nigeria with the drier inland areas of largely Muslim nomads. In general, British rule was based on informal control and tax collection through local African chiefs and village headmen.

The French made considerable investments in their western and equatorial African colonies, although the lands were mainly either rain forest or dry savanna and supported few people. Attempting to make their African lands and people part of France, they allocated a few seats in the French parliament. However, France's policy of differentiating between the few rich French citizens and the majority of Africans led to conflicts.

The other colonial countries—Belgium, Germany, Portugal, Spain, and Italy—regarded their African lands as sources of minerals and other wealth. Germany lost its colonies during World War I, but it was only after World War II that Portugal tried to introduce settler farmers to Angola and Mozambique.

Political Independence

After gaining its independence within the British Empire in 1910, South Africa's white population, dominated by the Boers, maintained a separation of black African, Asian, and European populations. In 1948 the **apartheid** policy was passed into law (see Figure 8.24), including a "petit apartheid" of separated ethnic groups and a "grand apartheid" of relocating black Africans to homeland areas. These Africans were regarded merely as laborers for the whites, and each was assigned to a homeland or temporary urban location.

Independence came gradually to the rest of the region, beginning with Ghana in 1957 and ending with the Portuguese colonies in 1975 and Namibia in 1990. Local agitation since the 1920s and virtual abandonment by European countries that were preoccupied in World War II stirred nationalisms within the colonial territories. In some countries, such as Ghana, the colonial power transferred control reluctantly but peacefully. In others, such as Kenya, Mozambique, and Angola, there were years of guerrilla warfare. In Southern Rhodesia (Zimbabwe) the European settler population declared independence in 1965,

although this was not accepted by the United Kingdom or other countries. External trade sanctions and internal guerrilla conflicts ended with independence under an elected African government in 1979.

Few new African countries were prepared for independence. Conflicts between chiefs and educated and commercial elites, who had frequently led the push to independence, mirrored the contrasts between rural areas dominated by the chiefs and the growing urban centers, where chiefs had little power and lawyers were prominent in politics. The authoritarian colonial experience foreshadowed the centralization of political power and dictatorships in the new countries. Oppositions were removed "in the national interest." Military interventions and takeovers replaced poor governments and corrupt leaders, though they often continued such practices. Such internal conflicts over political power hampered local and international support of economic development.

Growing and Mobile Populations

Population Distribution and Dynamics

The population distribution of Sub-Saharan Africa (Figure 8.9) reflects its cultural history, environmental challenges, and external influences. The highest densities are often found in lands where traditional empire organization was followed by colonial product extraction. Coastal areas and some inland locations such as central Ethiopia, northern Nigeria, interior Kenya and Uganda, and the Johannesburg region of South Africa have the highest densities. The desert margins of the northern edges and southwest areas of the region support few people. Two-thirds of the region's population remain rural and exist by traditional subsistence farming. Urban populations increase as people move to cities for better employment prospects in a money-based economy, although many find that there are few job opportunities in the formal sector. Better educational and health facilities also attract migrants to cities. Many Africans also see urban centers as more safe in times of civil war.

After independence, African countries had low to moderate population densities. Assisted by modern medical treatments, clean water, and sewage facilities, population numbers rose as a result of falling death rates and continuing high birth rates. The 380 million people in the region in 1980 rose to over 750 million by 2012 and could reach 1 billion by 2025. However, the annual rate of population increase of 3 percent was paralleled by agricultural production growth of only 2 percent. African countries did not take part in the Green Revolution that enabled other materially poor countries to escape undernutrition and periodic famines. The age-sex diagram for Nigeria (Figure 8.10) shows the continuing dominance of young age groups (those under 15 years old). Reducing population growth is a major key to Africa's future, but efforts to encourage smaller families seldom work. Large families are a sign of male virility, and many women prefer them as a basis for security in old age.

African population growth continues despite the widespread occurrence of long-term diseases such as malaria, sleeping sickness, and river blindness, the more recent onset of HIV/AIDS, and the prevalence of civil wars. Although life expectancies increased in many locations, they were reduced in countries with the highest incidence of HIV/AIDS.

Migration

Sub-Saharan Africa has a significant proportion of its people living outside of their country of birth. Some migrants move for economic reasons to places where they may be more likely to obtain paid employment. They include Western Africans from other countries moving into and out of Nigeria and Côte d'Ivoire, and millions moving to work in South Africa from the surrounding countries.

However, increasing numbers of migrants flee from violence against political oppositions or ethnic minorities. Movements across the borders of Rwanda and Burundi into the Democratic Republic of the Congo, Uganda, and Tanzania in the 1990s were associated with mobile militia groups, genocide, and property destruction. Movements from Somalia into Ethiopia and from Ethiopia into Sudan and Uganda are also composed of civilians displaced by wars. Civil war in Liberia spread into Sierra Leone, Guinea, and Côte d'Ivoire starting in 1989, causing over 1 million people to flee their home towns

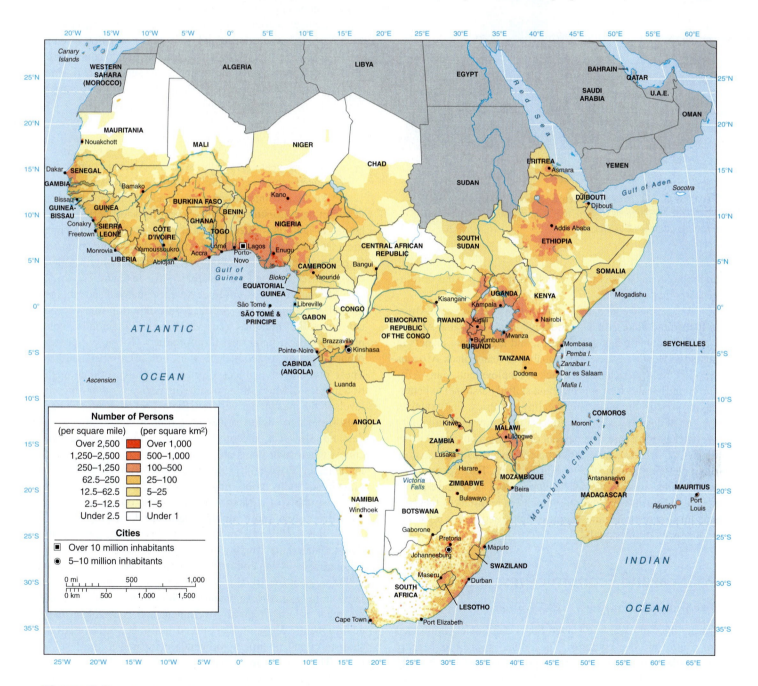

Figure 8.9 Sub-Saharan Africa: population distribution in the year 2012.

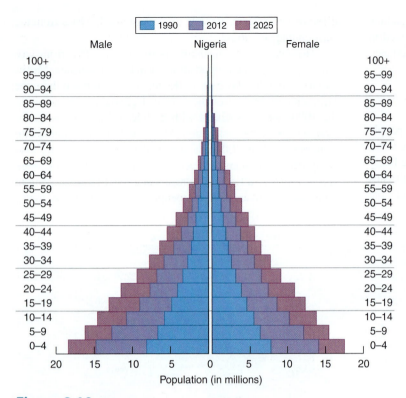

Figure 8.10 Nigeria, Africa: age-sex diagram. *Source: U.S. Census International Database: http://www.census.gov/population/international/data/idb/region.php.*

and villages and to become internally displaced persons or refugees in other countries.

As apartheid ended, many whites moved from South Africa to the United States, the United Kingdom, or Australia. Around half have higher education degrees, and many were leaders in business. However, there are probably as many black Africans who left South Africa due to a lack of employment. Many African countries also lose educated and skilled personnel to better-paying jobs in the materially wealthy countries. For example, the loss of doctors and nurses affects the quality and availability of African health care. Ten percent of Canadian hospital doctors are South Africans, and in 2005 there were more Ethiopian doctors practicing in Chicago than in Ethiopia. Few doctors who were trained in Zambia since independence have stayed. The upside of this outward migration includes remittances that bolster local economies and business links, and a few returning experienced professionals.

Political and Economic Pressures

Independence and the Cold War

During the Cold War, the United States and Soviet Union vied for control of other countries. They either propped up friendly governments or encouraged rebel groups against less friendly ones. After independence, many African countries had economies similar to materially poor countries in Asia and Latin America, but levels of internal conflict increased within the

countries, delaying economic development. For example, in Ethiopia the Soviet Union supported an unpopular Communist government, which oppressed some groups. In Angola, Soviet and Cuban troops supported government forces, while South Africa and the United States supported guerrilla oppositions. In Mozambique the guerrilla groups had Soviet support. In the Democratic Republic of the Congo, the United States supported the military dictator, attempting to ensure the continued production of important minerals from the interior. Supported leaders often used much of the financial assistance to buy armaments for protection. Many invested funds in personal accounts abroad. When the Cold War ended, internal conflicts fueled by the widespread availability of weaponry and even by the adoption of combative multiparty politics, often turned into ethnic strife. At the same time, the ending of the superpower rivalries also resulted in African issues of poverty and conflict being ignored until they further deteriorated.

Global Outsiders

By the year 2012, almost every African country had faced major economic and political problems. Heavy debts and internal conflicts made it difficult for them to attract new investment or aid, except for short periods when humanitarian crises directed world attention to specific countries. Of the foreign investment going to developing countries, 50 percent went to Asian countries, 20 percent to Latin American countries, and under 3 percent to African countries.

African countries continue to face many disadvantages in global trading. They produce and export primary products: metallic minerals (copper, bauxite, diamonds, gold, platinum), oil, commercial farm products (cocoa, coffee, tea, rubber, palm oil, vegetables, flowers, cotton), timber, and fish. These products are subject to price swings as world competition increases. Much of the improving African country income levels in the present derives from the export of such raw materials subject to higher prices at a time of increasing world demand. However, the most significant financial advantage comes when such items are processed and marketed, mainly in the materially wealthy countries. Apart from the local manufacture of bulky goods from cement to bottled drinks and some assembly of automobiles and trucks, only South Africa has a diversity of more sophisticated manufacturing.

In the first decade of the 2000s, political leaders of the world's materially wealthy countries in Europe, North America, and some Asian countries belatedly offered more support, including debt forgiveness and longer-term investments. But a large number of African countries with the greatest need could not meet criteria of good government and peaceful business environments. Most of the offered assistance was relatively short-term and of little significance compared to the level of need. The lack of agreement among the wealthier countries on such aspects as reducing their own tariffs on imported food products and commodities, such as cotton and sugar, also held back development prospects for African countries.

8.5 Geographic Diversity

The countries of Sub-Saharan Africa are grouped into four subregions (Figure 8.11). Each subregion contains a large country of over 50 million people. Further, each of the four largest countries made up half of the region's population in 2012 and produced almost 60 percent of the region's total GNI PPP. By contrast, over half of the countries in the Sub-Saharan Africa region had fewer than 10 million people each. Small countries and internal country regions illustrate further the geographical diversity within the region.

Subregion: Central Africa

In 2012, Central Africa (Figure 8.12a) had the fewest people (Table 8.1) and lowest population density of the African subregions, although rapid population increases may take it above Southern Africa by 2030. The subregion's development was slowed by its natural environment (Figure 8.13). The tropical rainy climate, tropical rain forest vegetation, and plateaus crossed by the Congo River and its tributaries with many rapids form a difficult environment. Humans, from precolonial times through colonization to political independence, always struggled with these conditions. Before European intrusions, these forested lands supported few people, and they were subject to external exploitation by Europeans from all sides.

Colonial development of mineral and crop exports was less extensive than in other subregions and, apart from the coastal countries of Gabon, Equatorial Guinea, Cameroon, and southern Democratic Republic of the Congo, economies developed slowly. Ocean trade with the rest of the world is still slowed by poor transportation, mainly due to the numerous series of rapids along the Congo and Niger rivers. Internal conflict divides and disrupts all countries in this subregion, and there is little, if any, intraregional cooperation.

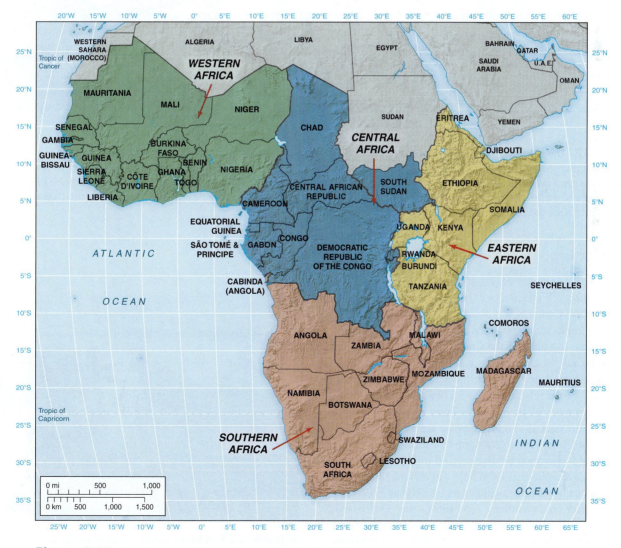

Figure 8.11 Sub-Saharan Africa: subregions and political boundaries.

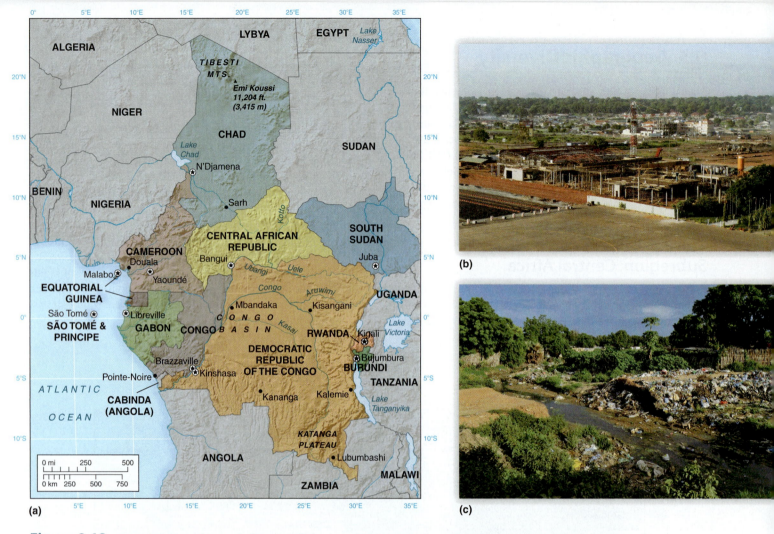

(a)

(b)

(c)

Figure 8.12 The subregion of Central Africa. (a) The subregion of Central Africa. (b) South Sudan, the world's newest country, is actively upgrading its infrastructure through profits from oil and other mineral deposits. Shown here in the capital city of Juba is a new international airport under construction. (c) South Sudan and other Central African countries have vast amounts of work ahead of them to build water and wastewater systems for their communities. Shown here is a dumping site next to a stream in Juba, South Sudan. *Photos: (b), (c) © Michael D. Kock/Gallo Images/Getty Images.*

Democratic Republic of the Congo

The combination of large size (largest country in the region) and large population (68 million, over half of the subregion total), centrality in the Congo River basin, many mineral resources, and huge hydroelectricity potential might be expected to make the Democratic Republic of the Congo (DRC) a focal country in Central Africa, or even in the continent. However, instead of forming a central driving force for development in Central Africa, DRC was largely taken over by feuding private armies, making government of the whole country nearly impossible. These events are central to understanding the geography of the country today. From 1998 to 2004, up to 4 million people died of war-induced starvation or disease, and millions more were displaced from their homes. Even huge aid income failed to lift GNI PPP per capita income, which was US$310 in 2012.

Soon after independence in 1960, the Democratic Republic of the Congo army under Mobutu Seko seized control from the Communist Prime Minister Patrice Lumumba. A catalog of confusion, corruption, and atrocities continued.

Figure 8.13 A fisherman steers his dugout canoe on the Congo River. *Photo: © Per-Anders Pettersson/The Image Bank/Getty Images.*

Table 8.1

Table 8.1 | **Sub-Saharan Africa:** Data by country, area, population, urbanization, income (gross national income purchasing power parity), ethnic groups

Country	Land Area (km²)	Population (millions) Mid-2012 Total	Population (millions) 2030 Est. Total	% Urban 2012	GNI PPP 2012 Total (US$ billions)	2012 Per Capita (US$)	Ethnic Groups (%)
CENTRAL AFRICA							
Burundi, Republic of	28	8.4	15.0	10	3.4	400	Hutu (Bantu) 79%, Tutsi (Hamitic) 20%
Cameroon, Republic of	475	19.6	25.5	57	44.4	2,270	200 groups: Fang, Barnileke, Fulani, Pahouin
Central African Republic	623	4.5	5.8	38	3.5	790	Baya 34%, Banda 27%, Mandjia 21%, Sara 10%
Chad, Republic of	1,284	11.2	17.5	27	13.7	1,220	Muslim groups in north, non-Muslim in south
Congo, Republic of the	342	4.0	6.2	60	13.0	3,220	Kongo 48%, Sangha 20%, Teke 17%, Mboshe 12%
Congo, Democratic Republic of the	2,345	66.0	107.5	33	21.4	320	Over 200 African groups: Mongo, Luba, Kongo
Equatorial Guinea, Republic of	28	0.6	0.8	39	—	—	Fang 80%, Bubi 15%
Gabonese Republic	268	1.5	1.7	84	19.8	13,180	Fang 60%, Mpongwe 15%, M'bete 14%, Punu 12%
Rwanda, Republic of	26	10.6	14.6	18	12.3	1,150	Hutu 90%, Tutsi 9%, Twa 1%
Totals/Averages	**5,419**	**127**	**195**	**41**	**114**	**2,818**	
WESTERN AFRICA							
Benin, Republic of	113	8.9	14.5	41	14.0	1,590	42 groups: Fon, Adju, Yoruba main
Burkina Faso, Democratic Republic of	274	16.5	23.7	16	20.7	1,250	Mossi 25%, Gourounsi, Senufo, Lobi, Bobo, Mande
Côte d'Ivoire, Republic of	322	19.7	27.4	48	35.8	1,810	60 groups: Akan, Kru, Mande, Senufo main
Gambia, Republic of the	11	1.7	2.2	54	2.2	1,300	Mandinke 42%, Fulani 18%, Woluf 16%, Jola 10%
Ghana, Republic of	239	24.4	33.1	48	40.5	1,660	Fanti, Ashanti, Ga-Adangbe, Ewe, Hausa
Guinea, Republic of	246	10.0	15.7	30	10.2	1,020	Fulani 35%, Malinke 30%, Susu 20%
Guinea-Bissau, Republic of	36	1.5	2.9	30	1.8	1,180	Balante 27%, Fulani 23%, many others
Liberia, Republic of	111	4.0	6.8	58	1.4	340	Bassa, Gio, Kpelle, Kru 95%, U.S. Liberians 5%
Mali, Republic of	1,240	15.4	20.6	31	15.8	1,030	Mande 50%, Peul 17%, Voltaic 12%, Tuareg/Moor 10%
Mauritania, Islamic Republic of	1,026	3.5	4.5	40	6.6	1,910	Mixed Moor/Black 40%, Moor 30%, Fulani, Wolof
Niger, Republic of	1,267	15.5	26.3	17	11.2	720	Hausa 56%, Djerma 22%, Fulani 8.5%, Tuareg 8%
Nigeria, Federal Republic of	924	158.3	204.9	47	344.2	2,170	Hausa, Fulani, Yoruba, Ibo total 71%
Senegal, Republic of	197	12.4	18.0	41	23.8	1,910	Wolof 44%, Fulani and Tutulor 24%, Serer 15%
Sierra Leone, Republic of	72	5.9	7.6	37	4.9	830	Mende, Temne, Limba, Creoles, others
Togo, Republic of	57	6.0	9.9	40	5.4	890	37 groups: Ewe, kabre, Gurma main
Totals/Averages	**6,135**	**309**	**418**	**39**	**539**	**1,307**	

(continued)

Table 8.1

Sub-Saharan Africa: Data by country, area, population, urbanization, income (gross national income purchasing power parity), ethnic groups *(continued)*

Country	Land Area (km²)	Population (millions) Mid-2012 Total	Population (millions) 2030 Est. Total	% Urban 2012	GNI PPP 2012 Total (US$ billions)	2012 Per Capita (US$)	Ethnic Groups (%)
EASTERN AFRICA							
Djibouti, Republic of	23	0.9	1.1	87	—	—	Somali 60%, Ethiopian 35%
Eritrea, Republic of	118	5.3	7.7	21	2.8	540	Tigrinya 50%, Tigre-Kunama 40%, Afar 4%
Ethiopia, Federal Democratic Republic of	1,104	83.0	108.7	16	86.1	1,040	Oromo 40%, Amhara, Tigrean 32%, Sidamo 9%, others
Kenya, Republic of	580	40.5	51.3	19	68.1	1,680	Kikuyu 21%, Luhya 14%, Luo 12%, Kalenjin 11%, Kamba 11%
Somali Republic	638	9.3	15.0	37	—	—	Somali 85%
Tanzania, United Republic of	945	44.8	57.4	25	62.6	1,430	Over 120 African culture groups
Uganda, Republic of	241	33.4	55.9	13	41.8	1,250	Ganda 18%, Nyankole 10%, Kiga, Soga, Iteso, Langi
Totals/Averages	**3,649**	**217**	**297**	**31**	**261**	**1,188**	
SOUTHERN AFRICA							
Angola, Republic of	1,247	19.1	26.2	57	103.1	5,240	Ovimbundu 37%, Mbundu 25%, Bakongo 15%, others
Botswana, Republic of	582	2.0	1.7	57	27.5	13,700	Tswana 75%, others
Lesotho, Kingdom of	30	2.2	1.7	24	4.3	1,960	Basotho 79%, Nguni 20%
Madagascar, Republic of	287	20.7	28.2	30	19.9	960	Merina 27%, Betsimisaraka 15%, Betsileo 12%, others
Malawi, Republic of	118	14.9	20.6	17	12.7	850	Chewa, Nyanja, others
Mozambique, Republic of	802	23.4	27.5	29	21.7	930	Makua-Lomwe, Yao, Makonde, Chewa, others
Namibia, Republic of	824	2.3	2.6	35	14.7	6,420	Orambo 50%, other black 36%, white 6%, mixed 7%
Republic of South Africa	1,219	50.0	51.5	59	517.9	10.360	Black 76%, white 13%, mixed 9%, Indian 2%
Swaziland, Kingdom of	17	1.1	1.0	24	5.7	5,430	Black 97%, white 3%
Zambia, Republic of	753	12.9	14.8	37	17.8	1,380	Black (over 70 ethnic groups) 98.7%, white 1.1%
Zimbabwe, Republic of	391	12.6	16.0	37	—	—	Shona 71%, Ndebele 16%, others, white 2%
Totals/Averages	**6,270**	**161**	**192**	**37**	**745**	**4,740**	
Region Totals/Averages	**21,473**	**809**	**1,102**	**37**	**1,676**	**2,513**	

Sources: *World Development Indicators*, World Bank (2009); *World Population Data Sheet 2008*, Population Reference Bureau; Microsoft Encarta 2005.

Elections set for June 2005 were postponed, and funds designated for development were redirected to military hardware. An election took place in mid-2006, surprising many that it could be held under such difficulties: the last census was in 1984, there is no experience of democracy, no functioning civil service, no registered voters, few electoral workers, and virtually no roads. In late 2006 Joseph Kabila was sworn in as president.

South Sudan

In 2011, the world's newest country was established with the independence of South Sudan and its capitol city of Juba. This was the result of several decades of civil war and tension with the country of Sudan, based on a religious division between Islamic fundamentalists in the northern part of the country and the Christian-tribal southern regions. The civil war depleted the country's natural and economic resources over time, with the result being one of the most poverty-stricken nations in the world. In the 1990s, substantial deposits of oil were discovered throughout Sudan, particularly in the southern areas. The promise of economic gain motivated southern Sudan to seek independence, and secession was put to a vote in 2011. Over 99 percent of the population of southern Sudan voted for independence, and it was formally recognized as a sovereign nation in July of 2011. However, economic and humanitarian challenges continue to plague this country, as most of its people are either farmers or nomadic herders without the skills needed to contribute to oil production or economic development. The prospect of continued war with Sudan is still a pressing issue, along with dependence on foreign aid for the building of the country's infrastructure needs.

Rwanda

In 1994, the genocide of nearly 1 million people brought Rwanda and Burundi, two countries in the heart of Africa, to the world's attention. More than 10 years later, Rwanda is quiet under the virtual dictatorship of Paul Kagame, who ended the massacres by force.

Rwanda is one of Africa's smallest countries in land area, although its 9.9 million people in 2012 gave it one of the highest population densities. It is known locally as the "Land of a Thousand Hills" with the highest points (often volcanic) in the west, where the steep slopes lead down to Lake Kivu in the rift valley. Eastward is a rolling, hilly upland, and there are swampy plains on the border with Tanzania. Although Rwanda lies just south of the equator, the hilly terrain and altitudes create a pleasant climate—warm temperate with temperatures around 70°F (25°C) throughout the year. Rain falling from frequent thunderstorms is heaviest in the west.

Landlocked, with poor surface transportation, some 90 percent of the population is dependent on **subsistence agriculture**. Most families produce a little coffee or tea for cash and export (in contracts, for example, with Starbucks), but there are few natural resources for a manufacturing base. Incomes are very low; few people live in towns.

Before the colonial period, the original population of Twa Pygmies (now only 1 percent of the total) was displaced by Bantu peoples—the majority Hutu cultivators, and the taller, thinner Tutsi cattle and sheep herders. Tribal hierarchy was never an issue, intermarriage was common, and many Hutus socially upgraded to become Tutsis. Unfortunately, the Belgian colonists created separation and resentment by deposing Hutu chiefs in favor of Tutsis and issuing ethnic identity cards. At independence in 1962, the Hutu majority won the elections and imposed a 9 percent quota on Tutsis (their proportion of the population) for salaried jobs. General Juvenal Habyarimana seized power in 1973 as one of "the majority people" and further discriminated against the Tutsis—many of whom left the country. In 1990, Kagame gathered these exiles and invaded from Uganda. Despite a 1993 peace accord, some members of the previous Hutu regime recruited and indoctrinated thousands of Hutu militiamen and armed them with machetes to meet the Tutsi "threat." When Habyarimana's plane was shot down in 1994, his most bigoted associates took control and set off the genocide of Tutsis through local gatherings. The property of those killed was given to enthusiastic murderers.

Kagame's Rwanda Patriotic Front (RPF), formed of exiled Tutsis, won the short war that ensued, killing many of the murderers and chasing them into the Zaire (now DRC) rain forests. The RPF continues to govern Rwanda, using massive foreign aid to rebuild the school and health care systems. Hundreds of thousands of Tutsis returned, bringing cash and skills to replace the lost middle class. Many had been born abroad, and they often returned with little knowledge of the local language. Rwandans returning home or to visit surviving friends and families find improved housing, shopping facilities, and transportation links without feelings of fear.

As part of the RPF's main objective is to maintain peace and involve all Rwandans, it tries to re-educate the murderers who survived. However, security is tight, with no press freedom or freedom of association, party loyalty is imposed, and Rwandan prisons, pronounced as "life-threatening" by the U.S. State Department, provide a major deterrent to opposition. In 2007, Kagame, in a gesture of defiance against the French forces, whom he accuses of being a bad influence in 1994, attended the British Commonwealth of Nations conference in Nairobi—also a reflection of increasing links with Eastern Africa.

Subregion: Western Africa

Western Africa's geography (Figure 8.14a) contrasts with that of Central Africa and makes it more open to the rest of the world. Instead of an interior focus on a major river system and a short section of coast, it has a long coastline and several rivers connecting the interior to the coast. European exploration and trade began in the mid-1400s, but colonial political control began in the late 1800s. France, the United Kingdom, Germany, and Spain were the colonial powers who left a legacy of country boundaries and common languages still used in commerce and government.

The colonial development of the subregion was based on local mineral, forest, and climate resources. The humid tropical rain forest in the south was cut for hardwood lumber, while plantations were established for cocoa, palm oil, rubber, and tropical fruits by multinational corporations, and the smaller plots of local farmers. These produced exports and income. Inland, cattle were raised in the savannas. Mining for gold, diamonds, and bauxite was mainly by international corporations and depended on world markets. Railroads joined inland centers and mines to ports. After independence, single-party government and military dictatorships became common, often tearing apart the countries on ethnic lines. Ports and inland historic cities experienced rapid population growth.

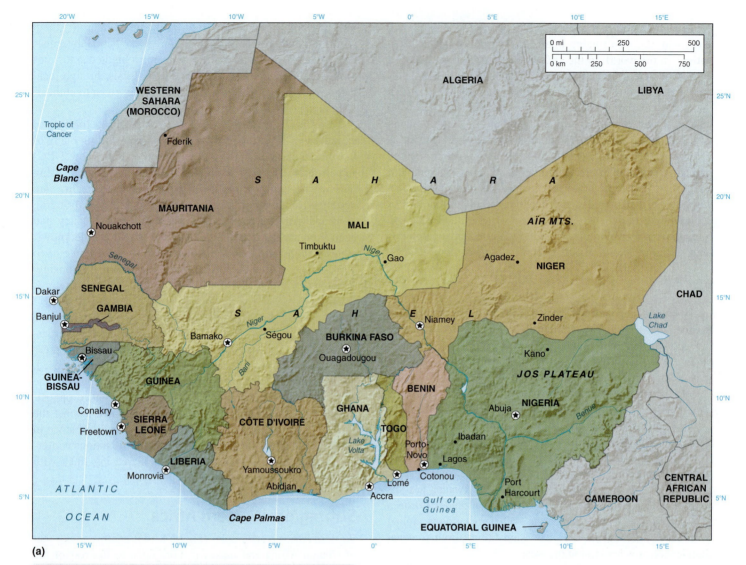

(a)

(b)

Figure 8.14 **(a) The subregion of Western Africa. (b) The Sahel region in the country of Mali, Western Africa.** This area has been heavily overgrazed due to political boundaries superimposed on the region by the Europeans in the 1900s, resulting in desertification and few grasslands left for grazing. *Photo: © Ariadne Van Zandbergen/Lonely Planet Images/Getty Images.*

Arising from their history as French colonies, many countries of Central and Western Africa continue financial links with France (Figure 8.15). After independence, the **Communauté Financière Africaine (CFA)** franc was shared by 14 African countries. Its value was tied to the French franc and guaranteed

by the Bank of France. High exchange rates allowed French companies to retain dominant positions in local contracting, but they made it difficult for French-speaking African countries to export goods, used up their foreign exchange reserves on imports, and made them increasingly dependent on France. In January 1994, the French government devalued the CFA franc by 50 percent. This raised the cost of imports, but France wrote off the debts of the poorer countries, and the World Bank made increased grants available. By 2011, most CFA countries saw some benefits in lower inflation and better foreign currency reserves.

The **Economic Community of West African States (ECOWAS)**, established in 1975, is Sub-Saharan Africa's most active regional organization. It has been prominent since the early

Figure 8.15 **The French influence in Western and Central Africa.** Former French colonies had a currency depending on the French franc and used French loans. The French government devalued the CFA franc in 1994 and adopted the euro in 2002, resulting in some confusion in the African countries, which were also part of the ECOWAS (Economic Community of West African States) zone.

1990s through peacekeeping operations in Liberia, Sierra Leone, and Côte d'Ivoire and opposition to a coup in Guinea-Bissau. The consultative ECOWAS parliament with 115 members was established in 2002. A Court of Justice arbitrates individual, corporate, and country-level complaints. ECOWAS has also promoted integration among countries, and many workers cross borders daily. Tariff barriers are being removed to allow free trade, an investment bank supports private sector enterprises, and a gas pipeline connects Nigerian oilfields to Benin, Togo, and Ghana. There is also scope for rationalizing air routes to make internal connections easy. A common currency, the "eco," is being developed for member countries. It will meet strict criteria, though will compete with the CFA franc in former French colonies. The problems of open borders include easier arms traffic across them and con artists taking advantage of a travelers' check initiative.

Nigeria

With half the subregion's population, Nigeria exemplifies many of the problems facing the other countries of Western Africa. Nigeria's geography includes coastal tropical rain forest and inland savannas, drained by the Niger River and its main tributary, the Benue River. The Muslim Hausa and Fulani people in the north are more numerous than the combined Christian and traditionalist Yoruba groups in the southwest and Igbo in the southeast. After the initial division of the independent federal country into three states, the conflicts resulting from the three-fold division and the presence of other ethnic groups in the center of the country led to the designation of 36 states within Nigeria.

Like the DRC, Nigeria failed to become a dominant force in its subregion. It experienced so much bad government from dictator generals in the 1980s and 1990s that it moved from an oil-rich, middle-income country to become one of the world's poorest. A lot of the problems arose from the continuing north–south ethnic and religious conflict. The northern Hausa and Fulani took control of the military and government, with the southern Yoruba being outvoted and the Igbos being widely disregarded after losing the 1960s civil war in which they had tried to establish their independence.

While the 1966–1979 military rule brought life to the economy and united peoples with very different interests inside the country, the military rule from 1984 to 1999 was a disaster. Despite huge controversy within the country, the northern Muslims made Nigeria a member of the Organization of the Islamic Conference, annulled the 1993 presidential election, and repressed the peoples of the oil-producing Niger River delta region. Cultural divisions deepened and the economy faltered through neglect and corruption. In 1998, it was estimated that three-fourths of official GDP was generated by the informal economy such as local street markets (Figure 8.16). Some US$12.4 billion of government funds was paid out without proper accounting, while the military dictator in the mid-1990s stole US$3 billion.

After the 1999 elections, Nigeria made a rapid transition from military to democratic government. The newly elected democratic government of Nigeria led by Olusegun Obasanjo faced the difficulties of a hastily devised new constitution and the legacies of a fragile federal system, an undermined judicial system, and a police force that was reduced in numbers in case it competed with the previous military. In 1999 and 2000, several states in the Muslim north challenged the new government

Figure 8.16 **Local market scene in the city of Lagos, Nigeria.**
Photo: © Eco Images/Universal Images Group/Getty Images.

by declaring that Islamic (sharia) law should take precedence over the common law order inherited from the colonial era. This raised questions about the relationships between public institutions and religious traditions throughout Nigeria.

A second challenge to the government comes from the delta of the Niger River, where most of Nigeria's oil is produced, and which is inhabited by several small, poor ethnic groups. The oil revenues, however, go to the federal government, which takes the largest share and divides the rest among the country's 36 states on the basis of population and area. The oil-producing areas demand that more funds be returned to them, and local people cause pipeline disruptions. The federal government scarcely listens and replies with repressive measures such as the murder of the activist Ken Saro-Wiwa, who protested oil company and government attitudes. Demands from the delta peoples increased when world oil prices rose sharply in 2004, 2008, and again in 2011.

A third challenge arose when the 17 southern states of Nigeria complained that the federal government made most of the important decisions without listening to the states. The states want more local power over the police force, education, revenues from resource exploitation, and infrastructure provision. However, the 19 northern states resist this proposal. By 2005, Obasanjo had achieved more democracy and had appointed young technocrats to key posts, and steered Nigeria toward greater economic stability. In 2007 he stepped down from power after serving for two terms as president, and a new regime rules Nigeria as of the year 2012.

Mali's Geopolitical Issues

Mali, a former colony of the French, is a country in Western Africa that often has been described as a model for democracy in the continent. However, in 2012, Mali experienced a military coup and takeover of its government after decades of stability. The conflict stemmed from the military's anger over the management of rebels in the northern regions of the country known as the Toureg people, who had attempted a rebellion the previous year. The Toureg are composed mainly of nomads who live in the southern Sahara Desert and established grazing and trading routes across much of present-day Mali, Niger, Libya, and Algeria. These people became separated from one another by the political boundaries superimposed upon them by European nations that controlled the continent in the early 1900s. Rebellions by the Touareg erupted in the neighboring countries of Niger and Libya in recent years, and it is widely thought that Muammar el Qaddafi, former leader of Libya, supported the Mali clan of Toureg people in raising a rebellion there in 2011, providing them with weapons to aid in their uprising.

Immediately after the military coup had commenced, rebel fighters organized and effectively split Mali into a northern and a southern entity, though the United Nations still recognizes Mali as one country. The rebels, mainly composed of Islamic radicals, captured the city of Tomboctou (Timbuktu) and declared independence for the northern half of Mali, destroying many ancient temples and other religious holy grounds in the process.

The capitol city of Bamako was eventually returned to civilian governmental control, though it has little ruling effect on the northern parts of the country today. The military coup left the army of Mali virtually unable to enforce any governmental laws in the rogue northern region, and as a result this part of the country emerged as a haven for terrorists. At this writing it was occupied by several radical factions of Islam, including al-Qaeda in the Islamic Maghreb, who imposed brutal Shariah law upon its people that included public beatings, whippings, amputations, and stonings. As this was being written Mali was being led by a temporary government and was formulating plans with other neighboring countries to recapture the north from the rebels occupying the region.

Subregion: Eastern Africa

Eastern Africa (Figure 8.17a) is dominated by plateaus and the Ethiopian highlands in the north, both broken by rift valleys. The subregion has mostly summer rainfall which is heaviest in the Ethiopian highlands, with increasingly arid climates on the coasts toward the north. The subregion became a crossroads where African groups from the Nile River valley met those from Western and Central Africa. Facing the Red Sea and the Indian Ocean, Arab influences were strong. Europeans (mainly British and Italian) brought new pressures after the 1869 opening of the Suez Canal. Superimposed on the African ethnic areas with their emphases on cropping or animal husbandry, British colonial commercial farming in Kenya and Uganda produced tea, coffee, and cotton. People from South Asia, introduced to construct and manage transportation, factories, and shops, added to the subregion's ethnic diversity.

The **East African Community (EAC)**, linking Kenya, Tanzania, and Uganda was launched in 1967 but soon collapsed. Revived in 2000, it began operations in 2005, with its headquarters at Arusha in northern Tanzania. The EAC faces concerns from those now used to their own countries' sovereignty: the Kenyans seek to dominate, the Ugandans see it as a source of conflict, and the Tanzanians are worried about keeping their distinctive identity.

Ethiopia

Ethiopia has the distinction of being the only African country that was never colonized by a European power. Although occupied by the Italian military from 1936 to 1941, it was soon returned to the Emperor Haille Selassie's rule by British forces at the outset of World War II. Ethiopia remained an isolated country, and Selassie was the last in a line of Ethiopian emperors dating back to the Coptic Church conversion in the AD 300s. He reigned from 1930 to 1974, when he was deposed and murdered in a military coup that brought to power a Communist government with Soviet Union support. After further coups, devastating droughts, and movements of refugees in and out of the country, rebel groups replaced that regime in 1991. A new constitution in 1994 led to multiparty elections.

After World War II, the United Nations combined Eritrea with Ethiopia. Disappointed at being reduced to a province

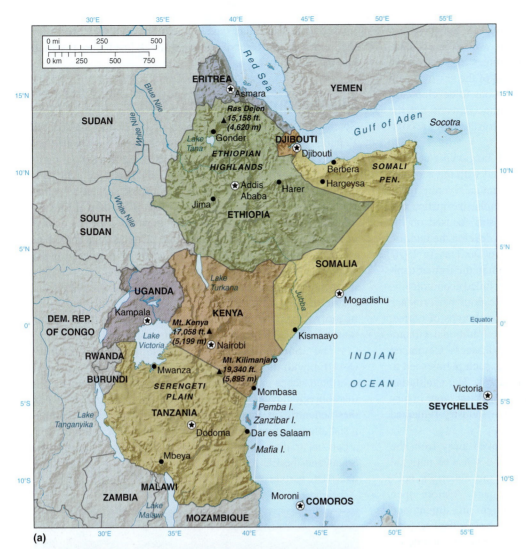

Figure 8.17 (a) The subregion of Eastern Africa. (b) An elephant makes a solitary journey across a game park in Kenya, through grassland environments known as savannas. *Photo: © Darrell Gulin/ Digital Vision/Getty Images RF.*

(b)

within Ethiopia, Eritreans fought Ethiopian and Soviet Union forces. After the Communist government in Ethiopia was deposed and the Soviet forces departed, Eritrea gained independence in 1993. Ethiopia is now landlocked and has difficult relations with surrounding countries. Tensions continue between Ethiopia and Eritrea over the placing of the border and access to an ocean port (i.e., on the Red Sea) for Ethiopia.

Ethiopia is greatly concerned with events in neighboring Somalia. The international boundary is poorly defined, and Ethiopia seeks to influence events in Somalia, while Somalis act as opposition groups within Ethiopia. A loose alliance of warlords based in Baidou, supported by Ethiopia, controls most of southern Somalia, but it was under pressure from militant Islamic groups who controlled the capital, Mogadishu, until early 2007.

Interior Tanzania

Inland Tanzania has a tropical plateau environment with variable moderate summer rains, low to moderate population density, and low income. Although the Olduvai Gorge in northern Tanzania is the site of many finds of the earliest humans, most of this area's history is unknown. Remnants of Khoisan (click-tongue) languages suggest that older groups of people were displaced by Bantu and Nilotic groups in the north. Some of the interior groups had well organized societies, especially in the northwestern lakes area, while others remained poor and combating a semiarid natural environment. They were largely untroubled by the coastal Arab and Portuguese developments. In the 1800s, European explorers, including David Livingstone (whose last mission was at Ujiji), passed through.

Germany indicated its colonial interest in Tanganyika in 1884 and made treaties with tribal chiefs, who accepted German protection, and with the British colonies to the north. In 1891 the German government took control of the territory with headquarters at Dar es Salaam and intentions to develop cash crops and transportation lines. Neither the Germans nor the subsequent British rulers in the post-1919 trust territory accomplished much development.

Julius Nyerere, a schoolteacher, led Tanganyika at independence in 1961, soon (1964) to be united with the island of Zanzibar as The United Republic of Tanzania. Nyerere introduced African socialism (*Ujamaa*) to emphasize justice and equality. While most economic activity still focused on the coastal area, major attempts were made to develop the interior, including moving the country's capital to Dodoma in 1996. However, many government offices remain in the former capital of Dar es Salaam. The Nyerere policies failed economically, politically, and socially. In particular, the poor performance of national boards to market the commercial crops of coffee, tea, and cotton, and the attempts to move traditional village-based groups into collective farms, were disasters. Combined with the failure to develop manufacturing, such policies led to increased poverty, famine, and indebtedness. The egalitarian socialist Tanzanian government resisted building large luxury hotels in the game parks, and the tourist industry remained small.

In the 1990s, a new government moved away from the Nyerere policies. In an attempt to forge a sense of national unity, Nyerere's successor detribalized the country's politics and made Swahili the official language. A more open economic policy now encouraged foreign investment and improvements in farming technology. However, the bad feelings between the mainland and Zanzibar remain; people on the mainland think Zanzibar has too much representation for its 1 million people, outmoded politics, and poor human rights, while Zanzibarians complain about the erosion of their previous sovereignty.

Having resisted the development of tourism for wealthy foreigners in Nyerere's equality-driven society, there was a major expansion of park resorts in the Serengeti area in the mid-1990s. Tanzania received nearly 900,000 visitors in 2012, up from 80,000 in 1980. Tourism made up over 50 percent of export earnings, replacing the crop sector as a dominant revenue source. The future depends on further improvements in government to restrict overspending and corruption.

Tanzania is known to possess large deposits of minerals, including gold and natural gas. Road construction is increasing to bring the interior into the orbit of global trade. The Tazara Railway connecting Zambia across Tanzania to Dar es Salaam was completed in 1975 with Chinese funding, in order to provide an outlet to the sea for countries that did not wish to trade through South Africa. The railway continues to function since the end of apartheid, and there is a tourist train from Cape Town to Dar es Salaam. However, competitive rates attracted Zambian copper and other trade southward again. Furthermore, the railway was built in a direct line that missed most Tanzanian development opportunities except for the Selous Game Reserve, passed through just before reaching Dar es Salaam.

Subregion: Southern Africa

Southern Africa includes the lands south of the Democratic Republic of the Congo and Tanzania (Figure 8.19a). The tropical summer rain climates grade into the southwestern arid and southern warm temperate conditions. The dominant plateaus with their pleasant climate provided many good farming areas for colonial Europeans, who displaced African groups. This subregion experienced the development of some of the world's largest deposits of valuable minerals (diamonds, gold, platinum, and important steel constituents). In the later 1900s, the apartheid policies of the white-ruled Republic of South Africa made enemies of the newly independent, black African-ruled countries in this subregion. However, since the mid-1990s the new black African-dominated majority government in South Africa improved the links of this most advanced African country with its neighbors.

The **Southern African Development Community (SADC)**, with its economic policies taking over from its anti-apartheid stance in the early 1990s, includes all the countries in this subregion, plus DRC and Tanzania. Although trade between these countries is free, South Africa, Botswana, and Namibia resist allowing free movements of people. Other countries fear a two-tier trading system as South Africa looks to the United States and the EU for trading opportunities without including its SADC partners.

Figure 8.18 **Tourists on a safari in Southern Africa observing giraffes, the tallest of all animals.** Several subspecies of giraffe exist in the world, though almost all live in Sub-Saharan Africa. *Photo: © Frank Herholdt/Stone/Getty Images.*

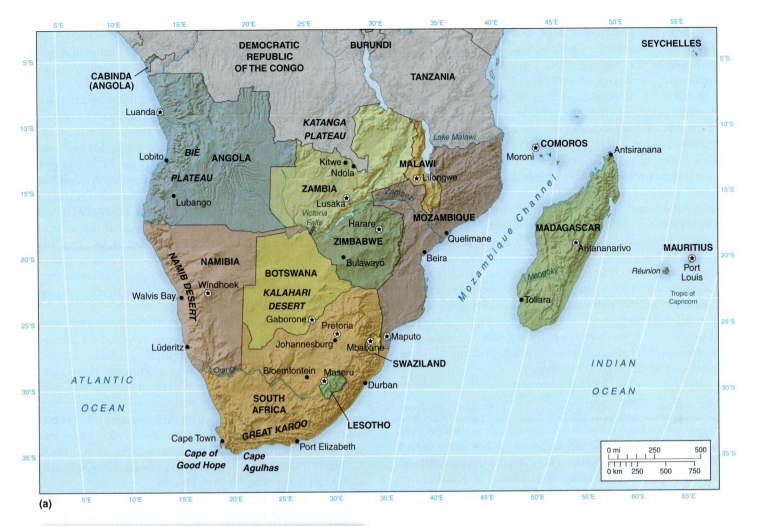

(a)

(b)

Figure 8.19 **(a) The subregion of Southern Africa. (b) Aerial view of Victoria Falls on the Zambezi River.** While the falls are neither the highest nor the widest in the world, during the flood season (February to April) they are the largest on Earth in terms of the amount of water carried over their steep cliffs. *Photo: © Jake Wyman/Photonica/Getty Images.*

Republic of South Africa

South Africa contains 6 percent of the population of Sub-Saharan Africa, yet its GNI PPP is 41 percent of the region. Its diverse economy of mining, manufacturing, and increasing services such as the tourism industry (Figure 8.18) continues to prosper following the peaceful transition from the apartheid regime to majority black government after the 1994 elections. The Government of National Unity, and the ability of the Truth

and Reconciliation Commission to deal with sensitive interethnic concerns, made this possible. The flight of some white professionals is partly balanced by more black Africans taking jobs in public services and private firms. This transition is changing the geography of South Africa as the abandonment of the homeland policies and open access to urban areas increase internal mobility. It also changes the orientation of neighboring countries from imposing sanctions against apartheid to welcoming economic association.

Much has been achieved since the 1994 elections. A lively culture was liberated, with open media and vigorous political debate. Throughout the 1980s, South Africa's distorted and underperforming economy grew at only 1 percent per year. Apartheid deprived nonwhite people of education and economic opportunities. In the 2000s, annual economic growth reached almost 4 percent, inflation was controlled, trade and exchange controls relaxed, and new external investment attracted. The government built nearly 2 million low-cost homes and created over 1 million new jobs. However, annual increases in the youthful labor force keep unemployment, particularly of unskilled people, high (30 to 40 percent), and one-fourth of the population lives on government handouts. Training programs and major public works projects to improve transportation and power supplies are helping the economy, but the high incidence of HIV/AIDS and restrictive labor practices slow further increases in well-being.

Debate over who controls South Africa continues. A new black middle class of some 15 million people that emerged after 1994 is fueling a consumer boom that continues through the present day. However, 5 million whites who remain here control 90 percent of the country's assets and major corporations. The Black Economic Empowerment (BEE) legislation of 2003 attempted to redress these inequalities. Private commercial companies give a percentage of shares to BEE consortia, contract to employ a percentage of blacks in senior positions, develop skills, and provide access to financial services for small companies. BEE has its critics, black and white, particularly for suspicions of "crony capitalism" in which government-favored people are placed in good positions.

Botswana

Botswana, situated on the northern border of South Africa, forms a contrast to other small African countries such as Rwanda. It has 20 times the land area, but only 1.8 million people, because around 70 percent of the land is dominated by the Kalahari Desert. It was formed as the British colonial protectorate of Bechuanaland following a request by the Tswana people in the face of a perceived threat from Boer farmers in the Transvaal. The country took its present name at independence in 1966.

Botswana is a stable democratic country, electing its National Assembly. An advisory House of Chiefs represents the main ethnic groups, and the chiefs preside over traditional courts (although citizens have the option to be judged under the British-based legal system). Traditionally the people are cattle herders, but Botswana has experienced major economic growth based on diamond mining. The government owns half of the Debswana mining company and uses the income to build infrastructure, diversify employment opportunities, and build foreign exchange reserves. However, present spending on economic development has been reduced in response to the rise in health care expenditures. Botswana has one the world's highest HIV infection rates (24 percent of the population aged 15 and over in 2012) and is spending money on free antiretroviral treatments and a program to reduce the mother-to-child transmission. Botswana also maintains a relatively high level of military spending, despite the low probability of internal or external conflict.

Industrialization in the Johannesburg Region

With 8 million people, Greater Johannesburg (Figure 8.20) is Sub-Saharan Africa's third-largest city (after Lagos, Nigeria, and Kinshasa, DRC) and this region's main economic hub.

Unlike many other major world cities, it is not situated on the coast or a major river. Nor is it the political capital of South Africa, although it is the capital of Gauteng, South Africa's wealthiest province.

Johannesburg occupies an area that yielded some of the oldest human remains. The long-term occupation by nomadic Bushmen gave way to Bantu farmer incursions from the north around AD 1060 and Boer farmers in the 1800s. The area was entirely rural until the discovery of gold in the 1880s and subsequent gold rush. Further exploration discovered the riches of the Witwatersrand Mountains, which became the center of development. Following the Boer War (1899–1902) and the declaration of the Union of South Africa (1910), a harsh racial system barred blacks and Asians from skilled jobs and instituted the migrant labor system. From the 1940s, people of non-European descent were removed to specified areas, such as the South Western Townships (Soweto), which often became sprawling shantytowns.

The abandonment of apartheid in 1990 and the 1994 elections led to freedom from discriminatory laws and the integration of black townships into the municipal government system. Although the suburbs became more multiracial, many businesses moved out of Johannesburg's central business district to the northern suburbs following perceptions of increased crime rates and serious traffic congestion. The north and northwestern suburbs are the wealthiest, with high-end retail shops and residential areas, including Houghton and the mostly black Sophiatown. The southern suburbs are low-income residential, including old townships such as Soweto.

The mining basis of Johannesburg's economy gave way to manufacturing and service industries. Today there is no mining within the city limits. Sanctions during the apartheid period led both to losses of multinational corporations

Figure 8.20 **Southern Africa: the central city of Johannesburg.** Its central business district is in the foreground, leading outward to the suburbs and old mines on the perimeter of the city. Photo: © Neil Beer/Photodisc/Getty Images RF.

and their replacement by local firms in many fields. Post-apartheid economic renewal brought back the multinationals. Steel and cement manufacture are paralleled by many medium and smaller concerns, making engineering products, chemicals, armaments, and food and drink. There are government branch offices and consular offices. Many new shopping centers are opening in the suburbs. Johannesburg is the center of South African media from newspapers (leading national Afrikaans, English language, and others aimed at black readers) to broadcasting. This city is also a center for higher education, with the University of Johannesburg and University of Witwatersrand having the reputation of having resisted apartheid.

As an inland center, Johannesburg was dependent on surface transport until the age of air transport. As a rail and road center it has the City Deep container "dry port" that takes 60 percent of cargoes arriving at Durban for further distribution. Although the large numbers of low-income people depend on bus and informal minibus taxi transportation, the many cars and trucks cause huge traffic jams, even on the 12-lane sections of the ring road. Johannesburg has South Africa's premier international airport, which serves as a hub for the subregion, and it is the center of tourist arrivals despite the area's industrial and crime-ridden reputation. Visits to Soweto, the Apartheid Museum, the world-famous zoo, and museums are often part of packages that use the airport as an exchange point for visits to Kruger National Park, Cape Town, and Namibia.

8.6 Contemporary Geographic Issues

HIV/AIDS Pandemic

Sub-Saharan Africa contains just over 10 percent of the world's population, yet in 2012 it had 68 percent of the people living with HIV/AIDS (Figure 8.21). In 2012, 1.4 million people became newly infected, a drop from 3 million in 2005. **HIV/AIDS** is a major threat to world health and especially to millions of people in poorer countries, where 90 percent of infections occur. The World Health Organization lists AIDS as the third main cause of global deaths. Its occurrence warrants **pandemic** status—a disease that has a long-term presence around the world. For geographers, one of the major issues arising from a study of the HIV/AIDS pandemic is how differently materially wealthy and poor countries approach it (Table 8.2).

People contract HIV through unprotected sexual contact with HIV carriers or through contact with HIV-contaminated blood or body fluids. Current medical research does not indicate that transmission can occur by other types of contact with HIV carriers. HIV infection can be passed from mother to baby. Patients become prone to many other sexually transmitted diseases and to other serious illnesses such as tuberculosis (TB), pneumonia, toxoplasmosis, fungal infections, and cancers. Available medical treatments are complex and expensive, and they require close monitoring. These do not cure HIV

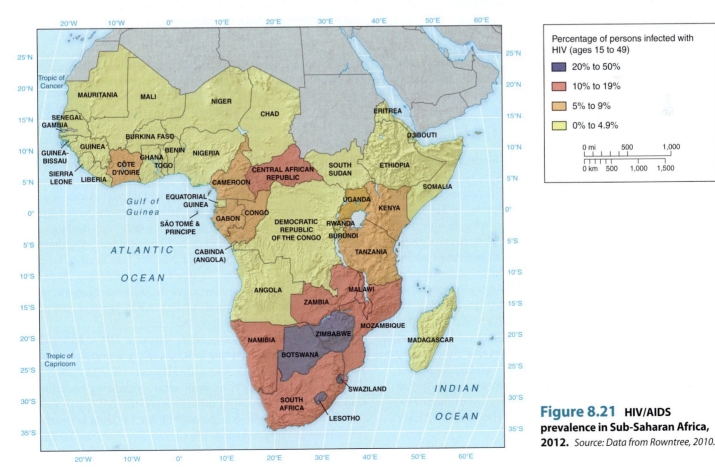

Figure 8.21 HIV/AIDS prevalence in Sub-Saharan Africa, 2012. *Source: Data from Rowntree, 2010.*

Table 8.2	Debate: HIV/AIDS in materially wealthy and poor countries

Materially Wealthy Country	Materially Poor Country
Retroviral drugs are having a significant effect in delaying the onset of AIDS, giving a sense that the disease is on the wane.	The drugs are designed in materially wealthy countries for the strains of HIV that are common there, and those produced by multinational pharmaceutical companies are too expensive for wider use in poorer countries. However, Brazilian and Indian producers sell them more cheaply, and the MNCs reduced the price when faced with the terrible outcome of no action. From the mid-1990s to 2011, drug prices had been reduced by 96 percent.
The greater investment in health care facilities makes vital monitoring centers widely available.	Monitoring centers are few and far between, making the distribution of drugs difficult. Access is being improved in middle-income countries such as Botswana and South Africa, but not in the poorest countries.
HIV rates are low, but are rising again after a period of careful control. In particular, the behavior of sexually active males is tending toward less protection, while some of the drugs are less effective.	HIV rates of infection and deaths from AIDS are high, although recent surveys show that previous estimates on less evidence were often too high.
Men have been affected more than women because of the connection with homosexual activities.	Women are affected more than men. Reasons include: male circumcision helps prevent infection; women are vulnerable to rape; older male sexual partners pick up infections; mobile miners and military personnel are most likely to be infected from prostitutes and to pass on infection to marriage partners and the resultant children.
Education programs are effective, together with provision of syringes for drug users, although few countries impose HIV/AIDS tests.	Effectiveness of education varies. Some countries did not take the threat seriously until the later 1990s, by which time HIV/AIDS was a major cause of death and social disruption. Even when testing is available, people avoid voluntary involvement because of stigma and social penalties. Botswana, with the highest rate of infection, reduced such high-profile testing to a routine action during doctor visits.
HIV/AIDS is not regarded as a major socioeconomic threat, although more significance is given to the worldwide situation. Cash invested in HIV/AIDS programs increased by 20 times from 1996 to 2003. UNAIDS reports that US$25 billion will need to be invested in HIV/AIDS services in low and middle-income countries in 2010.	The high incidence could lead to economic collapse in South Africa, which has more cases than any other country (5.7 million HIV/AIDS citizens out of a total population of 48 million). Arising from this threat, more countries are devising national plans to combat the disease. Although the epidemic appears to be stabilizing, HIV/AIDS is still taking a high toll in the region.
Most countries are now open in reporting cases of HIV/AIDS.	Many countries resist full reporting, giving a false view of the total picture. There is a particular difficulty in many Arab countries, where activities contributing to the spread of HIV/AIDS are illegal or not admitted and so there is little detailed monitoring. However, it is becoming clear that there is a high incidence among sex workers and drug users, and that the many migrant workers pose a considerable threat.

but can prolong life. First recognized in wealthier countries, HIV/AIDS is now a major plague in Sub-Saharan Africa, where most countries have increasing HIV prevalence and few show a decline. Southern Africa is the epicenter of the global HIV/AIDS pandemic, with the main countries having over 20 percent of the adult population infected. Western and Central Africa have lower HIV prevalence, under 10 percent of adults. Eastern Africa has the most evidence of declining prevalence, particularly in Uganda and Kenya, but HIV rates in other countries in that subregion remain at high levels (Figure 8.22).

The causes of the high levels of HIV/AIDS in Africa include poverty, the breakdown of traditional family support systems, the apartheid policy in South Africa that brought miners into male-only camps serviced by prostitutes, continuing promiscuity at a time when traditional polygamy gives way to the taking of sexual partners outside monogamous marriages, and mistaken government policies. HIV/AIDS spreads quickly in cultures that value male sexual prowess.

Although reduced (for a time) in Europe and North America in the 1990s by expensive triple-drug therapy monitored at special clinics, the disease diffused rapidly through Africa. The adoption of antiretroviral drugs in African countries is patchy. By mid-2005, a third of those needing the drugs received them in Botswana and Uganda, with up to 20 percent receiving them in Cameroon, Côte d'Ivoire, Kenya, Malawi, and Zambia. Low levels of provision occurred in South Africa, Ethiopia, Nigeria, and Zimbabwe. Progress requires the well-managed use of these drugs, but the high levels of HIV/AIDS are expected to continue in Eastern and Southern Africa.

In addition to the demographic impacts, HIV/AIDS has geographic social impacts that will take many years to change. Half the miners in South Africa are HIV carriers, and millions of orphans, often carriers as well, create a growing need for help in the region. By 2020, orphans with HIV/AIDS will rise from 2.0 to 2.7 million in Nigeria and from 5 million to nearly 6.5 million in Eastern Africa. Throughout Sub-Saharan Africa,

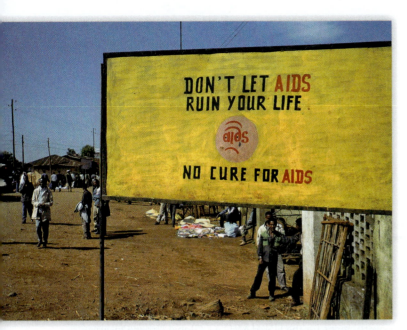

Figure 8.22 This sign in the village of Gimbii in Ethiopia reflects the serious issues the region has with HIV/AIDS prevalence, which in some locations can be as high as 50 percent of the population. *Photo: © Bruno Barbier/Robert Harding World Imagery/ Getty Images.*

people migrate into towns for perceived safety in times of civil war. However, the informal urban economy becomes the only means of livelihood for many people who are cut off from their home villages and subsistence food production. In the absence of the wealth common in Western countries, many Africans create ways of living that enable them to survive and even enjoy life. Items that might be regarded as waste become children's toys. Bicycles or walking are common modes of travel, while cheap rides on crowded vans and buses enable wider mobility. Commodities such as water and other goods and possessions are often transported on citizens' heads.

When placed alongside the fact that Africa is the only world region where extreme poverty is expected to increase (from 100 million people in 2000 to an estimated 360 million in 2020, mainly in towns), it is clear that there is an urban crisis. Although the rate of urban population increase is similar to that in Europe and North America in the late 1800s, with similar problems, few African cities benefit from the economic growth experienced in those regions.

the lives of young workers and their families are being shattered. Military personnel, migrant miners and their wives, and prostitutes have the highest proportions of infection. The lack of medical understanding and panic at not having the funds or expertise to do anything about the disease generate false taboos and myths, such as the one that men can be cured by having sex with a virgin girl—a basis for many child rapes. Although HIV/AIDS mainly occurs in urban areas, it also threatens rural communities in Southern Africa, where one-fourth of farm workers are infected. HIV/AIDS requires global action that is as vital as controlling terrorism, the armaments and drug trades, and slavery.

Exploding Cities

Two-thirds of the populations of most countries in Sub-Saharan Africa remain rural, linked to the dominance of subsistence farming in the economy. However, Africa is now the fastest urbanizing continent in the world: the growth of urban centers (Table 8.3) reflects the rising influence of global connections. Around 15 percent of Africans lived in towns in 1950, 28 percent in 1980, and 54 percent are expected to do so in 2020. The 300 million African urbanites in 2004 could rise to 500 million by 2015. In 1960, Johannesburg was the only city in the region with more than 1 million people. There were four such centers in 1970, 12 in 1990, and 27 in 2012.

Urban places account for high proportions of economic activities. They are perceived to contain more prospects for waged employment and to provide better educational and health facilities than rural villages. Urban populations also grow as

Table 8.3	Populations of Major Urban Centers in Sub-Saharan Africa	
TOP 10 MOST POPULATED URBAN CENTERS		
City, Country	**2012 (millions)**	**2025 (millions)**
1. Lagos, Nigeria	11.5	15.8
2. Kinshasa, DRC	9.0	16.7
3. Johannesburg, South Africa	7.6	9.1
4. Luanda, Angola	5.2	8.2
5. Abidjan, Côte d'Ivoire	4.4	6.0
6. Accra, Ghana	3.8	4.9
7. Dar Es Salaam, Tanzania	3.7	4.4
8. Nairobi, Kenya	3.5	5.9
9. Kano, Nigeria	3.4	5.1
10. Cape Town, South Africa	3.4	3.7
NUMBER OF MAJOR URBAN CENTERS IN SUB-SAHARAN AFRICA, BY POPULATION CATEGORY		
Population Category		**Number of Urban Centers**
10.00 million and greater		1
5.00 – 9.99 million		3
2.50 – 4.99 million		14
1.00 – 2.49 million		10

African cities have distinctive geographies with current changes imposed on past patterns. Accra in Ghana illustrates some common features. Many of the oldest African urban landscapes occur in Western Africa, including the trading centers of Timbuktu, Sokoto, and Kano at the southern end of trade routes across the Sahara. These Islamic cities typically have central markets, mosques, citadels, and public baths, and are still dominated by craft workshops rather than modern industry. Ibadan, the center of the Yoruba culture in southwestern Nigeria, preserves another type of pre-European urban landscape. The central palace and nearby market are at the focus of streets radiating toward other towns. Fortifications were added later to resist Muslim attacks. The modern populations of these older towns remain much more ethnically unified than those in many of the newer urban areas.

The colonial era was a time of building ports and interior control centers that left often grandiose buildings of European design and separated economic zones. After independence, local elites took over colonial properties. Many new homes, offices, and factories were built, but planning was often at a low level, and such city expansion was haphazard.

Shantytowns are slum areas that occur in all African cities, housing over 70 percent of the urban population (Figure 8.23). Shantytowns are constructed of any materials that come to hand—from wooden crates to cement blocks and corrugated iron—and basic in their services. Few are linked to piped water or sewerage systems, which leads to high infant and child mortality. HIV/AIDS incidence in South Africa is twice as high in slums as in the rest of the urban areas, and three times as high as in rural areas. Some shantytowns condemn families to a hopeless future of poverty. Their inhabitants are often involved in the informal economy. The high incidence of such poor housing, unemployment, street children, and the wide availability of small arms are linked to high levels of urban crime.

In many cases, however, people move into slums on arriving from a rural area but eventually find better accommodations. Governments may supply utilities and build schools, hospitals, and roads to integrate shantytowns with the rest of a large urban area, usually after a considerable period while these become established. Shantytowns were encouraged under apartheid in South Africa along with segregation of many public places such as beaches (Figure 8.24). The long process of improving housing conditions and implementing integration is well underway.

A combination of the United Nations Habitat Agenda's Sustainable Cities Project, the Millenium Development Goals, and the New Partnership for Africa's Development (NEPAD) is beginning to address this region's urban problems. There is an emphasis on improving water, sanitation, and housing in NEPAD cities. The first international ministerial meeting on land, housing, and urban development was held in Durban, South Africa, in early 2005. Improved governance, security of tenure, and better urban–rural linked regional planning are also on the agenda. Geographers are involved in producing local poverty maps for the UN-HABITAT Global Urban Observatory that will provide a basis for focused policies.

Global Intrusions and Local Responses

Global Connections and China's Role

Much of Sub-Saharan Africa continues to serve as either a plantation or a quarry, providing raw materials to the materially wealthy Western world—often a holdover from trade established in colonial times. Some African countries and cities are more integrated into the global economy than others because their ports and airports link commercial farms and mines to markets in wealthier countries. Cities such as Nairobi (Kenya), Johannesburg and Cape Town (South Africa), and Lagos (Nigeria) exhibit

Figure 8.23 **Shantytown slum in Cape Town, South Africa.** Scenes such as these are repeated far too often across the urban landscapes of Sub-Saharan Africa due to extreme poverty, lack of educational skills to obtain better-paying jobs, and in some cases, a clear policy of racial discrimination. *Photo: © Franz Aberham/The Image Bank/Getty Images.*

Figure 8.24 **Sign on a beach in Cape Town, South Africa, that clearly reflects the policy of apartheid throughout the country that lasted from the 1960s until the early 1990s.** *Photo: © Denny Allen/Gallo Images/Getty Images.*

some signs of becoming global city–regions as they provide headquarters for the subregional centers of commerce, multinational corporations, and NGOs.

However, many African countries are part of the global economic system as dependent debtors and recipients of aid. In their post-independence focus on autonomy to develop the well-being of their people, they often spent their small resources on what they perceived to be important in raising their national identity, through lavish new capital cities and military purchases. In some countries, poorly managed projects, together with the corrupt siphoning of funds into personal bank accounts abroad, slowed economic opportunities and led to international lenders and bankers imposing conditions that made it difficult to obtain further loans.

In the 1980s, the World Bank and the International Monetary Fund—the two major lending institutions for developing countries—established new guidelines for grants and loans. This was a response to the low rates of success achieved with previous loans and the large portions of loans that were absorbed by high exchange rates, the cost of internal government bureaucracies, and the corrupt mismanagement of funds. To date, these structural adjustment policies have had little success in the countries of Sub-Saharan Africa, although at times Ghana, Tanzania, Burkino Faso, Nigeria, and Zimbabwe came close to adopting the guidelines. The policies relied too much on exporting commercial crops instead of growing food for local use. The reduction in government employees reduced health care and education provision. At present, countries in Sub-Saharan Africa seem to lose out whether they decide to follow structural adjustment policies or not. If they do not adopt them, they lose access to funds from the World Bank, International Monetary Fund, and aid agencies. If they adopt the stringent policies, they often alienate their people.

For many African countries, the immediate future looks brightest through association with China. Africa may become a prime trade battlefield between China and the West. In 2004, one of the first foreign visits made by Chinese president Hu Jintao was to Africa. The Chinese keep human rights and politics out of business arrangements and in doing so, gain vital resources and new markets in this region. China trades with countries such as Ethiopia, Somalia, Equatorial Guinea, and Zimbabwe, which are shunned by European Union countries over human rights issues. China purchases oil (30 percent of their oil imports), iron ore, and other commodities, and invests billions of dollars in producing the commodities, along with tourism, agricultural, infrastructure, and health (such as anti-HIV/AIDS) projects. China is a major supplier of military hardware to the region. Between 2005 and 2012, African trade with China increased from US$40 billion to US$125 billion. China is currently offering a prospect of free trade for African countries.

Local and Global Connections: Cell Phones

The development of telecommunications is indispensable to Africa's growth and domestic stability. In 2005, the United Nations launched its Digital Divide Fund to help reduce the technology gap between richer and poorer worlds. However, the establishment of rural telecenters (communal telephone, Internet, and other computing facilities) is an example of top-down development requiring capital and often government funding. The bottom-up spread of cell phones is faster because their use is a matter of private demand. The world's poorest people often share or rent cell phones by the call, and thereby overcome the high cost and shortage of landlines and electricity supplies. Cell phones have the greatest impact on development, raising long-term growth rates. Cell phones cut transaction costs, widen trading networks, and reduce travel costs. They can also access the Internet and provide a clock, camera, and music download source—bringing an information revolution. An extra 10 phones per 100 people in a developing country increases annual economic growth by up to 1.5 percent.

In Africa, while only 3 percent of the population had access to a landline telephone in 2001, there were already 50 million cell phone subscribers, with numbers increasing by one-third each year. By 2012, South Africa had over 84 cell phones per 100 people, and South African providers were expanding their facilities across the region. Cell phone growth rates doubled in 2012 in Nigeria, Angola, Ghana, Liberia, Tanzania, Zambia, Guinea-Bissau, and Chad. Urban areas are well provided, and rural users are being targeted. Ethiopia has the lowest usage, with just one cell phone network under the prevailing state telecommunication monopoly.

Multinationals and the World Trade Organization

At the time of independence in the 1950s and 1960s, Western economists claimed that the new countries would grow economically by following the world's wealthier countries in moving from the primary sector into manufacturing and services. Few African countries achieved this progression.

Multinational mining companies and makers of coffee, tea, and chocolate products continue to buy African raw materials. For example, multinational aluminum manufacturers helped arrange funding for the Volta River project in Ghana that generated hydroelectricity to refine bauxite, the ore of aluminum, as cheaply as possible. However, reliance on producing raw materials and low levels of processing keeps local incomes low.

Many countries reliant on primary product exports are kept poor by low world prices and restrictive practices in the wealthier countries over farm products. Thus Nigerian and Ghanaian coffee producers receive around 50 cents per pound. Each cappuccino served in U.S. coffee chains takes about one ounce (4.5 cents) of coffee, yet customers pay over US$2. Most of the markup goes to the coffee traders, blenders, grinders, and retailers in the United States (and other wealthier countries). Purchasers of African raw materials often maintained low world prices by opening up new areas of production in other parts of the world as growing markets absorbed what established areas produced.

Soaring raw material prices in the 1970s (minerals) and in the early 1980s (beverages) were short-lived, but they often enabled the producer countries to take out loans for economic development projects. When the prices fell, these countries faced debts that they could not repay. In 2004, the World Trade

Organization championed the cause of African products to help them gain wider access to world markets.

Tourism

Tourism is the world's largest industry, with international tourism arrivals in 2012 exceeding 950 million persons. Tourism in Sub-Saharan Africa grew by 10 percent until mid-2012. Tourism provides many African countries, especially those in parts of Eastern and Southern Africa, with a major potential for earning foreign currency. In 2012, 7 percent of all employment in the region was in tourism. The International Council of Tourism Partners aims to triple the numbers of visitors to Africa by 2020.

The commitment of some governments to the conservation of designated game and national parks resulted in Sub-Saharan Africa having a higher proportion of such land uses than any other continent. Unfortunately, the governments have little money to spend on maintaining the parks and their wildlife, and they cannot finance realistic management policies. Other tourist attractions include the historic slave-trading centers and Robben Island in South Africa, where Nelson Mandela was a long-term inmate before his release in 1990.

An example of the growing significance of tourism in Africa is provided by the Great Limpopo Transfrontier Park that connects the Kruger National Park (South Africa) with the extended Limpopo National Park (Mozambique) and the Gonarezhou National Park (Zimbabwe). Formalized in 2002, the three-country initiative was taken forward in 2006 by the presidents of the three countries opening a border post. The combined park area of over 41,000 km^2 (16,000 mi.^{2}) has both ecological and economic goals: animals will be able to range widely and the already prosperous tourist industry will be extended. In particular, the park was a major draw to the hundreds of thousands of visitors in South Africa for the 2010 Soccer World Cup.

Tourism does have adverse impacts, since tourist revenues fluctuate with global security crises and changes in demand. It puts pressure on land, water, and power resources, diverting demand from often poorly provided residential and industrial consumers. However, Africa's unique cultural and natural environments, along with the rewards from well-planned tourist facilities, will continue to make this industry significant.

Local Emphasis in a Globalizing World

Globalization places an emphasis on export goods and foreign trade, but most people in the countries of Sub-Saharan Africa rely on the local economy. Consumer goods remain unusual or communal (Figure 8.25). Many rural Africans live their lives with little reference to the global economy, even though it brings the use of motor vehicles or clothes made of synthetic fibers into the remotest villages. Many small towns experience a growing mobility of people, increased levels of commercial exchange, and rising demands for the consumer goods advertised on global TV channels. Some villages mushroom into small service centers with rapidly built shops and market stalls where food and consumer goods are sold and buses bring people from surrounding rural areas.

In some countries, a trend toward interaction with the global economy has been reversed. For example in Zimbabwe, which was one of the main African growth countries in the early 1990s, President Robert Mugabe encouraged the takeover of large, white-owned commercial farms by war veterans. This changed the farming emphasis from tobacco and vegetable crops for export to the growing of corn for local consumption. This, in turn, cut off much of Zimbabwe's foreign exchange and reduced many black former farm workers to unemployed poverty.

Culture Shock

Human Rights and Women's Roles

Sub-Saharan Africa has a very poor record on human rights. Few people can earn what Westerners classify as a decent standard of living. Many Africans with formal employment are paid low wages and are exploited by their firms. Personal security is often threatened by arbitrary arrest and violence, especially in civil wars that create millions of refugees (Figure 8.26). Injustice is widespread, illustrated by the apartheid culture that affected South Africa for many years and is proving difficult to change. Discrimination by gender, race, ethnicity, and religion is rife, as examples earlier in the chapter show.

Women in Africa are expected to bring up children, draw water, collect wood, raise crops, and cook meals. Few avoid such a life, and hardly any attain high political office. Only 13 percent of African members of parliament are women. Before the colonial period, ethnic groups such as the Kongo people in Central Africa had matriarchal inheritance, but the colonial powers ended many female institutions and reduced women's rights.

Unequal access to education has led to a major deprivation of rights for many women, especially in the northern, Muslim parts of this region. In 2012, substantial efforts were being made to correct this situation. Female genital mutilation is common, with recent estimates that 100 over million women are affected, mainly in the northern countries such as Nigeria (25 percent of women) and Mali (90 percent). Complications from the cutting include bleeding that encourages HIV/AIDS, painful intercourse, and childbirth difficulties.

In contrast to the lot of most African women, a few women wield immense power. For example, wives of dictators often make their presence felt. The wives of Nigerian dictators in the 1980s and early 1990s built their own personal fortunes. In Rwanda, the wife of the dictator from 1994 is suspected of links to the groups who carried out the genocide. In Gabon and Zambia, estranged leaders' wives returned to embarrass their husbands as pop stars or in court cases. The wife of the leader of Liberia in 2003 claimed that she, not her husband, was in charge.

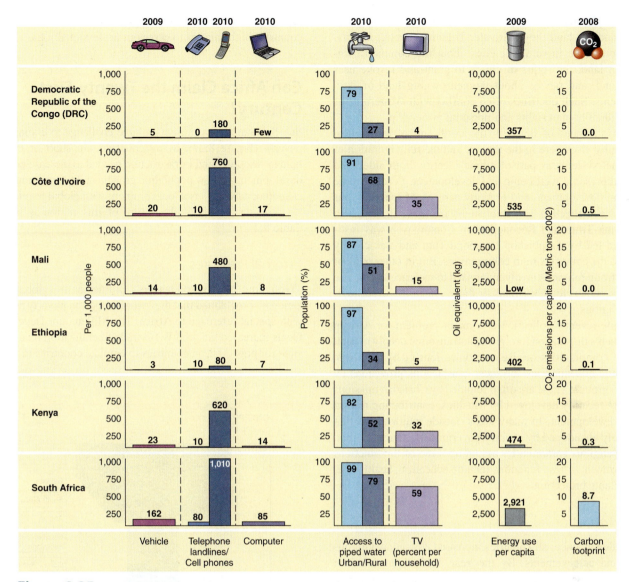

	2009	2010	2010	2010	2010	2010	2009	2008
	Vehicle	Telephone landlines	Cell phones	Computer	Access to piped water Urban/Rural	TV (percent per household)	Energy use per capita	Carbon footprint
	Per 1,000 people				Population (%)		Oil equivalent (kg)	CO₂ emissions per capita (Metric tons 2002)
Democratic Republic of the Congo (DRC)	5	0	180	Few	79 / 27	4	357	0.0
Côte d'Ivoire	20	10	760	17	91 / 68	35	535	0.5
Mali	14	10	480	8	87 / 51	15	Low	0.0
Ethiopia	3	10	80	7	97 / 34	5	402	0.1
Kenya	23	10	620	14	82 / 52	32	474	0.3
South Africa	162	80	1,010	85	99 / 79	59	2,921	8.7

Figure 8.25 Sub-Saharan Africa: indicators of well-being. The ownership of consumer goods is low in almost all countries, and access to piped water and energy varies greatly among the countries. Compare this diagram with those for other regions. *Source: Data from* World Development Indicators, *World Bank (2009);* Population Data Sheet, *Population Reference Bureau (2007).*

Figure 8.26 **Refugee camp in the Democratic Republic of the Congo (formerly known as Zaire), just across the border from Rwanda.** Civil war in Rwanda resulted in millions of people fleeing the country into this and other neighboring countries. *Photo: © Wesley Bocxe/Photo Researchers, Inc.*

Religion: A Controversial Role

Religion has always been important in Africa and continues to be so, often in new ways. Many people profess membership in a religious organization, with animism, Islam, and Christianity being the most widespread in the region.

For Africans, traditional religious doctrines are about an invisible world linked to the visible, as spiritual beings or forces communicate with them and influence their daily lives. People relate to a spirit world, linking them to other people and the land they cultivate. Traditional religious systems loom large in rural societies about the common ownership of land, the rights of chiefs to grant and revoke uses of the land, and taboos about women owning land or farm implements. Such doctrines often conflict with Western ideas of individual human rights and personal ownership.

The advances of Islam in the north and of evangelical Christianity elsewhere in the region, and the reinvigoration of traditional systems as part of ethnic identity, both aided and hindered personal well-being and development. Peace is basic to human development, and in many parts of Africa, religious groups work together to establish or maintain it. In South Africa, the Truth and Reconciliation Commission was instituted and led by Archbishop Desmond Tutu and was closely linked to the country's faith communities. But in other parts of Africa strong religious involvements led to violence as extreme Islamists persecuted Christians or those practicing tribal (animist) religions.

While revenue collection is a major problem for African governments that rely on foreign aid and are often deeply in debt, religious communities survive on monies donated by members. Health and education are the most conspicuous provisions by religious groups. These are often founded by Christian missions and more recently developed by Muslims, contributing toward human development. In many areas outside major towns that are largely abandoned by country governments, there is a trend toward religious communities assuming some of the functions of government. These functions include education, health care, and vigilante protection.

Better Education

After independence, all countries increased educational enrollments and achievements. By the year 2012, countries such as Botswana, Cameroon, Kenya, South Africa, Zambia, and Zimbabwe had nearly 100 percent enrollment in elementary schools—a substantial and significant increase from 50 percent in 1965. Burundi, Chad, and Mauritania made major strides by increasing primary education from under 20 percent in 1965 to nearly 80 percent in the present. However, Burkina Faso, Guinea, Mali, and Niger in the northern Muslim belt and Ethiopia still have only half of their children in elementary school, and female education lags behind male.

Increasing numbers of people in the region also have opportunities to become fully literate in secondary school and to earn higher academic qualifications. It is often disappointing to many who gain higher qualifications that there are few jobs available in their home countries. African doctors, lawyers, and airline pilots are increasing in numbers, but many enter the brain drain and find their employment abroad in the world's wealthier countries. Emigration from Africa to the United States more than doubled since the 1990s, disproportionately in the professional, managerial, and technical occupations, and continues to grow. Many Africans living in the wealthier countries may benefit their home countries by sending money to their families. Some of the skilled and experienced personnel return.

Meanwhile, African countries pay expatriates from wealthier countries high salaries to carry out professional jobs.

Can Africa Claim the Twenty-First Century?

Sub-Saharan Africa presents a huge challenge to ending global poverty (Figure 8.27). Many countries are short of the basic human resources and infrastructure needed to increase economic development, to slow population growth, and to encourage political democracy. This is a local problem with global implications.

The debate about the future of this region is posed in Table 8.4.

External Pressures

Throughout 2005, world leaders—politicians and global organizations—proclaimed that they would focus on poverty reduction with special reference to Africa. The Millennium Development Goals increased aid, and WTO trade liberalization made it possible for poor African countries to emulate countries in Asia.

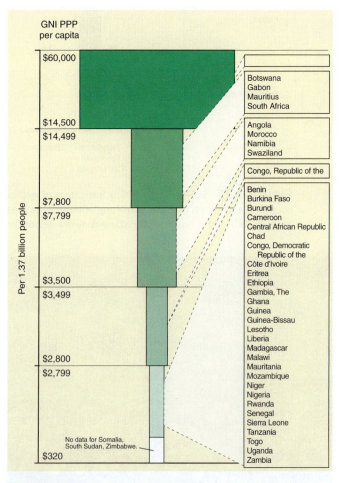

Figure 8.27 **Distribution of incomes in the Sub-Saharan Africa region.** *Source: Data from 2012 World Development Indicators, World Bank.*

Table 8.4	Debate: The future of Sub-Saharan Africa
Africans Can Do It	**No Hope for Africa's Future**
There are signs that Africans are moving toward a better future. The peaceful transition in South Africa, Uganda's success with AIDS, fewer wars, and increasing democracy are examples of good trends.	Such signs are few and temporary. Most things get worse. Civil strife springs up in new places, and most democracy is a façade. These are basic reasons today for crippling African development.
Younger "born frees" (since independence) are better educated and more inclined to expect good leadership from the current leaders rather than blaming the past.	The present problems stem from the colonizers and their racism. Whatever Africans do to put them right is futile. (This is a common complaint of older people born into colonial rule.)
African countries can produce goods in addition to the crops and minerals that others want to buy if trading terms with the wealthier countries are improved. Some countries are growing rapidly, some as a result of oil windfalls, and a few others such as Mozambique, Rwanda, and Uganda have seen a decade of economic growth after decades of strife and poverty.	Few leaders place much importance on sustained economic growth, and their actions tend to reduce the ability of governments to develop education, health, and employment prospects. Few African leaders allow widespread involvement in government that reduces their powers of patronage and their ability to make arbitrary decisions.
Smart businessfolk can do well, as in the oil companies on the west coast, the mining corporations, and those making cheap luxuries such as bottled drinks and soap powder. Cell phones increased from almost none 10 years ago to nearly 100 million today. The business climate could improve with more privatization in areas such as utilities (telephones, electricity) and with shorter periods of business registration.	Too many countries place difficulties in the way of foreign corporations wishing to do business in African countries, including the continuation of the bribe culture and lawlessness. The past has left too many examples of squandered opportunities.
The rise in the numbers of urban Africans is leading to wider political participation and, hopefully, to demands for better government. Better government is particularly important at this time, but it has to be widely wanted and supported. Power is based in country governments. But improved political involvement has seldom been linked to greater prosperity.	There is still too much rule and control by a few "big men" who turn the law and finances to their own ends. President Mugabe of Zimbabwe is a well-known example. His once fairly prosperous country is now among the poorest in the world. Too many governments are predatory, and few are competent. There are still few leaders who leave after electoral defeat, compared with those overthrown by war or coup.
Aid agencies are now putting more research into pre-funding activities. More philanthropic donors are needed, overseeing their giving in relation to criteria such as "saving the maximum number of lives at minimum cost" (Bill Gates).	In the past, too many aid agencies left behind more harm than good. Dependency on outside resources, and spending aid such as World Bank grants in profligate ways, do not lead to local entrepreneurial actions.
Land reform is occurring, although there is a need to ensure that people who can farm get title to farming land and government favorites do not take over productive land.	Most countries suffer from a lack of security in property rights. People with communal or tenant rights cannot use that land to underpin bank financing.
African countries are at last realizing that they have to deal with the HIV/AIDS problem, and there are encouraging signs of stabilization of the epidemic.	HIV/AIDS continues to result in devastating social problems, including a generation of orphaned and infected children. Even temporary palliatives are too expensive, and there is little infrastructure to monitor their use.
South Africa acts as an example of democracy, modernization, involvement of black talent in major corporations, free press, strong labor unions, independent judiciary, and large middle class.	South Africa does little to help other African countries improve their systems, partly because it is also struggling. Its recent dominance by a single political party (African National Congress) could result in greater corruption.
Africans are beginning to pay more attention to marrying local and global trends.	The change from traditional approaches, ways of doing things, and expectations to the trappings of modernization has been too great.

Source: Data from The Economist Newspaper Group, Inc. (2004).

Critiques of Africa's problems from the developed world highlighted four areas of action:

1. *Improved governance and conflict resolution* is the most basic need. To date, reductions in poverty have resulted as much from better domestic government in a few countries, as from external aid. Civil conflicts impose huge costs at home and in neighboring countries through deaths, maimings, property destruction, and refugee migrations.

2. *Investment in people.* The vicious circle of high fertility and mortality, low enrollments in education (especially of girls in some countries), high numbers of dependent children and youths, slow action against HIV/AIDS, and low family savings lie behind much of Africa's static or declining development.

3. *Economic diversification* makes countries more competitive in world markets. To date, Sub-Saharan Africa has been a loser in the global economy. The countries of this region produce only 1 percent of global GNI PPP. Most people have little access to the consumer goods that are the signs of material well-being, or lack the financial ability to develop entrepreneurial skills or engage with global connections. The region's industries need new products and better terms of trade, new incentives, and wider access to markets in wealthier countries. The perceived risks of investing and doing business in Africa make job creation slow.

Internal reforms needed include reducing corruption, improving infrastructure and financial services, and providing better access to the information economy.

Countries do not have sufficient all-weather roads or other forms of internal transportation. Ports are poorly equipped and expensive. People lack clean water supplies and adequate sanitation. There are shortages of electricity and telecommunications.

4. ***Reduced aid dependence, debt, and stronger intra-regional partnerships.*** Africa remains the world's most aid-dependent and indebted region. Programs of debt relief have become more significant since the 1990s. By the end of 2012, more than half of the region's countries had debt burdens eased by US$35 billion under the Heavily Indebted Poor Countries initiative that commenced in 1996. Aid donors still insisted on approved development policies to avoid corruption. The World Trade Organization tries to help African and other developing countries improve their access to markets for agricultural products and to reduce farm subsidies in wealthier countries, but the United States and the European Union make small concessions and increase their own farm subsidies to maintain their farming communities.

Internal Efforts: African Union and Regional Links

In July 2001, the heads of African governments meeting in Lusaka (Zambia) changed the name of the Organization of African Unity (OAU) to the **African Union (AU)**, with headquarters in Addis Ababa, Ethiopia. The purpose was to enable African countries to compete better in a tough global environment by creating strong African institutions—including an executive assembly, a fixed parliament, a central bank with a single currency, and a court. However, by 2012 only moderate progress had been made, and few countries had contributed their share of costs. The Central Bank of Africa was still perceived as "decades away." The main contribution of the AU is through its Peace and Security Council involvements in conflicts within countries. The **New Partnership for Africa's Development (NEPAD)** operates within the AU to foster integrated socioeconomic development in the region.

At the initial conference, President Thabo Mbeki of South Africa proposed the Millennium Action Plan with a twofold thrust. First, it restated the policies previously urged by Western countries and institutions: better government, more democracy, respect for human rights, market reforms, and recognition of the advantages of globalization. Second, the plan highlighted the need to reduce poverty by improving education and public health. It asked for continuing, more accountable aid together with the removal of trade barriers and agricultural subsidies in richer countries. However, subsequent actions suggest that the world's wealthier countries demand the first part but contribute little to the second. When the wealthiest (G8) countries met in 2002 and discussed African needs, they offered US$1 billion of the US$64 billion requested at a time when the United States increased its own farm subsidies by US$190 billion.

African countries also try to work with each other through regional trading groups along subregional lines. Unlike the EU or NAFTA, the African groups are loosely organized and often overlap. They commonly set out with enthusiasm but then become dormant or achieve little for want of political support from members, credibility among the wealthier countries, or difficulties in administrating their activities.

By 2012, soaring prices for oil and minerals gave Sub-Saharan Africa's encouraging annual economic growth a boost, and not just in the mineral-rich countries. Most country economies are better with lower inflation, tighter government accounting, and less violent politics. Economic blocs of Sub-Saharan Africa are shown in Figure 8.28. The improvements attracted foreign direct investment, which trebled since the low point of the 1990s. Yet the real income per person has scarcely risen—by only one-fourth from 1960 to 2012, whereas it multiplied by over 50 times in some Asian countries. Half the population continues to live on less than a dollar a day, compared to reductions in this criterion to 30 percent in South Asia and 17 percent in East Asia. Aid and investment in Sub-Saharan Africa have been uneven in the past, and the region, apart from South Africa, continues to rely on commodity exports, paying less attention to diversifying economies and developing human resources.

Amid the complexities of the challenges facing Africans today, it is clear that many of the materially poor are not poor in the broadest sense. Particularly in rural areas, there are isolated communities with clear leadership, community festivals, music, songs and stories, herbalist "doctors," practical education, and the satisfaction of basic needs. Conflicts cause more rural poverty than droughts. Africa in general faces major difficulties in the transition from an isolated and largely rural economy to one in a modern industrial and urban setting.

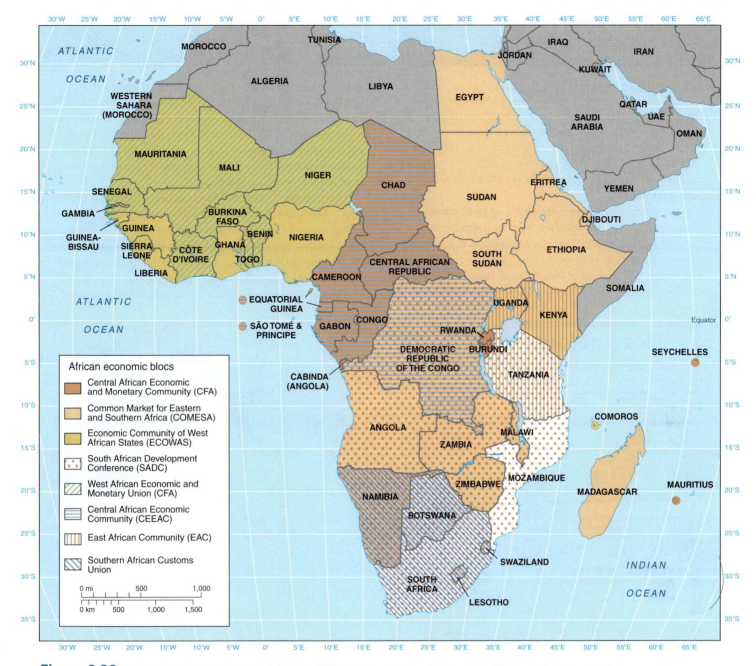

Figure 8.28 Sub-Saharan Africa: economic blocs in 2012. *Source: Data from The Economist Newspaper Group, Inc. (2012).*

Geography at Work

Mapmakers and GIS Analysts

Elio Spinello and Steve Lackow are Californians who are the authors and managers of AtlasGIS, a software package that combines mapping with a geographic information systems approach. In the 1990s, the huge GIS software company, ESRI, took over AtlasGIS, but it eventually placed marketing and development back with Elio and Steve's company, RPM Consulting.

As an example of their work, Elio completed a project that was an epidemiological assessment of river blindness (onchocerciasis) in Mozambique. He was commissioned by Aircare International, which provides small aircraft to help local groups in Mozambique. The company wanted to know the most important areas for delivering medical care. The disease does not kill, but it is chronic and widespread and affects the lives of many people. The African blackfly inhabits areas with fast-flowing streams, and the female carries the river blindness from an infected person to an uninfected one. Fibrous nodules form in the infected persons, producing microfilariae that attack skin pigmentation, causing skin atrophy and blindness. Some surgery is needed to remove the nodules, but new medicines make it possible to treat many more people without bad side effects.

Elio began by mapping the factors that encourage a concentration of African blackflies: the density of population, the concentration of rivers, and the occurrence of steep slopes that give faster river flow. He put together a composite index of these factors, giving greater weight to the population and waterway density than slope steepness (Figure 8.29). The results highlighted districts of high risk (yellow), making it possible for Aircare International to concentrate its delivery of medical support.

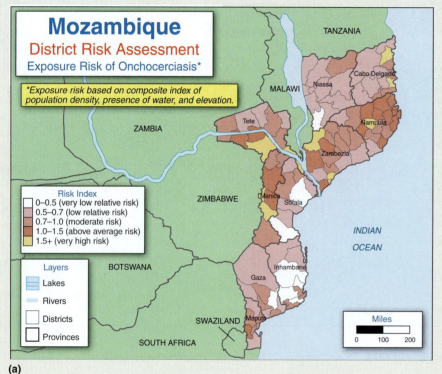

(a)

(b)

Figure 8.29 **Mapmakers and GIS.**
(a) Mapping the risk of river blindness in Mozambique. (b) Elio Spinello. *Source: (a) Data from Elio Spinello, RPM Consulting; Photo: (b) Courtesy Elio Spinello.*

Sydney, Australia. *Australia's largest city is situated on a picturesque natural harbor. Visible are Sydney's central business district, harbor-front parks, and the world-renowned Sydney Opera House. Photo: © Corbis RF.*

LEARNING OBJECTIVES

After reading this chapter, you should be able to:

- Understand the maritime situation of the countries of the region.

- Identify and describe the diverse climates and unique ecosystems of Australia, Oceania, and Antarctica.

- Contrast the environmental problems of the region's countries. Describe Antarctica's unique environmental issues.

- Describe the economies of Australia, New Zealand, and their changing global linkages.

- Contrast the diverse economies and political governance of the islands of Oceania, and explain the role of external aid for the material poverty in the subregion.

- Discuss Antarctica's unique status in the world, and explain the continent's economic potential.

Australia, Oceania, and Antarctica

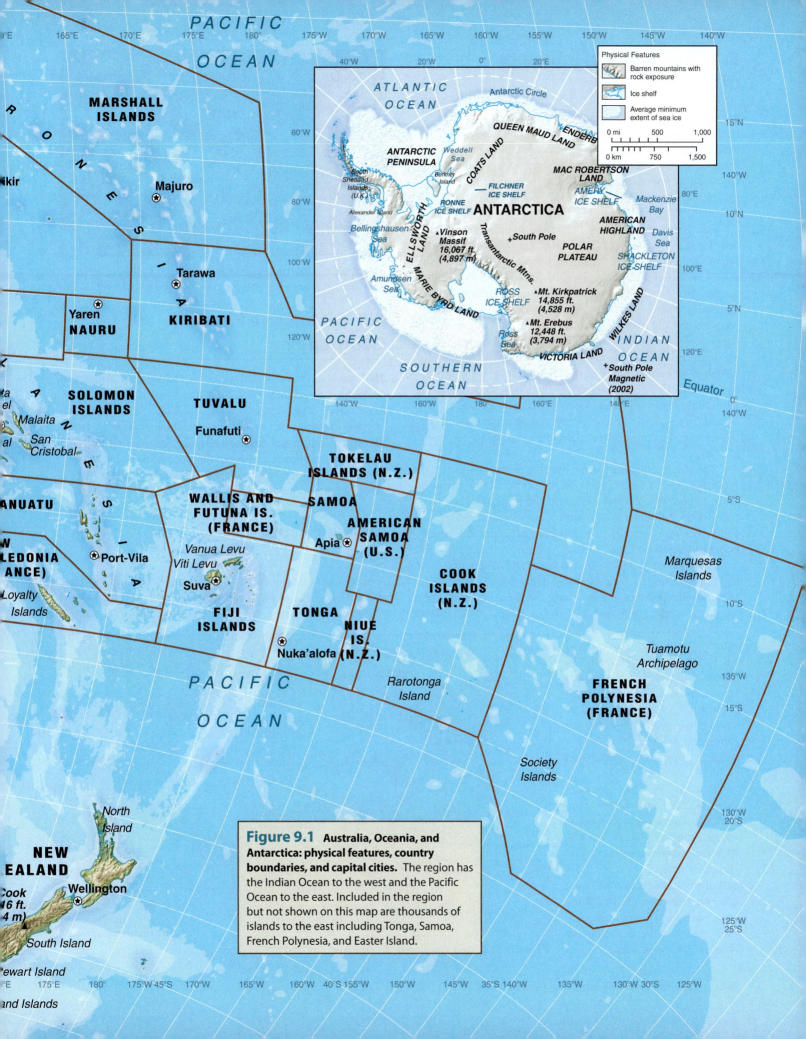

PACIFIC OCEAN

MARSHALL ISLANDS

Majuro ⊛

MICRONESIA

Tarawa ⊛

Yaren ⊛
NAURU

KIRIBATI

SOLOMON ISLANDS

Malaita
San Cristobal

VANUATU

Port-Vila ⊛

CALEDONIA
(FRANCE)

Loyalty
Islands

TUVALU

Funafuti ⊛

WALLIS AND
FUTUNA IS.
(FRANCE)

Vanua Levu
Viti Levu

Suva ⊛

FIJI
ISLANDS

TOKELAU
ISLANDS (N.Z.)

SAMOA

Apia ⊛

AMERICAN
SAMOA
(U.S.)

TONGA

NIUE
IS.
(N.Z.)

Nuka'alofa ⊛

COOK
ISLANDS
(N.Z.)

Rarotonga
Island

Marquesas
Islands

Tuamotu
Archipelago

FRENCH
POLYNESIA
(FRANCE)

Society
Islands

PACIFIC

OCEAN

NEW
ZEALAND

North
Island

Cook
6 ft.
4 m)

Wellington ⊛

South Island

tewart Island

and Islands

Physical Features
Barren mountains with rock exposure
Ice shelf
Average minimum extent of sea ice

0 mi   500   1,000
0 km   750   1,500

ATLANTIC
OCEAN

Antarctic Circle

QUEEN MAUD LAND
ENDERB

ANTARCTIC
PENINSULA

Weddell
Sea

COATS LAND

MAC ROBERTSON
LAND

South
Shetland
Islands
(U.K.)

Berkner
Island

FILCHNER
ICE SHELF

AMERY
ICE SHELF

Mackenzie
Bay

Alexander Island

RONNE
ICE SHELF

ANTARCTICA

AMERICAN
HIGHLAND

Davis
Sea

Bellingshausen
Sea

ELLSWORTH
LAND

▲ Vinson
Massif
16,067 ft.
(4,897 m)

Transantarctic Mtns.

✛ South Pole

POLAR
PLATEAU

SHACKLETON
ICE SHELF

Amundsen
Sea

MARIE BYRD LAND

ROSS
ICE SHELF

▲ Mt. Kirkpatrick
14,855 ft.
(4,528 m)

WILKES LAND

PACIFIC
OCEAN

Ross
Sea

▲ Mt. Erebus
12,448 ft.
(3,794 m)

INDIAN
OCEAN

SOUTHERN
OCEAN

VICTORIA LAND

✛ South Pole
Magnetic
(2002)

Equator

Figure 9.1 Australia, Oceania, and Antarctica: physical features, country boundaries, and capital cities. The region has the Indian Ocean to the west and the Pacific Ocean to the east. Included in the region but not shown on this map are thousands of islands to the east including Tonga, Samoa, French Polynesia, and Easter Island.

9.1 The Regional Influence of the Sea

Australia, Oceania, and Antarctica encompass a wide range of historical influences, religious traditions, linguistic diversity, and economic goals. The region is a geographic transition zone from the Indian Ocean on Australia's west coast to the Pacific Ocean on its eastern shores, and from the warmth of the tropical Pacific latitudes of Oceania to the harsh polar climates of Antarctica (Figure 9.1).

European colonial efforts reached Australia, New Zealand, and the Pacific islands of Oceania in the 1800s. The British established port and trade facilities on the coasts of Australia and New Zealand and attempted to reproduce their Western culture. However, British development in the region did not materially engage Australia and New Zealand's indigenous peoples, nor did Britain establish strong links between this region and their rapidly developing colonial enterprises in East and Southeast Asia. Nevertheless, the local impact was nearly complete Westernization (Figure 9.2). Culture, geographic situation, and established economic relations kept the postcolonial Australia and New Zealand closely linked with Europe and the United States. Their beaches, reefs, mountains, dynamic cities, and stability make them desirable destinations for world travelers.

The Pacific islands attracted little international attention or sympathy, apart from exploitative intrusions, from other parts of the region. Whalers and missionaries from the United States and Britain infiltrated many islands in the region during the 1800s, but no colonial power formally integrated the islands into the global economy.

Contemporary Australia, Oceania, and Antarctica is a region containing extreme diversity in material wealth, where countries like Australia are among the wealthiest and others are severely challenged by poverty. Many of the island countries of Oceania are almost entirely dependent on external aid. Population distribution in the region varies from a few crowded but relatively small island-based populations to the coastal orientation of Australia's low-density population to Antarctica, which lacks a permanent population (Table 9.1).

Situated at the southern extent of the global sphere, Antarctica comprises a very unique and geographically distinctive site. Although Antarctica has never had permanent settlements, many countries claim segments of it and fund scientific research on it. Their findings contribute to global research in such areas as ozone depletion and oceanic fishing resources.

Figure 9.2 **Australia's Western culture.** Australian children wave flags at an international cricket match. *Photo: © Corbis RF.*

9.2 Distinctive Physical Geography

The natural environments of Australia, Oceania, and Antarctica range from those influenced by expansive land mass areas to those influencing hundreds of small islands. Arid continental interiors contrast with humid coasts and harsh polar climates. The region's physical formations range from young volcanic islands and coral reefs to some of the world's oldest landscapes. The long-term geographic isolation of some lands in this region has given them a unique flora and fauna.

Geologic Activity

Australia the Ancient Continent

Australia was once joined to Africa, Antarctica, South America, and the peninsula of India in the continent of **Gondwanaland** (Figure 9.3), but it now lies on the eastern portion of the Indian

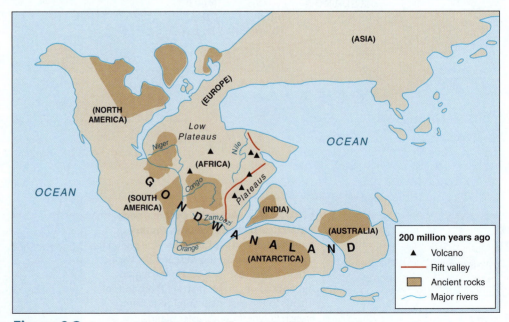
Figure 9.3 **Gondwanaland.** Earth's landforms 200 million years ago.

| | | Table 9.1 | Australia, Oceania, and Antarctica: Data by country, area, population, urbanization, income (gross national income purchasing power parity), ethnic groups | | | | |

Table 9.1 — **Australia, Oceania, and Antarctica:** Data by country, area, population, urbanization, income (gross national income purchasing power parity), ethnic groups

Country	Land Area (km²)	Population (millions) Mid-2012 Total	Population (millions) 2025 Est. Total	% Urban 2012	GNI PPP 2010 Total (US$ billions)	2010 Per Capita (US$)	Ethnic Groups (%)
AUSTRALIA AND NEW ZEALAND							
Australia, Commonwealth of	7,713,360	22.0	26.27	82	823.0	36,910	White 95%, Asian 4%, Aborigine 1%
New Zealand	270,990	4.4	5.1	86	121.3	28,100	White 73%, Maori 12%, Asian 5%, Pacific Islander 4%
Totals/Averages	**7,984,350**	**26.4**	**31.3**	**84**	**944.30**	**32,505**	
SOUTH PACIFIC ISLANDS							
Fiji, Republic of	18,270	0.8	0.9	51	—	—	Fijian 50%, South Asian 45%
Kiribati, Republic of	730	0.1	0.1	44	—	—	Micronesian with Tuvaluan minority
Marshall Islands, Republic of	179	0.1	0.1	68	—	—	Micronesian, German, Japanese, U.S., Filipino
Micronesia, Federated States of	702	0.1	0.1	22	—	—	Chuukese 41%, Pohnpeian 26%, other groups
Nauru, Republic of	20	0.01	0.01	100	—	—	Naurean 58%, other Pacific 26%, Chinese 8%
Palau, Republic of	458	0.02	0.02	77	—	—	Malay/Melanesian/Polynesian, Filipino, Chinese
Papua New Guinea, Independent State of	462,840	7.0	9.1	13	16.6	2,420	Melanesian 98%
Solomon Islands	28,900	0.6	0.8	20	—	—	Melanesian 93%, Polynesian 4%
Tonga, Kingdom of	750	0.1	0.1	23	—	—	Tongan 98%
Tuvalu	30	0.01	0.01	47	—	—	Polynesian 96%
Vanuatu, Republic of	12,190	0.3	0.4	24	—	—	ni-Vanuatu (Melanesian) 94%, white 4%
Western Samoa, Independent State of	2,840	0.2	0.2	21	—	—	Samoan (Polynesian) 93%, Euronesian mixtures and others 7%
Totals/Averages	**527,909**	**9.3**	**11.8**	**42.5**	**16.6**	**2,420**	
Region Totals/ Averages	**8,512,259**	**35.7**	**43.1**	**63**	**960.9**	**17,462**	

Source: *World Population Data Sheet 2012*, Population Reference Bureau; Microsoft Encarta 2005.

plate at a distance from the region's major plate margins. The relief features of the western half of the Australian continent are low plateaus and plains on the ancient shield rocks. Mountain ranges formed over 600 million years ago were worn down by erosion to form landscapes of little relief today.

The **Great Dividing Range** is Australia's one significant mountain chain running just inland along the continent's east coast. This mountain range is formed of rocks that were deposited from the erosion of the ancient continent and then uplifted and broken into blocks by faulting. Apart from the steep edges of these eastern uplands, Australia has fewer major relief contrasts than other large landmass areas. The world's largest continuous chain of coral forms the **Great Barrier Reef** off the northeastern coast of Australia.

Plate Movements, Mountain Ranges, and Volcanic Activity

The boundary between the Indian and Eurasian plates extends horizontally across the western portion of the region, staying north of the Australian continent. The Indian and Pacific plates form an active collision zone through Papua New Guinea and islands to the east. A **transform plate margin**, where plates are sliding horizontally along one another, marks the eastern boundary of the Indian plate and then angles southwestward through New Zealand.

New Zealand is the product of plate margin activity. Its two main islands are part of a fragment of the southern continent that broke away from the main mass of Gondwanaland around

100 million years ago (along with Australia, South America, and Antarctica). The uplift of the land keeps pace with the glaciers and rivers that wear it down.

Many volcanic islands also occur farther out in the Pacific Ocean, away from plate margins. These islands form above areas of crustal heating and may have a volcanic core surrounded by fringing coral reefs, a volcanic island with a wide lagoon between it and the coral reef or **barrier reef** (Figure 9.4), or a central lagoon surrounded by coral reef but no volcanic island (atoll). It is thought that such coral reef formations represent a sequence that begins with active eruptions building a volcanic island and ends with the island sinking beneath the waves under its own weight after the volcanic eruptions cease. Coral reefs colonize the islands when tiny animals secrete limy structures (limestone) at or just below sea level. The coral animals live near the tropical ocean surface, where it is warm enough and where algae on which they feed have access to light and oxygen.

Antarctica: The Land of Ice and Snow

The divergent plate margin between the Indian and Antarctic plates separates Australia and Antarctica. The Antarctic continent formed the ancient rock core of Gondwanaland. After the combined continent broke apart, Antarctica remained at the South Pole. The **Transantarctic Mountains**, which divide the continent into West Antarctica and East Antarctica, are one of the world's longest continuous mountain chains and are an extension of South America's Andes Mountains. The rocks of the **Eastern Antarctic Shield** are over 4 billion years old, placing them among the world's oldest known rocks. East Antarctica primarily consists of high plateau. The majority of Antarctica's terrestrial extent lies in East Antarctica, while the Antarctic Peninsula, numerous islands, and the ice-covered Weddell and Ross Seas lie to the west of the mountains in West Antarctica.

Figure 9.4 **South Pacific: coral island.** A typical South Pacific island with a hilly volcanic core, an outer coral reef, and a shallow lagoon between the reef and mainland. The village is situated where a channel leads from a breach in the reef front. *Photo: © Patrick Ward/Corbis.*

Extreme Climate Variations

Australia's climates are dominated by its arid continental interior (Figure 9.5). One-third of Australia is arid, and another one-third is semiarid. Water shortages are permanent characteristics of much of the continent. The aridity of the interior is occasionally tempered by storms and flash floods, which may temporarily replenish stores of fresh water.

The Western Pacific is subject to frequent **typhoons** (hurricanes). The tropical Pacific has high water temperatures throughout the year, supplying moisture and heat to the air above, which fuel intense tropical storms and typhoons. Tropical disturbances and typhoon generation occurs most frequently between 10 and 25 degrees of latitude north or south of the equator (although more frequently north of the equator).

The coastal areas of Australia tend to have more regular rainfall. Winter rains of the Mediterranean climatic environment, brought by the **midlatitude cyclones** of the southern oceans, are characteristic of the southwestern corner of the continent and the Adelaide area. **Monsoon** or seasonal summer rains and tropical cyclones occur along Australia's tropical northern coasts.

Australia is near the western end of the oceanic and atmospheric circulations that cause the El Niño fluctuations off western Latin America. When Peru is dry, the western Pacific has plentiful rains and vice versa. During the major 1997–1998 El Niño, an area stretching from Australia and Oceania suffered intense drought and forest fires, while Peru had unusual deluges of rain.

New Zealand's mild and humid midlatitude climate is similar to the British climate that many of its settlers left behind. Frequent storms bring rain to the islands and snow to the higher-elevation mountains. Temperatures vary with latitude and altitude. North Island is warmer on average than South Island. The mountains on New Zealand's South Island create a rain shadow to their east, necessitating the irrigation of some of the farmland.

Tropical Ocean Climates

The islands of Oceania are nearly all situated in the tropical latitudes. Poleward of 10 degrees north and south, the trade winds blowing from the northeast (Northern Hemisphere) or southeast (Southern Hemisphere) are nearly constant factors. The windward east-facing sides of mountainous islands experience **orographic** enhanced precipitation. Low-lying **coral atolls** (fringing reef) lack sufficient hilly relief to cause uplift and support precipitation production. Many are arid with little annual rainfall. Temperatures in Oceania are consistently warm throughout the year. Oceania's temperature regimes are only slightly modified by winds, storms, and elevation.

Antarctica's Harsh Polar Climate

Residents of many Southern Hemispheric countries assert that the Antarctic landmass is surrounded by the **Southern Ocean**, fourth largest of the Earth's five oceans. Map makers based in the Northern Hemisphere and geographers rarely acknowledge the Southern Ocean and simply extend the Atlantic, Pacific, and

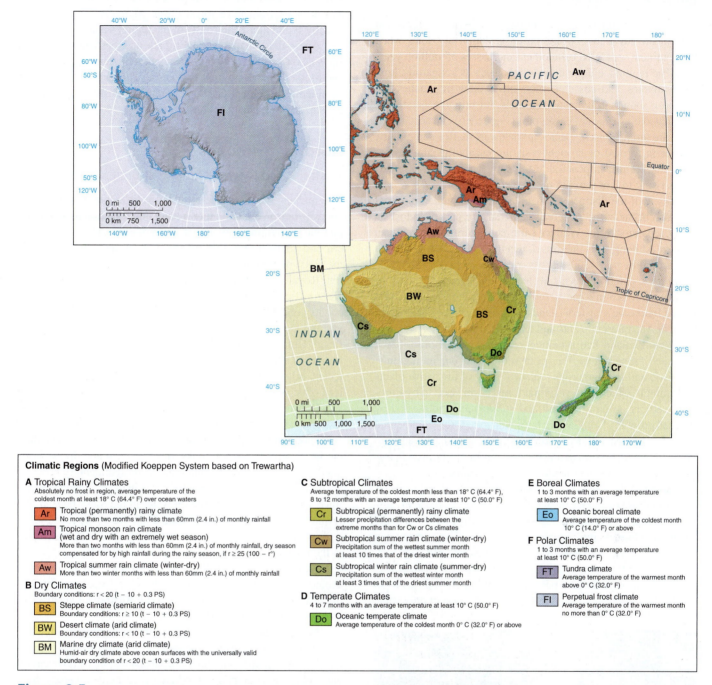

Figure 9.5 Climate patterns for Australia, New Zealand, the islands of Oceania, and Antarctica.

Climatic Regions (Modified Koeppen System based on Trewartha)

A Tropical Rainy Climates
Absolutely no frost in region, average temperature of the coldest month at least 18° C (64.4° F) over ocean waters

Ar Tropical (permanently) rainy climate
No more than two months with less than 60mm (2.4 in.) of monthly rainfall

Am Tropical monsoon rain climate
(wet and dry with an extremely wet season)
More than two months with less than 60mm (2.4 in.) of monthly rainfall, dry season compensated for by high rainfall during the rainy season, if r ≥ 25 (100 − r°)

Aw Tropical summer rain climate (winter-dry)
More than two winter months with less than 60mm (2.4 in.) of monthly rainfall

B Dry Climates
Boundary conditions: r < 20 (t − 10 + 0.3 PS)

BS Steppe climate (semiarid climate)
Boundary conditions: r ≥ 10 (t − 10 + 0.3 PS)

BW Desert climate (arid climate)
Boundary conditions: r < 10 (t − 10 + 0.3 PS)

BM Marine dry climate (arid climate)
Humid-air dry climate above ocean surfaces with the universally valid boundary condition of r < 20 (t − 10 + 0.3 PS)

C Subtropical Climates
Average temperature of the coldest month less than 18° C (64.4° F), 8 to 12 months with an average temperature at least 10° C (50.0° F)

Cr Subtropical (permanently) rainy climate
Lesser precipitation differences between the extreme months than for Cw or Cs climates

Cw Subtropical summer rain climate (winter-dry)
Precipitation sum of the wettest summer month at least 10 times that of the driest winter month

Cs Subtropical winter rain climate (summer-dry)
Precipitation sum of the wettest winter month at least 3 times that of the driest summer month

D Temperate Climates
4 to 7 months with an average temperature at least 10° C (50.0° F)

Do Oceanic temperate climate
Average temperature of the coldest month 0° C (32.0° F) or above

E Boreal Climates
1 to 3 months with an average temperature at least 10° C (50.0° F)

Eo Oceanic boreal climate
Average temperature of the coldest month 10° C (14.0° F) or above

F Polar Climates
1 to 3 months with an average temperature at least 10° C (50.0° F)

FT Tundra climate
Average temperature of the warmest month above 0° C (32.0° F)

FI Perpetual frost climate
Average temperature of the warmest month no more than 0° C (32.0° F)

Indian Oceans southward to the coasts of Antarctica. In either case, the continent is surrounded by water. While the majority of Antarctica's landmass is situated within the Antarctic Circle (66° 33′ south latitude), the **Antarctic Peninsula** and several surrounding islands extend north of the circle to near 60°N latitude (some 600 miles south off the Southern Cone of South America). The majority of the continental landmass and the surrounding islands are covered by ice. The climate of Antarctica's high eastern plateau is frigid, with average annual temperatures below freezing. Coastal areas of Antarctica are affected by lati-

tude, with slightly warmer temperature regimes existing in the lower latitudes than in the high-latitude, high-altitude continental interior. The mildest climate exists along the Antarctic Peninsula's west coast and in the surrounding islands where summer high temperatures in January average above 32°F.

The high plateau of Eastern Antarctica is extremely arid with average annual precipitation of around 2 inches of rainfall. Higher amounts of precipitation occur along the coasts, where coastal low pressure systems bring heavier snows and wind along with an occasional summer rainfall.

Distinctive Ecosystems

Australia's mammals were at an early stage of evolution that did not include a long womb-based gestation of babies when the continent separated from other landmasses 35 million years ago. The region's unique **marsupials**, such as kangaroos, koalas, wallabies, and possums, raise their young in pouches and compose about half the native animals. Australian vegetation is dominated by species of eucalyptus (Figure 9.6) and acacia, and there are unique desert species in the dwarf mallee community. **Mallee** is formed of drought-resisting eucalyptus shrubs that grow into almost impenetrable thickets of many close-spaced stems rising to 8 or 9 m (25 to 30 ft.) high.

Some of the distinctive Australasian plant and animal species also occur in the eastern islands that are part of Indonesia today, separated by the **Wallace Line** (see Figure 9.1) from Asian species. Although this line was drawn by botanist Alfred Russell Wallace in the mid-1800s, it was over 100 years before the origin of the separation was made clear. The line marks the edge of plate tectonic action some 10 million to 15 million years ago that forced Indonesia's eastern islands against its western islands and brought Australasian plants and animals with them.

New Zealand had a unique, temperate, rain forest–based flora and fauna before European settlement, but immigrants' domestic animals and crops replaced many of the native plants and animals. Much of the original forest cover was cut by European settlers on North Island and subsequently reforested by quicker-growing Douglas fir trees and pines for commercial uses. Introduced reindeer damaged trees and shrubs and are now contained. South Island contains stands of old-growth forest, much of which is protected by the government.

Some of the larger Pacific islands are forested with species closer to those of Indonesia and East Asia than to those of Australia. Palms are particularly numerous. Few islands have many animals, although bird species are diverse. The surrounding waters contain a wealth of tropical fish varieties, but each is in relatively small numbers, and they are too easily overfished. Vegetation on many of the South Pacific islands of Oceania grows in relatively nutrient-poor sandy soils that may support only some

Figure 9.6 **Australia: eucalyptus forest in humid Victoria state.** *Photo: © Corbis RF.*

scrub vegetation and coconut palms. Islands with heavier rains and higher terrain have forest cover. Mangrove areas are frequently found along the coasts of the region's islands.

Antarctica's harsh polar climate and permanent ice and snow cover support very little vegetation beyond algae, lichens, and mosses. There are only two known flowering plants on the continent; these exist on the Antarctic Peninsula and some of the surrounding islands.

Natural Resources

Australia and New Zealand Resources

The ancient rocks of Western Australia contain many large deposits of iron ore and other metallic ores such as nickel, gold, platinum, uranium, and copper. Natural gas fields exist in eastern and southeastern parts of the country and offshore along the coast of Western Australia. A joint project between Chevron, Royal Dutch Shell, and Exxon Mobil to exploit gas in the Gorgon field along the coast of Western Australia makes the country one of the world's largest exporters of liquefied natural gas and is an important part of Australia's trade relations with Japan, China, and other Pacific Rim countries. The rocks that form the Great Dividing Range in eastern Australia are of more recent origin and contain significant deposits of coal, silver, lead, zinc, and copper ores.

Between the shield rocks and the Great Dividing Range, the lowlands are drained by the Murray-Darling River system in the south and form the Great Artesian Basin in the north. Rain falling on the eastern mountains soaks into the rocks and drains westward and downward, accumulating in the sedimentary rocks of the basin (Figure 9.7). Because the rocks in the mountains remain filled with water, the pressure on water in the rocks beneath the lowlands is so great that wells drilled into the rocks cause water to flow out onto the surface without pumping. The northern part of the lowlands was named for these **artesian wells**.

Australia is the world's largest producer of opals, contributing more than 90 percent of the global opal supply. The majority of the opals are mined in South Australia, Queensland, and northern areas of New South Wales.

New Zealand relief and rainfall combine to make fresh water and hydroelectric power generation a significant local natural resource. Some minerals and more substantial deposits of natural gas add to the country's limited natural resource diversity. Sustainable forestry is a growing part of the New Zealand resource economy. Although the country is a relatively small contributor of forest products on a global scale, it supplies nearly 20 percent of the value of forest products in Asia-Pacific markets, including significant exports to Australia, China, and Japan.

Pacific Islands

Some of the Pacific islands along the plate collision zone have mineral resources. The copper deposits on Bougainville, an offshore island that is part of Papua New Guinea, contain one of the world's largest copper reserves, which had been intensively mined and dominated the Papua New Guinea economy

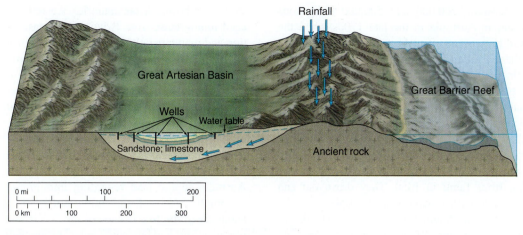

Figure 9.7 **Australia: Great Artesian Basin and the Great Barrier Reef.** The artesian basin of northeastern Australia is created by the geologic conditions that produce a major groundwater source as rain falls on the Great Dividing Range, seeps into the rocks, and flows downward to replenish the water in the rocks beneath the basin. The water in the deepest rocks flows out under pressure from the water moving down from the hills. The Great Barrier Reef is one of the world's largest developments of coral reef and forms a feature just off the Queensland coast.

until terrorist activities halted production. Efforts are currently underway between the Australian-based company that owns the mines and the government of Papua New Guinea to reopen the mines and make them operable by 2015. New Caledonia is the world's third-largest producer of nickel ore. Most of the larger islands have a covering of rain forest, but some of the drier and flatter islands have only sparse vegetation.

Antarctica

The Transantarctic Mountains and other areas of the continent are rich in mineral resources. It is believed that significant deposits of coal, ores, copper, nickel, silver, and tin are present. Scientists also assert the existence of significant deposits of oil and natural gas, especially along Antarctica's continental shelf. The harsh climate and ice cover combined with political treaties restricting resource exploitation will keep most of Antarctica's resources in the ground for the next several decades.

Natural Hazards

Threats from typhoons, floods, droughts, earthquakes, and volcanic eruptions affect various parts of the region. The 1995 eruption of Mount Ruapehu on the North Island of New Zealand led to the temporary closure of airports and highways. A 6.3 magnitude earthquake occurred 6 miles from Christchurch, New Zealand's second-largest city, in February 2011. The quake killed more than 170 people and injured more than 1,000. It is considered the second-deadliest natural disaster on record in New Zealand.

Ice melt from anthropogenic climate change on Antarctica could open the continent for settlement and resource exploration, but the corresponding sea level rise may inundate low-lying coastal areas around the world. Many low-lying inland countries in Oceania literally could be submerged. The highest parts of some of the coral atoll islands are only a few meters above that level, and many

have a concentration of settlement on low-lying coasts. Australia's Great Barrier Reef faces environmental and corresponding tourism economic disaster as warming waters and rising sea levels continue. Much of the region's coral, which serves a coastal protective function and acts as a major tourist attraction, is under threat from rising sea levels, predators, and pollution.

New Zealand, which is known for strong environmental awareness and activism among its citizenry and its tough environmental legislation, still faces issues of soil erosion and air and water pollution. Farming practices, industrial effluent, automobile emissions, and controversy over the construction of new dams for hydroelectricity generation remain environmental challenges for New Zealanders in the early 2000s.

The common perception of the South Pacific islands as a pleasant and untouched part of the world is spoiled by the dumping of oil and other materials from commercial ocean vessels. In addition, several of the islands are uninhabitable after being mined for copper, nickel, or phosphate, or after nuclear testing. Nauru was left as an uncultivable skeleton following the extraction of its phosphate deposits.

Waste from the resident scientific community and increasing numbers of tourist visitors to Antarctica coupled with fuel discharge and occasional leaks from commercial cruise ships are currently developing environmental issues for the polar continent.

9.3 Distinctive Human Geography

Indigenous Peoples

The indigenous people of Australia, New Zealand, and the Pacific Ocean islands include ethnic and culture groups whose origins are still debated. The populations of the islands are distributed over vast ocean distances and contain mixtures of Melanesian, Polynesian, Micronesian, and mainland Asian heritage.

There were between 200,000 and 500,000 indigenous **Aborigines** present in Australia in the late 1700s when the Europeans arrived (Figure 9.8). The Aborigines were nomadic hunters and gatherers living in communities or clans spread across the continent and speaking 200 different languages. Rock paintings, based on their **animistic** (nature worship) religious beliefs and social organization, are part of their legacy (see "Contemporary Geographic Issues: Aborigines and the "New" Australians," p. 286).

The **Maoris** of New Zealand came from the wider South Pacific around the AD 800s, with some of the final waves of people migrating from Tahiti in 1350. They drove out and replaced the Moriori, an earlier, dark-skinned people.

The inhabitants of the South Pacific oceanic islands are distributed over a vast expanse of ocean territory. They are grouped in three broad geographic categories—the **Melanesian** ("black islands," so named by Westerners because of the presence of dark-skinned people), **Micronesian** ("small islands"), and **Polynesian** ("many islands") people.

Geographic Identity

Claims were made in the early 1900s that "empty" Australia could accommodate over 200 million people. British geographer Griffith Taylor asserted in the 1920s that Australia would be significantly challenged in sustaining more than 20 million inhabitants. After more than 80 years of immigration, natural increase, and economic development, Australia's current population is only slightly more than Taylor's 20 million figure.

Most Australian immigrants came from the British Isles until the mid-1900s. In the early days of the Dominion of Australia, an informal **white Australia policy** encouraged the acceptance of European immigrants and discouraged immigration

Figure 9.8 Australia: Aborigines. Aborigines after shopping at a store in Nangalala, Northern Territory. Tensions arise between traditional ways and Western lifestyles. *Photo: © Penny Tweedie/Corbis.*

from neighboring Asian countries. Australia especially encouraged immigration from Britain to increase its overall population and maintain cultural and economic links with the United Kingdom. As that source became insufficient, more immigrants came from other parts of Europe. Australia ended its "whites only" immigration restrictions in 1972. Immigration trends in Australia in the early 2000s include significant migration streams from New Zealand (which became the single largest contributor of immigrants in the early 2000s) and numerous Southeast and East Asian countries. Australian voters and business leaders are occasionally at odds over which course Australia should take regarding immigration. The situation is complicated by illegal immigration, which increased substantially in the first decade of the 2000s. Many pro-business organizations and business leaders promote open immigration policies in conjunction with increasing trade links with East and Southeast Asian countries and in their desire to secure adequate labor for growth industries. Many Australian workers express concerns that immigrant communities will take jobs away from Australian-born citizens. The Australian government's approach toward immigration continues to evolve. The country instituted a "point" system for foreigners requesting residency visas and since 2007, the government decreased the overall number of permanent residency visas while increasing visas for skilled workers qualifying with the most "points" under Australia's system. Points are granted based on age, English speaking proficiency, education, and prior work experience (among other criteria). A positive cultural trend that may be the result of the increased diversity in Australia's immigrant sources is that the increasingly multiracial, multiethnic, and multicultural population of the country seems to be becoming more tolerant of the rights of Aborigines and cultural diversity in general, as evidenced by recent changes in related legislation.

Although high unemployment in the 1990s led to a halving of immigrants allowed into Australia, Australian businesspeople argue for greater numbers in the light of less-than-replacement levels of births and predictions of falling population totals and skills levels. Australia also aims to give 12,000 refugees a year the right to permanent residence, but in 2001, its insistence on preentry interviews offshore earned criticism from human rights groups such as Amnesty International as the Australian navy intercepted boat people and transferred them to the isolation of detention camps on Nauru.

New Zealand's cultural landscape continues to be more British in origin and allegiance than that of Australia. Although many young, skilled New Zealanders migrate to Australia in search of economic opportunities, others remain because of economic opportunities in various growing service industries, pride in the country's culture, and attraction to its unique physical landscape. New Zealand's population growth is sustained by natural increase and immigration from many of the island countries of Oceania. The indigenous people, the Maoris, take a more significant part in New Zealand life than the Aborigines do in Australia. They total around a half million people, most living in North Island. Auckland has over 100,000 Maoris, many of whom have professional jobs.

Population Dynamics and Natural Increase

Dramatic contrasts in the regional population distribution are evident in Figure 9.9. People in Australia are primarily concentrated in urban centers situated in coastal areas around the country's periphery. Australia's vast interior space has some of the world's lowest population densities. New Zealand's North Island has higher population concentrations and a higher percentage of people than South Island. Only the largest Pacific islands show up on the map at this scale, but densities are often high on small land areas.

The South Pacific islands have small total populations, with the largest numbers on those islands nearest to Australia

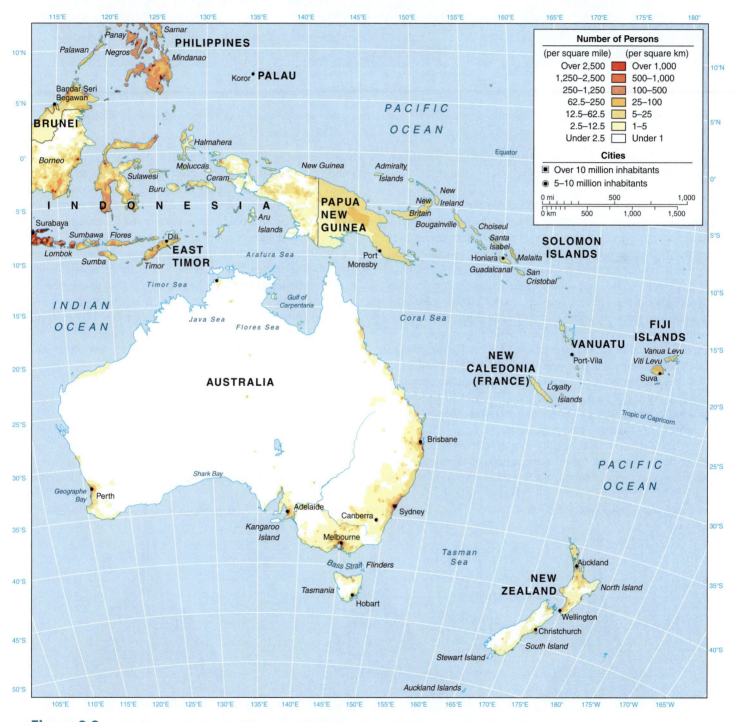

Figure 9.9 **Australia, New Zealand, and Oceania: distribution of population.** Australia's population is almost entirely distributed near the coasts. New Zealand's North Island has higher population densities than does its South Island. Papua New Guinea has high densities of people throughout much of its central and eastern territory. Not shown are numerous South Pacific islands north and east of the map's coverage and Antarctica, which has no permanent resident population.

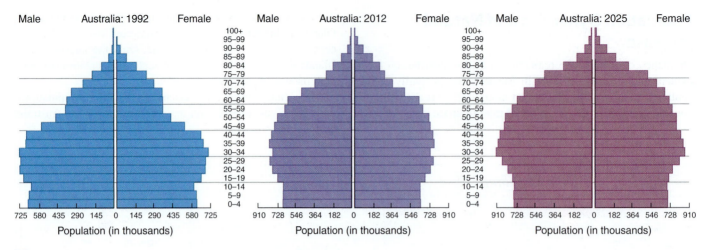

Figure 9.10 **Australia: age-sex diagram.** *Source: U.S. Census International Database: http://www.census.gov/population/international/data/idb /region.php.*

and New Zealand. In 2012, the islands had a total population of 9.3 million people, 7 million of whom lived in Papua New Guinea and 800,000 on Fiji.

Antarctica has no permanent resident population. The inhabitants of the subregion are primarily scientists studying Antarctica's geography, geology, and climate, and government representatives present to assert a political connection between several external sovereign countries and the region's territory and resources.

Australia's 2012 population of 22 million was more than double that of its 1950 population. As with other materially wealthy countries, Australia's rate of natural increase is slow, with total fertility rates of less than 1.9. Immigration results in annual population growth rates of over 1 percent, compared to less than 0.5 percent in European countries. Both Australia and New Zealand are well advanced in the demographic transition process. Australia's age-sex diagram (Figure 9.10) reflects a slowing in births after a baby-boom period from the mid-1950s to around 1970 and an aging population.

New Zealand's total fertility rate of 2.2 remains slightly above that of Australia, but its immigrants keep the annual population rate of increase at just over 1 percent. The population of

New Zealand in 2012 was 4.4 million. After gaining dominion status, New Zealand had to survive some years of labor force losses from its loyal support of Britain in two world wars. Today, many younger New Zealanders in particular migrate to Australia for professional jobs in various service industries.

Population growth continues on many of Oceania's South Pacific islands due to high total fertility rates. The age-sex diagram for Papua New Guinea resembles those for poorer countries elsewhere and implies a large proportion of young people in the total population (Figure 9.11). A combination of poverty, the consumption of high-fat-content foods in part from imported convenience foods, and genetic tendencies toward obesity contribute to health issues and relatively low life expectancies for many island residents.

Rapid Urbanization

Intraregional and interregional trade fueled the growth of numerous city-regions in Australia and, to a lesser extent, in New Zealand. These city-regions are centers of international business and are the global or regional headquarters for numerous multinational

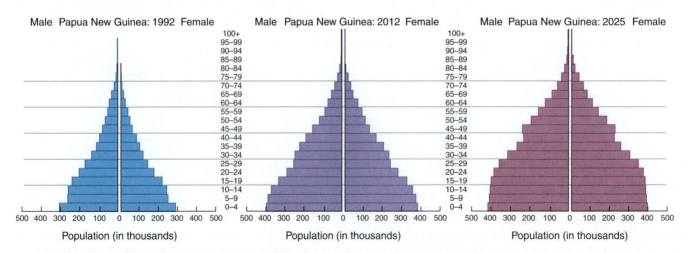

Figure 9.11 **Papua New Guinea: age-sex diagram.** *Source: U.S. Census International Database: http://www.census.gov/population/international /data/idb/region.php.*

corporations. Among global city-regions, Sydney (Australia) is the region's most connected city, while numerous other cities are increasing their connectivity.

Australia is one of the world's most urban countries, with 82 percent of its people living in urban environments in 2012 and almost two-thirds of the total population living in Australia's five largest cities (Table 9.2). All of Australia's largest cities are situated along the country's coastlines. The coastal distribution of urban centers reflects a reaction to the less habitable arid interior of the continent and the related historic development of ports within the separate colonies and subsequent states of the Australian Federation. Most of Australia's major cities retained their established port functions in addition to becoming established centers of finance, culture, and tourism. Melbourne grew further in importance by specializing

in container trade. Australia's cities have downtowns that combine state government and financial sectors (Figure 9.12) with major cultural facilities such as the Opera House in Sydney.

In 2012, 86 percent of New Zealanders lived in towns and cities. New Zealand's main cities on both islands began as ports that acted as "hinges," giving access to the interior and back to the main overseas markets of the colonial homeland. The contemporary distribution of New Zealand's cities along the coasts of its two main islands reflect historic settlements around port facilities and the challenges to inland habitation resulting from the rugged interior terrain, especially on South Island. Processing industries concentrated in the seaports became market, commercial, financial, and government centers. Auckland, with 1.4 million people in 2010, is the center of an urbanized area that stretches along the peninsula to its north (Figure 9.13). Wellington, with one-third of Auckland's population, is a major port and the center of government. Christchurch and Dunedin are the main towns of South Island. New Zealand towns reflect histories similar to those in Australia, including the most recent growth of manufacturing and service industries. All have port facilities.

Table 9.2	Populations of Major Urban Centers in Australia and Oceania	
TOP 10 MOST POPULATED URBAN CENTERS		
City, Country	**2012 (millions)**	**2025 (millions)**
1. Sydney, Australia	4.5	4.8
2. Melbourne, Australia	3.9	4.3
3. Brisbane, Australia	1.9	2.2
4. Perth, Australia	1.5	1.8
5. Adelaide, Australia	1.2	1.4
6. Auckland, New Zealand	1.2	1.5
7. Newcastle, Australia	0.5	0.6
8. Canberra, Australia	0.4	0.5
9. Wellington, New Zealand	0.4	0.5
10. Christchurch, New Zealand	0.4	0.4
NUMBER OF MAJOR URBAN CENTERS IN AUSTRALIA AND OCEANIA, BY POPULATION CATEGORY		
Population Category	**Number of Urban Centers**	
10.00 million and greater	0	
5.00 – 9.99 million	0	
2.50 – 4.99 million	2	
1.00 – 2.49 million	4	

Figure 9.12 **Sydney, Australia.** Australia's largest city is situated on a picturesque natural harbor. Visible are Sydney's central business district, harbor-front parks, and the world-renowned Sydney Opera House. *Photo: © Corbis RF.*

Figure 9.13 **New Zealand: urban landscape.** Auckland harbor, North Island, is the main commercial center of New Zealand. Its waterfront has extensive yacht docking facilities and oil terminals close to the high-rise city center offices. *Photo: © Corbis RF.*

Oceania's Small Towns

Few of the South Pacific islands are large enough to have major cities. Most island towns began as colonial ports and continued as political capitals in the postcolonial independent countries (Figure 9.14). Most towns in Oceania's South Pacific islands have small populations of a few thousand people. South Pacific towns contain a mixture of some basic economic facilities dating from the colonial era and limited postcolonial additions to the infrastructure. Town landscapes include some processing industries, port facilities, and shantytown areas. Many islands are linked by air or sea transport to the cities of New Zealand and Australia or to the Hawaiian Islands for greater access to world transportation routes. Suva on Fiji and Port Moresby in Papua New Guinea, with their international airports, are the major exceptions.

Globalization and Local Change

Figure 9.15 depicts the wide range in per capita income between Australia and New Zealand and the South Pacific island countries of Oceania. The countries at the top of the diagram in the figure have the highest incomes, the highest consumer goods ownership (Figure 9.16), and are the most involved in the global economic system.

Both Australia and New Zealand were settled by Europeans who comprised the majority of the work force and who maintained economic and cultural contacts with Europe. Global connections created in the islands of the South Pacific were fewer and did not have as lasting an impact on contemporary patterns present there today.

The majority of Australia and New Zealand's current trade is oriented toward East Asia and the United States. Contemporary globalization patterns resulted in less emphasis on a closed self-sufficiency with limited external market dependency and a more open attitude toward neighboring people, markets, and trade for both countries. Although many Pacific islands now have better connections to the rest of the world, few are so closely involved in the global economy, apart from their involvement in the global tourism industry, as are Australia and New Zealand.

Tourism

Although the physical environments and natural and human resources vary significantly among the countries of the region, nearly all play host to growing numbers of international tourists. The number of foreign visitors increased from around 1 million in the early 1980s in Australia to nearly 5.9 million in 2010. Over half of these visitors came from Asia and are attracted by the beaches, golf courses, and the theme parks in metropolitan Sydney and in Queensland's Gold Coast. It is clear that the industry's prosperity depends on growing affluence in Asia. Many visitors come from Japan and Taiwan partly to purchase goods at lower prices than are available at home. Tourists from North America and Europe travel to experience Australian culture, its cities, beaches, barrier reefs, unique flora and fauna, and outdoor adventures in the Australian Outback. An English-speaking populace, combined with Western perceptions of safety and stability, contribute to tourism revenues from American and European visitors.

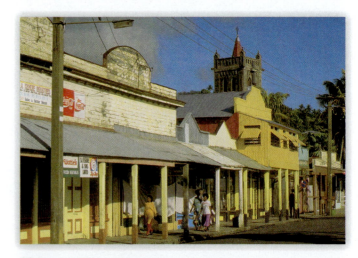

Figure 9.14 **Oceania: urban landscape.** In Levuka, Fiji, where colonial touches remain in the landscape, a British-style church tower and wooden store buildings line Beach Street. *Photo: © Robert Holmes/Corbis.*

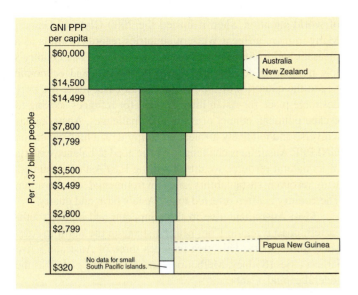

Figure 9.15 **Australia and Oceania: country per capita incomes compared for 2010 data.** The countries are listed in the order of their PPP GNI per capita. *Source: Data (for 2010) from* World Development Indicators, *World Bank 2012.*

New Zealand expanded its tourist industry through its outdoor attractions on both North Island and South Island. Improved and cheaper airline access overcame the great distances from the main sources of wealthier tourists. In 2010, New Zealand had almost 2.5 million tourist visitors, almost triple its 1990 number. New Zealand also serves as a gateway for scientific expeditions and tourist journeys to Antarctica, and the country's economy captures some revenue as Antarctic travelers spend some of their predeparture or post-polar adventure time there.

Oceania's tourism industry is one of the few economic vehicles people of the subregion have to increase their incomes. Some countries are favored over others by tourism operators, transportation carriers, and the tourists themselves. Tourism is important on some islands that are on international air routes, have particular scenic or cultural appeal, or have well-organized tourist facilities, such as Fiji, Guam, the Marshall and Northern Mariana Islands, Toga, Vanuatu, and Western Samoa. It has little or no impact on those that lack such facilities, such as Kiribati, Micronesia, Nauru, New Caledonia, Palau, the Solomon Islands, or Tuvalu. Tourism can, however, destroy or merely exploit the last remnants of traditional culture, and it often pays low wages to local people. Tourism in this region remains dependent on the retention of an attractive environment and political stability, and thus it may encourage good environmental practices.

Tourism on Antarctica grew slowly from its inception in 1958 to the early 1990s and then began to grow more quickly through the early 2000s. The International Association of Antarctica Tour Operators reported more than 33,000 visitors made their way to the continent's shores in the 2010–2011 tourist seasons. The numbers increased with the arrival of sturdy ice-breaking ships that add landing possibilities and as global warming reduced the extent of regional ocean ice cover. Most tourists reach Antarctica on commercial cruise ships sailing from Southern Hemispheric ports in Australia, New Zealand, or Argentina. Visitors are attracted to the scenery, wildlife, and scientific research stations. Many simply want to be able to say they visited the "bottom of the world." As more tourists arrive, however, the dangers of environmental damage increase. At present, the Antarctic Treaty system does not have a code regulating the tourism industry.

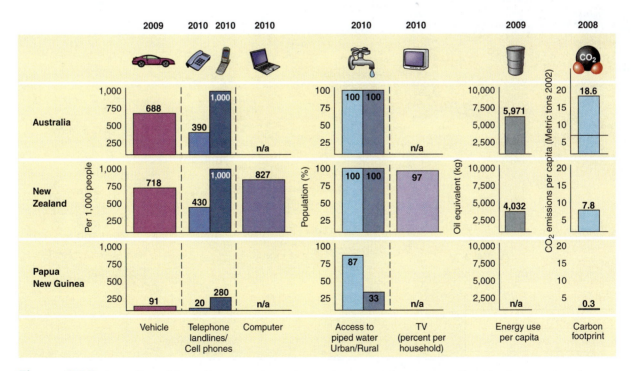

Figure 9.16 **Australia and Oceania: consumer goods, water access, and energy use.** Note the sharp contrasts in consumer goods ownership and access to water and energy use between Australia and New Zealand, and Oceania's Papua New Guinea (no energy data for Papua New Guinea). *Source: Data (for 2010) from* World Development Indicators, *World Bank 2012.*

9.4 Geographic Diversity

Subregion: Australia

The Commonwealth of Australia came into being as a federation of the states in 1901 that emerged from what had been separate colonies under the British Crown. Sydney and Melbourne competed for the role of national capital, but the question was resolved by building a specially planned city in a new Australian Capital Territory centered on Canberra, almost midway between them. The federal government moved there in 1928.

Australia (Figure 9.17) is inhabited almost entirely by European people, who transferred their politics, cultural traditions, and economies to the opposite side of the world (see Table 9.1). The ties that bound this country to the United Kingdom in particular were gradually severed, and trade interests were reoriented to Asia and the wider Pacific. Australia became a focus of world attention when it hosted the 2000 Olympic Games. Many in more affluent Western countries regard Australia as a pleasant and exciting place to visit or relocate. The country's low population density, high level of income, and economic ties to the countries of the Pacific Ocean basin make it a very desirable place for Asian migrants and for refugees seeking to escape political, ethnic, or economic challenges. Australia has a diverse and relatively affluent economy with a high per capita GNI PPP. Australia is challenged by the need to balance exports of its natural resource wealth to willing East Asian consumers with resource sustainability and environmental conservation. The country is often referred to as "Asia's farm and quarry."

Most Australians live in the temperate and humid southeastern coastal region from near Brisbane in the north to Adelaide in the south. This area includes three of Australia's five largest cities (Sydney, Melbourne, and Adelaide), as well as the federal capital, Canberra.

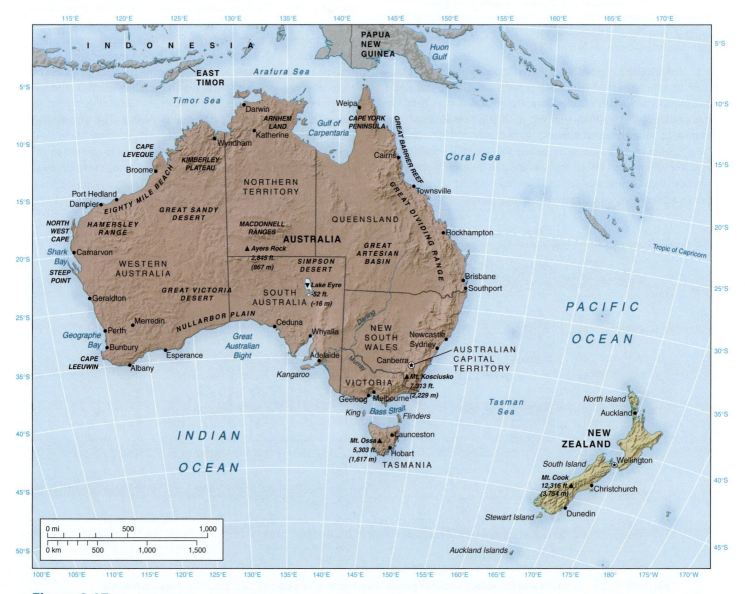

Figure 9.17 **Australia and New Zealand: major geographic features.** The subregion's political divisions, physical features, and major cities.

Much of the southeastern coastal region is bound inland by the hills and plateaus of the Great Dividing Range, which formed a barrier to inland movement during the formative years of colonial migration and development. Three states and their dominant cities occupy this area. Sydney, the largest city, lies at the heart of New South Wales. There are several industrial cities near the New South Wales coast, and the main concentration of Australia's sheep farming is concentrated on the western side of the Great Dividing Range. Human activity in the smaller state of Victoria is focused on its largest city, Melbourne. Its inland farming area is concerned mainly with irrigated farming for grapes, fruit, and grain crops. South Australia has its main city, Adelaide, in an agricultural zone at the southern edge of the state. Much of the rest of the state is desert.

From Brisbane northward, the more tropical climate and the offshore Great Barrier Reef—the world's largest continuous coral reef formation—form one of Australia's most significant tourist destinations. Tropical Queensland is the second-largest state in area with its main city, Brisbane, in the southeast corner. Queensland contains the majority of Australia's cattle ranching, has major mining operations, and is developing tourist attractions.

Northern Territory is the least populated of Australia's primary political divisions. In the far north, the tropical climate brings monsoon rains along the coasts. Most of the few remaining Aboriginal people who live in traditional ways are situated in this territory.

Inland of the Great Dividing Range, the Murray-Darling River lowlands have a subhumid climate that supports a major farming region producing cattle, sheep, grain, and fruit in western Queensland, New South Wales, and northern Victoria. Westward of this sparsely populated farming region, the virtually empty **Great Australian Desert** covers almost all of the remainder of the continent. In the southwestern corner, rains come in winter in a Mediterranean climatic regime and provide a basis for agricultural settlements centered on the state capital, Perth.

Western Australia is the largest state in area but is mostly desert, and its main settlements around Perth are separated by over 1,600 km (1,000 mi.) of uninhabited land from the large cities of southeastern Australia. Perth has farming and mines in its immediate hinterland and large iron ore mines farther north. Western Australia has globally significant deposits of natural gas along and offshore of its central and northwestern coastal areas. Parts of South Australia and Northern Territory are also included in the vast Australian interior.

Economic Trade and Development

By the mid-1900s, Australia had a two-tier economy. The first tier depended on exports of agricultural products and mined minerals to provide income to pay for imported goods. The second tier consisted of protected manufacturing and government-owned enterprises such as airlines, railroads, and docks that were often inefficient but had a role in spreading the wealth from the export industries' income over a wider group of people. Australia's practice of **import-substitution manufacturing**, where the government protected and encouraged domestic industries through tariffs and restrictions on certain imported goods, did not reap the policy's intended benefits. Protectionist policies collapsed in the 1970s and 1980s as exports of agricultural and mineral commodities commanded lower prices in world markets and the expansion of the European Union cost Australia its markets in Britain and Europe.

The Australian economy rebounded in the late 1980s and began to boom in the 1990s. Economic growth was fueled by supplying the rapidly expanding Asian markets with coal, iron ore, and other minerals and some grain and livestock products. Service industries increasingly contributed to Australia's GNI through the first decade of the 2000s. Some of the country's fastest-growing and financially significant industries in 2012 included banking, mining, telecommunications, retail, real estate, and niche-market shipbuilding (high speed catamarans and government service vessels).

Problems of Trade Dependence

Australia's economy is increasingly tied to Asian markets. As it accepts more imports of manufactured goods from Asian countries, it needs to export more to them. Australia wishes to add value to its exports by processing its agricultural and mining products in Australia, but Japan and other Asian countries retain import restrictions on processed materials. Certain specialty products created their own growing markets abroad, but Australia is a relatively small market for Asian goods.

Australia's Dominant Mining

Mining provides nearly half of the value of goods exported by Australia. It is the world's largest exporter of coal, and the world's largest supplier of high-quality (and low-polluting) coking coal. Plans for expansion could increase this level of production if world demand for coal remains high and the production of iron and steel increases. In addition to being a major supplier of coal to world markets, Australia supplies 10 percent of the world's uranium. Other minerals of importance to Australia include iron ore, of which it is the second-largest world producer, bauxite (first), nickel (second), and gold (third). Australia opened the world's largest diamond mine in the 1990s, and the country remained a significant diamond exporter into the 2010s. Developers who desire the significant expansion of Australia's mining capacity and production conflict at times with environmental activists and Aborigine communities who assert the need to proceed cautiously and to respect lands that are environmentally sensitive or historically significant to indigenous peoples.

New energy resources include globally significant natural gas fields along Australia's west coast and a growing industry which converts the gas from its natural state to liquefied natural gas for export to Japan and other East Asian markets. Such projects attracted investment from several large oil companies like Chevron, which helps Australia compete with other nearby producers in Malaysia, Brunei, and Indonesia.

Australia's Farm Output

Two percent of Australia's agricultural land is cultivated and the remainder is devoted to livestock grazing. Australian farmers export 80 percent of their production, with most sales going to Asian markets. Wheat, oilseeds, beef, veal, and wine are the main exports (Figure 9.18). Occasional interruptions of market access to Japan and the United States and currency devaluations in countries such as Brazil and Argentina are among the global challenges Australia faces in maintaining and increasing a stable market for its agricultural products.

Both Australia and New Zealand are among the 21 member economies (countries) of the Asia-Pacific Economic Cooperation, or APEC, and both seek to expand opportunities within the wider Pacific economic region. Australia also attempts to engage the Southeast Asian economic trade bloc, ASEAN, but the ASEAN members seem to be more interested in trading with East Asian countries. The **South Pacific Forum**, established in 1971, links 13 regional countries: Australia, New Zealand, Papua New Guinea, the Solomon Islands, the Cook Islands, Fiji, Kiribati, Nauru, Niue, Tonga, Tuvalu, Vanuatu, and Western Samoa. Its aim is to develop regional political cooperation. This agency has had some success in confronting regional problems for the islands in particular. Australia has championed a number of causes, but these often bring it into confrontations with Asian countries, such as when it opposed the exploitative Malaysian logging of Papua New Guinea.

Subregion: New Zealand

New Zealanders (who sometimes refer to themselves as "Kiwis") take pride in the natural beauty of the country's two main islands. Dramatic coastal fiords, rugged mountainous terrain, abundant rain and snowfall, and lush vegetative cover combine to provide an extremely wide range of environments and ecosystems. Residents enjoy active outdoor lifestyles and recreational pursuits. New Zealanders have nicknamed their country the "adventure (and adventure sports) capital of the world." The unique and diverse physical environments and landscapes of

Figure 9.18 **Australia: farming.** Harvesting grapes.
Photo: © Corbis RF.

the country likely contributed to New Zealanders' heightened awareness of environmental conservation and stewardship.

Internal rivalries in New Zealand are less apparent than those within and between some parts of Australia. North Island is distinguished by its central upland area, which is affected by volcanic activity. It is home to nearly all the Maoris, most of the people of European origin, the largest and most cosmopolitan city, Auckland, and the country's capital, Wellington. South Island has fewer people and a much more rugged physical environment. South Island's high western mountains, the Southern Alps, are geologically active and are occasionally rocked by earthquakes. The eastern plains provide good farming, herding, and grazing lands. New Zealand is home to far more sheep than it is to people.

Economic Strengths: Natural Resources and Agriculture

New Zealand's economy is primarily based on natural resources like that of Australia, the export of farm and forest products to Europe, and tourism. New Zealand's main economic products entering world markets are wool, lamb, and dairy products. Over half of New Zealand is in pasture, and over one-third of its exports is from livestock (Figure 9.19). In New Zealand, sheep outnumber people some 12 to 1. The government of New Zealand does not formally subsidize agriculture, making the country one of the world's least subsidized agricultural exporters and exposing the industry almost entirely to global market forces. New Zealand's farmers do not want government subsidies; they assert that their removal makes agricultural practices in the country among the world's most efficient. While the number of beef cattle stabilized in the first decade of the 2000s, increased demand for dairy products resulted in growing numbers of dairy cattle, and an expanding venison industry resulted in more than 3,000 deer farms in New Zealand by 2012 which produce nearly half of the world's venison supply.

Pastureland replaced forest cover during the formative and expansion years of agricultural production. The New Zealand government replanted large areas of forest with Radiata pines and Douglas firs, softwoods that thrive in New Zealand's climate and environment, from the early 1900s. This **afforestation** policy provides an increasing harvest that finds markets in Asian countries with limited or rapidly diminishing forest cover. **Sustainable forestry** is now as profitable as farming in New Zealand, and the state-owned forests have nearly all been privatized, albeit with strong conservation laws.

The contributions of agriculture and manufacturing to its economy fell as that of services rose to account for around two-thirds of GNI. In the 1980s, the New Zealand government instituted economic reforms in the 1980s, which resulted in economic growth in the following decades and produced a trade surplus, lower unemployment, and inflation. Tariffs and restrictive port practices were removed and government spending reduced as a proportion of GNI. Lower labor, transportation, and utility costs brought increases in manufacturing productivity. New Zealand's GNI per capita fell behind that of Australia and Western European countries in the early 2000s, although its high quality of life continued through 2012.

Figure 9.19 **New Zealand: farming.** Sheep show in Rotorua with 20 breeds of sheep, each having advantages and many originating in Scotland (U.K.). *Photo: © David C. Johnson.*

New Zealand now sells more goods to Japan than to Australia, the United States, and the UK combined. Most of its imports still come from Australia and the United States. Like Australia, New Zealand is a member of the South Pacific Forum, and ties with the Pacific islands continue that originated when New Zealand was the United Nations agent in those island countries before their independence.

The unprecedented success of the motion picture trilogy *The Lord of the Rings* (and, to a lesser degree, the film *Whale Rider*) brought significant global attention to New Zealand. *The Lord of the Rings* films, which set global box office records, in part showcased the spectacular scenic beauty of New Zealand. Some aspects of New Zealand's tourism grew in direct response to the success of the film trilogy. Also stemming in part from the successes of *Whale Rider* and *The Lord of the Rings* is positive growth in **New Zealand's film industry**, where a number of internationally viewed television series were already filmed (Figure 9.20).

Subregion: Oceania

The South Pacific islands of Oceania stretch from Papua New Guinea (an independent country situated on the island of New Guinea) to hundreds of tiny islands with few inhabitants positioned thousands of kilometers to the east (Figure 9.21). The smaller islands are grouped in independent countries or colonies. Their small sizes and markets place them at a disadvantage in any political, social, or economic negotiations with Asian and European countries, Australia, or the United States. Issues arise frequently between the islands and external countries regarding timber felling, mineral extraction, agricultural production, fisheries exploitation, and military testing in the waters of the South Pacific. The living standards of Oceania's inhabitants suffer in part due to the subregion's lack of authority or influence in the global community (see Table 9.1). Occasionally, local tension garners global media attention, as in the case of the eruption of violence on the French Polynesian island of Tahiti over the French nuclear testing in mid-1995. Although these islands were perceived by early travelers as a "Garden of Eden," day-to-day life is not as idyllic as romantic travel writers depicted (Figure 9.22).

The Pacific Islands Forum (formed as the South Pacific Forum in 1971) is a political grouping of 16 sovereign Pacific states, including Australia and New Zealand. The Forum's administrative headquarters, The Secretariat, are in Suva, Fiji, and the organization's overarching concern is with the regulation and management of the resources in the region. The growing problems faced by the small islands brought closer cooperation on an economic and political agenda that included achieving independence for the remaining European colonies, increased investment, and trying to stop French nuclear testing.

Island Countries

Most of the South Pacific islands gained their independence in the 1970s. Western Samoa achieved it in 1962 and the Marshall Islands in 1991. The Micronesian group of islands remains a United Nations trust territory. New Caledonia and French Polynesia remain under the sovereign authority of France.

Although independence appeared desirable to many of the islands, economic difficulties, internal tensions, and dependence on continuing economic aid and protection keep them linked economically to the United States, France, the United Kingdom, Australia, or New Zealand. Kiribati, Tonga, Tuvalu, Western Samoa, and the Solomon Islands remain among the world's poorest countries, with few products sought by global consumers.

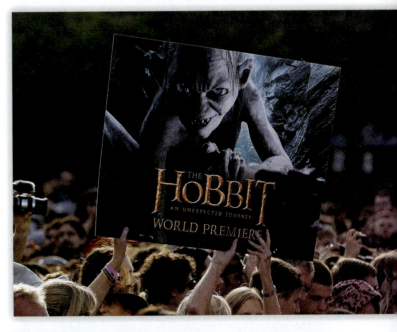

Figure 9.20 *The Hobbit* **world premiere.** Fans hold up signs during the world premiere of *The Hobbit* movie in Courtenay Place in Wellington, New Zealand, on November 28, 2012. *The Hobbit* films build on the success of Peter Jackson's *Lord of the Rings* film trilogy, which contributed significantly to the growth of New Zealand's motion picture and tourism industries. *Photo: © Marty Melville/AFP/Getty Images.*

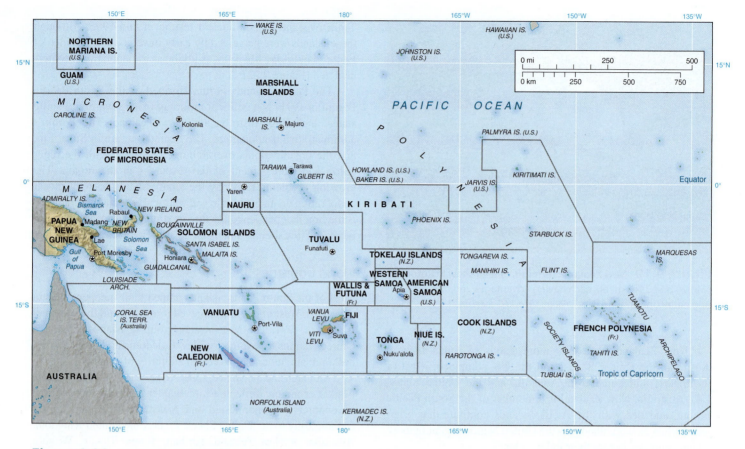

Figure 9.21 **Oceania's South Pacific islands: major geographic features.** The country groups and main cities. These islands are distributed across huge longitudinal distances.

The southwestern island group approximately coincides with former Melanesia and includes Papua New Guinea, the Solomon Islands, Vanuatu, Fiji, and New Caledonia. These are the largest and most populated islands and are the nearest to Southeast Asian markets and connections to Australia and New Zealand. Papua New Guinea is situated on the eastern part of the island of Irian Jaya and on several surrounding smaller islands (the western part of Irian Jaya belongs to Indonesia). The northwestern group, roughly coinciding with former Micronesia, includes Guam, the Northern Mariana and Marshall Island groups, Nauru, Palau, and the Federated States of Micronesia. The largest of the three groups in terms of ocean area but the poorest in economic development contains the islands of the central Southern Pacific such as Kiribati, Tuvalu, Wallis and Futuna, Tonga, Tokelau, Western and American Samoa, Niue, Cook Islands, Pitcairn, and French Polynesia.

The islands vary considerably in their well-being, as the consumer goods ownership in Papua New Guinea shows (see Figure 9.16). The French colonies of French Polynesia and New Caledonia are heavily subsidized by France and have above-average incomes. Others, such as the Marshall Islands,

Palau, Micronesia, Guam, and the Northern Marianas, have U.S. support.

Economic Development

Island communities had complex economies prior to colonization, often linked across wide ocean expanses. Exchanges took place among the better-watered islands where vegetables and fruit were grown and the arid islands that relied on fishing. Commercialization of inter-island trade occurred in the early postcolonial years. Some of the islands entered the world economy with sales of coconut palm products such as **copra**—the dried white meat that lines the inside of the coconut shell and provides an oil used in soaps and candles. Manufactured goods from Europe and the United States and expensive diesel oil for power were imported. Many islands needed subsidies from former colonial powers and remained materially poor. Some islands, including Tuvalu, relied on external aid for nearly 100 percent of their income.

Many South Pacific islands faced economic ruin in the early 2000s. Some see external aid as a compensation for past colonial exploitation; others see it as demeaning and a

(a)

(b)

Figure 9.22 **Oceania.** (a) An idyllic view of Ralarea, French Polynesia, which is close to French nuclear testing sites. (b) Traditional dress (plus sneakers) in Papua New Guinea. *Photos: (a) © Corbis RF; (b) © Mission Aviation Fellowship.*

mechanism for stagnation. Some islanders see tourism as an income prospect, while others view it as a further form of economic colonialism.

Farm, Forest, and Mine Products

Apart from Papua New Guinea (copper and gold), New Caledonia (nickel), and Nauru (phosphate), few islands have natural resources except for a warm climate, nutrient-rich volcanic or coral limestone soils, and the surrounding ocean. The small areas available for agriculture allowed few islands to diversify into commercial crops beyond coconuts and copra. Fiji produces sugar and Tonga produces bananas and vanilla in addition to coconuts.

World attention was drawn to the cutting of tropical hardwoods on Papua New Guinea and the Solomon Islands by a statement at the South Pacific Forum in Brisbane in the mid-1990s that the islands were being "ripped off." The imposition

of stricter environmental laws in Malaysia and Indonesia turned the attention of their logging companies to the islands that had fewer regulations. Cutting is now at levels that could remove all the timber, and little replacement planting is taking place.

Similar problems face local fishermen as fishing fleets from Japan, the United States, South Korea, and Taiwan take fish from this region to make up nearly 40 percent of the world catch. Although the island countries of the South Pacific have declared 200-mile exclusive economic zones and the foreign fishers often pay access fees, it is impossible for the whole area to be patrolled.

Strip mining for phosphate rock (used in fertilizers) left a barren, jagged surface over much of the island of Nauru (only 21 km^2 [8 mi.^{2}] in size) as mining companies cleared the majority of the island's tropical vegetation from the mid-1900s through the island's independence in 1968. The Nauru government bought out the phosphate company and secured investment from Japan and South Korea, which injected income into schools and free medical treatment facilities, and gave Nauru one of the region's highest per capita incomes in the 1970s. Few Naurians needed to work, and the wealthy conditions attracted Australian managers, Chinese shopkeepers, and miners from other Pacific islands, who now make up one-third of the island's 12,000 inhabitants. Air Nauru bought five Boeing 737s.

Surplus revenue was invested in Pacific Rim properties, but in the 1990s, the value of these assets dropped from over US$1 billion to $130 million. In the 1990s and early 2000s, the depleted phosphate resources and falling world prices also reduced Nauru's income. The Nauru government's strategy in the first decade of the 2000s for overcoming its previous financial crisis included allowing offshore banking that encouraged the deposits of unrecorded Russian mafia accounts (estimated at more than US$70 billion in the early 2000s), leasing land to Australia for a detention camp for asylum seekers, and supporting Russia's official recognition of the sovereignty of Abkhazia and South Ossetia in exchange for an estimated US$50 million in aid from Russia in 2009. Such sources of income are temporary, and the once-affluent Naurians suffer perceived indignities, including rationed electricity. After an Australian court permitted the Export-Import Bank of the United States to seize the airline's remaining assets in 2005, the government renamed the airline "Our Airline" and reinstated flight service after the purchase of one Boeing 737 commercial aircraft. In 2012, the government-owned Our Airline continued operations with just the one aircraft and approximately 65 employees.

Subregion: Antarctica

Antarctica is larger than either Europe or Australia, occupying 10 percent of Earth's land surface (Figure 9.23a). Seven countries—Argentina, Australia, Chile, France, Norway, New Zealand, and the United Kingdom—assert territorial claims to parts of the continent (Figure 9.23b). The United States and Russia do not recognize any political claims, but both openly state a reserved right to make future claims of their own. Although no permanent population exists on the continent, there is a constant human presence. Scientific teams representing numerous research and political interests maintain a continuous existence on Antarctica, stationed at various research facilities on the continent. Increasing numbers of tourists also visit during the region's summer months and contribute to an evolving human impact. Although the numbers of people visiting and studying Antarctica are steadily increasing, the aggregate amount remains relatively small, especially in relation to the continent's vast territorial expanse. Correspondingly, the majority of Antarctica remains virtually empty. Low population densities and limited human activity are likely to change in the decades to come. The continent's mineral wealth, oil and gas potential, and an enhanced accessibility resulting from anthropogenic climate change make Antarctica an increasingly important item on the political, economic, and strategic agendas of the world's wealthiest and most powerful countries.

Antarctica's Global Status

Geographic expeditions to Antarctica, beginning with Captain James Cook's voyages to the Southern Oceans in the late 1700s, concentrated on discovering what existed beyond the foggy, ice-filled waters. The interest of furthering geographic science combined with prestige politics in the late 1800s and early 1900s as governments raced to be the country that was "first"—first to reach the South Pole or first to cross the continent (see "Geography at Work: The Geographic Exploration of Antarctica," p. 288). Early expeditions and discoveries gave way to international political claims on sections of the continent's territory during the mid-1900s and the establishment of fixed scientific research stations. In 1961, 39 countries signed the **Antarctic Treaty** as a basis for nonmilitary scientific cooperation, environmental safeguards, and international control. In 1991, the **Wellington Agreement** banned commercial mining activities and introduced protection regulations. Signatory countries of the Antarctic Treaty have become, in effect, the governing body through the **Antarctic Treaty System**, or ATS. The ATS prohibits any weapons testing or military use of Antarctica. Some countries did not sign this agreement because they did not have an interest in the continent or did not agree with the restrictions imposed. The ATS encourages scientific research and exchange among interested signatory countries. Research conducted from the mid-1900s through the early 2000s significantly contributed to climate, oceanographic, geologic, and glaciological studies.

Antarctica's vast natural resources are not exploited, and the international agreements stipulate that they should not be. Only the future will tell whether this protective situation will last or whether Antarctica will turn into an open-pit mine for the rest of the world. The main interest in Antarctica in the early 2000s concerns its important role as a laboratory for monitoring global climate change and the study of the development of the ozone hole above it, the extent of ice on the continent and the surrounding oceans, and the bounty of the ocean ecosystems. Research carried out in Antarctica finds atmospheric pollutants that indicate a decline in world environmental quality. In their small but increasing manner, however, research stations also degrade the local environment by pouring untreated sewage into the ocean and dumping oil drums on sites that seabirds use for nesting.

Antarctica and the Southern Oceans

The frozen continent forms its own cold climate with a heating deficit throughout the year. In the frigid and dark winter months, ice coverage in the oceans around Antarctica increases as the sea surface freezes; in the summer months, glacial ice calves off icebergs into the surrounding ocean. Perhaps Antarctica's most significant boundary is between 50 and 60 degrees south, where cold and warm ocean water and atmosphere converge. The contrast in the atmosphere above creates a sharp boundary between Antarctic air and warmer midlatitude air, generating a frontal zone and a succession of midlatitude cyclonic storms with high winds.

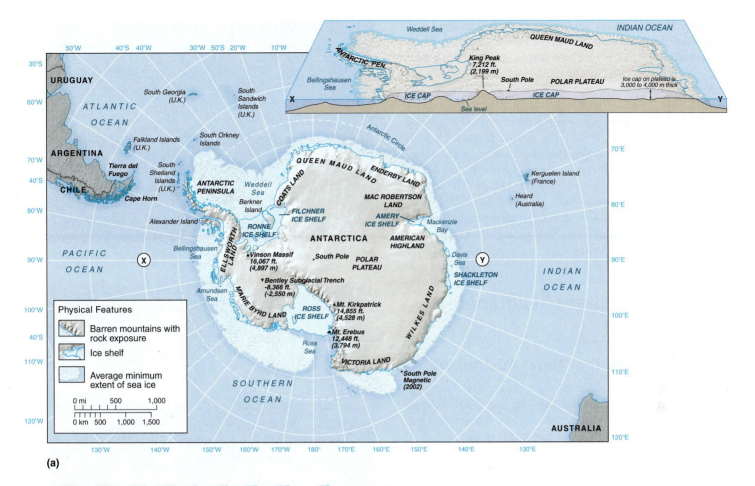

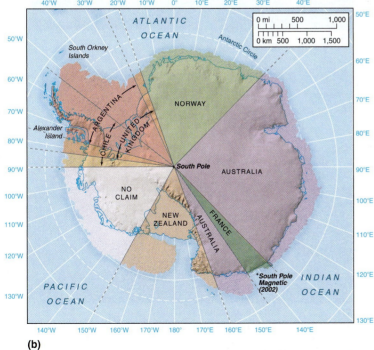

Figure 9.23 **Antarctica: the ice-covered continent.** (a) The continent's toponyms and physical features and a cross-section through 90°E and 90°W. (b) Political claims: the United States and Russia, among other countries, refuse to make such claims and do not recognize other countries' claims.

Despite the barrier to climatic elements entering Antarctica's atmosphere, small quantities of chlorine gases penetrate from lower latitudes to produce the **ozone hole** above the continent at the end of the Antarctic winter in October. This "hole" is a thinning of the protective ozone layer high in Earth's atmosphere that filters out harmful ultraviolet radiation from the sun. The hole enlarged in the 1980s and 1990s as the chlorine gases reacted with the ozone; however, it appeared to stabilize around 2000.

During the late 1990s and early 2000s, some of Antarctica's ice shelves, where ice flowed off the continents to cover the ocean surface, melted and broke up—seen by many as a sign of global warming.

Antarctica's Resources

Antarctica is not a country and so does not have an economy of its own. The ATS prevents the known deposits of coal and other minerals from being exploited. Oil and natural gas deposits beneath the continent's shelf also remain unexploited and somewhat protected by international agreements. As the regional ice continues to melt, political and economic pressure may dramatically change the international position regarding such resource exploitation.

Antarctica's surrounding oceans draw fishers from all over the world. The oceans surrounding Antarctica are among the world's most profuse life zones. Antarctica's living organisms are dominated by a huge variety of sea birds, including penguins that rely on the rich ocean life of plankton, fish, seals, and whales (Figure 9.24a, b, c). The Antarctic oceans are an important basis for wider ocean food chains and are being studied to gain an understanding of sustainable levels of fishing, sealing, and whaling.

In 1982, as commercial fishers from around the world increased their exploitation of the marine resources, it was agreed internationally to regulate such fishing. Fish stocks, such as cod, together with some groups of whales, were declining. Some fishing fleets claim ignorance of sensitive areas or ignore regulations or international concerns. Additionally, it is almost impossible to monitor fishing in this extensive area of ocean that has few ships passing through and is not the responsibility of a particular country.

9.5 Contemporary Geographic Issues

Aborigines and the "New" Australians

Long before Captain James Cook surveyed the Australian coast and claimed it for the British in the late 1700s, Aboriginal communities were thriving throughout the varied landscapes of this Southern Hemispheric continent. Archeological evidence suggests that Aboriginal communities have existed in Australia for as long as 50,000 years. Aborigines lived traditionally as hunter-gatherer communities based on extended family clans. They did not have a concept of land ownership—they believed the land was given to them to use by their creator, and the relationship they had with the land was a spiritual one. Aborigines did not buy or sell land or its resources, but their concept of place and territory would be forever changed with European settlement in Australia.

Shortly after Captain Cook's expedition along the coasts of Australia, the British began settling the continent in 1788. The impact of the Europeans was extremely degrading to the social fabric of the continent's indigenous communities. As the "new" Australians began their continental conquest, they surveyed and appropriated much of Australia's land and resources. Some Aborigines died from exposure to European diseases, while others were increasingly marginalized and literally moved out of the way. If indigenous peoples resisted, they were imprisoned or killed. Nearly the entire population of Aborigines on the island of Tasmania was wiped out by the settlement efforts of the British colonists in the 1800s. Only those living in the most remote or inhospitable parts of the Australian mainland were left alone to maintain their traditional lifestyle. By the early 1900s, many Aborigines were living on government-created indigenous reserves or in material poverty in and around Australia's growing cities.

Throughout the 1900s, the Australian government assumed control of freshwater resources, fisheries, and land considered sacred by and historically significant to the Aborigines. Land through which Aboriginal families migrated freely for hunting

(a)

(b)

(c)

Figure 9.24 Antarctica. (a) Penguins are part of Antarctica's unique biodiversity. The animals are both an important part of the continent's food web and a favorite "attraction" for tourists. (b) A tourist cruise ship plies the waters along the Antarctic coast. (c) A Danish scientific research station. *Photos: (a) © Corbis RF; (b) & (c) © Ian Coles.*

and gathering was also acquired by the government. Mining and other primary sector economic activity took precedence as the growing economy of Australia evolved to compete in the global marketplace.

Aborigines did not have the same legal or property rights as European Australians. They did not have the right to vote, and they were completely marginalized from the political process until the 1960s (Table 9.3). The lack of political empowerment for the indigenous communities contributed to their dire material poverty and the harsh legal judgments passed on them in the Australian legal system. It was not until the latter part of the 1960s that some officials in Australia began to publicly acknowledge past wrongs in terms of the treatment and exclusion Aborigines received at the hands of the European–Australians.

In addition to the appropriation of land for mining, tourism development played a role in the appropriation of Aboriginal sacred space. Prominent sites such as central Australia's Uluru (known by most Westerners as "Ayers Rock") became popular tourist destinations. Australian and foreign tourists flocked to Uluru, one of the world's largest monoliths and a World Heritage site, to see the unique rock formation, its colors, and the surrounding landscape. Visitors used to climb all over Uluru as part of their experience. Uluru is considered sacred ground by the central Australian Aborigines, and they believe people should not touch it or climb on it. Aborigines gained control of Uluru in 1985, and then leased it to the Australian National Parks and Wildlife Service, which currently operates the park.

The struggle for equality and recognition continued during the later decades of the 1900s. By the 1990s, much of "mainstream" Australian society began to appreciate the injustice forced on Australia's native peoples. The promotion of awareness of Aboriginal culture began to be emphasized in education programs, the arts, and through tourism. Kevin Rudd, prime minister of Australia from 2007 to 2010, issued a formal apology to Aboriginal communities on behalf of the country's government. Though significant progress toward equality continues, some Aborigines still lived in poverty in 2012, and it will likely take decades more for them to completely overcome the injustices of the past. Many Aborigines enjoy legal and social equality. Many have assimilated into Australia's urban culture, while others continue to live the traditional lifestyle of their ancestors in remote parts of the continent. While some see the recent appreciation of Aboriginal culture as a marketing tool for tourist or other commercial purposes, others believe true progress has been made and that movement toward egalitarian coexistence is in the right direction.

Water

Australia is often referred to as the driest inhabited continent in the world. Desert and steppe biomes dominate the territorial extent of the continent. The wettest climate in Australia lies in the country's sparsely populated northern, tropical latitudes, where nearly two-thirds of Australia's surface water runoff occurs. The majority of Australia's population is distributed along the country's east and southeast coasts, where the seasonal temperate rains do not produce significant water surpluses suited to fulfilling local exploitation needs. This geographic pattern exists in many areas around the world, where the highest population densities often do not neatly coincide with the natural resources, agricultural, or climatic conditions best situated for human consumption and societal well-being.

The changing needs of the growing population of Australia, along with climate change model projections, suggest that Australia's water challenges will intensify in the coming decades. Australia's population has grown steadily over the past three decades. The most significant factor in the country's population growth is the immigration of people from East and Southeast Asian countries. As Australia continues to foster increased ties to countries like Japan, residents are finding the climate, living standards, and stability of the country to be strong migration pull factors. The majority of immigrants to Australia are settling into the country's urban network along its east and southeast coasts. Growing urban centers place a steadily increasing strain on Australia's water supply.

Table 9.3	Debate: Aborigines, assimilation, and equal rights
Perspective of Some Aborigines	**Perspective of Some Non-Aborigine Communities**
Aborigines assert thousands of years of history and occupation of the Australian continent.	Some European–Australians assert that their hard work and a high price in human resources went into building Australia as safe and healthy living space.
Some Aborigines claim that true equality eludes them in terms of social and economic opportunities.	Some European–Australians claim that many Aborigines won't take responsibility for their own lives and actions, that they live off the hard work of others, and blame everyone else for their mistakes.
Aborigines want other Australians and tourists to respect sacred sites such as Uluru (Ayers Rock).	Uluru, or Ayers Rock, belongs to all Australians and should not be controlled by a very small percentage of the overall Australian population.
Aborigines want control of historically significant territory, and many call for the termination of mining and other industrial land-use practices on such territory.	Mining and other resource-based industries provides thousands of direct and indirect jobs in Australia and are an important source of foreign revenue for the country.

Australia's primary export industries are mining and agriculture, both of which have also grown steadily in the past three decades, and both have increasingly drawn on the country's water supply. Agriculture is the biggest consumer of fresh water in Australia, consuming some 70 percent of Australia's water resources.

Water-related research projects suggest that the past two centuries of land use and anthropogenic land-cover changes are having a direct impact on soil salinity and water vapor flows for Australia. Settlers in Australia began to modify the country's vegetation in the 1800s. The practice of clearing woody vegetation and replacing it with agricultural crops and pasture for animal grazing, which continued through the 1900s, altered the land-cover landscape in many parts of Australia's west, southeast, and north. The native tree vegetation in Australia is suited to the country's climate patterns, and thus the trees evolved to release much of their stored water into the regional lower atmosphere through evapotranspiration. The elimination of native woody vegetation and its replacement with crops and grassland has significantly decreased water vapor levels (through decreases in evapotranspiration). Decreased water vapor can directly translate to decreases in regional rainfall. The alteration of regional land cover appears to contribute to increased soil salinity in parts of Australia. High soil salinity can impair soil productivity and decrease crop yields.

Climate change models suggest that rainfall amounts will decline on average in the coming decades in southern Australia, where the majority of the population lives and where the majority of irrigated agriculture is located. If this projected trend evolves, it will further challenge Australian efforts to adequately supply water to its growing cities and to its economically significant agricultural industry.

Added to the challenge of growing demand on a potentially diminishing supply of water in the most populous areas is the additional challenge of degrading water quality. Industrial and agricultural use, coupled with diminished flow due to diversion for irrigation and other consumption habits, are contributing factors in the steady decline in water quality in the drainage basins of most of Australia's rivers in the east and southeast.

As Australia continues to welcome migrants from East Asia, Southeast Asia, and New Zealand, as its tourism industry continues to grow, and as the country expands mining and agriculture trade with Asian countries, it may face increasingly difficult challenges in adequately supplying clean, fresh water to its citizens. Decreasing evapotranspiration from land-use practices, coupled with climate change projections, will further complicate Australia's water demands in the coming decades.

Geography at Work

The Geographic Exploration of Antarctica

Early geographers set out to explore new lands and waters, to dispel rumors, and to further geographic science and awareness. Fables of a vast southern land, *Terra Australis*, persisted in the world's geographic consciousness for much of the 1600s and 1700s. Captain James Cook, a British cartographer, explorer, and navigator, sailed for the Southern Oceans on the first of several voyages in 1768. His journey led to Europe's "discovery" of New Zealand and Australia, which he claimed for Britain. Cook's second voyage began in 1772 and lasted for three years. He reached farther south than any known previous explorations, and though he did not reach land in Antarctica, he theorized that the large icebergs he encountered could only have come from land.

European and Russian explorers, along with with whalers and fishermen from various countries, significantly furthered global knowledge of the waters around Antarctica in the decades after Cook's voyages. Whalers and seal hunters plied the region's waters throughout the 1800s. Lured by numerous species of whales and seals, the hunters penetrated Antarctica's ice-filled waters in pursuit of their catch. They began to learn about the region's seasonal changes, wind patterns, and the different types of pack ice and ice flows. Their reports sparked great interest among many countries of the world, especially those in Europe already immersed in competitive pursuit of geographic exploration and discovery, to send expeditions to Antarctica's treacherous waters to advance science, build awareness, and garner prestige.

Delegates to the 6th International Geographical Congress, held in London in 1895, declared Antarctica to be the most important geographical exploration yet to be achieved. The race to explore, discover, and make "firsts" heated up over the next two decades. Numerous expeditions sponsored by geographical societies, governments, and private donors ensued from Britain, Belgium, Germany, and Norway, among others. Explorations led by British naval officer Robert Falcon Scott, and later Ernest Shackleton (who accompanied Scott on his first venture to the pole), continued to push farther toward the South Pole, furthering knowledge of local conditions, but without success in reaching it. In December 1911, Norwegian explorer Roald Amundsen led the first successful expedition to the South Pole. The expedition's success captivated the attention of the world. His conquest was followed in 1914 by Ernest Shackleton's Royal Geographical Society expedition to cross the Antarctic continent. Shackleton's ship, *Endurance*, became trapped in pack ice in the Weddell Sea within sight of the Antarctic continent. The crew eventually had to abandon ship, first camping on the ice, and later making their way in small boats to Elephant Island and some on to South Georgia Island before being rescued. Shackleton's ill-fated journey is chronicled in his memoir *South* (published in 1919).

The legacy of early geographic explorations to Antarctica continues today on the icy continent as contemporary geographers conduct climate research at numerous scientific stations located in places that bear the names of their predecessors in honor of their bravery and contributions.

Samba parade in Rio de Janeiro. *The world's largest Carnival celebration takes place each year in Rio de Janeiro just prior to the beginning of the Catholic season of Lent. Millions of locals and international visitors join in daily parades and other festivities in what has become the most iconic festival experience for the city of Rio and the country of Brazil.*
Photo: Alex LeFevre.

Latin America

LEARNING OBJECTIVES

After reading this chapter, you should be able to:

■ Understand the diversity of and the relationship among the region's wide range of climate types, major landforms, river basins, vegetation, and natural resources.

■ Identify regional natural hazards and understand the impact on and relationship with the region's human geography.

■ Recognize spatial population patterns throughout the region and understand the impact demography, migration, and urbanization have on creating the region's population distribution.

■ Compare and contrast Mexico and the countries of Central America in terms of their economic development, political stability, and relationship with the United States.

■ Explain the diverse development paths of the countries of the Caribbean Basin subregion in terms of their colonial histories and resources.

■ Link the export-based economies of the Northern Andes countries to the modest economic development of the subregion and dependence on the drug trade.

■ Examine the relationship between the history of Brazil's coastal settlement and economic development and the country's contemporary urban, environmental, and economic challenges.

■ Compare and contrast the demographic characteristics and economic development of Southern South America with the other subregions of Latin America.

Figure 10.1 **Latin America: physical features, country boundaries, and capital cities.** Latitude, altitude, mountain barriers, islands, proximity to water, river basins, and geographic situation all affect the human geography of the region.

10.1 Latin America: Dramatic Contrasts

Latin America is a vast region situated from temperate latitudes north of the equator in northern Mexico, to the subarctic conditions of southern South America (Figure 10.1). Latin America is considered to be a world region based on the lasting impact of Iberian colonization and the influence of the global economy on Latin America's human and physical geographies. The countries of Latin America range in size from population giants such as Brazil (194.3 million people in 2012) and Mexico (116.1 million), to the Caribbean Basin, where many countries have fewer than 100,000 people (Table 10.1). Latin American countries range from high to low incomes, from dependence on a single economic product to a diverse and integrated economic base, and from highest levels of involvement in, to isolation from, the global economic system.

Table 10.1	**Latin America:** Data by country, area, population, urbanization, income (gross national income purchasing power parity), ethnic groups						
		Population (millions)		**% Urban**	**GNI PPP 2010**	**2010**	
Country	**Land Area (km²)**	**Mid-2012 Total**	**2025 Est. Total**	**2012**	**Total (US$ billions)**	**Per Capita (US$)**	**Ethnic Groups (%)**
MEXICO, CENTRAL AMERICA							
United Mexican States	1,956,200	116.1	131.0	77	1,627.0	14,340	Mestizo 60%, Native American 30%, Euro 9%
Belize	22,960	0.3	0.4	44	—	—	Mestizo 44%, Creole 30%, Maya 11%
Costa Rica, Republic of	51,000	4.5	5.1	62	52.5	11,270	Mestizo and Euro descent 95%
El Salvador, Republic of	21,040	6.3	6.8	63	40.6	6,550	Mestizo 90%, Native American 9%
Guatemala, Republic of	108,890	15.0	19.7	50	66.8	4,650	Mestizo 56%, Native American 44%
Honduras, Republic of	112,090	8.4	10.5	50	28.6	3,770	Mestizo 90%, Native American 7%
Nicaragua, Republic of	130,000	6.0	6.9	57	16.1	2,790	Mestizo 69%, white 17%, black 9%, Native American 5%
Panama, Republic of	75,520	3.6	4.2	65	44.9	12,770	Mestizo 70%, Native American, mixed black 14%, white 10%
Totals/Averages	**2,477,700**	**160.2**	**184.6**	**59**	**1,877**	**8,020**	
THE CARIBBEAN BASIN							
Antigua & Barbuda	440	0.1	0.1	30	—	—	Black African 96%, white 3%
Bahamas, Commonwealth of	13,880	0.4	0.4	84	—	—	Black African 85%, white 12%
Barbados	430	0.3	0.3	45	—	—	Black African 90%, mixed 4%, white, other 6%
Cuba, Republic of	110,860	11.2	11.4	75	—	—	Mixed white/black 51%, white 37%, black 11%
Dominica, Commonwealth of	750	0.1	0.1	67	—	—	Black, Carib natives
Dominican Republic	48,730	10.1	11.6	66	89.6	9,030	Mixed 73%, white 16%, black 11%
French Guiana	91,000	0.2	0.3	81	—	—	Mixed white, Native American, black, Arawak natives
Grenada	340	0.1	0.1	40	—	—	Black African, South Asian, white
Guyana, Cooperative Republic of	214,970	0.8	0.8	29	—	—	Asian Indian 51%, black and mixed 43%, Native American 4%
Haiti, Republic of	27,750	10.3	11.9	47	11.1	1,180	Black 95%, mixed and white 5%
Jamaica	10,990	2.7	2.8	52	19.7	7,310	Black African 76%, mixed black/white 15%, Asian Indian 3%
St. Kitts & Nevis, Federation of	360	0.1	0.1	32	—	—	Black African
St. Lucia	620	0.2	0.2	28	—	—	Black African 90%, mixed 6%, South Asian 3%

(continued)

Country	Land Area (km²)	Population (millions) Mid-2012 Total	Population (millions) 2025 Est. Total	% Urban 2012	GNI PPP 2010 Total (US$ billions)	2010 Per Capita (US$)	Ethnic Groups (%)
St. Vincent & The Grenadines	390	0.1	0.1	40	—	—	Black African 82%, mixed 14%, whites, South Asian, Carib native
Suriname, Republic of	163,270	0.5	0.6	74	—	—	South Asian 37%, Creole white/black 31%, Indonesian 15%, Maroon 10%
Trinidad & Tobago, Republic of	5,130	1.3	1.3	13	32.3	24,050	Black African 41%, South Asian 40%, mixed 17%
Totals/Averages	**689,910**	**38.5**	**42.1**	**50**	**153**	**10,393**	
NORTHERN ANDES							
Bolivia, Republic of	1,098,580	10.8	12.5	66	46.0	4,640	Quechua 30%, mestizo 25–30%, Aymara 25%, white 5–15%
Colombia, Republic of	1,138,910	47.4	52.4	76	419.6	9,060	Mestizo 58%, white 20%, mixed white/black 14%
Ecuador, Republic of	283,560	14.9	17.2	66	114.0	7,880	Mestizo 55%, Native American 25%, white 10%, black 10%
Peru, Republic of	1,285,000	30.1	34.4	74	259.6	8,930	Native American 45%, mestizo 37%, white 15%
Venezuela, Republic of	912,050	29.7	35.1	88	350.2	12,150	Mestizo 67%, white 21%, black 10%, Native American 2%
Totals/Averages	**4,718,100**	**132.9**	**151.6**	**74**	**1,189**	**7,802**	
Brazil, Federative Republic of	8,511,970	194.3	210.1	84	2,144.9	11,000	White 55%, Black African 11%, mixed white/black 32%
SOUTHERN SOUTH AMERICA							
Argentine Republic	2,766,890	40.8	46.9	91	629.3	15,570	European descent 85%, mestizo, Native American, other 15%
Chile, Republic of	756,950	17.4	19.3	87	250.5	14,640	European and mestizo 95%, Native American 3%
Paraguay, Republic of	406,750	6.7	8.2	59	32.8	5,080	Mestizo 95%
Uruguay, Eastern Republic of	177,410	3.4	3.5	94	45.7	13,620	European descent 88%, mestizo 8%, black 4%
Totals/Averages	**4,108,000**	**68.3**	**77.9**	**83**	**958**	**11,982**	
Region Totals/Averages	**20,505,680**	**594.2**	**666.3**	**70**	**6,322**	**9,839**	

Source: *World Population Data Sheet 2012*, Population Reference Bureau; Microsoft Encarta 2005.

The region is one of dramatic physical contrasts. The Andes Mountains, the second-highest mountain range in the world, contrast with the huge, low-lying basin of the world's largest river system, the Amazon. The Earth's largest tropical rain forest, the Amazon rain forest, is situated across the mountains from one of the world's driest deserts, the Atacama. The region's great range in latitude produces a variety of climate regimes, which are further altered dramatically by changes in elevation and proximity to mountain ranges. Some countries, such as Mexico, have both tropical beaches and snow-capped mountain peaks.

The contrasts present in Latin America also exist within the region's human geography and are especially evident inside Latin America's large cities and between the growing cities and their hinterland rural areas. Mexico, Brazil, Argentina, Peru, and Chile contain large urban-industrial areas around their major cities. The São Paulo and Mexico City metropolitan areas each contain more than 20 million people and are two of the world's largest urban centers. Extreme contrasts between the region's materially wealthy and materially poor are most dramatic within Latin America's cities.

10.2 Distinctive Physical Geography

Although the Latin America region extends through almost 90 degrees of latitude—the greatest north-south distance of any major world region—the majority of Latin America lies within the tropical latitudes. When coupled with high mountain altitudes, the latitudinal expanse of the region results in a wide variety of climates, natural vegetation types, and soils.

Tectonics and Landforms

The major relief features of Latin America were formed by a combination of tectonic plate interaction and precipitation runoff in major rivers and mountain glaciers. Along the west coast, the South American plate overrides the Nazca plate (see Figure 1.5). The convergent plate margin marks the line of the Andes Mountains and causes earthquakes and volcanic eruptions. The tectonic pattern is more complex in Middle America (Mexico, Central America, and the Caribbean Basin), where the North American, South American, Caribbean, and Cocos plates meet.

Insular and Mainland Middle America

High-altitude plateau lands between the eastern and western Sierra Madres dominate northern Mexico. The plateau rises over 2,000 m (6,000 ft.) and contains shallow basins. The western slopes facing the Pacific Ocean are steep, but those on the east are less steep with wide coastal plains (see "Geography at Work: Agricultural Change in Indigenous Land in the Amazon," p. 328). Prominent mountains continue southward from Mexico through the Isthmus of Panama.

South of the Tehuántepec isthmus in Mexico, the rapidly moving Cocos plate subducts under the slower eastward progression of the Caribbean plate, forming a single spine of mountains along and parallel to the geologically active Pacific coast. There are very narrow coastal plains and some areas without any flat land between mountain and ocean. A large limestone platform emerged from the seafloor to form the Yucatán Peninsula.

Middle America is subject to earthquakes and volcanic eruptions where tectonic plates collide. A major earthquake centered off the west coast of Mexico devastated Mexico City in 1985, and earthquakes twice leveled the Nicaraguan capital city of Managua in the 1900s. There are 25 active volcanoes between northern Mexico and Colombia that periodically spew lava and ash on surrounding areas.

Insular Middle America is also affected by clashing tectonic plates. The eastward-moving Caribbean plate is overriding small portions of the North and South American plates along the Caribbean's eastern boundary. Periodic earthquake and volcanic activity significantly disrupt the lives of island residents. A series of eruptions of the Soufrière Hills volcano on Montserrat buried towns and forced the evacuation and movement of significant percentages of the island's residents in the

Figure 10.2 **The Presidential Palace, Port-au-Prince, Haiti.** Men stand outside the ruins of the Presidential Palace in Haiti's capital city of Port-au-Prince. The city's infrastructure and that of surrounding areas was devastated by a strong earthquake in January 2010. The earthquake killed or injured hundreds of thousands of people. *Photo: © Design Pics/Reynold Mainse/Perspectives/ Getty Images.*

1990s. The volcano began another series of eruptions in 2008, which interrupted the economic recovery from earlier events and attracted the ongoing attention of scientists who continue to closely monitor the volcano. The transform boundary along the northern margin of the Caribbean plate, where it slides past the westward-moving North American plate, was the site of a devastating earthquake in Haiti in 2010. The quake destroyed much of the urban infrastructure of Haiti's capital city, Port-au-Prince, and killed or injured hundreds of thousands of people (Figure 10.2). The Bahamas and an outer group of islands including Anguilla, Barbuda, and Barbados are flat limestone islands constructed of coral reefs on top of subsiding former volcanic peaks (Figure 10.3).

Figure 10.3 **Caribbean Basin: coral islands.** The Exuma Cays, Bahama Islands, are formed of coral reef limestone and rim the Great Bahama Bank for over 150 km (100 mi.). *Photo: © Bruce Dale/ National Geographic Image Collection.*

Mountains and Plateaus

The Andes Mountains have an impact on most of the physical environment of South America. The collision of the South American and Nazca plates produced a volcanic and earthquake-prone western mountain range and a folded and faulted eastern range. The Andes rise to over 6,500 m (20,000 ft.) in Argentina, Chile, Peru, and Ecuador. In Bolivia, Peru, and Ecuador, the central Andes have two main ranges, the Cordillera Occidental (west) and Cordillera Oriental (east) (Figure 10.4). Between the two ranges, a high plateau, the **Altiplano**, is widest in Bolivia and narrows northward into Peru. In Peru, rivers cut deep gorges as they flow northward and eastward to join the Amazon River tributaries.

In Colombia, the Andean ranges divide into three cordilleras—the Occidental, Central, and Oriental. In northern Venezuela, the Cordillera Oriental branches into a further series of lower ranges. The mouths of the Colombian rivers, Lago de Maracaibo, and the Orinoco River delta provide limited areas of lower coastal land between the ranges along the Caribbean coast. Islands such as Trinidad and the Dutch Antilles are extensions of mainland geologic structures.

The southern Andes Mountains dominate the landscapes of Chile and the western parts of Argentina, with their highest points constituting the border between the two countries for most of its length. The snow line on the Andes gets lower toward the southern tip of the continent, where glaciers descend to sea level (Figure 10.5).

On the Chilean side, the Andes come close to the Pacific Ocean in the north. Southward, a coastal range is separated from the main Andes ranges by a series of basins and then a wide continuous valley south of the Chilean capital, Santiago. The coastal range and the valley get lower in height alongside the main Andes range and are ultimately situated below sea level south of Puerto Montt. On the Argentine side of the Andes, deep, river-carved valleys break the front ranges in the north. Their eastern margins have large alluvial fans formed by the deposition of rock material eroded from the mountains. These fans mark both sides of the Andes throughout Chile and Argentina.

Broad plateaus and wide river valleys dominate Brazil's physical environment. Locally, relief is sharp near physical transition zones from plains to plateaus and ridges. The main relief features of Brazil consist of the ancient rocks of the Brazilian Highlands, which are topped in the southeast by layers of lava flows, and the similar ancient rocks of the Guiana Highlands on the Venezuelan border to the north. The Guianas have low coastal plains that were formed by the deposition of sediment brought to the Atlantic Ocean by the Amazon River and then moved westward along the coast by longshore and offshore currents. Inland, these countries rise to the Guiana Highlands plateau. In southern Argentina, the Patagonia Plateau is cut deeply by rivers draining eastward from the Andes.

Tropical to Polar Climates

Middle America

Nearly all of Middle America lies within the tropics (Figure 10.6). East-to-northeast trade winds dominate much of the Caribbean Basin for about two-thirds of the year, contributing to the area's consistent warmth and humidity. Temperatures average around 30°C (86°F) through most of the year. The northern Caribbean islands, Mexico, and even parts of northern Central America are occasionally affected by modified cold air masses, locally referred to as "*nortes,*" that move southward from continental Canada and the United States during the Northern Hemisphere's winters. In Middle America the rainfall generally increases southward, from the arid region that straddles the Mexico–U.S. border toward the coasts of Nicaragua, Costa Rica, and Panama, which receive over 2,600 mm (100 in.) of rain each year.

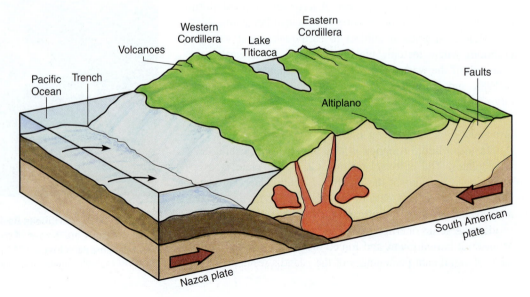

Figure 10.4 **South America: Andes Mountains.** The major features of the central Andes Mountains relate to their formation along an active plate margin. The Nazca plate plunges beneath the South American plate, causing volcanic activity in the Western cordillera along the coast and uplift of the eastern cordillera and the high plateau (altiplano) between the two ranges.

Figure 10.5 **Southern South America: Andes Mountains.**
The Moreno glacier, Glacier National Park, Santa Cruz province, Argentina. The glacier is 5 km (3 mi.) wide, and its front melts on entering Lago Argentino. Such glacial features increase in significance in the southernmost parts of the Andes Mountains.
Photo: © James P. Blair/National Geographic Image Collection/Getty.

Most rains fall in the summer and fall, when heating of the lower atmosphere causes humid air to rise, cool, and condense into clouds. On Caribbean islands with mountains, windward slopes facing the east or northeast trade winds force the moist air upward into cooler atmospheric conditions. This increases precipitation totals. Where the air descends down the leeward slopes, it warms, becomes less humid, and stabilizes, thus producing significantly less rain in a localized, or **microclimate** area, referred to as the **rain shadow**. For example, in Jamaica the windward northeast coast receives over 3,300 mm (130 in.) a year, while the leeward southern coast requires irrigation for farming in areas where the annual rainfall is less than 750 mm (31 in.).

Middle America is a region of annual hurricane activity. The North Atlantic hurricane season runs from June through November. Early in the season, hurricanes form in the Western Caribbean and Gulf of Mexico. During the seasonal peak in August and September, hurricanes form in the eastern Atlantic near the Cape Verde Islands west of Africa. The storms move westward into the Caribbean Basin and curve toward either the northwest or north (affecting the U.S. mainland or shipping channels) or continue westward, striking the northern countries of Central America or the Mexican Yucatán Peninsula. The southernmost countries of Middle America lie outside the hurricane zone. Hurricanes cause both a dramatic loss of life and extensive damage to crops, livestock, personal property, and the vital regional tourism industry. Recovery is often slow in small countries with few resources.

Brazil

Northern Brazil experiences an equatorial climate in which the temperatures hover in the low 30s°C (80s°F), humidity is high, and rain falls in all seasons. On the Brazilian and Guiana Pla-

teaus, the rains are more variable and more seasonally concentrated in the high-sun period, when evaporation and rising air currents are most intense. The variability is particularly marked in the northeastern corner of Brazil, where severe and prolonged drought occurs periodically. In southernmost Brazil, the climate becomes temperate, with cool winters and a shorter growing season.

Andean South America and the El Niño Southern Oscillation

The climate in the southern Andes countries ranges from arid regions in northern Chile and parts of Argentina to one of the world's stormiest and wettest regions in southern Chile. The Andes ranges affect the climates of the lands on either side in differing ways.

In northern Chile, the Atacama Desert is situated in the rain shadow along the coast between the Andes Mountains and the cold Pacific Ocean. Winds blow almost parallel to the coast or offshore, pushing the cold Peruvian current northward along the coast. Heavy, dense cold air over the cold ocean water cannot rise and cool in the atmosphere, and thus it does not produce precipitation; this results in the region's coastal deserts. In southern Chile, by contrast, the midlatitude westerly winds bring precipitation throughout the year. Between the desert and the stormy southern region of Chile is a transition zone of warm, dry summers when the arid climate moves south and wet winters when the storm tracks move north. Agriculture and a booming wine industry thrive in this Mediterranean type (subtropical winter rain) climate zone in Chile.

On the eastern side of the Andes is another contrast between northern and southern lands. In the north, the warm Brazilian current in the Atlantic Ocean allows high rates of evaporation offshore, and trade winds blow humid air into the continent. The Patagonia region of southern Argentina is arid. Westerly airflow descends after crossing the Andes, warms up, and becomes drier. The strong winds blowing across Patagonia, after crossing the Andes, often produce very dry rain shadow conditions at the surface.

The **El Niño Southern Oscillation (ENSO)** is regarded as a significant feature in explaining connections among worldwide climatic environments (Figure 10.7). The major 1997–1998 El Niño event brought drought to parts of Middle America and northern South America, and exceptionally heavy rains to the deserts of Peru and Chile. The same El Niño event was blamed for unusual weather around the world. Droughts in Indonesia and Australia coupled with drought-based fires in Florida and unusually hot weather in southern Europe were indirectly linked to the event. No two El Niño events are the same, and thus the impact on global climatic conditions varies.

In general, the basic features of El Niño are understood. Every two to five years, there is a decrease in the easterly tropical trade winds that usually circulate cooler waters from the Peruvian current westward across the tropical Pacific (where it warms as it progresses westward). Warmer water from the western Pacific thus flows eastward, eventually reaching the coasts of western

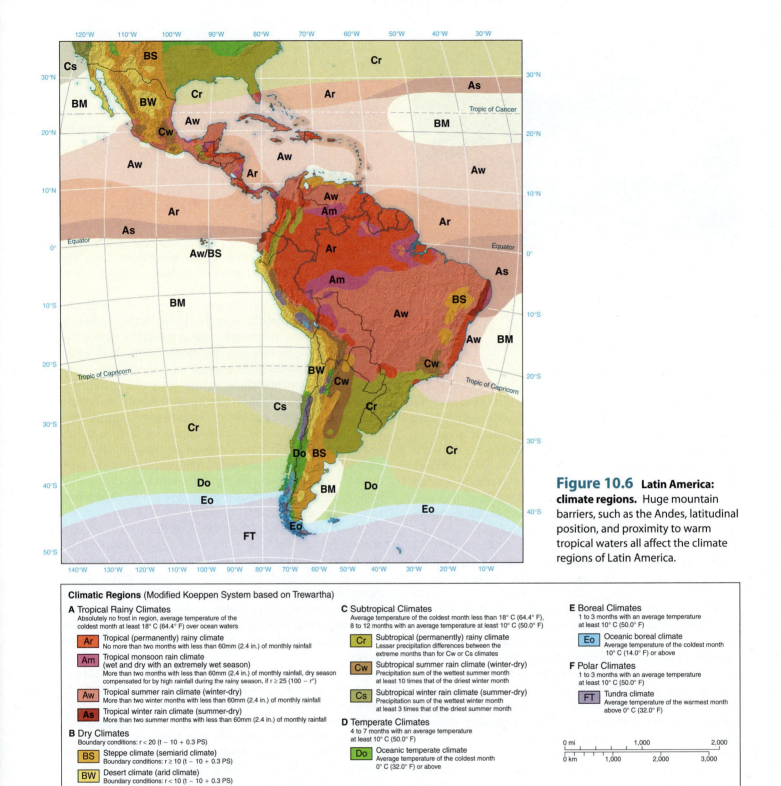

Figure 10.6 **Latin America: climate regions.** Huge mountain barriers, such as the Andes, latitudinal position, and proximity to warm tropical waters all affect the climate regions of Latin America.

Climatic Regions (Modified Koeppen System based on Trewartha)

A Tropical Rainy Climates
Absolutely no frost in region, average temperature of the coldest month at least 18° C (64.4° F) over ocean waters

Ar Tropical (permanently) rainy climate
No more than two months with less than 60mm (2.4 in.) of monthly rainfall

Am Tropical monsoon rain climate
(wet and dry with an extremely wet season)
More than two months with less than 60mm (2.4 in.) of monthly rainfall, dry season compensated for by high rainfall during the rainy season, if r ≥ 25 (100 − r⁰)

Aw Tropical summer rain climate (winter-dry)
More than two winter months with less than 60mm (2.4 in.) of monthly rainfall

As Tropical winter rain climate (summer-dry)
More than two summer months with less than 60mm (2.4 in.) of monthly rainfall

B Dry Climates
Boundary conditions: r < 20 (t − 10 + 0.3 PS)

BS Steppe climate (semiarid climate)
Boundary conditions: r ≥ 10 (t − 10 + 0.3 PS)

BW Desert climate (arid climate)
Boundary conditions: r < 10 (t − 10 + 0.3 PS)

BM Marine dry climate (arid climate)
Humid-air dry climate above ocean surfaces with the universally valid boundary condition of r < 20 (t − 10 + 0.3 PS)

C Subtropical Climates
Average temperature of the coldest month less than 18° C (64.4° F), 8 to 12 months with an average temperature at least 10° C (50.0° F)

Cr Subtropical (permanently) rainy climate
Lesser precipitation differences between the extreme months than for Cw or Cs climates

Cw Subtropical summer rain climate (winter-dry)
Precipitation sum of the wettest summer month at least 10 times that of the driest winter month

Cs Subtropical winter rain climate (summer-dry)
Precipitation sum of the wettest winter month at least 3 times that of the driest summer month

D Temperate Climates
4 to 7 months with an average temperature at least 10° C (50.0° F)

Do Oceanic temperate climate
Average temperature of the coldest month 0° C (32.0° F) or above

E Boreal Climates
1 to 3 months with an average temperature at least 10° C (50.0° F)

Eo Oceanic boreal climate
Average temperature of the coldest month 10° C (14.0° F) or above

F Polar Climates
1 to 3 months with an average temperature at least 10° C (50.0° F)

FT Tundra climate
Average temperature of the warmest month above 0° C (32.0° F)

North, Central, and South America. Warmer and more humid air masses off the west coasts of the Americas alter the regional climate. Warm water also dramatically changes the marine ecosystem, depleting the usually cold water of its nutrients, altering the marine food chain, and producing dramatic reductions in fish catches for the countries of Andean South America.

Major River Basins

Three major river basins between the high mountains and lower plateaus dominate South America. Tributaries of the Orinoco River primarily drain the largest areas of lower land in Venezuela and Eastern Colombia. The Amazon River tributaries flowing from the Andes are muddy "white water" rivers in contrast to

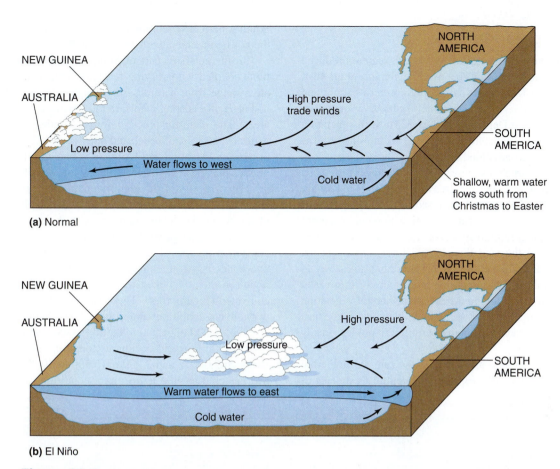

NEW GUINEA

AUSTRALIA

Low pressure

High pressure
trade winds

NORTH
AMERICA

SOUTH
AMERICA

Water flows to west

Cold water

Shallow, warm water
flows south from
Christmas to Easter

(a) Normal

NEW GUINEA

AUSTRALIA

Low pressure

High pressure

NORTH
AMERICA

SOUTH
AMERICA

Warm water flows to east

Cold water

(b) El Niño

Figure 10.7 **The El Niño effect.** During an El Niño event, strong tropical easterly winds (a) diminish, permitting warm water in the western Pacific to flow eastward (b) toward the Americas. *Source: From Bradshaw et al.,* Contemporary World Regional Geography, *3rd ed. Copyright © 2009 by McGraw-Hill Global Education Holdings, LLC. Used with permission.*

the "black" rivers, which contain little sediment as they flow from the plateaus. The contrast is visible where the "black" Rio Negro joins the muddy Solimões just below Manaus in the center of the Amazon basin (Figure 10.8). The Amazon River is navigable well into Peru. In Manaus, Brazil, 2,500 km (1,500 mi.) from the ocean, the Amazon River is 15 km (10 mi.) wide, and over 50 m (160 ft.) deep.

In southern Brazil, rivers drain south to the Paraná-Paraguay River system. The southward-flowing rivers present great hydroelectricity potential in the waterfalls at lava plateau breaks.

Figure 10.8 **Brazil: Amazon rain forest.** A LANDSAT satellite view of the area around Manaus, Brazil, in July 1987, approximately 150 km (100 mi.) across. Red areas indicate tropical rain forest cover. The wide Rio Negro, which contains little silt, is the large dark area in the top half of the image. The blue river is the Solimões branch of the upper Amazon, which brings large quantities of silt from its upper reaches in the Andes Mountains. Manaus is the colored area just west of the Negro-Solimões confluence, with radiating roads leading to and from it north and south of the river. Small white areas with shadows to the west and northeast are clouds. *Photo: Courtesy of Space Imaging, Thornton, CO, USA.*

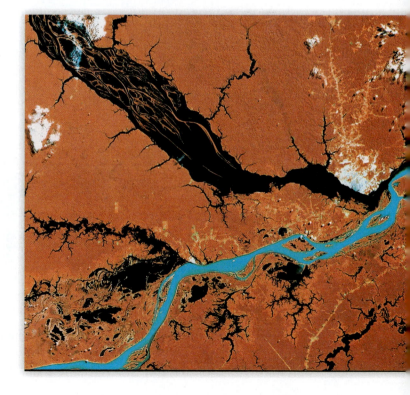

Natural Vegetation

Tropical rain forest is the natural vegetation regime produced where warm, humid conditions persist through most of the year, as in the Amazon River basin, along the northern Pacific and Central American coasts, and on some Caribbean islands (see "Tropical Forests and Deforestation," p. 323). Such vegetation spreads several thousand meters up the east-facing slopes of the Andes (where it becomes cloud forest). The Amazon basin contains the world's largest expanse of tropical rain forest. Soils beneath the forest vary, but the areas of good soils are small, apart from the flooded areas close to sediment-carrying rivers.

Tropical grasslands or shrub vegetation communities dominate where tropical rainfall is seasonal or significantly lower on average during the year than in rain forest areas. On the Brazilian Highlands, dense deciduous woodland gives way to more open woodland with increasing proportions of shrubs and grasses. Soils are generally poor beneath the natural vegetation and need fertilizers to support agriculture, but some of the lava flows capping the plateau in southeastern Brazil produce easily worked soils.

Cold ocean currents, arid air, and the rain shadow from easterly winds produce deserts along the central west coast of South America. Vegetation ranges from tropical plant species in the north to temperate types in the south.

In the southern part of South America, temperate conditions coupled with varying levels of annual precipitation create a range from bare desert, through semiarid bunch grasses and drought-resisting plants, to tall grasses and forest in humid areas. The plentiful precipitation of southern Chile supports natural vegetation of beech and pine forests. The pampas region of central Argentina and Uruguay is named after the tall, lush grasses that grew there before the region was plowed.

The majority of the high mountain ranges of Latin America are located within tropical latitudes. The latitudinal position of the Andes Mountains coupled with their very high altitude results in a series of vertical zones with distinctive climate and vegetation regimes. The greatest number of distinctive vegetative zones is at the equator. The **altitudinal zonation** in the region is directly linked to the variety of crops that may be cultivated (Figure 10.9). The lowest 1,000 m (3,000 ft.) have warm to hot conditions and tropical forest in the *tierra caliente*. The next 1,000 m (3,000 to 6,500 ft.) have mild to warm temperate conditions and deciduous forest in the *tierra templada*. Between 2,000 m and 3,000 m (6,500 to 10,000 ft.), the *tierra fría* has cold to mild temperature conditions and pine forests.

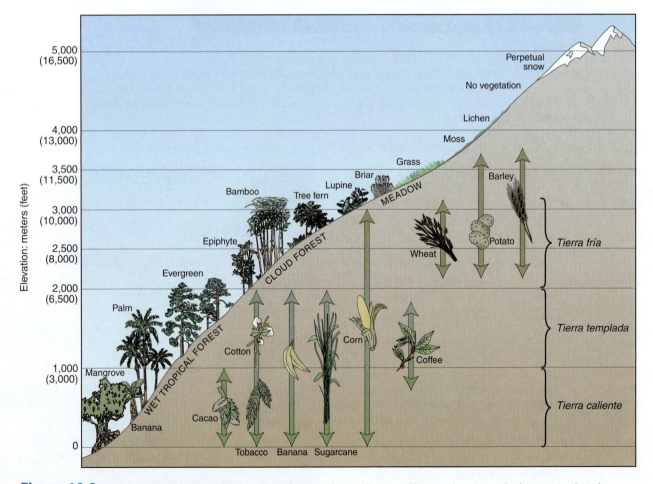

Figure 10.9 **Northern Andes: altitudinal zoning of vegetation and crops.** Climate changes with changes in altitude (combined with latitude), which affects the type of vegetation present and the type of agricultural crops that may be grown. Altitudinal zones often support numerous different types of crop production over relatively short land distances.

Above this zone, there is grassland, and at the greatest heights, in the *tierra helada*, snow cover is present all year, even on the equator.

Natural Resources

The natural resources of Latin America include minerals, soils, forests, water, and marine life. The Andes Mountains and the ancient plateau rocks contain considerable resources of metal ores that were among the early attractions for European settlers. Sedimentary rock basins between the mountains and in offshore areas became petroleum and natural gas reservoirs, especially in eastern Mexico, northern Venezuela, Colombia, the offshore areas of northeastern Brazil, and along the eastern slopes of the Andes in Ecuador, Bolivia, Peru, and Argentina. The alluvial soils in parts of the Amazon River basin, the weathered lava surfaces in southern Brazil, and the pampas of Uruguay and central Argentina form large areas with predominantly good soils for farming.

The water resources of Latin America are huge, including the world's largest river by volume, the Amazon, which carries more than twice the amount of water of the next largest river, the Congo in Africa. Many Latin American countries currently generate a large proportion of their power as hydroelectricity, and they continue to explore the expansion opportunities of this power source. Marine life includes some of the world's richest fisheries along the western coast of Ecuador, Peru, and northern Chile. Other fishing grounds in the Caribbean and off southern Argentina are being developed. Fish production for North American markets is a growing source of income to many countries in Central America and the Caribbean. Such resources are vulnerable, however, to human overuse and natural phenomena such as El Niño.

Environmental Problems

Natural hazards such as earthquakes, volcanic eruptions, and hurricanes bring destruction and death to the Andean and Middle American mountain ranges and to the Caribbean islands. Many environmental problems, however, are related to European colonial patterns and subsequent decades of political instability and corruption. Two of the more serious environmental issues in Latin America are soils and air quality. Perhaps the most serious and globally controversial environmental issue centers on deforestation in the Amazon tropical rain forest (see "Tropical Forests and Deforestation," p. 323).

Soil Erosion

Soil erosion is a major problem in many countries of Latin America, especially where growing populations place additional pressure on subsistence lands. The primary subsistence crop is corn, which is commonly grown in rows on sloped terrain, which contributes to soil erosion during episodes of heavy rainfall. Economic and population pressures increasingly force subsistence farmers to cultivate crops on hillsides, eliminating natural vegetation and removing one of nature's mechanisms for holding soil in place on steep slopes.

Small Caribbean islands colonized for intensive commercial agriculture became notoriously susceptible to soil erosion and other forms of environmental degradation. The combination of agricultural practices and hotel development along hilly areas above the coasts accelerated soil erosion. Clearing natural vegetation on the more hilly islands of the Caribbean causes soil to wash rapidly down the hills and into the surrounding sea, resulting in the accumulation of sediment in the coral reef areas surrounding many of the islands. The **sedimentation** of a reef system may ultimately kill the coral. Fertilizers used to compensate for lost soil nutrients from rapid runoff also enter the sensitive coral reef system, further degrading the health of the reefs.

Air and Water Pollution

Air and water pollution results from mineral extraction and refining, and from the concentration of human activities in urban areas. Mexico City suffers more than any other metropolitan area in Latin America from air pollution. Mexico City is a very densely populated urban center situated in a bowl-shaped depression known as the Central Valley of Mexico. The city sits on the valley floor at approximately 2,200 m (7,000 ft.) above sea level, surrounded by much higher mountains which virtually form walls around the urban expanse. The air in and just above the city is often trapped in the valley and becomes extremely stagnant. Vehicle exhaust, coupled with industrial and domestic pollutants, collects in the valley's stagnant air masses and settles near the valley floor where millions of people breathe. Naturally occurring **temperature inversions**, or periods during the winter months when cold, dense air remains "trapped" at the surface under warmer air for several days or even weeks, further complicate the air pollution phenomenon of Mexico City. Although migration and natural population increase are now slower than they were a few decades ago in the Central Valley of Mexico, population numbers continue to challenge pollution reduction efforts.

10.3 Distinctive Human Geography

Ethnocultural Diversity

Three groups dominate Mexico's population: Native Americans (around 30 percent), those of European heritage (9 percent), and people of mixed Native American and European ancestry known as mestizos (60 percent) (see Table 10.1). Native Americans are particularly numerous in the southern province of Chiapas, where language dialects of Maya origin are commonly spoken in preference to Spanish, the official language. Almost 90 percent of Mexicans are nominally Roman Catholic.

The populations of Central American countries are comprised of racial and ethnic groups that became social classes. Native American, or indigenous peoples, remain in large numbers in western Guatemala (Figure 10.10) and comprise significant minorities in the other countries of the subregion. Many indigenous groups in Guatemala and other Central American

Figure 10.10 **The peoples of Central America.** Indigenous people comprise a significant portion of the population of Guatemala. Patterns and colors present in clothing signify the home village or region for many Guatemalans. *Photo: © Bill Gentile/Corbis.*

locations do not speak Spanish as their first language. Differences in language, social customs, and livelihood marginalize such communities from those with political power and economic wealth. Several indigenous groups in Honduras and Panama have successfully achieved a measure of autonomous status within the federal systems of those countries. Costa Rica has the smallest indigenous population of the subregion.

Along the east coast of Nicaragua and the northern and eastern coast of Honduras are many communities of English-speaking Afro-Caribbean peoples, mixtures of Miskito Indians and blacks from Jamaica and other Caribbean islands, and even pockets of English-speaking whites. The Spanish largely ignored coastal activity in this region during the colonial era, which opened the coasts to decades of influence and control by the British. The British forced the migration of blacks from the Caribbean to work primarily in agricultural production. The historical combination of African and Caribbean cultural influence coupled with the imprint of British rule created a very distinctive cultural landscape region that remains strong today. Belize also has English-speaking Afro-Caribbean peoples.

Very little of the indigenous population of the Caribbean Basin survived the early days of European settlement, succumbing especially to the diseases brought across the Atlantic. Although a few Caribs live on Dominica and St. Vincent and some Arawaks on Aruba and in French Guiana, the vast majority of the regional population traces its origin to the colonial occupation. Residents of the Caribbean speak a variety of languages, from French and English to regional hybrids of Patois and Papiamento. Religious practices are equally diverse.

The Andean countries contain racial and ethnic contrasts that have important economic, social, and political effects. In Colombia and Venezuela, Native American groups live primarily in highland and interior areas that were not taken over by Hispanic or mestizo groups. In Ecuador, Peru, and Bolivia, indigenous peoples also tend to be concentrated in the highlands. Smaller numbers of different Native American groups occur in the Amazon River lowlands, totally isolated from

European intrusions until the 1950s. Current settlement for European descendants and mestizo groups reflects colonial occupations of upland basins and coastal locations that provided overseas links to Spain. Workers from Japan, China, and Africa immigrated to work in the coastal irrigated farmlands of Peru and established neighborhoods near their work sites. The economic and social differences between European-origin and Native American peoples continue to raise tensions within the countries of the Northern Andes.

Although the large Brazilian population is extremely varied ethnically, the citizens of the country maintain a strong sense of national pride. The original Native American population is much reduced: in the Amazon River basin, it was probably around 3.5 million in AD 1500 but is now closer to 200,000. People of Portuguese and other European heritage make up a major proportion of the current population, with the highest concentrations in the southern states and elite areas of the cities. The descendants of African slaves form a significant minority proportion of the population along the northeastern coast and have spread into the southern cities. Brazilians of African decent have had a major impact on Brazilian art, food, music, and dance. Immigrants over the last 30 years include large numbers of Japanese who have easily insinuated themselves into Brazilian business and political communities. Japanese Brazilians form the largest Japanese community living outside of Japan. Mixtures of people of African, European, and Native American descent comprise a sizable and growing portion of the Brazilian population (Figure 10.11).

The proportions of immigrant population give different emphases to the racial and ethnic mixes of the Southern South American countries. In Chile, around 40 percent of the population claims to be of European heritage, with the remainder considered to be mestizo. In Paraguay, virtually all the people are mestizos apart from a few communities of European or Asian origin. Argentina and Uruguay have the largest proportions of European heritage and the smallest proportions of mixed populations.

Figure 10.11 **Brazil: the people.** Brazilians watch a World Cup soccer match between their country and the United States on TV. The crowd includes a variety of peoples with origins in Africa, Europe, Asia, and the Americas, together with those of mixed ancestral heritage. *Photo: © John Maier, Jr./The Image Works.*

Latin American Population Patterns

The Spatial Distribution of People

The population of Mexico is concentrated in the country's central region, from Guadalajara in the west through Mexico City to Veracruz in the east (Figure 10.12). This central plateau and the valleys cutting into it formed both an indigenous and a colonial hearth, and more recently became the center of industrial development and government functions focused on Mexico City. In Central America, the main concentrations of people are in and around the largest cities—often in the highlands and closer to the west coast, where temperatures and soils are better for cultivation. Rural-to-urban migration continues in Panama where the capital, Panama City, is undergoing rapid growth. Haiti and Puerto Rico have very high population densities, while other islands are more sparsely populated. In countries of the northern Andes Mountains, population distribution reflects the Spanish pattern of colonial settlement and the utilization of harbors and fertile valleys. The distribution of population in Brazil is a combination of both historical and contemporary regional development goals. The highest densities are in the southeast, around and inland of São Paulo and Rio de Janeiro. Moderate densities occur in a band parallel to the coast

Figure 10.12 Latin America: population distribution. High population densities are evident in the plateau areas of central Mexico, throughout the highland areas of Central America, the western islands of the Caribbean, and along the coastal periphery of South America. Population patterns reflect indigenous and colonial settlement patterns, as well as climate, soil conditions, and proximity to water.

from the southeast around Pôrto Alegre to west of Fortaleza in the north. Farther inland, the very low densities of the Amazon rain forest area create a major geographic contrast within the country. The main population centers in Southern South America are around the Río de la Plata estuary (Argentina, Uruguay) and in central Chile. Smaller centers occur in the irrigated farming oases of northern Argentina and around Asunción, capital of Paraguay.

Natural Increase and Migration Patterns

Although fertility rates and birth rates in Mexico have been declining for a couple of decades, the country's 2012 population of 116.1 million is projected to rise to nearly 131.0 million by 2025 (see Table 10.1). The age-sex diagram (Figure 10.13) shows a decline in births that was achieved through a well-developed family planning program, halving total fertility rates from about 6 in 1976 to under 2.3 in 2012. The very low death rate and increasing life expectancy, however, maintain a gap between births and deaths that keeps population totals increasing.

Central America's population is growing rapidly, and the 2012 subregion population of 44.1 million is expected to increase to 53.6 million by 2025. Such growth will place increasing stress on the already pressured natural environment and the political, economic, and social systems of these countries. Very high fertility rates in the 1960s and 1970s—over 6 children per woman, apart from Costa Rica (fewer than 5)—continue to impact the current population growth within the subregion. Fertility rates significantly decreased for the majority of the countries of Central America by 2012, but they remain high relative to those in the United States, Canada, and most European countries.

The total population of the Caribbean Basin more than doubled from 17 million to 38.5 million people from 1965 to 2012. Population pressure is already severe in many islands, with densities of over 600 people per km² (1,500 per mi.²) in Barbados and over 200 in many of the other islands. Only in the Guianas does population density fall below 5 per km².

In the 1960s and 1970s, the Northern Andean countries had annual rates of population increase that were among the highest in the world at 3 to 4 percent. In the 1980s and 1990s, total fertility rates fell from over 5 to around or below 3 by 2012, except in Bolivia. These countries lag behind others in South America with 2012 fertility rates ranging from Colombia's 2.1 to Bolivia's 3.3. The total population of 132.9 million people in mid-2012 may grow to nearly 151.6 million by 2025.

Brazil's population is moving out of a period of rapid growth, which produced the relatively young population of today and a population total that rose rapidly to the mid-2012 figure of 194.3 million people (see Table 10.1). Although Brazil is the world's largest Roman Catholic country and that church in Brazil has a conservative hierarchy, its local priests are often progressive regarding birth control. Moreover, a high proportion of women choose to give birth by a cesarean operation followed by sterilization, a service available free from the government. Studies suggest a significant impact from television on family size choices for many Brazilians. Television programs depicting happy and stable situations for small families seem to influence Brazilian family planning choices. Life expectancy in Brazil remains similar to most other South American countries, at 74 years.

The population of Southern South America is growing more slowly than in the other subregions of Latin America. In 1930, this subregion had nearly 20 percent of Latin America's total population, but it now has less than 12 percent. The 68.3 million people who lived there in mid-2012 could rise to around 77.9 million by 2025 (see Table 10.1). Rates of population growth and fertility are low.

Regional Urban Geography

Mexico's population is highly urban. From 1970 to 2012, estimates of the proportion of the Mexican population in towns increased from 59 to 77 percent (see Table 10.1). Mexico City grew from a population of 500,000 in 1900 (2.5 percent of Mexico's population) to 20.1 million in 2010 (17 percent of Mexico's total population). Mexico City is a classic example of a primate city. In Mexico, the stimulus for movements of people to cities is the increasing growth of urban-based manufacturing and service jobs. The official rates of unemployment, however, remain high, and many jobs are in the informal sector (see "Urban Pressures in Mexico and Brazil," p. 324).

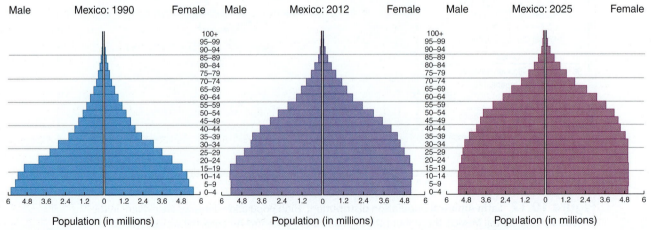

Figure 10.13 **Mexico: age-sex diagram.** *Source: U.S. Census International Database: http://www.census.gov/population/international/data/idb/region.php.*

Total urban percentages are lower in Central America than in Mexico, ranging from a low of 44 percent urban in Belize to 65 percent in Panama. Most countries have a single primate city with populations that are several times that of the second-largest city. Central America's largest cities are the political capitals of each country.

The Northern Andean countries vary in their levels of urbanization. Venezuela was 88 percent urban in 2012, while Colombia and Peru were at 76 and 74 percent respectively (see Table 10.1). Ecuador and Bolivia, however, had lower urban percentages of 66 percent. All these figures show increases of 10 to 20 percent since 1970. Peru's primate urban expanse is the Lima–Callao metropolitan complex, where one-third of the country's population resides. This urban complex extends along the Rimac River valley from Lima to the port of Callao. In Bolivia, metropolitan La Paz has 20 percent of the country's total population.

Leading urban centers in Venezuela include Caracas, Maracaibo, and Valencia (Table 10.2). Recent economic growth in the Pacific coast city of Guayaquil, Ecuador, positions this metropolitan region ahead of the established highland capital of Quito. Colombia has a number of major cities that grew up in relative isolation from one another, producing urban regional

cultural identities that challenge national cohesion. Bogotá, the capital and largest city, has extensive manufacturing industries as well as the national government bureaucracies. Medellín is another major manufacturing center for Colombia, and Cali is linked to one of Colombia's busiest port areas. Both Cali and Barranquilla are rapidly growing urban centers.

Brazil has two of the three largest metropolitan areas in Latin America and a highly urban population overall. The proportion of the population living in towns or urban settings increased from 56 percent in 1970 to 84 percent in 2012 (see Table 10.1). São Paulo is one of the largest metropolitan areas in the world with 19.7 million people in 2012 (see "Urban Pressures in Mexico and Brazil," p. 324). The state of São Paulo produces nearly half of Brazil's GDP and two-thirds of its manufacturing output. In the late 1990s, new industries tended to be sited elsewhere, often farther out in São Paulo state to avoid the high costs of traffic congestion, pollution, and strong union control (high wages and restrictive working conditions). The provision of housing, schools, and health facilities could not keep up with the metropolitan area's growth. Brazil's second-largest metropolitan center is the Rio de Janeiro complex. Rio de Janeiro first grew as a port city for early interior gold mine development, becoming Brazil's largest port and capital city in 1763. Today, Rio de Janeiro is a contemporary culture hearth for Brazilian nationals and a major draw for international visitors to the country.

Southern South America is the most urban subregion in Latin America. Except for Paraguay (59 percent urban in 2012), all the countries of Southern South America had from 87 to 94 percent of their populations living in urban environments. As in many countries of Latin America, each country has a primate metropolitan center. Buenos Aires grew from a population of 170,000 in 1870 to around 13.4 million people, one-third of the Argentine total, in 2012. Montevideo, Uruguay, Santiago, Chile, and Asunción, Paraguay, all contain a significant percentage of their respective countries' total populations. These cities are the centers of government, manufacturing, and service industries. Few other cities in any of the four countries approach the size or important role of these primate cities.

Colonial Legacy

Christopher Columbus embarked from Spain in 1492 on the first of four trans-Atlantic voyages. Within 50 years of the initial voyage of Columbus, much of the region was conquered and occupied. In 1494, Spain and Portugal signed the Treaty of Tordesillas, which established a demarcation line (approximately 46°W longitude) between their global spheres of interest in the Americas, giving the eastern quarter of South America to Portugal (Figure 10.14).

The Spaniards conquered most of the region in the 1500s and imposed a high degree of control on their colony of New Spain by establishing an oppressive system of agricultural production and tightly connected urban settlements and ports. Several forces hindered the Native American inhabitants from forming strong resistance to Spanish intentions in the Americas. The indigenous people's lack of immunity to European diseases, such as smallpox, drastically reduced their populations. Many surviving indigenous communities were forced to convert to Catholicism by

Table 10.2	Populations of Major Urban Centers in Latin America	

TOP 10 MOST POPULATED URBAN CENTERS

City, Country	2012 (millions)	2025 (millions)
1. São Paulo, Brazil	20.2	22.8
2. Mexico City, Mexico	19.5	21.4
3. Buenos Aires, Argentina	13.6	14.7
4. Rio de Janeiro, Brazil	12.0	13.5
5. Lima, Peru	9.1	10.2
6. Bogotá, Colombia	8.7	9.6
7. Santiago, Chile	6.0	6.5
8. Belo Horizonte, Brazil	5.5	6.7
9. Guadalajara, Mexico	4.5	5.1
10. Monterrey, Mexico	4.2	4.7

NUMBER OF MAJOR URBAN CENTERS IN LATIN AMERICA, BY POPULATION CATEGORY

Population Category	Number of Urban Centers
10.00 million and greater	4
5.00 – 9.99 million	4
2.50 – 4.99 million	10
1.00 – 2.49 million	39

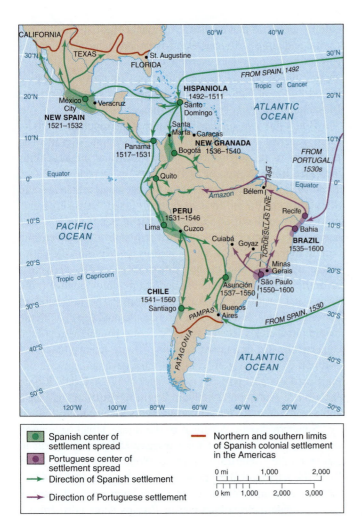

was applied throughout the Spanish colonies for more than 200 years. In addition to the subjugation of native peoples, the Spanish eventually imported more than 1.5 million people from the African continent to labor as slaves in mining and agriculture.

Spanish colonial agriculture developed large production estates, or *haciendas*, to cultivate crops and livestock products for local or domestic markets. Landholders often advanced credit to laborers, causing them to be indebted to the hacienda. Coastal areas around the Caribbean were home to plantations, which produced agricultural products for export to Europe and elsewhere outside of the region.

Inefficient and corrupt colonial administration of the region established a foundation for pervasive and lasting tension between various socioeconomic groups. The native peoples were dispossessed of land and rarely afforded educational opportunities. At the other end of the socioeconomic scale, the *Peninsulares*—Spaniards born in Spain who were working in the New World—took the highest offices and largest land grants. *Criollos*—Spaniards born in the colonies—and the increasing numbers of *mestizos* (mixtures of European and Native American ancestry) had fewer privileges and became resentful as their links to Spain weakened.

The Portuguese installed a governor general in Brazil in 1549. The city of Salvador became the first capital and northeastern hub of Portuguese Brazil. São Paulo was established much farther south to solidify Portuguese territorial claims and to settle southern lands. The Portuguese established a settlement along Rio de Janeiro Bay after expelling French settlers from that portion of the region. The bayside settlement of Rio de Janeiro became the Portuguese colonial capital in the late 1700s. In the 1600s, the discovery of gold inland of Rio de Janeiro led to increased Portuguese immigration. This in turn set off expeditions to explore and claim the interior. In 1750, Spain agreed to permit Brazilian interior expansion westward of the Tordesillas line—which was really an expression of Spain's inability to prevent what was already occurring.

The Portuguese process of occupying Brazil resembled that of Spain in its colonies, although it had some distinctive features. Colonial Brazil was economically and socially controlled by an elite class who purchased or captured millions of African slaves and transported them across the Atlantic to work on large sugar plantations along the northeastern coast. The Portuguese were responsible for the forced migration of more people from the African continent than any other single colonial power. Estimates suggest that more than 4 million Africans were shipped to Brazil to fill the labor needs of economic development. Large numbers of present-day Brazilians trace their ancestral heritage to African slaves.

French, Dutch, and British attempts to colonize Latin America came later and were largely resisted by the Spanish and Portuguese. There were limited incursions on lands controlled by the dominant colonizers, primarily in and on the periphery of the Caribbean Basin.

Independence

The initial wealth that attracted Europeans came from exports of high-value minerals such as silver, gold, and gemstones from

Figure 10.14 **Latin America: colonial conquest and settlement.** During the 1500s, Spanish and Portuguese adventurers established bases and rapidly conquered vast territories. The areas of initial conquest became centers of administration and development. Spain and Portugal signed the Treaty of Tordesillas in 1494, which established a demarcation line separating their colonization of Latin America.

the powerful Roman Catholic Church. Native peoples were also forced to give up their subsistence agricultural practices to provide crops and livestock for the food needs of the foreign city dwellers. Mining settlements in the Andes Mountains used various forms of slavery to coerce the local people, which further challenged the strength and survival of the region's indigenous people.

The colonial period in Latin America left a legacy of underdevelopment because it primarily focused on mining for export and the imposition of large landed estates with a strict feudal system. Native peoples and many of mixed ancestry were relegated to being landless laborers. By the end of the Spanish colonial period, antagonisms between privileged and underprivileged groups were ingrained. Subsequent history is largely a record of such antagonisms at work.

Spain granted large land areas to nobles, soldiers, and church dignitaries, in conjunction with the responsibility for political control. These landowners were given jurisdiction over native peoples to use them as laborers. This feudal *encomienda*, or tribute system,

Latin American colonies to Europe. When the flows of such wealth slowed in the 1700s, combined with the devotion of Iberian resources to fight the Napoleonic conquest of Spain in the 1790s, the colonial powers of both Spain and Portugal weakened considerably. The dilution of Iberian strength enabled many Latin American countries to gain independence in the early 1800s. The new countries largely emerged from administrative divisions within the Spanish viceroyalties.

Countries newly independent from Spain were poorly prepared to capitalize on their sovereignty. The next 150 years after independence were marked by political instability punctuated by short periods of economic growth. Spain left behind a geographic system based on mineral exploitation, transportation to a limited number of ports, and large estates engaged in raising livestock. The post-independence period created rivalries and boundary disputes between countries. Rigid social class stratification fueled deep and lasting internal resentments. In most former Spanish colonies, government control alternated between the elite conservative landowning groups, often allied to the military, and the more liberal *criollos* desiring more democratic governance.

Independence came in the 1820s to Brazil, which had a stronger foundation in place for modernization than other post-colonial territories. In 1807, the Portuguese royal family evacuated from Europe to Brazil when Napoleon threatened to take their country. Rio de Janeiro became the temporary capital of Portuguese government. Although the royal family returned to Portugal after Napoleon's defeat, the prince regent returned to Brazil in 1816 at a time of flourishing revolutions in the Spanish colonies. In 1822, he proclaimed Brazil's independence from Portugal and himself king as Pedro I. Under King Pedro II (1840–1889), the Brazilian economy grew rapidly with the construction of railroads and ports and the expansion of mining in the east-central parts, commercial farming of coffee in the south, and rubber collecting in the Amazon River basin. In 1889, a military coup made Brazil a republic, but falling prices of coffee and rubber led to widespread unrest and a period of dictatorship.

The United States first asserted its perceived right to influence affairs in Latin America by formulating the **Monroe Doctrine** in 1823, in which it asserted its role as the geopolitical leader of the Western Hemisphere. From the late 1800s, the United States intervened in the affairs of Cuba, the Dominican Republic, Haiti, and Panama (among other regional countries) on several occasions. Increasing economic ties throughout the 1900s culminated in the 1990s with the acceptance by Mexico, the United States, and Canada of the **North American Free Trade Agreement (NAFTA)**. In the early 2000s, a unique countertrend eroded some of the dominant unidirectional flow of influence between the United States and Latin America. Growing Latin American immigrant communities in the United States increasingly asserted their political voice while the payments, or **remittances** they sent to family members in Latin American countries, continued to significantly influence regional economies. In the 1990s and early 2000s, Latin American popular culture began to rival U.S. homegrown popular culture. Latin American influences in the U.S. restaurant industry thrived, and Latin American popular music topped the U.S. charts.

Numerous elections in the early 2000s in Latin American countries brought center-left to far-left political parties to power in places such as Venezuela, Bolivia, and Peru. Outspoken leaders, such as Venezuela's Hugo Chavez, vehemently opposed U.S. economic influence in the region. Venezuela, Bolivia, and Cuba signed the **Bolivarian Alternative for the Americas** (known by its Spanish acronym—**ALBA**) agreement, a political and economic pact based in part on the exclusion of the United States. Bolivia nationalized its oil and gas industries in May 2006, a move that hurt investors such as Brazil's government-owned oil and gas company, Petrobras. Brazil's territorial expanse and economic size created an internal perception in recent decades that Brazil was the regional leader in South America's political and economic affairs. Venezuela's oil reserves and current political influence, coupled with moves such as the Bolivian oil and gas nationalization, present formidable challenges to Brazil's desire for regional leadership.

Economic Geography

Geopolitics and the Global Economy

Import Substitution The global economic depression of the 1930s and World War II caused many decision makers in Latin American countries to strive to be more internally self-sufficient. The countries of the region established the goal of becoming less dependent on selling unprocessed or unrefined raw materials in exchange for high-priced manufactured goods from industrial countries. Under import substitution, Latin American countries attempted to use their raw materials in their own internal production of various manufactures for domestic markets. Governments established high tariffs, quotas, and bureaucratic barriers on goods arriving from countries outside of the region. The various country governments owned many of the new industries, and others, established earlier by foreign interests, were nationalized, or taken over by the government of the Latin American country in which they operated. The process proved to be costly for many Latin American countries as they incurred large debts to fund the construction of industry and the purchase of manufacturing equipment. Import substitution policies contributed to the rapid growth of one or two major urban centers in each country by the 1970s. Because of their larger home markets, the most populous countries (Argentina, Brazil, Colombia, and Mexico) produced the most under this system. Economic growth continued to be uneven.

Oil Crisis and Debt In the 1970s, geopolitical events in the Middle East contributed to rapidly rising oil prices and diminished supply to world markets. Oil producers initially incurred increased revenues from the higher oil prices. They invested their revenues in European and U.S. banks. These **petrodollars** were urged on Latin American countries in the form of loans. Such loans were used to pay for oil imports in the nonproducing countries and major infrastructure projects in oil-producing countries such as Brazil, Mexico, and Venezuela—and to improve some government officials' overseas bank accounts.

High oil prices contributed to a global recession in the more materially wealthy countries, weakened markets for products from Latin America, and resulted in much higher interest rates on the loans. The combination of debts and falling export income caused many Latin American countries to default on debt payments by the mid-1980s, resulting in the 1980s distinction as the "**Lost Decade**." The Brazilian government had to devote its large overseas trade balance, generated by import-substitution industries, plus sales of mineral and farm products, to servicing its extensive international debts. Brazil's debt servicing reduced its ability to invest in domestic production, thereby dramatically diminishing economic progress. The high interest rates caused foreign investment and aid to dry up in the late 1980s, and debt liabilities forced countries to emerge from reliance on their internal markets.

Protectionism and Structural Adjustment

The current global connections of many economies in the region are partially a legacy of past economic institutions and international relations and partially a response to 1990s changes in governmental approaches to participation in the world economy. Elected governments were in power everywhere (except Cuba), and emphasis shifted from inward-looking policies to the need for cooperation. The policies that focused on government-run industry and the protection of domestic products through tariffs, quotas, and red tape, or **protectionism**, involved high levels of government intervention and highlighted the lack of capital available for internal investments. These policies gave way to those of structural adjustment, urged by the World Bank and the International Monetary Fund, which encouraged less government spending, more foreign investment, and more export industries. The new policies also opened markets to foreign products and privatized government corporations. Structural adjustment policies were partly forced on Latin American countries as a means of reducing debt burdens incurred during the 1970s and 1980s. As with other simplistic approaches to development, this economic restructuring superimposed on the Latin American regional culture was subject to potential disasters, as Mexico found in 1994–1995 when its economy opened too rapidly, sucking in imports and capital investments and creating a huge trade imbalance.

Trade

NAFTA remains the largest and most globally competitive trade bloc in the Americas. Regional trade blocs, such as the **Caribbean Community Common Market (CARICOM)** cannot compete with much more globally significant organizations such as NAFTA, the EU, and ASEAN. Although the **Common Market of the South (MERCOSUR)** was among the world's largest trade blocs in 2012, its global significance remained challenged by a lack of cohesion among the countries of South America. Some South American countries have no affiliation with the trade bloc, while others maintain associate membership and prioritize their roles in other regional trade associations. Venezuela joined MERCOSUR in 2012, which created tension among some members and caused concern regarding the impact Venezuela's con-

troversial leader, Hugo Chavez, would have on trade relations with the United States and Europe. Attempts by the United States and several Latin American countries to create what would be the world's largest free trade bloc, a **Free Trade Area of the Americas (FTAA)**, were unsuccessful and unrealized in 2012 due to a lack of ratification of FTAA agreements. The newly formed ALBA trade bloc presents a further challenge to MERCOSUR and other regional associations. Membership in ALBA requires a rejection of formal trade relations with the United States.

Mexico

Mexico is by far the most economically developed of the countries of Middle America and one of the strongest economies of the Latin American region. The country's GNI PPP is among the highest in Latin America (Figure 10.15), but its economic growth and stability are challenged by millions of Mexicans who live below the poverty line. In 2012, Mexico had a total GNI PPP that accounted for over 80 percent of Middle America's total GNI PPP. Mexican GNI PPP per capita is one of the highest in Latin America and near the top of the World Bank's upper-middle income group. Ownership of consumer goods is well above that in the countries of Central America apart from Costa Rica, but it is below that of many regional countries such as the Bahamas, Brazil, and Chile (Figure 10.16). Mexico's HDI and GDI are much higher in rank than Brazil's, but not as high as Argentina's.

Farming employs one-fourth of Mexico's labor force. Manufacturing, combined with jobs in service and government, account for most of the rest of Mexico's labor. Tourism

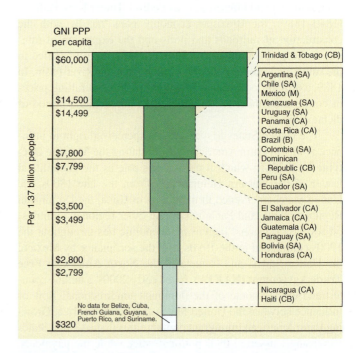

Figure 10.15 **Latin America: country average incomes compared.** The countries are listed in the order of the GNI PPP per capita for 2007. M=Mexico; CA=Central America; CB=Caribbean; B=Brazil; SA=South America. Many small Caribbean Basin countries are not included because of unavailable data. *Source: Data (for 2007) from* World Development Indicators, *World Bank 2008.*

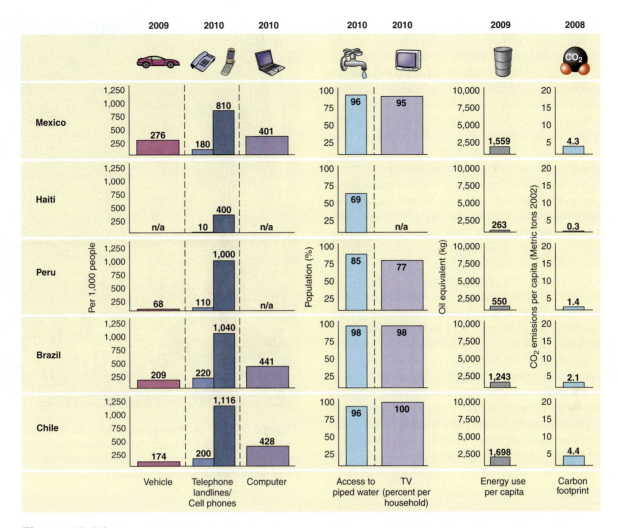

Figure 10.16 Latin America: ownership of consumer goods, access to piped water, and energy usage. Contrast conditions in the materially wealthy, moderately wealthy, and poor countries. *Source: Data from* 2012 World Development Indicators, *World Bank 2012.*

is a major and growing source of income and employment for Mexico, with 22.3 million visitors in 2010 (up from 17 million in 1990). Mexico is a regional leader in developing a tourism industry; it is one of the first Latin American countries to have a separate ministry of tourism with dedicated funds. Mexico City, Pacific coastal resorts such as Cabo San Lucas, Mazatlán, and Acapulco, well-preserved ruins of Mayan city-states, and the Caribbean coastal resorts of the Yucatán Peninsula are major tourist attractions for Mexicans, Latin Americans, and visitors from North America, Europe, and Asia.

While diversified urban-industrial economies develop in Mexico's center and north, the east coasts facing the Gulf of Mexico supply much of Mexico's wealth from large oil and natural gas fields. Oil production had a major impact on the economy of the coastal area between Tampico and Campeche, and recent finds extended this oil-producing area inland. Although Mexico was among the top 10 oil producers in the world in 2012, a growing domestic demand, and government control of Pemex, Mexico's government-owned oil company, contributed to a lack of growth in foreign revenue from oil sales.

Central America

In contrast to Mexico's diversifying economy, the Central American countries have a much narrower economic base, resulting in slower economic and social infrastructure development. Guatemala's capital, Guatemala City, has manufacturing industries, but recent influxes of people swamped the job markets, and half of the working-age population was unemployed at the time of this writing. The lowland Petén area in northern Guatemala is forested and less populated than parts of the country to the south. Although dramatic archaeological relics of the Maya civilization attract tourists (nearly 1.9 million in 2010) and scholars from all global regions, concerns over kidnappings and crime challenge tourism developers' ability to attract larger numbers of foreign visitors. Although tourism brings some revenue to the Petén, much of the forest resource is being cleared, which could ultimately hurt the tourism industry.

Honduras is one of the most materially poor countries in the Western Hemisphere. Barely 25 percent of its land can be used for farming. The ranching economy established in colonial times persists in parts of the west, and large banana and pineapple plantations dominate the Ulua River valley and the northern coastal plain. Coffee became a primary export in the 1960s. Shrimp farming on the Pacific Ocean coast grew rapidly during the 1990s making Honduras one of Latin America's largest shrimp exporters. Although shrimp sales to U.S. markets provide Honduras with foreign income, the water-based farming of shrimp is degrading to important mangrove ecosystems along the country's Pacific Coast. Tourism is a growing industry, attracting 896,000 visitors in 2010, centered on Maya ruins in Copán, nature tourism-related mountain hikes and river rafting, and a relatively thriving scuba diving and beach attraction along the north coast and in the Bay Islands. Much of the former forest cover was logged, and even the secondary growth of pines is being removed rapidly. Manufacturing was encouraged after 1950, and a growing maquiladora industrial belt is forming around the city of San Pedro Sula.

Costa Rica has a more prosperous and diversified farming industry than the other countries, based mainly on coffee production along the Pacific slopes, bananas and cacao on the eastern coast, and a range of new crops such as tropical fruits, ornamental plants, cut flowers, and tropical nuts. Areas of former rain forest are used for livestock production. Costa Rica has many successful industries, including Microsoft facilities, textiles and pharmaceuticals for export, and cement, tires, and car assembly for the domestic market. Costa Rica has the subregion's most developed tourism infrastructure and attracts more visitors each year than any other Central American country. More than 2 million foreign tourists visited Costa Rica in 2010.

Caribbean Tourism and Other Economic Activity

Caribbean Basin tourist destinations cater to a range of visitor interests from the elite to the mass markets. The gleaming beaches and turquoise water of the region are its primary attraction (Figure 10.17). The islands of the Caribbean also provide unique historical and cultural experiences, and a wide range in topography from the flat Cayman Islands to the rugged hills of Jamaica and Dominica. The United States now provides two-thirds of the tourists, while most of the rest come from Canada, Europe, and Japan. Puerto Rico, the Bahamas, and the Dominican Republic attract the greatest number of visitors, while the Virgin Islands, Antigua, Guadeloupe, and Martinique continue to grow in significance.

Initial tourism development consisted largely of foreign-owned and managed resorts, many of which were all-inclusive. Foreign control of tourist infrastructure resulted in severe rates of **economic leakage** for many areas. Such leakage occurs when tourism flows to the foreign owners/investors and foreign employees of tourism businesses (hotels, restaurants, gift shops, bars, and dive shops) and thus "leaks" from the local economy. Although many foreign-owned, all-inclusive resorts remain in the Caribbean, some governments are increasingly taking control of tourism infrastructure and marketing and are trying to involve local communities in tourist infrastructure investment and management.

The number of cruise ship passengers visiting Caribbean ports continues to rise annually. Although a cruise ship may

Figure 10.17 **St. Thomas, Virgin Islands.** While tourists enjoy the sea, sand, and sun of the Caribbean, much of the revenue generated from their vacations may not reach the local economy or community. *Photo: © Neil Rabinowitz/Corbis.*

unload thousands of money-spending tourists into a port town, many of the cruise visitors participate in prepackaged events where revenues generated go to the cruise lines and only a couple of local businesses fortunate enough to be working with the ship's activity planners. Island residents view resort and cruise tourism with mixed emotions. Some see the industry as economically beneficial, providing jobs, housing, and schools, while others do not reap any direct economic benefits and resent the crowds, crime, resource consumption, and pollution that may accompany tourism growth in many areas.

Tourism is extremely susceptible to external forces. Economic recessions in the wealthier countries from which visitors originate, destination-based crime or terrorism, or media reports of natural disasters such as volcanic eruptions or hurricane damage all may produce immediate reactions in the form of dramatic declines in tourist visits. Although tourism brings in foreign currency, much of it is used to buy the foods and equipment expected by the tourists.

Agriculture is another major industry in the Caribbean Basin. Commercial cultivation of sugarcane (and corresponding rum industries) and bananas are among the region's dominant crops. Cattle ranching and coffee cultivation persist in upland areas of islands with topographic relief.

Mining is prominent in the Dominican Republic and Jamaica, as is manufacturing in Puerto Rico. A booming oil-based economy exists in Trinidad and Tobago. Natural gas and oil production, together with the refining of oil from Southwest Asia, is a major sector of Trinidad's industrial base. Manufacturing industries include petrochemicals and steel, and they employ 15 percent of the labor force, while government and other service jobs provide 50 percent of employment. Tourism is not yet a rapidly developing industry.

Andean South America

Andean economies range from materially impoverished Bolivia to some of the continent's more materially wealthy countries such as Colombia and Venezuela. Although oil and coffee increasingly factor into the subregion's formal economies, the informal economy remains pervasive among the urban poor and through the illegal cultivation, processing, and trafficking of cocaine.

Colombia has a more diversified economy than the other Northern Andean countries, and it was the only country in the subregion to increase its income per capita during the first decade of the 2000s. Medellín, and the capital, Bogotá, became Colombia's primary urban industrial centers for textiles, steel, agricultural equipment, and domestic goods. In the lower valleys and near Colombia's northern coast, large plantations raise tropical cash crops, including cotton, sugarcane, bananas, cacao, and rice for export. Colombia is the world's third-largest producer of coffee, after Brazil and Vietnam. Coffee, Colombia's largest legal export, thrives in the optimum conditions of well-drained mountain slopes, fertile soil, and sufficient rainfall. The country began to grow new varieties of corn and rice in the mid-1990s, which were bred for higher yields in the poor soils of the eastern savannas. These crops, coupled with livestock and locally grown cattle feed, are meeting with success in this sparsely settled area. Colombia is the world's largest producer of cocaine (see "The Northern Andes, Mexico, and the International Drug Trade," p. 326). Colombia is also a mining country with considerable iron, coal, oil, natural gas, gold, and emerald resources. Overseas trade, however, is regularly hampered by continuous internal political strife.

Venezuela's income is directly related to global demand for its oil and mineral exports. The region around metropolitan Caracas is the focus of Venezuela's manufacturing and service industries. Southern Venezuela is a source of mineral wealth from iron and bauxite mines. The mines provide a basis for a manufacturing zone in Ciudad Guyana on the Orinoco River. The manufacturing zone is powered by hydroelectricity that is generated in the Guiana Highland valleys. This region has potential for future economic development but remains isolated by poor transportation facilities and is affected by Venezuela's shortage of investment capital.

Economic growth in Ecuador occurs primarily on the coast around Guayaquil, where the expansion of fishing for tuna and whitefish, together with shrimp farming, is adding to an increasingly diversified base of light manufacturing, commercial farming for sugarcane, bananas, coffee, cacao, rice, and tourism. Oil and related petroleum product production make up nearly half of Ecuador's exports. New laws opened mining prospects for foreign corporations from Canada, South Africa, France, and Belgium that are testing the viability of gold, silver, lead, zinc, and copper deposits.

Global attention remains focused on Ecuador's Galápagos Islands, where unique animal and plant species attract scientists and tourists from all parts of the world (Figure 10.18). The government of Ecuador is caught between the desire to fully exploit the tourism potential of the islands and the need to preserve a world-class natural heritage site, as well as to maintain the conditions that make Galápagos an attractive tourist destination. Visitor quotas, established by the government to protect the ecological balance of the islands, have been increased several times and are often exceeded. The economic potential of the islands could easily be destroyed if too many visitors degrade the health of the animals and natural resources that make the Galápagos such a desirable destination. An even greater challenge to the Galápagos environment may be taking shape as Ecuadorian migrants move from the mainland coast to the islands in search of tourism jobs or to illegally fish in the marine reserve's waters. Uncontrolled population growth and the fishing of protected resources may pose a more destructive and rapid threat to the fragile balance of the Galápagos ecosystem. Ecuador attracted slightly more than 1 million foreign visitors in 2010, a number that likely would have been significantly higher without quotas and limits for the Galápagos.

Peru dramatically opened its economy to external investment in the 1990s, creating a boom in mining exploration. In 1992 Peru opened Latin America's largest gold mine at Yanacocha near Cajamarca. Mining production grew through the first decade of the 2000s, and Peru's economy grew significantly from increased mining output and expanded exports of fish meal, agricultural products, textiles, copper, and molybdenum (a type of metal used as a steel alloy). Peru also remains one of the world's largest producers of cocaine. Revenue generated from cocaine may cushion some of Peru's economic challenges, but distribution of wealth related to the drug trade is extremely uneven. Prominent indigenous sites, such as Macchu Picchu, continued to attract tourists from around the world

Figure 10.18 Galápagos. The volcanic landscape of the Galápagos Islands. *Photo: © David Zurick.*

and contributed to bringing 2.3 million foreign visitors to Peru in 2010, making Peru the leading Andean country in terms of attracting foreign visitors.

Bolivia's mining output of tin and silver is less in demand on world markets than it used to be. The country lacks an ocean outlet since a war over phosphate deposits in the 1800s resulted in Chile taking over its route to the sea at Arica. Bolivia has South America's second-largest natural gas reserves (Venezuela has the continent's largest natural gas reserves). Natural gas is a highly contentious and politically charged natural resource in Bolivia, where elections are often determined by the candidates' positions regarding the natural gas industry. Exports of silver and tin dominated the Bolivian economy for decades. A 1980s fall in world tin prices hurt Bolivia's limited economy. The Bolivian economy today depends on other metallic ores, such as gold, which comprise up to 50 percent of Bolivia's exports. New development is occurring in the eastern interior plains where nutrient-rich soils and improved transportation make it possible to move more soybeans from field to consumer each year. The crops are trucked across the Andes to Peruvian ports for export. Oil from the eastern Andean slopes makes Bolivia self-sufficient in energy. Bolivia is the world's second-largest producer of cocaine. United States pressure on the Bolivian government to crack down on production in the Chaparé Valley east of Cochabamba has met with little success due to internal political disagreement and an economically challenged population. Bolivia's 2006 nationalization of its oil and gas industries intensified preexisting investor fears, which may further challenge Bolivia's ability to attract foreign capital.

Brazil

The Brazilian government continues to encourage agricultural production, but agriculture's proportion of GNI fell as that of manufactured goods rose. Coffee and sugarcane remain important exports, but coffee, once the mainstay of the economy, now makes up less than 10 percent of agricultural exports. New developments, such as the growing of oranges and other citrus in the south and tropical fruits and nuts farther north, added to the diversity of commercial farming. The efficient commercial production of soybeans rapidly positioned Brazil as the world's top soybean producer (Figure 10.19). The largest manufacturing sector in Brazil is the production of automobiles and trucks, concentrated in the southern São Paulo suburbs.

The Brazilian government owns many large corporations that employ thousands of citizens. Electrobras is a nationalized corporation that produces and distributes electricity. Small hydroelectricity projects on the plateaus in the east gave way to huge projects in the interior. The world's second-largest hydroelectricity project (China's Three Gorges Dam is the largest) is at the Itaipu Dam on the Paraná River at the border with Paraguay (Figure 10.20). The dam, which opened in 1983 under the joint ownership of the governments of Brazil and Paraguay, produces more than one-fifth of Brazil's power and more than three-fourths of Paraguay's. Brazil's famous beaches and tropical rain forest, as well as its unique culture, food, and music, attracted more than 5 million international tourists in 2010.

Figure 10.19 **Brazil: western Mato Grosso.** A space shuttle photo of new cattle ranches and soybean farms in a 60 km² (40 mi.²) area. The plateau surface has been cleared and only the steep, floodable, intervening valleys remain forested. Lands along the Rio Sangue (bottom) are flooded up to 10 m (40 ft.) deep for three summer months. *Photo: NASA.*

Figure 10.20 **Brazil and Paraguay: hydroelectricity.** The spillway of the Itaipu hydroelectricity station on the Paraná River between Brazil and Paraguay. Eighteen massive turbines make this Brazil's largest hydroelectric investment. Although the electricity is shared between the two countries, Paraguay sells nearly all of its quota to Brazil, where there is greater demand. *Photo: © Michael Bradshaw.*

Economic Activity in Southern South America

Agricultural products such as apples, soft fruits, grapes, and wine, together with fish, lumber, and new minerals, are growing components of the Chilean economy. The agricultural products and lumber come from central Chile, where a large proportion of the population lives. Copper mining is important in the Atacama Desert of northern Chile, with the world's two largest mines at Chuquicamata and Escondido. After the highest-grade copper deposits were mined out by the early 1900s, Chile used U.S. technology to mine lower-grade ores. Its companies now lead the world in refining technology.

Argentina has a larger manufacturing base than Chile. Most of the Argentine factories are based in and around the Buenos Aires metropolitan area. The farm products of the pampas are still important to Argentine trade, but oil and gas resources fuel new industries, mainly on the coast. Argentine service industries developed as part of this growth. In 2010, more than 5.3 million foreign tourists visited centers on the coast and in the mountains. Some older settlements such as the northern cities of Tucumán, Córdoba, and Mendoza, where the economy is still based on irrigation agriculture, do not grow as rapidly because of the dominance of Buenos Aires in attracting manufacturing

investment. Some inland centers, such as Jujuy near the Bolivian border, remain poor and overwhelmed by Bolivian migrants. In the late 1990s, the Mexican crisis and fears of open competition with Brazil led to a slowing of growth in Argentina, which contributed to Argentina's economic crisis of 2001 and subsequent currency devaluation in 2002. Although the country's economy rebounded and grew strongly from 2002, a dramatic increase in federal spending and higher domestic prices for goods due to higher foreign demand for Argentine products combined to produce inflationary increases through the end of the first decade of the 2000s.

10.4 Geographic Diversity

Mexico

Mexico gained its independence from Spain in 1821 through a series of rebellions (Figure 10.21). Political instability resulted in numerous power struggles and dozens of government turnovers in the decades after the end of Spanish rule. Stability came in 1871 with the harshly repressive dictatorship of Porfirio Díaz

Figure 10.21 Mexico: a diversity of regions. Northward, the climate is drier, but economic opportunities increase close to the U.S. border. Southward, the land narrows and localities become increasingly remote from the influence of the Central Valley of Mexico.

at the cost of many social freedoms. Oppression during this period, referred to as the Porfirioto, planted the seeds for revolution that would shape Mexico's political culture for most of the 1900s. Revolutionaries who overthrew the Porfirian government in 1911 created a new land tenure system with the purpose of easing rural poverty. The plots of land created in 1917 became known as *ejidos*, which were state-owned rural cooperatives developed to provide landless peasants with a degree of control over the land they farmed. The government granted the right to use land to individual *ejido* farmers, or *ejiditarios*, as well as the theoretical right to pass land use to their offspring. However, ownership of the land remained with the government. The *ejido* system resulted in the fragmentation of landholdings as the government took control away from some landowners and gave it to *ejiditarios*. In many cases, the government placed already impoverished farmers on marginal land, exacerbating their situation. Although rights to individual property ownership were defined more clearly in the 1990s, many rural, materially poor farmers continued to struggle while large-scale agribusiness took control of more land.

Another long-lived institution arising from the Mexican revolutionary period (1910–1920) is the Institutional Revolutionary Party (*Partido Revolucionario Institucional*, PRI). The PRI grew out of the revolutionary movement into the most powerful and lasting political party in the history of Mexico, dominating Mexican politics for more than 70 years. The party controlled the government, nationalized industries, and allegedly fixed many political votes to remain in power.

Serious debt problems in the 1980s contributed to a reversal of extreme protectionist government policies for Mexican industries, leading to policies that allowed foreign competitors to enter Mexican markets. The Mexican government significantly reduced tariffs and trade restrictions, cut inflation, tried to reduce deficit spending, privatized telecommunications, banking, and agriculture, and reduced a wide range of central government controls. Mexico signed the North American Free Trade Agreement (NAFTA) with Canada and the United States in 1992, and after passage in the respective legislatures, the pact was formally implemented in 1994.

In July 2000, a historic election overturned the PRI's dominance of Mexican politics with the election of Vicente Fox, who represented the National Action Party (*Partido de Acción Nacional*, PAN). Fox and the PAN inherited a vastly corrupt system supported in part by graft and money from illegal drug smuggling to the United States. Mexico's voters went to the polls in July 2006 to elect a replacement for President Fox as he neared the end of his second and final term (ending December 1, 2006). The election results were extremely close (within 0.6 percent) and forced a recount between the PAN, which claimed victory, and the leftist Democratic Revolution Party (*Partido de la Revolución Democrática*, PRD). On December 1, 2006, PAN candidate Felipe Calderon assumed the presidency of Mexico.

Southern Mexico is less developed than northern areas of the country and more reminiscent of the countries of Central America. Most of the region is remote and is populated by Native American subsistence farmers who grow corn on the hillsides and some wheat on the valley floors. Overgrazing by

sheep is common. Poor transport facilities slow development in this region. Contrasting with this poverty, the government-planned tourist resorts such as Acapulco bring tourists to the coast, and new highway links connect the resorts to airports.

Indigenous identity is strongest in the Mexican states of Oaxaca, Yucatán, and Chiapas. Rapid tourism development in parts of the Yucatán Peninsula is dramatically altering the human geography for some while furthering a feeling of isolation for others (Figure 10.22). Over a million Native Americans in Chiapas do not speak Spanish. Local controversy surrounds government-sponsored oil and gas operations, power generation, and the construction of a road connecting Chiapas to Guatemala. The road, which is supported by the Mexican government, the World Bank, and the Plan Puebla Panama, passes through the Sepultura Biosphere Reserve, cutting the reserve in two. Many members of indigenous communities assert that there has been a lack of local involvement in decisions made concerning land and resources in their state. Indigenous communities argue any financial benefit bypasses the residents of Chiapas while it reaches more influential Mexican citizens in the Central Valley. Chiapas also houses thousands of refugees who fled earlier civil disturbances in Guatemala and El Salvador. The combination of dispossessed refugees and local disillusionment with the Mexican treatment of these states contributed to the Zapatista rebellion and regional unrest.

The 1994 challenge mounted by the Zapatista National Liberation Army in Chiapas drew attention to the social costs of economic reform. The Zapatistas made the point that the Native Americans, who comprise one-third of the Chiapas population and were already very materially poor, were further disadvantaged by the economic restructuring and NAFTA. The Native Americans in Chiapas are only a part of the 13.5 million Mexicans who live in "extreme poverty" and a further 23.6 million who are "poor." In the mid-2000s, the Chiapas situation remained unresolved. Reforms promised in 1996 bought time for the government but were not implemented. In response, Zapatistas set up their own autonomous municipalities in defiance of government orders. The Mexican government maintains a large military presence in the region but does little to stimulate dialogue.

Maquila

Towns along the U.S. border, from Tijuana in the west to Matamoros in the east, experienced very rapid growth. The Mexican government enacted the Border Industrial Program (BIP) in the mid-1960s to stimulate growth in northern Mexico and to relieve growth and population pressures from Mexico City and the surrounding Central Valley of Mexico. The BIP, more commonly referred to as the **maquiladora** program, made it possible for foreign-owned factories (*maquila*) situated in Mexico to import components for assembly without customs duties. Such factories utilized significantly cheaper Mexican labor and took advantage of Mexico's less stringent labor and environmental regulations to assemble goods that could be exported, duty free, across the Mexican border and back into the producing country (most often the United States).

Figure 10.22 **Mexico: Cancún resort development.** Tourism growth supports the diffusion of resort hotel development along the Caribbean coast of Mexico's Yucatán Peninsula, while economic conditions worsen in some interior villages. A significant percentage of tourism revenue generated in Cancún and nearby resorts leaks out of the area and into the coffers of foreign investors. *Photo: © Corbis RF.*

U.S. corporations involved in textiles, apparel, electronics, and wood products built their factories in these towns, which continued to grow through 2012. Asian and European corporations joined U.S. facilities in producing under *maquila* laws in the northern zone. Ford, GM, Nissan, IBM, Whirlpool, Eastman Kodak, and Caterpillar are just a few of the hundreds of major foreign companies operating within Mexico's borders.

The concentration of economic activity along the border brought increased air and water pollution. Medical research continues to indicate a strong correlation between the industrial expansion of the maquiladora region and increased rates of cancer and other diseases among the people on both sides of the Mexico–U.S. border. Both Mexico and the United States are implementing programs to reduce industrial pollution in the region. Violent crime, gang violence, and drug trafficking have also flourished in border towns throughout the *maquila* zone across northern Mexico (see "The Northern Andes, Mexico, and the International Drug Trade," p. 326).

Central America

Central America is a land bridge, or **isthmus**, that narrows in width from the Mexican border in the northwest to the Colombian boundary in the southeast (Figure 10.23). The Central American isthmus provides a narrow physical boundary separating marine environments of the Caribbean Sea/Atlantic basin to the east and the Pacific Ocean to the west. Seven independent countries occupy the land of Central America. Belize was a British colony from the late 1800s until 1981 and has a relatively high percentage of people who trace their ancestry to the African continent. Its mainland location, increasing use of the Spanish language, and Roman Catholic populace link Belize to Central America rather than to the Caribbean Basin.

Regional independence from Spain came with a brief union between contemporary Mexico, El Salvador, Guatemala, Honduras, Nicaragua, and Costa Rica as the Mexican Empire from 1821 to 1823. In 1823, the incipient Central American countries separated from Mexico as the United Provinces of Central America, which lasted until 1838 when they each became independent. Panama existed as a province of Colombia until a U.S.-brokered canal deal created a separate country in 1903. Belize remained within the British sphere until 1981.

The four largest countries—Guatemala, El Salvador, Honduras, and Nicaragua—had 2012 populations ranging from Nicaragua's 5.7 million to Guatemala's 13.7 million (see Table 10.1). Coffee plays an important role in the cash crop export-oriented economies of each. Civil unrest and natural disasters have wreaked havoc on the social fabric, physical infrastructure, and economies of all four countries. Deadly conflict, stemming in part from dramatic inequalities in land ownership in Guatemala, El Salvador, and Nicaragua, focused attention on disparity issues but did little to bring about an egalitarian solution.

Figure 10.23 Central America: major geographic features, countries, and cities.

Civil unrest in El Salvador, Guatemala, and Nicaragua eased considerably in the 1990s. The three countries made a slow progress repairing the physical and social damage inflicted by years of conflict and attempting to build more diverse economies in the first decade of the 2000s. Although it remains difficult for the governments of these three countries to attract significant external investment in part because of their violent histories, Nicaragua developed a successful tourism resort, San Juan del Sur, along its Pacific coast in the first decade of the 2000s. In 2012, San Juan del Sur continued to attract foreign visitors, investors, and people from the United States and Europe seeking an affordable vacation property or retirement home along the beach.

Belize, Costa Rica, and Panama have the smallest populations of the Central American countries. Belize is by far the smallest country in the region, having only 300,000 citizens. The country became independent in 1981 after a long period of British colonial rule. Although years of British tenure established an English language base and some cultural elements different from those of neighboring countries, the Hispanic influence of the region exerts a strong force on Belize and is causing the development of a more regionally uniform human geography. Many Guatemalans assert a historic right to the lands and waters of Belize.

Costa Rica, with 4.5 million people in 2012, is the only country in the region that has had a long-term democratic government. This may be related to the legacy of its initial colonial settlement, when there were few Native Americans to be dominated and the people of European origin took up medium-sized farms. As a stable democracy, Costa Rica attracted manufacturing and free trade zones financed by U.S. and Taiwanese corporations. Costa Rica has little civil strife. It does not have an army, but the militarized police and relatively good quality of life help to maintain internal peace. Costa Rica attracts more international tourists than any other Central American country. Costa Rica's beaches, volcanoes, and culture helped the country to attract 2.1 million foreign visitors in 2010. An extensive system of national parks incorporates a variety of ecological habitats and microclimate zones (through altitudinal zonation) and protects a vast array of flora and fauna. Beginning in the 1990s and continuing in 2012, Costa Rica became home for U.S. retirees seeking to take advantage of its climate, stability, quality health care, and low cost of living compared to the United States. In 2012, there were more than one-half million U.S. retirees living in the country, and Costa Rican developers were increasingly targeting aging baby boomers in the United States to sustain Costa Rica's growth in this market.

Countries

Although El Salvador experiences frequent earthquakes, land reform conflict in the 1980s proved far more costly to the country than any natural disaster. Civil war erupted when landowners of vast coffee estates challenged changes imposed by a reformist government. Thousands of people migrated in an attempt to escape the bloodshed. Such large numbers of war refugees moved to the capital, San Salvador, that the economy and infrastructure of the city virtually collapsed. Thousands of Salvadorans fled the country during the civil war and took up residence in the United States, where they established lasting expatriate communities. These foreign-based communities send regular payments to relatives remaining in El Salvador. Such remittances are currently El Salvador's largest source of foreign revenue.

Nicaragua's population is concentrated in an earthquake-prone depression around the lakes of western Nicaragua and the capital city, Managua. Other parts of the country have poor transportation links to this region. Decades of oppression under the Somoza family dictatorship in Nicaragua instigated revolutionaries to overthrow Somoza control. The Marxist-oriented Sandinista revolutionaries took control in 1979 and inflicted another form of oppression and corruption on the people of Nicaragua for the next 11 years. The United States imposed a trade embargo on the Sandinista government, which added to the country's economic disruption and already impoverished social conditions. The United States also backed armies (Contras), partially comprised of former Somoza military personnel, who were fighting against the Sandinistas. The Sandinista party was eventually voted out of office in Nicaragua in 1990. The U.S.-imposed trade embargo devastated the country's economy.

Panama became a separate country when the United States facilitated its independence from Colombia in 1903, in preparation for the construction of the Panama Canal. The newly independent country had an immediate economic and military dependence on the United States, creating a relationship in which Panama functioned more like a U.S. colony than as an independent country. The construction of the canal brought migrants from the Caribbean islands, Asia, and Southern Europe. The canal created a dramatic dichotomy in the human landscape for Panama, which exists to varying degrees today. The corridor along the canal, known as the Canal Zone, is relatively wealthy and developed in contrast to areas of Panama to the east and west. Panama City, the country's capital, retains a cosmopolitan population and an international role in trade and finance. The United States virtually ran the country until 1979 and then invaded in 1989 to protect strategic and economic interests. Control of the Panama Canal reverted entirely to the Panamanian government in 1999, although the United States reserved the right in its treaty with the government to protect the canal should the United States perceive it to be threatened. The United States attempts to maintain close ties to Panama in an effort to combat the northward progression of narcotics trafficking through the region.

In 1991, Taiwanese investors and Panamanian officials established a free trade zone in Panama's Caribbean port city of Colón (the Colón Free Trade Zone), in which some 8,000 people were employed in bulk retail and value-added manufacture of consumer goods by 2012. The free trade zone serves regional customers, such as buyers for stores in Colombia and Venezuela, as well as clients from other world regions such as Europe. The free trade zone enjoyed considerable growth since its inception, and is currently undergoing a major expansion. To the east of the Canal Zone, the Central American Highway tapers off in Panama's impenetrable Darién, an area of dense vegetation between the canal and the Colombian border that hinders the completion of effective north-south highway links. Many believe completion of the highway would dramatically increase the flow of illegal drugs northward from the Northern Andes region of South America.

The Panamanian government is currently undertaking the expansion of the Panama Canal in order to facilitate the traffic of ships, referred to as **post-Panamax**, that are too large to pass through the existing locks (Figure 10.24a, b). The expansion will create a new channel for part of the transit, and two new sets of locks (current transit requires passage through three sets of locks). Environmentalists are concerned that expansion of the canal and increases in ship traffic will deplete fresh water from the Chagres River watershed (Panama's primary freshwater source), while the Panama Canal Authority asserts the expansion design will conserve and recycle a significant amount of the fresh water needed to fill the locks for each ship's transit. The expansion project will create thousands of temporary jobs in Panama and will likely result in increased immigration of workers from Central American and Caribbean countries. Expansion-related immigration will create new human geographies in Panama.

(a)

(b)

Figure 10.24 **Panamax vessels transit the locks of the Panama Canal.** The ships pictured transiting the Pedro Miguel (a) and Miraflores (b) locks of the Panama Canal are known as Panamax vessels, which indicates they are the maximum size of vessel that will fit through the canal's locks. The world's largest retailers, such as The Home Depot, utilize vessels that are currently too big to transit the Panama Canal (known as post-Panamax vessels). The government of Panama won the approval of the country's citizens in the fall of 2006 and is moving forward with a canal expansion project that will enable post-Panamax vessels to transit the canal. *Photos: (a) & (b) Joseph P. Dymond.*

The Caribbean Basin and Environs

The contemporary human and physical geography of the Caribbean Basin reflects dramatic transformations imposed on the region by European countries during the colonial era. The loss of indigenous people to disease, the forced migration of Africans to the basin, and the establishment of extensive sugar cultivation on the islands laid the foundation for the social, environmental, and economic conditions present in the Caribbean today. The Caribbean Basin consists of a few large islands, several small islands, and three political units on the South American mainland (Figure 10.25). In 2012, more than 38 million people lived in the Caribbean Basin. The political geography of the Caribbean is one of the most diverse of any subregion. Caribbean countries vary in size and population from Cuba, with 11.2 million people in 2012, to numerous tiny islands with few inhabitants (see Table 10.1). There is a wide range in the basin's economic diversity from poverty-stricken Haiti to the more affluent Bahamas. The majority of the Caribbean's residents live on one of the four largest islands, known as the Greater Antilles: Cuba, Hispaniola (consisting of two countries: Haiti and the Dominican Republic), Puerto Rico, and Jamaica. The smaller Caribbean islands are commonly referred to as the Lesser Antilles. The Lesser Antilles chain is an arc of small islands around the eastern edge of the Caribbean Sea, with the Leeward Islands in the north (Virgin Islands to Guadeloupe) and the Windward Islands in the south (Dominica to Grenada). Many of the small island countries of the Caribbean have only a few thousand residents.

North of the Caribbean are the Bahamas, which include some 700 islands situated to the east of Florida. Another group of islands, including Trinidad and Tobago, lies along the northern shore of South America. Also included in this subregion are the three Guianas—Guyana, Suriname, and French Guiana. Although situated on the South American mainland, their histories, people, present economies, and cultural expressions are more connected to those of the Caribbean Basin than to those of other mainland South American countries.

Figure 10.25 The Caribbean Basin. Numerous island countries and some overseas colonies vary significantly in size and population.

National Identity

The people of the Caribbean have a very strong sense of belonging to the individual island or country of their birth, as opposed to strong feelings of regional unity or identity. Although the majority of the Caribbean residents are of African ancestral heritage, most people identify with their fellow citizens and perceive readily apparent differences between their nationality and those of neighboring countries. A Jamaican would identify with other Jamaicans, and see her or his national people as different from those of Barbados, rather than perceive any unity from ancestral heritage. National pride is often expressed through rivalries with neighboring island countries. For example, sporting events such as cricket and soccer among the people of Jamaica, Barbados, Trinidad, and Guyana are very competitive, prestigious, and emotional affairs.

Colonial Farming Heritage

The Caribbean colonies of Spain, France, Britain, and the Netherlands assumed a huge significance in the global economy when the Caribbean Basin took over the leading sugarcane production role from Mediterranean countries in the early 1600s. From 1500 to the 1800s, some 10 million African slaves were shipped to the Americas, of which half went to the Caribbean Basin, 39 percent to Brazil, and under 5 percent to British American colonies (including the United States). African peoples brought their own social customs, religious beliefs, handicraft skills, and various forms of artistic expression. At the height of sugar cultivation prosperity in the 1700s, the average plantation unit was 200 acres, used 200 slaves, and produced 200 tons of sugar each year.

Island and Country Distinctions

Puerto Rico is not an independent country, nor is it a state within the U.S. political system. An official relationship between Puerto Rico and the United States began in the early 1900s and became more formally structured in 1952, with the establishment of commonwealth status. Puerto Ricans are considered to be U.S. citizens, yet they do not have true representation in Washington, D.C. Puerto Rico's representatives in the U.S. Congress do not have the right to vote on the passage of bills. Puerto Ricans have open access to migration into the United States, and many have taken advantage of this since the 1950s. U.S. statehood remains an issue, with distinct political views within the commonwealth regarding Puerto Rico's status. The smallest group would like Puerto Rico to become an independent country, free of any U.S. authority. Another group supports statehood. The largest group currently is content to maintain commonwealth status. Although the Puerto Rican government declared Spanish to be the official language in 1991, English as a second language is mandatory in the Puerto Rican public school system.

Puerto Rico has a much higher per capita GNI than other Caribbean islands. Although its sugar industry expanded, its economy remained narrow until after World War II, when farming shifted to dairying and manufacturing increased rapidly. Machinery, metal products, chemicals, pharmaceuticals, oil refining, rubber, plastics, and garments provide a diverse product mix and have grown under special U.S. tax laws for the island. Tourism became a leading economic sector for Puerto Rico, with 3.7 million visitors in 2010. The geography of drug trafficking from Northern Andes cultivation to U.S. markets dramatically changed the landscape of Puerto Rico in the first decade of the 2000s. While the U.S. government focuses its resources on combating drug trafficking across the U.S.-Mexico border, illegal narcotics made inroads into the United States through Puerto Rico, and drug-related crime increased significantly through 2012 on the island.

Haiti, occupying the smaller western part of the island of Hispaniola, is the poorest country in the Americas. Life expectancy is low, infant mortality high, and 50 percent of its adults are illiterate. Over three-fourths of the people are crowded onto poor-quality lands, the result of an imbalanced division of land holdings. Independence in 1804 was followed by the political instability that continues today. The sugar plantations deteriorated, and economic stagnation and decline set in. The United States occupied the country from 1915 to 1934 and again in 1994 to end military takeovers of the government. In between these dates, corrupt and repressive governments produced little economic development, leaving most people dependent on their subsistence plots of land. Some commercial farming for coffee and cacao produced export income, but a 1980 hurricane destroyed many trees, and recovery was slow. In the 1970s, tax incentives attracted U.S. corporations to set up factories in Haiti. Exports of clothes, electronics, and sports equipment became more valuable than farm products. Haiti's mineral resources, however, remain undeveloped. Tourism is not developing in Haiti as it is in other Caribbean countries due to the extreme material poverty, political instability, and a high incidence of HIV/AIDS. Following the military coup against a democratically elected government in 1992, an international embargo on Haitian products caused the economy to collapse, including the loss of vital aid, default on its public debt, and closure of the new assembly industries. Exports and imports halved. A democratic government was reinstated in 1994 after U.S. intervention. Economic growth in Haiti as of 2012 was further challenged by drug-traffic-related corruption and a very slow recovery from a devastating earthquake in the winter of 2010.

The former French colonies of Martinique, Guadeloupe, and French Guiana are political subdivisions of France, known as overseas departments. Their residents are French citizens and members of the European Union. The former Dutch colonies of Aruba, Bonaire, and Curaçao off the northern coast of Venezuela are low-lying and arid, obtaining their water supplies from desalination plants. They remain administratively linked to the Netherlands today. Although Aruba has had a separate status since 1986, it is moving back toward closer political ties with the Dutch government. The people are exceptionally cosmopolitan, including descendants of Africans, Indians, Dutch, Portuguese, Danes, and Jews. Residents are able to speak an old trading language, Papiamento, as well as Dutch (the language of their government), English (the language of tourist visitors), and Spanish (the language of influential neighbor Venezuela). Oil refineries linked to the Dutch development of Venezuelan oil, along with tourism and offshore banking, are among the main sources of income.

Guyana (former British Guiana), Suriname (former Dutch Guiana), and French Guiana all have small populations on relatively large areas of land. Dutch drainage engineers in the early 1800s made the coastal plains of the Guianas habitable. Indentured laborers were shipped in from South Asia to work on the vast sugar plantations near the coast. In both Guyana and Suriname, living standards declined after independence as the result of civil strife. Guyana and Suriname export bauxite from inland mines, while French Guiana exports timber to the EU. The reopening of gold mines by Canadian companies in Guyana caused exports to quadruple through the early 2000s.

United States' corporate ownership of the sugar industry dominated Cuban economic geography in the early 1900s. When Fidel Castro led a Communist takeover in 1959, ties with the United States and U.S. companies were severed, and dependency shifted to the former Soviet bloc countries. Large state farms and cooperative farms replaced confiscated private plantations. Although some farms diversified to produce citrus fruit for Eastern Europe, overall productivity remained relatively low. Manufacturing included import-substitution units for locally needed goods such as textiles, wood products, and chemicals. Cuba exported sugar, tobacco (cigars), and some strategic minerals to the Soviet bloc in exchange for oil, wheat, fertilizer, and equipment. During the 1980s, 85 percent of Cuba's trade was with the Soviet bloc, but this close link was broken with the 1991 dissolution of the Soviet Union. Cuba suffered, as did other Communist countries, from the breakup of the Soviet bloc. From 1989 to 1994, Cuban GNI fell by 34 percent, and its sugar crop fell from 8.4 million tons in 1990 to 3.4 million in 1995. Most farm workers found growing fruit and vegetables for the black market to be more profitable. The United States maintained its trade embargo through the early 2000s. Tourism, however, began to grow as a source of foreign currency, with 2.5 million visitors in 2010. Canadian and other non-U.S. mining companies took over Soviet-instigated mining

projects, primarily for nickel, cobalt, and gold. Cuba formed close economic and political ties in the early 2000s with oil- and natural gas-endowed countries such Venezuela and Bolivia, a move which provided some short-term economic relief from losses related to the Soviet dissolution. After nearly five decades of rule, Fidel Castro suffered serious health challenges in the summer of 2006, and he turned over the reins of government to his brother.

Northern Andes

The Andes Mountains are a dominant feature in all five countries of the Northern Andes subregion, which consists of Bolivia, Colombia, Ecuador, Peru, and Venezuela (Figure 10.26). The world's second-highest mountain range creates a multitude of local environments at different heights throughout the subregion. The geologically active Andes are in the process of uplift, resulting in frequent earthquakes, landslides, and

Figure 10.26 **Northern Andes: main geographic features.** Note the position of the Andes Mountain ranges in each country and how they isolate interior lowlands from the main centers of population and trade.

occasional volcanic eruptions. The dramatic topography of the Andes Mountains isolates large interior sections of the Orinoco and Amazon River basins from global connections.

Mountain-dwelling culture groups exhibit centuries-old social customs in agricultural towns tucked into the valleys of the region's steep terrain. Cash from the illegal drug trade creates a unique elite with dramatically different lifestyles from those in rural mountain villages. Political stability is occasionally challenged by powerful criminal elements and intermittent border disputes, such as the disagreement between Ecuador and Peru over Amazonian territories.

Countries

Although rugged terrain and high relief are common features in each of the Northern Andes countries, topographic and environmental variety is present throughout. The high altitude of the Andes in the western part of the subregion contrasts with and isolates the lower lands to the east. Eastern areas are largely covered by tropical rain forest vegetation, and they experience high rainfall throughout most of the year. The interior reaches of Colombia, Venezuela, and Bolivia experience more seasonal rains and a tropical grassland vegetative cover. The most dramatic contrast of the subregion exists on the Pacific coast of Peru, where air circulation patterns, cold currents, and the rain shadow of the rugged Andes combine to produce one of the most arid climates in the world.

The informal and illegal economies of the countries of the Northern Andes, based on a globally significant role in the production of coca and cocaine, grew rapidly in the 1970s. Increasing demand from the wealthy U.S. market fostered dramatic growth and continues to support coca cultivation and cocaine production. Government efforts thus far have not been successful in eradicating coca production. The number of routes from Colombian processing centers through the Caribbean Basin to the United States is rapidly increasing. Coca-growing areas are diffusing from the eastern slopes of the Andes into Brazil.

Drug-related instability in Colombia improved during the early 2000s. The hardline administration of Álvaro Uribe is credited with dramatically reducing the country's murder and kidnapping rates. Additionally, Uribe offered amnesty to all paramilitary who would turn themselves and their weapons in—a move that significantly decreased the role and influence of the paramilitary. While many credit Uribe with effecting positive changes, others argue that the administration committed gross human rights violations in rural areas of the country in order to achieve its goals. Although Colombia was safer and less violent by 2012 than it was in the 1990s, the guerrilla group known as the Revolutionary Armed Forces of Colombia (FARC) remains a formidable threat to the government and security of the country. Conflict and violence remain, and some parts of the country continue to function outside of the effective control of the government (see "The Northern Andes, Mexico, and the International Drug Trade," p. 326). The United States infuses large amounts of money, primarily aimed at reducing the cultivation of coca. Geographic analysis of aerial photos and satellite imagery has indicated that the overall amount of territory used in coca cultivation has changed very little.

Colombia's cities are extremely contemporary, and urban life is similar to that in cities of the United States and Europe. Bogotá boasts dozens of large shopping malls, museums, multiplex cinemas, thousands of restaurants, and modern recreation facilities. The city government encourages outdoor recreation and has built hundreds of miles of paved bike paths or designated bike lanes through commercial and residential areas (Figure 10.27). Efforts to increase biking in Bogotá serve multiple purposes; civic leaders hope to reduce both traffic and pollution by encouraging greater bike use in the metro area. Bogotá started a program known as *ciclovia* in the early 2000s where major car routes are closed to automobile traffic on Sundays and holidays and are used by families to bike throughout the city. The program has been copied in cities in Australia, the United States, and other Latin American countries. Although life in Bogotá is statistically much safe in 2012 than it was just 15 years ago, parks, malls, and restaurants in upper-income areas remain protected by numerous visibly armed guards.

Brazil

Brazil is the largest country in both area and population in all of Latin America (see Table 10.1). Brazil has three times the area of Argentina, the second-largest South American country. In 2012, Brazil's population exceeded 194.3 million. Politically, Brazil is divided into states that are part of a federal government system (Figure 10.28). The people of Brazil have a strong sense of national pride and a passion for cultural expression. Although the Portuguese language and Roman Catholic Church provide some national homogeneity, the diverse contemporary human geography

Figure 10.27 Formal bike paths: Bogotá, Colombia. Bike paths such as this one were built throughout central and northern Bogotá in the first decade of the 2000s. Bogotá's city government has a very progressive attitude toward biking and other outdoor recreation. This bike path and adjacent paved walkway are located in the median of a very broad shopping avenue in the city. *Photo: © Joseph P. Dymond.*

Figure 10.28 Brazil: major geographic features. The states, rivers, and major cities.

Periodic droughts devastated the interior of this region (as in the late 1990s), causing many to emigrate to other parts of Brazil. The most significant economic development in Brazil occurred in the southeast around Rio de Janeiro, which was the colonial capital and the original national capital for 200 years, and São Paulo. The early industrial focus of the region centered on mining and commercial agriculture (based on coffee cultivation). Today, this region is a major center of manufacturing and financial service activities for Brazil.

The Amazon River network drains 60 percent of Brazil's territory. Much of this area is covered by tropical rain forest. Although the Amazon River basin has relatively low population densities, development efforts led by the Brazilian government are increasingly encroaching on tropical forests of the region, and deforestation is rampant in some parts of the Amazon basin (see "Tropical Forests and Deforestation," p. 323).

The three southern states of Paraná, Santa Catarina, and Rio Grande do Sul became centers of growing population and agriculture from the 1930s. After 1950, the coffee crop exhausted the soils of the area west of São Paulo, causing cultivation to spread to the west-southwest along new railroads into northern Paraná. Cattle and a variety of temperate and subtropical agricultural products, including oranges, are produced in these states, which were settled largely by immigrants who were encouraged to move from Germany, Italy, and Japan. Hydroelectricity generated on the Paraná River and its tributaries powers manufacturing in the area. Inland of São Paulo, straddling the drainage divide between the Amazon and Paraná-Paraguay River systems, is an area of rapid farming expansion known as the *Cerrado*. This vast stretch of savanna to the east and southeast of the Amazon basin became one of the world's main soybean producers, contributing more than 60 percent of Brazil's soy harvest in 2012.

and cultural expression of Latin America's largest country are tangible extensions of the historical mixture of European, African, and indigenous influences. The physical expanse of Brazilian territory incorporates a vast array of topography, climate, and vegetative cover, and holds a diverse and abundant supply of natural resources. Brazil presents an immense economic market, has the greatest economic diversity of any Latin American country, and was among the world's ten largest economies in 2012.

The Portuguese language and unique blend of Portuguese, African, Caribbean, and indigenous traditions give Brazil a national flavor that is distinctively different from Spanish Latin America. The Portuguese plantation economy forced the migration of millions of Africans to Brazil. The contemporary human mosaic includes the largest African and Afro-Caribbean heritage of any Latin American country.

The northeastern coastlands and plateau were the first settled areas of Brazil during the colonial era. The establishment of sugar plantations there began the course of European-induced human and physical geographic change experienced in so many locations in Latin America. Sugarcane plantations prospered until competition with the Caribbean Basin began in the later 1600s.

Modern Mining

The national iron ore company is developing the iron ore mining Carajás Project in the eastern Amazon River basin. A railroad built in 1985 takes the ore to a port near São Luis on the northern coast for export to Japan, the United States, and Europe, while the Tucurui Dam generates hydroelectricity for mining and industrial needs. The controversial project required the clearance of 3 million hectares (7.4 million acres) of rain forest, and much of that area is now used for ranching and small farms.

Other Brazilian mining developments include the production of manganese (used in hardening steel), tin, and bauxite. One-third of the world's bauxite resources occur east of Manaus in the Amazon River basin. In the upper reaches of the basin, small deposits of gold attract thousands of independent miners, but the use of mercury to separate the gold pollutes the rivers, and this activity forms a major intrusion in the lives of Amazon tribes such as the Yanomami.

Brazil was 90 percent dependent on foreign energy sources in the mid-1970s. The oil consumers' crisis of the 1970s severely challenged Brazil's economy and directly contributed to Brazil's incurrence of foreign debt that would last for several decades. An immediate response was the development of an ethanol fuel program (from sugarcane) in 1975. Brazil dramatically transformed its energy needs, gaining complete fuel energy independence by 2006 and becoming an energy exporter. The government steadily increased the acreage devoted to sugarcane cultivation and continued to invest in sugar mills and other ethanol production technologies and facilities during the 1990s and early 2000s. Nearly 100 percent of new autos sold in Brazil by 2010 were flex-fuel vehicles capable of running on gasoline or ethanol blends, and all gasoline sold was a blend of a minimum of 20 percent ethanol. Exports of ethanol to India, South Korea, the United States, and other foreign markets grew steadily into the first decade of the twenty-first century. Petrobras, the Brazilian state oil company, was originally established to import, refine, and distribute oil products but now finds itself an oil producer with offshore wells along the eastern and northern coasts and major reserves in the western Amazon.

Connections to the global economy thrive in the **free trade zone** established in 1966 in the Amazon River city of Manaus (Figure 10.29). In this zone, foreign companies may import materials for assembly without tariffs and export the assembled products to other countries. Manaus is now Brazil's fastest-growing city and the second in manufactured goods value after São Paulo. The local human and physical landscapes are dramatically different from outlying forest areas. There are more than 6,000 factories in the huge industrial parks of Manaus, many of which are the production facilities of foreign-owned companies, including Honda, Sharp, Kodak, Olivetti, Toshiba, Sony, and 3M. Manaus was transformed by the number of visitors from elsewhere in Brazil coming to purchase goods that they were not allowed to import into Brazil. The "electronics bazaar" occupies a maze of streets in the city center. People come to buy foreign-made computers and electronic goods, products that Brazilian companies, protected by high tariffs, make poorly and sell at high prices.

Southern South America

Argentina, Chile, Paraguay, and Uruguay form the southernmost part of South America and are sometimes called the "Southern Cone" because of their combined shape on a map (Figure 10.30). Physically, this long and narrow subregion extends far south into the midlatitudes. Average temperatures are cool, and rainfall amounts increase as one goes from the north to the south in the subregion. The Hispanic cultural imprint is a strong common element of the human geography of this subregion. The majority of the population of each country speaks Spanish as their first language and adheres to Roman Catholicism.

Southern South America has the highest percentage of European descendants among the subregions on the South American continent. Sparse indigenous populations and climatic environments more reminiscent of Europe led to the establishment of pervasive European-based populations in Argentina, Chile, and Uruguay. Argentina's 2012 population of 40.8 million people dominates the four countries of the subregion, with almost 60 percent of Southern South America's people living here.

The diverse physical and human geographic spatial patterns of Southern South America relate in part to the subregion's extensive latitudinal coverage from north to south, a relatively narrow width, the rugged Andes Mountains and geologic instability of the west, and the influences of the cold Pacific and warm Atlantic Ocean currents. All the physical properties of the region support a wide range of temperature and precipitation regimes, resulting in varied settlement and land use patterns. The physical geography of this subregion, coupled with the historic goals of Spain, created a vibrant contemporary human mosaic. The Andes Mountains proved an important dividing factor between types of settlement.

Figure 10.29 Brazil: Amazon River. The city of Manaus is the largest along the Amazon River. It has an opera house that was built early in the 1900s to cater to the 2,000 or so rubber barons who lived there and controlled the valuable rubber trade. The refurbished opera house is in the center of a city of over 1 million people, with high-tech industries, a thriving free trade zone, and improved communications and transportation by river, road, and air. *Photo: © Michael Bradshaw.*

Figure 10.30 **Southern South America: major geographic features.** The countries, Andes Mountains, rivers, and major cities.

The eastern part of Paraguay is more developed, with forest products being diversified by commercial agriculture. Livestock products and some industrial crops are exported. Paraguay gets its power from hydroelectricity and earns foreign exchange from its share of the Itaipu project on the border with Brazil by selling most of its power to Brazil. Transportation connections with Brazil have improved, and that is now the main direction of trade. Paraguayan border towns such as Ciudad del Este sell cheap consumer goods to Brazilians and are involved in smuggling between Brazil and Argentina.

Uruguay has the highest urban percentage of any mainland country in Latin America. The majority of the country's urbanites live in metropolitan Montevideo, the country's capital and largest city. Uruguay has one of the region's highest literacy rates, and it is one of the only countries in the Western Hemisphere to offer a free college education to its citizenry. Agriculture, in the form of livestock raising, and tourism are the most important components of the economy. Uruguay's warm climate and picturesque beaches attract tourists from around the world. Government-controlled hydroelectricity production provides more than 75 percent of the country's energy needs.

In the 1990s, the Argentine economy was very robust, and the citizens of the country were enjoying a relatively prosperous standard of living. A series of both local and global actions combined to devastate the economy in Argentina by the end of 2001 and to bring the middle classes into the streets of Buenos Aires in protest. The combination of government policies, global trade relations, and government corruption necessitated the devaluation of the Argentine currency. Economic recovery after 2003 was halted by the global economic crisis of 2008 and 2009. In 2010, the Argentine economy began to rebound substantially, and it appeared to be in a material stage of recovery by 2012.

Paraguay remains a materially poor country. The larger portion of the country to the west of the Rio Paraguay, the mainly semiarid Gran Chaco, remains lightly settled apart from military camps and lumber operations. The quebracho ("ax-breaker") tree is its commercial timber product. The tree is very hard and is in demand for railroad ties and its tannin extract.

Buenos Aires, Argentina

The rapid growth of Buenos Aires, Latin America's third-largest city in 2010 (13,370,000), followed patterns that are common to other cities in Southern South America. After slow growth in its early history, Buenos Aires expanded in the late 1800s as commercial farming took hold in its pampas hinterland. The built environment developed a trading and industrial waterfront and a central thoroughfare at right angles along the route inland. Rapid growth occurred in both the economy and the immigrant population around 1900. This resulted in a middle class of skilled workers and office workers moving out to new suburbs, while the poor and most affluent residents remained in the inner city. Amenities came slowly to the suburbs. During the later 1900s, increasing rates of population growth produced squatter settlements and rising inner-city population densities as apartment blocks replaced mansions. Population growth stagnated in the coastal industrial areas. From the 1960s, much of the commercial and industrial activity moved out of central Buenos Aires to the city edges. Attempts were made to divert the overcrowding in the Buenos Aires metropolitan area by placing new projects and development in other centers farther inland from Buenos Aires, but they have not been successful.

10.5 Contemporary Geographic Issues

Tropical Forests and Deforestation

Deforestation, or the permanent clearing of forest vegetation, is a centuries-old land modification practice. Societies in all world regions and in both temperate and tropical latitudes used forest resources for fuel, shelter, and transportation. Dramatic increases in the permanent clearing of tropical forests, and especially tropical rain forests, in recent decades is causing significant alarm among a growing global body of scientists, medical researchers, government officials, and environmentalists. Global communities concerned with the potential ecological and human health impact from tropical forest clearing call on governments practicing or permitting tropical deforestation to cease. Locally, where deforestation is taking place, governments, business communities, and farmers assert their sovereign rights to resource use.

Tropical rain forests exist at latitudes where high temperatures and high levels of humidity year-round produce a fairly consistent precipitation and vegetation pattern each season. The three most significant locations of tropical rain forest in the world are in Southeast Asia (see Chapter 5), Central Africa (see Chapter 8), and the largest, the Amazon River basin in South America (Figure 10.31a, b).

Tropical rain forest is the dominant vegetation and climate regime in the Amazon River basin of Brazil and adjacent parts of its neighboring countries, including eastern Colombia, Ecuador, Peru, Bolivia, southern Venezuela, and the Guianas. The tropical rain forest of the Amazon basin is the largest in the world. Although other locations in Latin America contain similar vegetation and climatic conditions, including the Pacific coast of Colombia, along the eastern coasts of the Central American countries, and down the northeast coast of Brazil, the Amazon dominates in size and global impact.

Tropical forests contain the highest plant and animal species diversity per unit of land of any ecosystem or biome in the world. Seventy percent of the world's known plant and animal species reside in tropical forests, with hundreds of distinct species existing in relatively concentrated areas. Contemporary global health care depends on the species diversity of tropical rain forests. Existing treatments and promising cures for various forms of cancer come from tropical forest species. Numerous human diseases and aging conditions may be treated,

Legend:
- Rain forest
- Former rain forest area
- Asphalt roads passable all year-round
- Gravel roads
- Roads under construction
- Planned roads
- ▲ Indian reservation
- P National park
- ■ Major hydroelectric project

(a)

(b)

Figure 10.31 Brazil: Amazon rain forest. (a) Map of the Amazon River basin in Brazil and the extent of tropical rain forest. (b) Aerial view of the tropical rain forest in Rondonia state, where the forest is cut into as farms are established along the roads. *Photo: (b) © Michael Bradshaw.*

cured, or slowed through medicines derived from the tropical rain forest plants. Medical research communities assert the need to explore the unknown potential hidden in the diversity of tropical rain forest vegetation. International pharmaceutical research, development, and sales generate hundreds of millions in revenue from products related to tropical rain forest species. The potential benefit of hundreds to thousands of species present in the tropics has yet to be identified. Permanent clearing could eliminate countless medicinal cures and treatments yet undiscovered.

Trees absorb carbon dioxide from the atmosphere and return oxygen to the atmosphere. Tree respiration takes place in the troposphere, the lowest level of the Earth's atmosphere, where humans live and breathe. Removing huge tracts of tropical forest eliminates a primary source of oxygen and a mechanism for removing carbon dioxide at the same time. Burning forests, or drowning them under dammed waters, both common practices in Latin America, contributes to the release of carbon dioxide and methane, both considered to be culprits in global warming.

Tropical forests provide a number of local environmental stabilizers, holding moisture and releasing it into the local and downwind environment, and regulating stream flows and preventing downstream flooding. Tropical rain forests provide a source of water when global cycles bring drier periods to a region. Through the process of **evapotranspiration**, tropical rain forests can provide humidity and rainfall to the region they inhabit as well as to areas far downwind. Tropical soils are not nutrient rich like many temperate soils. The vegetation in the tropical forests stores the ecosystem's nutrients, so clearing the forest removes the local nutrient base and diminishes the potential for future species diversity. The vegetation keeps the local soil intact. In its absence, soil will wash through the watershed and out through the drainage system. Trees on poorer soils maintain their existence by circulating the nutrients without letting them enter the soil: when leaves fall, insects and fungi soon break them down so that roots near and above the surface can capture the chemicals again.

Government debt, high population densities in the east, millions of materially impoverished and jobless people, farmers with no land to farm, and a wealth of unexploited natural resources prompted the Brazilian government to promote the development of the Amazon basin. The Brazilian government is hoping huge mining projects located in forested areas will bring in foreign capital and ease Brazil's foreign debt. The government claims sovereign rights to exploit its resources. Brazilian political leaders assert the need to develop their resource potential fully in order to decrease their international debt. Brazilian officials point to countries such as the United States, where unencumbered resource exploitation led to material wealth and prosperity. Many local and international environmental groups are calling for international reduction of Brazilian debt to help the country shift its development efforts away from the Amazon.

Critics argue that the development of large mining operations in the Amazon basin is leading to an influx of people to provide services. As supporting service and other industries grow, more people will move to the region. As roads are cut into the forest, farmers and ranchers settle along both sides. Poor farming practices coupled with the nature of the soils and vegetation often lead to rapid nutrient depletion, and the farms are either sold to large-scale enterprises or abandoned for new land. As new roads are built, new settlers move in, existing farmers move to more productive land, and the rain forest vegetation slowly disappears.

The government claims such projects produce jobs that employ people who would otherwise have no work or income. People working the mines and providing services to the developing regions are grateful for the opportunities presented. Some farmers claiming land in the region assert their pride in land ownership and in their ability to be self-employed. In places where better soils coincide with an understanding of the local environmental conditions, there are very successful commercial crops. Soy production, which is thriving in Mato Grosso state, provides significant income to the country, but it is a major culprit in forest clearing.

The future of the tropical rain forest in Latin America will be determined by the ability of local and global governments to understand the ecosystem better and to cooperatively seek egalitarian uses that may serve the needs of countries like Brazil while sustaining the global ecological and human health value of the rain forest system. Table 10.3 provides a matrix with some of the positions taken by those urging tropical rain forest conservation and those who believe resource exploitation in the rain forest is in the best interest of local communities and nations.

Urban Pressures in Mexico and Brazil

Mexico City is by far the largest urban agglomeration in Middle America. Its site was a major population center before the arrival of the Spaniards and even before it became the Aztec capital. It continued to be the focus of road and rail networks within New Spain and independent Mexico. Mexico City is the political capital and media center of the country, and it has three-fourths of Mexico's manufacturing industry and nearly all of its commercial and financial establishments.

During the 1950s, 1960s, and 1970s, up to 1 million people per year migrated from rural areas in Mexico to metropolitan Mexico City. Housing supplied by government efforts, combined with that from private developers, could not nearly provide adequate accommodation for almost one-third of the rapidly growing city's population. The city attracted rural migrants with the expectation of greater economic opportunities, better education, more diverse recreation and cultural choices, and more substantial health care services. Squatter settlements exploded in many parts of the Central Valley of Mexico in and around the city. Nezahualcóyotl (Neza) is a sprawling municipality with more than 1.5 million materially impoverished people living on flood-prone lands, adjacent to the northeastern border of Mexico City. The municipality grew quickly during the 1960s and 1970s as poor rural

Table 10.3	**Debate:** Tropical deforestation

Conserve Tropical Rain Forest	Use Tropical Rain Forest Resources
Tropical rain forest (TRF) resources provide a "sink" for carbon dioxide. Burning TRF vegetation adds carbon dioxide to the Earth's atmosphere. TRF areas are a source of oxygen in the lowest level of the Earth's atmosphere, where humans live and breathe.	There is incomplete carbon dioxide data for the Earth's atmosphere. Large portions of the Earth's surface are unreported. Ocean exchanges with the lowest levels of the Earth's atmosphere are more significant than TRF exchanges.
There is tremendous biodiversity in the plant life present in TRF ecosystems.	There is no conclusive evidence that TRF clearing will permanently change the biodiversity of the Earth as a whole.
Many medical treatments are derived from TRF products, and many disease cures come from TRF products, including current treatments and potential cures for cancer patients. Destruction may eliminate many undiscovered cures and treatments. TRF-derived pharmaceuticals earn billions internationally each year.	Medical treatments come from many sources. Many treatments and cures may be synthetically generated in laboratories and do not require the use of naturally growing species from TRF.
Governments permit the rapid clearing of TRF resources and sell them internationally, claiming rights to destroy domestic resources that impact the entire Earth. Yet the same governments may be corrupt and wasteful of other resources.	Debt-ridden and impoverished countries need to and have the right to use their natural resources for their own best interest. The wealthier countries of the world obtained high material living standards by depleting much of their own and others' resources as they grew. Now those countries want to hold back countries that have TRF resource wealth.
Indigenous tribes and local people are displaced by TRF clearing. In some cases, bloody conflicts ensue while government officials turn a blind eye.	Growing countries need to push their frontiers and develop their resources. "Productive" members of society have a right to use land in a manner that will benefit them and their country.
TRF resources provide an increasing tourism revenue potential. Although some governments claim to balance resource clearing for export sales and development goals with conservation for tourism growth, few have shown a true commitment to achieving such a balance.	Governments have the right to determine how they will earn revenue from their resources. Governments of TRF resource-wealthy countries assert their ability to balance resource depletion and extraction with conservation and replenishment.
TRFs provide a natural habitat for species found only in this biome. Removing the TRF would eliminate habitat and cause permanent loss to global species diversity. Loss of species could alter the ecological balance of the Earth.	A good source of income in a debt-challenged country with a large materially impoverished segment in its population is far more important than the conservation of a bird or a tree.

migrants settled by the tens of thousands each year in make-shift, self-built housing. Parts of Neza were given municipal services like piped water and electricity, while other areas remain without amenities. Garbage collection is conducted in part by donkey cart, and Neza has one of the highest crime rates in Mexico.

The overall result of such rapid growth is a combination of overcrowding, congestion, and air and water pollution. The physical geography of the Central Valley, coupled with the intensely crowded living conditions, make the urban center one of the most polluted, in terms of air quality, in the world. Although residents of metropolitan Mexico City have greater access to health care than rural Mexicans, infant mortality rates and other social health indicators are among the worst in the country due in large part to poor air and water quality. Depletion of underground water resources during the past several decades presents another problem for the city. Many areas within the urban system are subsiding, some more than 5 meters. Engineers struggle to stabilize historic structures that have been slowly sinking for decades. As job markets also failed to cope with the population explosion of the city, the informal sector of the economy grew from around 4 percent to 26 percent of people of working age from 1980 to the 1990s. Government efforts from the 1980s through the early 2000s to attract jobs and migrants to other regions in Mexico helped to slightly alleviate population pressure in Mexico City, but overcrowding and

unemployment remain formidable challenges in the Federal District.

In Brazil's two largest cities, São Paulo and Rio de Janeiro, shantytowns, known as *favelas*, house millions of materially poor people who cannot be accommodated by the formal patterns of housing construction, infrastructure, and service provision (Figure 10.32a, b). It is estimated that over 7 million people in São Paulo and up to 6 million people in Rio de Janeiro live in favelas. Favelas vary in their character. In and around the city centers, they occupy gaps in the built environment, in which families erect their own minimal accommodations. It is common for these favelas to lack potable water, electricity, waste disposal, or paved roads. Old apartment blocks in the city centers, which were abandoned and taken over by impoverished peoples, resemble favelas in many ways, with several families occupying each room. On the outskirts of the cities, different types of favelas are pushing the built-up area outward at a rapid pace. Some favelas spring up almost overnight and grow by thousands of people within months. Once established, some favela areas may be provided with some basic roads and may soon acquire utilities, shops, and schools. Crime is rampant in many favelas. Gang violence and kidnappings severely challenge police authority and the justice system in Brazil's large cities. Strong allegations were made in the late 1990s and early 2000s that police officers were either bribed or frightened away from

(a)

(b)

Figure 10.32 Rio de Janeiro. The dramatic physical geography of Rio de Janeiro contributes to the strong pride Brazilians feel for the city. (a) The harbor, beaches, and marina areas seen from the hilltop statue of Christ. (b) Contrasting with the tourist's view and expensive hotels and condominiums along Copacabana beach, the Rocinha favela crowds thousands of materially poor Brazilians into marginal structures on a steep hillside. *Photos: (a) © Stephanie Maze/Corbis; (b) © Corbis RF.*

patrolling the streets of some favela areas where drug dealers and gang leaders took control of the streets. In an effort to prepare the city for the 2014 Soccer World Cup and the 2016 Olympics, city officials aggressively tried to reduce crime in some favela areas. Reports of material change vary, with some parts of Rio's favelas appearing to have less crime, while others are experiencing more.

The favelas are also the most common places where thousands of Brazilian street children base their nomadic and often shelterless existence. Economic, political, and social globalization forces have neither eliminated nor meaningfully diminished the conditions that combine to place hundreds of thousands of Brazilian youth in jeopardy. Some argue that multinational corporations, international banking and development institutions, and intergovernmental trade relations have increased Brazilian poverty rates rather than reduced them. Brazilian government debt and resource-export economic goals figure prominently in the social environment that produces so many children of the streets. The socioenvironmental conditions produced by such domestic and foreign relationships include extreme disparities in wealth, high poverty rates, relatively high death rates for women in childbirth, high rates of teenage pregnancy, high illiteracy rates, and high unemployment. Street children either literally live on the streets with no shelter or work the streets at a young age, trying to earn money for their survival or that of their families. The street children often end up in prostitution or in an illegal narcotics trade, and many contract life-threatening diseases. The most challenging situation is the reported periodic murder of groups of street children (often believed to be the work of vigilantes), whose bodies are abandoned in ravines or concealed areas. Although local and global human rights groups are persistent in their attempts to attract significant attention to the plight of the street children, the problem remains very serious.

The Northern Andes, Mexico, and the International Drug Trade

The countries of the Northern Andes comprise one of the world's primary drug-producing regions (Figure 10.33a, b). The first system in the subregion centered on cultivation areas in Peru and Bolivia, with processing or production centers based in Colombia. The expansion of Bolivian cultivation slowed in the early 1980s, and the Peruvian coverage declined in output following a fungus infestation of the crop. Colombian cultivation and production increased; it includes coca, opium poppies, and marijuana, and is now greater than that in Bolivia and Peru.

The human and physical geography of Colombia enable drug cultivation, processing, and shipment to thrive. Colombia has three rugged mountain ranges that run roughly in a north-south axis through the country. Inland cities developed in relative isolation from one another, contributing to the establishment of more local identities within Colombia at the expense of a strong unified national identity. A history of political turmoil,

(a)

(b)

Figure 10.33 **Drug-producing regions.** (a) This map shows coca-growing areas in the Northern Andes. The dominance of Peru and Bolivia in the 1980s gave way to rapid expansion of cultivation in Colombia in the 1990s and early 2000s. (b) Federal police officers and forensic workers inspect the crime scene where five people were shot to death on March 6, 2011, in the northern border city of Ciudad Juarez, Mexico. Cities such as Ciudad Juarez increasingly became zones of violent crime in the first decade of the 2000s due in large part to the increased influence of Mexican drug cartels on the cross-border drug trade from Mexico into the United States. *Photo: (b) © AP Photo/Raymundo Ruiz.*

including intense violence between left-wing and right-wing ideologues, further fragmented national cohesion. Climate, soils, altitude, and shade conditions are well suited to coca cultivation. A lack of transportation infrastructure, combined with dense vegetation over vast amounts of land, facilitate a thriving cocaine industry.

The global demand for cocaine from consumers in the United States, Europe, and Asia has a devastating impact on the local politics, agriculture, and the social fabric in the Northern Andes. An excess of production over market demand in the early 1980s caused the price of cocaine to fall by 75 percent. The huge drop in price dramatically opened the international market to those with less disposable income, resulting in a subsequent rapid increase in demand. The United States led efforts to prevent drug production in these countries, but its supply-based (rather than market-based) strategies were diluted by limited government cooperation, corruption, coercion, and challenges in crop eradication. United States investment in combating narcotics trafficking from the region makes Colombia the third-largest recipient of U.S. financial aid (after Israel and Egypt) in the world. Although Bolivian farmers planted

other commercially valuable crops on former coca-growing land as a result of U.S. aid, the acreage under coca remained the same. Peruvian government attempts to halt cocaine production are limited in part by fears of the resurgence of guerrilla groups among disaffected former coca growers. Subsistence and small-scale farmers are attracted to working on large coca farms where average wages are far in excess of what they would otherwise earn. Other farmers are forced by threats of violence to become part of the drug labor force, and in many cases, their farms are forcibly taken over by the drug cartels.

The combined efforts of the U.S. and Colombian governments significantly reduced the influence of drug cartels and gangs in Colombia in the first decade of the twenty-first century. Drug-related crime dramatically decreased in Colombian cities from the late 1990s through 2012. Cocaine and other narcotics cultivation in Colombia, however, remained unchanged. Primary control of the trafficking and trade of narcotics migrated from Colombia to Mexico, which shares a large border with the world's largest cocaine market, the United States. Mexican cartels and drug gangs dominate drug flows from the Northern Andes to the United States and other global market areas. Drug-related crime skyrocketed during the first decade of the 2000s in Mexico, especially in northern areas close to Mexico's border with the United States. Kidnappings, murder, intense fighting between drug cartels, and gang violence are commonplace in many northern cities in Mexico, which prompted the U.S. Department of State to issue a travel warning for northern Mexico late in the decade which remained in place in 2012.

Agricultural Change in Indigenous Land in the Amazon

Agricultural expansion, a major proximate cause of land cover change, is estimated to account for about 90 percent of all forest clearing in the tropics and temperate latitudes. Scientific understanding of the relationships between agricultural land use and land cover improved substantially in the past few decades. Because this body of knowledge is still incomplete, research projects like that of Dr. Santiago López (Figure 10.34) are needed in order to advance our understanding of human–environment interactions in a variety of regions, including a broad diversity of human groups and land-use strategies. Dr. López is an assistant professor of interdisciplinary arts and sciences at the University of Washington, Bothell. Current research by Dr. López analyzes agricultural expansion in areas managed by indigenous peoples in the Amazon region. Specifically, his research focus is on the lower watershed of the Pastaza River basin in the Ecuadorian Amazon. The study area encompasses approximately 20 percent of the Pastaza megafan, an area of 60,000 km^2 that is considered by some to be the largest alluvial fan in the world.

The indigenous groups who live in the area are the Shuar, the Achuar, the Shiwar, the Zápara, and the Andoas. The Shuar, Achuar, and Shiwar belong to the same dialect group, which is part of the Jívaro linguistic family. The Jívaro are probably one of the largest and most homogenous indigenous groups of the Amazon basin, and their spatial distribution covers different eco-zones of the Pastaza River basin that include the foothills, lowlands, and inundated evergreen humid forests. Most of these groups energetically—and very often violently—rejected any attempt by nonindigenous people to access their ancestral lands until the late 1960s. These groups continue to live in relative isolation from cities and the so-called Western world. They live in semipermanent settlements called *centros*. There is no accessibility by roads, and the most common means to reach these communities is by single-engine aircraft. For these and other reasons, their cultural and ecological heritage has remained almost intact.

Dr. López's research involves the collection of spatial and socioeconomic data through the integration of geotechnologies

Figure 10.34 **Analysis of swidden agriculture in the Ecuadorian Amazon.** Dr. Santiago López analyzes data he collected while conducting field work on slash-and-burn (swidden) agriculture in the watershed of the Pastaza River basin in the Ecuadorian Amazon. *Photo: Courtesy of Santiago López.*

(i.e., remote sensing, geographic information systems, and global positioning systems) and field surveys to accurately characterize shifting slash-and-burn cultivation areas. Field research activities often involve the acquisition and *in situ* interpretation of submeter resolution aerial images and the ground truthing of these interpretations to accurately map how indigenous households use their land. Some land uses, such as hunting and foraging (which are still very important for subsistence societies), are impossible to identify through remotely sensed data, so GPS receivers are used to georegister these areas. Dr. López's team has currently mapped more that 800 kilometers of hunting trails in the area. Because this type of data has not been collected before in this area, Dr. López's project will help to establish baseline information that will be used to determine the demand for land and forest resources of these populations. The results of this project will be used later to establish conservation policies in the region.

Cocoa Beach, Florida, United States. *American pride is shown in this photograph of the sunrise at Cocoa Beach, with beach volleyball courts adjacent to the Atlantic Ocean.*
Photo: © California CPA/Flickr/Getty Images RF.

LEARNING OBJECTIVES

After reading this chapter, you should be able to:

- Compare and contrast the physical landscapes of North America with their respective climate zones.

- Understand the spatial history of the region as it pertains to European colonization and historical settlement patterns in the region over time.

- Identify and describe the natural hazards and environmental situations facing North America today.

- Understand the spatial distribution of population densities in North America and reasons for agglomeration.

- Describe the role of migration in North America, and the geopolitical issues facing the region today.

- Compare the different periods of industrialization and urban settlement in North America, from its agricultural roots through the Industrial Revolution and into the twenty-first century's global economy.

- Locate the subregions of the North American region, and describe and contrast the major qualities of each.

North America

Figure 11.1 North America: physical setting, political boundaries, and major cities.

Elevation (ft.)

below sea level 0 500 1,000 2,000 5,000 10,000

-10,000 -5,000 -500 0

0 mi   500   1,000   1,500   2,000

0 km   500   1,000   1,500   2,000   2,500   3,000

Lambert Azimuthal Equal-Area Projection

11.1 Defining the Region

The North American region is composed of two countries, Canada and the United States (Figure 11.1). Greenland, being a political unit of Denmark, is not considered in this section. These countries are the world's second-largest (Canada) and third-largest (United States) in terms of land area. They both possess a wealth of natural resources, are highly urbanized, and their citizens enjoy some of the highest standards of living in the world. Canada's huge land area is sparsely inhabited with a 2012 population of 34.9 million (Table 11.1). The United States is the world's third-largest country in terms of population with 2012 estimates of 313.9 million living in the country. By comparison, this is still substantially smaller than the Earth's two population giants, China and India, each with more than 1.2 billion residents. Almost 80 percent of North America's residents live in urban areas.

The very high standards of living in Canada and the United States attract migrants from all over the world, who come to North America seeking prosperity and stability. The majority of North America's immigrants congregate in the region's vast urban areas where opportunities are greatest, particularly in the southwestern and northeastern United States, and in the Canadian cities of Vancouver and Toronto (Figures 11.2a, b). Internal migration patterns further add to the growth of the region's cities and, coupled with international immigration, create increasingly plural yet often spatially segregated metropolitan centers. Illegal immigration and the continued prevalence of undocumented workers in the United States is creating a growing controversy surrounding job competition, along with the exploitation of national financial and social resources. Technology improvements such as seawater desalination are providing economical solutions to environmental issues such as groundwater depletion in North America (Figure 11.3).

Table 11.1	North America: Data by country, area, population, urbanization, income (gross national income purchasing power parity), ethnic groups						
		Population (millions)		% Urban	GNI PPP 2012	2012	
Country	Land Area (km²)	Mid-2012 Total	2030 Est. Total	2012	Total (US$ billions)	Per Capita (US$)	Ethnic Groups (%)
Canada	9,976,140	34.3	39.1	81	1,170.7	35,500	British origin 35%, French origin 25%, other European 20%, First Americans 3%
United States of America	9,809,460	313.5	364.8	80	13,827.2	45,840	White 83%, African-American 12%, Asian 3%, Native American 1%

Sources: *World Development Indicators 2007*, World Bank (1, 5, 6); *2012 World Population Data Sheet*, Population Reference Bureau (2, 3, 4); Encarta 2007, Microsoft (7, 8, 9).

(a)

(b)

Figure 11.2 **North America: urban and rural landscapes.** (a) The urban setting of Toronto, Canada. (b) The contrasting rural landscape of Quebec: Gatineau Park, located just north of Ottawa (Ontario), Canada's national capital. *Photos: (a) © Joseph P. Dymond; (b) © Brian Phillpotts/Photolibrary/Getty Images.*

Figure 11.3 **The Tampa Bay Desalination Plant, located just south of Tampa, Florida.** This facility has the ability to convert 50 million gallons of seawater daily into 25 million gallons of fresh water for use in the region, alleviating pressure on the local aquifer system.
Photo: © The Tampa Tribune.

Canada and the United States are closely linked culturally and economically. However, their political systems differ, even with both being free, federal governmental systems. For example, in Canada voters do not directly elect their prime minister—to vote for this person at the national level, you have to vote for his or her political party at the local level, and then the vote trickles upward as a vote for their candidate at the highest level.

In terms of physical features and landscapes, North America is relatively isolated from most of the world. Canada and the United States are bound to the east by the Atlantic Ocean, separating them from Europe and Africa by thousands of miles. Canada's northern border is the Arctic, while the southern Canadian boundary serves as the entire northern border of the conterminous United States. The western border of both countries is the Pacific Ocean, which also puts thousands of miles between the region and East and Southeast Asia. The U.S. southern border is divided between the landmass of Mexico and the Gulf of Mexico.

11.2 Distinctive Physical Geography

Mountains and Plains

The continent of North America rests almost entirely upon the **North American plate**, except for the extreme western boundaries that are a part of the Pacific plate (see Figure 1.5). New plate material forms along the divergent margin at the Mid-Atlantic Ridge in the Atlantic Ocean, forcing the plate to move northwestward and clash with the Pacific plate at the convergent and transform plate margins along the western coast of North

America. Eastern North America is "along for the ride" on this plate, and as such does not feature very much active geologic activity such as earthquakes. However, the west coast of North America is affected by frequent earthquakes and periodic volcanic activity, due to its convergence and/or lateral movement with the Pacific plate. The oldest rock in North America is found in the **Canadian Shield**, an area of exposed rock outcrops that cover roughly half of Canada's surface area and contain major deposits of mineral ores such as copper, diamonds, and iron ore.

The western third of Canada and the United States is mainly composed of two chains of high mountains, with the Basin and Range physical region located in between these. The highest peak in the region is Mount McKinley in the Alaskan Range (6,194 m, 20,231 ft.). The Rocky Mountain ranges extend from Alaska and northwestern Canada southeastward to New Mexico. To their west in the United States are extensive high plateaus from the Colorado Plateau in the south to the lava layers of the Columbia Plateau just south of the Canadian border. West of the plateaus are the Sierra Nevada Mountains of California and the Cascade Ranges (northern California through Oregon and Washington to the U.S.-Canada border).

Lowlands with few topographic changes in elevation dominate southern Ontario, the Prairie Provinces, and the central United States. Layers of sedimentary rock cover the shield rocks, forming plateaus and escarpments such as that over which the Niagara Falls plunge. In the U.S. Midwest and in Ontario, deposits from the melting ice sheets of an earlier age cover most of these rocks. Similar deposits occur over part of the Canadian prairies. However, in areas not covered by the ice sheets, distinctive bedrock outcrops and hundreds of thousands of lakes were formed by the glaciers as they melted and retreated in the last major glacial age of around 12,000 years ago.

East of the Rockies, the Mississippi drainage basin (the world's third-largest watershed) empties an area of 3,225,000 km² (1,245,000 mi.²). The northern lowlands within the Mississippi watershed were covered by rock fragments and particles deposited by the ice sheets blanketing the area during the Pleistocene Ice Age. Other regions within the watershed have deposits of windblown silt, or **loess,** which accumulated on the river floodplains as far south as Louisiana and Mississippi.

East of the Mississippi lowlands, the Appalachian Mountains form a continuous chain of rolling hills and mountains extending from northern Georgia into the Adirondacks of New York, the Green and White Mountains of New England, and the Atlantic Provinces of Canada. Few Appalachian mountain ridges exceed 2,000 m (6,500 ft.). The rocks were deformed and uplifted by ancient plate movements around 220 million years ago, and subsequently have been exposed to heavy wind and water erosion. Geologists hypothesize that these mountains at one time may have been taller than 35,000 feet high! West of the Mississippi River, the Ozarks and Ouachitas form similar upland areas and were once connected to the Appalachian Highlands. However, erosion by the Mississippi River over millions of years has resulted in isolation from the rest of the mountain chain. In the northeastern United States and eastern Canada, the glaciers scraped away much of the surface rock and soil and carried it southward, depositing it along the coast in the low ridges (moraines) that form much of Long Island and Cape Cod.

Glaciation, Major Rivers, and the Great Lakes

During the Pleistocene Ice Age that occurred on and off from 2 million to 12,000 years ago, advancing ice sheets covered most of Canada along with the northern and midwestern United States.

The southern boundary in North America correlates closely with the locations of the Missouri and Ohio rivers, which were formed as rivers of glacial meltwater at the end of these massive ice sheets. During this time period, tremendous amounts of dirt and debris were ripped off the landscapes in Canada and carried along by glaciers, being dropped in place across the midwestern United States as these started melting and retreating. This is the reason for the exposure of bedrock throughout most of Canada, as the glaciers were a mile-high bulldozer that tore apart anything in their path, scouring sediment down to base level.

The Great Lakes formed as large pieces of ice from the last glacial age melted in place. Specifically, as the ice sheet retreated northward at the end of the last glacial phase, meltwater accumulated in the depressions that became the present-day Great Lakes. Today, the Great Lakes function as an inland seaway with coastlines, industrial ports, and recreation areas. To facilitate transportation of raw materials and manufactured goods to the rest of the world, the Canadian and U.S. governments worked together to open the St. Lawrence River Seaway in 1959 (Figure 11.4). Ocean-bound ships from as far inland as Duluth, Minnesota, and Thunder Bay, Ontario, utilize this water passage. Raw iron ore from Minnesota is transported via boat to the great steel mills south of Chicago in the Gary, Indiana, area.

North America's surface **hydrology** (surface water drainage pattern) has a major impact on both the physical and human geographies of the region. The major rivers of the continent along with the Great Lakes altogether provide sources of fresh water and transportation routes. The Mississippi River was the basis of early interior transportation, and it continues to play a role in the transport of bulk materials. Its second-largest tributary, the Ohio River, was a transportation route at the heart of manufacturing developments in the United States during the mid-1800s. The Missouri River, its largest tributary, served as a

Figure 11.4 **North America: water-based transport.** The St. Lawrence River, which connects the Great Lakes to the Atlantic Ocean, is an important waterway for recreation and commerce in Canada. This container ship is docked alongside a wharf in Montréal. *Photo: © Joseph P. Dymond.*

conduit for exploration and settlement into the Great Plains. The Colorado River in the arid Southwest is harnessed for power and irrigation water. The Columbia River in the Pacific Northwest is used to irrigate farmland and generate hydroelectricity, along with being a major source of salmon. In northwest Canada, the Mackenzie River is a major summer transportation route that facilitates the movement of minerals such as oil and diamonds from mines in Alberta and the Northwest Territories to the rest of the world.

A series of dams and levees were constructed in the early 1900s in an effort to control the flooding of the Mississippi River floodplain, including the Ohio, Missouri, and Tennessee rivers. Today, nearly 100 percent of all rivers longer than 10 miles anywhere in the southeastern United States have some sort of man-made structure built upon them, whether dams or levees. The flood protection system prevents most excess runoff from spilling onto surrounding land. However, extreme rain events, such as the summer flooding event of 1993 (Figure 11.5) and Hurricane Katrina in 2005, prove to be too much for the levees and dams to handle.

Tropical to Polar Climates

The majority of North America's climatic environments are temperate midlatitude in type (Figure 11.6b). The western coast is dominated by an oceanic temperate climate that is generally characterized by dry summers and sunshine, giving way to more northern climates such as those of Seattle, Washington, in which it is wet all year and mild. East of the Cascade Mountains, the climate becomes very dry and arid, represented by the great deserts of the United States in Nevada and Arizona. East of the Rockies, the Great Plains are characterized by dry grassland climates that vary from being brutally cold in Sasketchewan and Manitoba to North Dakota, to extremely hot in Oklahoma and Texas. These climates transition into the Midwest region with more frequent rainfall, allowing for a wider variety of crops to be grown yet containing four seasons. The Gulf Coast

Figure 11.5 **The historic 1993 flooding of the Mississippi River in the United States caused some of the greatest flood devastation in modern history.** This photo shows miles of inundated farmland along the banks of the river near St. Louis, Missouri. *Photo: © Les Stone/Sygma/Corbis.*

from Texas to Florida and the Atlantic Coast as far north as the Virginia tidewater have subtropical conditions that bring warm spring and fall temperatures, and hot, humid summers. North of this region, the northeastern United States and Canada are four-seasons climates that receive varying amounts of precipitation in the form of both rainfall and snowfall. Finally, from eastern northern Canada westward to Alaska, the long and harsh Arctic winter gives way to a brief, mild climate from late May to early August.

The central plains and lowlands enable bursts of Arctic air to move southward in winter, bringing extreme cold especially to the Midwest and northeastern United States but also extending in shorter bursts to the Gulf Coast. Warm, dry southwestern air or humid air from the Caribbean and Gulf of Mexico moves northward in spring and summer. The frequent confrontation of cold Arctic air and humid subtropical air in the central lowlands leads to the formation of active cold fronts, with thunderstorms that can produce violent tornadoes in the spring and summer. Due to ideal climatic conditions for their formation in the Great Plains, the United States experiences more tornadoes per year than any other country in the world.

Natural Vegetation and Soils

Figure 11.6a depicts the spatial distribution of major vegetation types in North America. The hot deserts of the Southwest support mainly drought-resistant varieties such as cactus and low shrubs. The continental temperate Great Plains and prairies of the western Mississippi River basin and south-central Canada historically contained prairie grasslands that are now largely plowed to make use of the underlying black earth soils. Eastward, the more humid conditions, from tropical southern Florida to the more temperate Northeast, supported **deciduous** (broadleaf) forest, which gave rise to brown earth soils of moderate to good fertility. North of the deciduous forests and the prairies, a wide band of **coniferous** (needle-leaf) forest has poor podzolic soils and gives way to the tundra along the Arctic Ocean shores.

Today, the eastern mountains have thin soils on steep slopes. Before extensive clear-cutting by loggers in the late 1800s and early 1900s, the eastern mountains supported a profusion of trees that contained a much greater variety of species than those in Western Europe. The best farming soils formed under forest and grassland in the interior plains, where the old glacial and windblown materials combined with the moderate amounts of precipitation to produce rich brown (forest) or black (grassland) soils. In the southeastern United States, the frequent presence of sandy soils with low nutrient content caused some areas to be dominated by pine trees. Along the western coast of the United States, north of San Francisco, huge firs and cedars, growing over 100 m (300 ft.) tall, formed a massive timber resource.

Natural Hazards

A greater range of natural hazards than in any other country in the world marks the natural environments of North America. This region is affected by hurricanes, severe thunderstorms with accompanying tornadoes, lightning, hail, earthquakes,

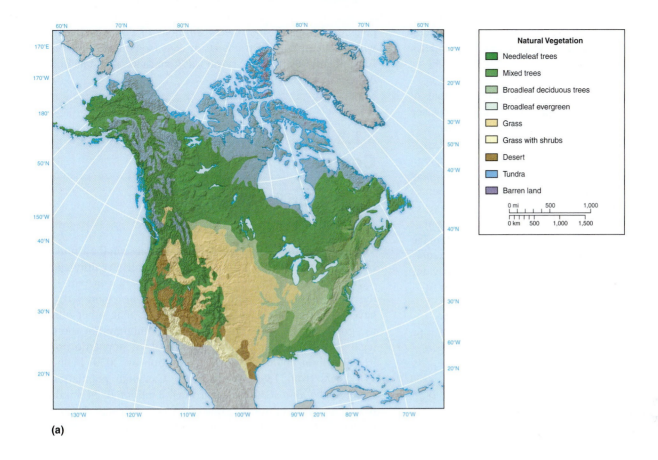

(a)

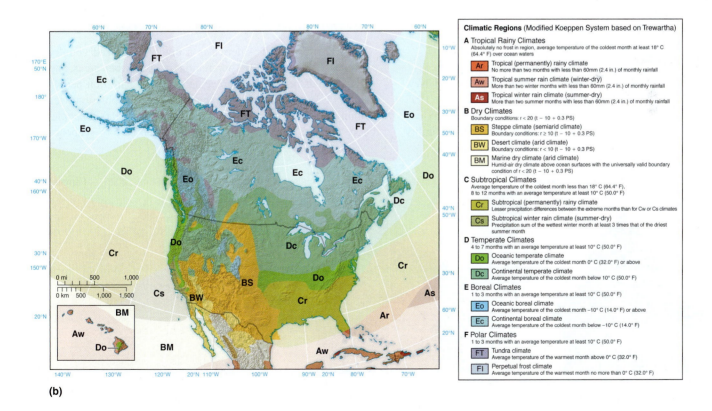

(b)

Figure 11.6 North America: vegetation and climate. (a) Relate this distribution of natural vegetation to the distribution of the region's climatic environments. (b) The climate regions of North America.

volcanoes, floods, blizzards, ice storms, and wildfires. Specific examples of these hazards and geographic locations include the following:

- Hurricanes threaten the Atlantic and Gulf Coast states mainly from July through September, when ocean water temperatures are warmest, providing fuel for these storms.
- Tornadoes are most common in the Great Plains region, where the necessary climate ingredients meet to support their formation. Tornadoes also frequently affect the southeastern United States, including Florida.
- River floods occur most commonly in the Mississippi River valley, following either the spring snowmelt in the Midwest or heavy summer rains. Flash flooding occurs in all regions of the United States and is particularly dramatic in arid regions in the West after sudden rains fill dry streambeds.
- Lightning is a year-round event across the southern part of the United States and is especially prevalent in Florida, with Tampa being the "lightning capital of the world."
- Severe winter storms affect the Great Plains states, the Midwest and Northeast, and the western mountains of the United States.
- Earthquakes and volcanoes occur mainly along the west coast from Alaska southward through California. Volcanic activity is nearly continuous in Hawaii, and it also occurs in the Cascade Mountains from central California northward through Oregon, Washington State, British Columbia, and into Alaska. The Aleutian Islands (i.e., Alaska's "tail") are entirely composed of active volcanoes and experience some of the world's greatest earthquakes.

The Canadian provinces of Québec, Ontario, and New Brunswick were devastated by an ice storm in January 1998 when 7–11 cm (3–4 in.) of ice accumulated over a six-day period. The heavy ice brought down trees, power lines, utility poles, and transmission towers. Millions of Canadians went without power and electric heat for days (some had no power for a month). The 1998 ice storm was Canada's costliest natural disaster to date.

11.3 Distinctive Human Geography

Population Patterns

Population Change: Natural Growth

The relatively rapid annual population increase in the United States is unusual among the world's materially wealthy and technologically advanced countries. This is due in part to the young immigrant communities within the country that have larger families and higher birth rates than other ethnic groups, along with strong levels of immigration (both legal and illegal) into the region. The U.S. population of 313.5 million could rise to 370.7 million by 2030. In the 115 years from 1891 to 2012, the Canadian population increased by almost 30 million, from 4.8 million to 34.3 million. Canada's total fertility rate of 1.9 is less than that of the United States, though its population growth rate is similar due to immigration (Figure 11.7). Canada has lower death rates than the United States, and its continued immigration rates almost match those of the United States.

Population Distribution: Increasing Urban Density

The United States of America is a highly urbanized country with 80 percent of its 2012 population living in urban areas, and just over half in metropolitan centers of over 1 million people (Table 11.2). More than 90 percent of the U.S. population lives within a two-hour drive of a large city of over 300,000 people. More than 35 million people live along the eastern seaboard

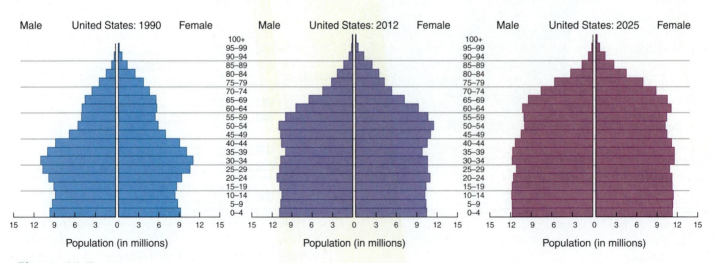

Figure 11.7 **United States: age-sex diagrams.** *Source: U.S. Census International Database: http://www.census.gov/population/international/data/idb/region.php.*

Table 11.2 — Populations of Major Urban Centers in North America

TOP 10 MOST POPULATED URBAN CENTERS

City, Country	2012 (millions)	2025 (millions)
1. New York City, USA	20.5	21.2
2. Los Angeles, USA	14.9	15.7
3. Chicago, USA	9.1	10.4
4. Toronto, Canada	6.1	6.5
5. Dallas–Fort Worth, USA	5.9	6.3
6. San Francisco, USA	5.8	6.6
7. Miami, USA	5.6	6.5
8. Philadelphia, USA	5.5	5.8
9. Houston, USA	5.4	5.9
10. Atlanta, USA	4.7	5.5

NUMBER OF MAJOR URBAN CENTERS IN NORTH AMERICA, BY POPULATION CATEGORY

Population Category	Number of Urban Centers
10.00 million and greater	2
5.00 – 9.99 million	7
2.50 – 4.99 million	10
1.00 – 2.49 million	24

in an urban corridor from Washington, D.C., to Boston, Massachusetts, known as the **Megalopolis**, and another 24 million people live in the metropolitan areas of the western Manufacturing Belt and Midwest. In addition to the Megalopolis, urban clusters exist around the Great Lakes from Chicago to Detroit, in Florida, and westward along the Gulf Coast to southeastern Texas, and along the Pacific coast in the west (Figure 11.8). Large cities are the essential feature of the country's contemporary geography. Urban population clusters are more numerous in the eastern contiguous United States. Vast areas of the western United States away from the Pacific coast contain very low population densities. Alaska, the the largest U.S. state in terms of land area, has the lowest population density in the country.

Canada has approximately 10 percent of the population of the United States yet on a larger land area, giving it a much lower average population density than that of the United States. More than 90 percent of its people live within 100 miles of the U.S. border, and over 60 percent of Canada's population lives in the southern Ontario and St. Lawrence River region of Québec. It is necessary to further consider the distribution of people in Canada in order to better understand Canada's population geography.

The vast majority of Canadians are concentrated in a belt across the southern part of the country nearly parallel to the border with the United States. Thus, portions of this populated belt across southern Canada have high population densities, while most of the remainder of the country is virtually empty. Canada is experiencing a trend toward metropolitan expansion and has 81 percent of its population classified as urban in 2012. The two largest metropolitan areas, Toronto (5.4 million) and Montréal (3.8 million), rival many U.S. metropolitan areas in size. Together with Vancouver on the west coast and Ottawa, the national capital, these cities contain one-third of Canada's population. Winnipeg, Edmonton, and Calgary are other major Canadian cities, found in the Great Plains region, and they owe their existence to both agricultural and mining economies. Canadians are far less spatially divided by ethnic segregation than Americans, with their diversity being due to language differences primarily.

Patterns of Migration in the United States

The continuation of relatively high rates of population increase for the United States is a result of immigration, which has been vital to the country's economic growth since the days of European exploration. Immigration currently accounts for one-third of U.S. population growth. British, Irish, and German people immigrated in significant numbers during the 1700s through the middle of the 1800s. The highest numbers of immigrants came from southern Europe, particularly Italy, in the later 1800s, and at the turn of the century, large numbers came from the Slavic countries of Eastern Europe, the Balkans, and Russia.

Peoples from the African continent were forced to migrate as slaves to North America from the 1700s until the early 1800s. The African-American population continued to grow through natural increase after the end of the Atlantic slave trade in the 1860s.

Latin American peoples from Middle and South America, and Asian peoples from South, Southeast, and East Asia became major sources of immigrants from the mid-1900s through the present day.

The dynamics of population geography in the United States involve internal migrations as well as international immigration (Figure 11.9). From the early to mid-1900s, large numbers of African-Americans moved from the rural and urban South to northern cities. Those from the Atlantic coastal plain mostly moved to Washington, D.C., Baltimore, Philadelphia, New York, and Boston; those from the Mississippi Delta moved primarily to Cleveland, Cincinnati, Indianapolis, and Chicago. Smaller numbers of African-Americans moved to the West Coast. In 1900, nearly 90 percent of African-Americans lived in the South, though this proportion declined to just over half by the 1990s. The movements also stimulated the dissemination of cultural features of African-American society, such as the varied types of blues music.

Most of the movements of African-Americans out of the South were completed by the 1970s. After that, smaller **countermigration** balanced or exceeded the northward flow, spurred by retirement and the advent of air conditioning that made living conditions in these regions much more bearable. Movements of

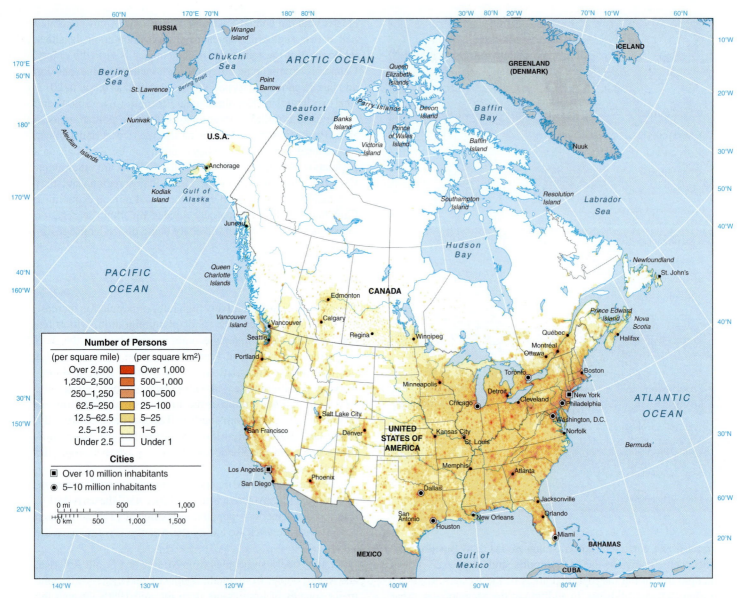

Figure 11.8 **North America: population distribution.** Compare the distribution of population in the eastern and western halves of the United States and note the proximity of most Canadians to the U.S. border.

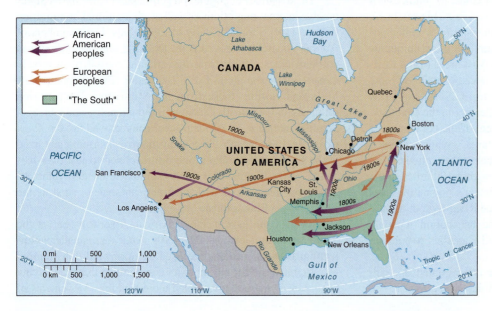

Figure 11.9 **United States: The major movements of European and African-Americans in the 1800s and 1900s.**

European Americans from the Northeast to both the West Coast and the South continued into the 1980s, although recessions in that decade slowed southern economic growth and reduced such migrations.

Canadian Patterns of Ethnic Integration

Government policy protecting multiculturalism and well-known high living standards combine to make Canada an extremely attractive destination for peoples emigrating from their homelands. The Canadian federal government instituted policies outlawing ethnic discrimination and protecting the rights of immigrants. For several decades, Canada has been one of the leading immigrant-receiving countries of the world. Contemporary immigration trends for Canada exhibit significant increases in the numbers of people migrating from Asia and the Americas (including the Caribbean), along with an overall percentage decrease in the European immigrant population.

The large urban centers of Toronto, Vancouver, and Montréal are experiencing dynamic cultural processes that increase the diversity of each metropolitan area. Canada's largest city, Toronto, attracts more immigrants from all reaches of planet Earth than any other Canadian city. Asian migration is most notable along the west coast of the country. The city of Vancouver and its hinterlands in the province of British Columbia are the hosts to two types of immigrants from Asia: those who settle year-round, and those who use the city and environs as a second home or home for their families while they "commute" to work in Hong Kong.

Contributing History

Indigenous groups of people, who are referred to today as **Native Americans** in the United States and **First Nations** in Canada, inhabited North America for many centuries before 1500. The first Americans probably migrated from Siberia to present-day Alaska over 20,000 years ago. By 1500, these people lived in hierarchically structured ethnic groups, commonly called tribes. The combination of natural resources, environmental conditions, and a sedentary settlement structure enabled groups such as the Ojibwa (Michigan) and Algonquin (Québec) to flourish for centuries.

After the arrival of the Europeans, many Native Americans perished from the introduction of diseases such as smallpox and influenza, to which they had no immunity. The survivors were increasingly pushed to the most marginal lands in the region. This process began in the east and southwest in the 1600s, and lasted into the 1800s as natives were herded into the central and northwestern parts of North America on government-designated lands known as **reservations**.

In the time period after Spain's initial exploration of the Americas, the region became a series of colonies and occupied territories governed by the French, Spanish, British, Dutch, Russians, and Swedes. The British became the dominant power in North America by the mid-1700s. Almost immediately, settlers south of the St. Lawrence River valley fought to become the independent United States of America by signing the Declaration of Independence in 1776, eventually forcing Britain to recognize U.S. sovereignty.

Canada arose mostly out of the colonial territories first established along the St. Lawrence and the Great Lakes, and remained a British colony far longer than the United States. It achieved independence through the **British North America Act** on July 1, 1867. However, Canada maintained legal ties to Britain until 1982, when it gained sole control of its constitution.

U.S. Development

Three primary economic and sociostructural patterns emerged during the British colonial development of the eastern portion of the United States. After independence, areas along the Atlantic coast formed springboards for thousands who moved westward within the United States as new lands were acquired. The New England community subculture diffused into areas around the Great Lakes. The Pennsylvania system of individual family farms became the basis of farming in the southern Midwest. The plantation system that was first established in the Virginia tidewater, along with the accompanying use of slaves, was extended south and then southwestward along the Gulf lowlands to Texas. The North–South tensions that culminated in the Civil War of the 1860s arose from the spatial diffusion of these early differences.

The United States tripled its area by 1850, buying land from France and Spain, and acquiring the western third by negotiation and military conquest (Figure 11.10). To ensure the rapid occupation of these new areas, land was surveyed and sold by the U.S. government at lower and lower prices. The **Homestead Act of 1862** provided families with very inexpensive and at times even free farmland. Settlers spread from the eastern seaboard into the vast interior plains. Families able to overcome the challenges of starting farms in the interior, such as building roads and clearing vegetation from acquired lands, took advantage of the fertile soils and the warm, moist summer climate. This combination proved ideal for growing corn, soybeans, and wheat, and raising cattle and pigs. The United States exported a range of farm produce to Europe from the early 1800s, competing directly with European farmers in bulk grain markets, as well as continuing to export crops such as cotton that could not be grown in northern Europe.

Gradual Canadian Changes

The settled region along both sides of the St. Lawrence River (part of modern Canada) was a materially poor and often environmentally hostile remnant that was left to British authority after the United States declared independence. British loyalists who voluntarily emigrated or were forced out of the incipient United States joined small numbers of Native Americans and French-speaking European settlers in Canada. Those loyal to the British Crown mostly settled on the east coast or farther inland around the northern and western shores of Lake Ontario, due in part to a concentration of French settlers along the lower St. Lawrence who resented British rule.

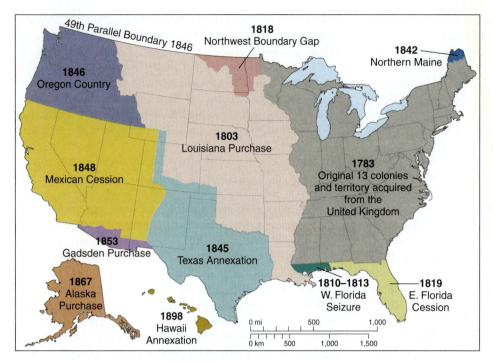

Figure 11.10 **United States: rapid expansion of territory.** This map shows the lands occupied by European colonial empires acquired from the Native Americans who inhabited the continent until their arrival. The eastern lands were acquired from the United Kingdom by the year 1783, just a few years after America declared independence in 1776. The 1803 Louisiana Purchase acquired lands that France had just gained from Spain. Texas claimed independence after defeating Mexico in the early 1830s, and then joined the United States in 1845. Subsequent lands in the southwestern United States were acquired through the 1848 Mexican Cession after a fast military victory.

Westward of the Great Lakes there was little settlement in an area still then largely administered by the Hudson's Bay Company, which was created by a Royal British Charter in 1670 to administer the development of nearly one-third of present-day Canada. The interior remained the realm of indigenous peoples and isolated groups of French settlers attempting to escape the watchful eye of British rule. When the Hudson Bay Company's fur trappers began to compete with Americans for land along the lower Columbia River near the coast of the Pacific, negotiations in 1846 set the boundary between the two countries west of Lake Superior at the 49th parallel of latitude.

Growth in U.S. Manufacturing

The first Industrial Revolution began in Western Europe (see Chapter 2), coinciding with the United States' independence from Great Britain. Independence and British hostility motivated Americans to establish their own manufacturing industries, providing goods that were at first not obtainable from Britain. Textiles, metal goods, and leather goods were among the first industries to develop, often powered by water mills in New England, New Jersey, and eastern Pennsylvania. By 1850, Pennsylvania's coal replaced charcoal in iron making, and ensuing steel production in Pittsburgh set off further industrial developments in the region. The northeastern United States soon emerged as a dominant manufacturer of the country, hous-

ing the bulk of the population with growing material wealth and related political power based on different types of industries.

The primary symbol of the early steel-based Industrial Revolution—the railroad—spread across the country to the Pacific coast. The Manufacturing Belt of the northeastern United States, which initially developed based on the regional resources of coal, iron, and the agricultural produce of the Midwest, later drew in mineral resources from other parts of the country. Railroads linked the mines of the western mountains to the eastern economic and political core. The mining of gold, silver, copper, and zinc in the new western lands provided capital for further development and widened the scope of the late-1800s metal industries.

In the 1900s, the availability of such natural resources, utilization of rivers for irrigation, navigation, and hydroelectricity, and discoveries of oil and natural gas drew more people and manufacturing corporations southward and westward from the country's core. Although most Americans viewed natural resources in terms of the potential for economic gain, influential groups in the late 1800s and early 1900s persuaded the federal government to set aside large areas of attractive wilderness or places of historic significance as national parks, often before much settlement had occurred.

The United States also sought to develop conditions in which its population could flourish by making the most of their freedoms under the democratic Constitution. It established compulsory education much earlier than European countries. Its emphasis on technology transfer began in the late 1800s with land grants for establishing engineering and farming universities in many states, such as Michigan State University in East Lansing, Michigan.

These natural and human resources, along with innovative manufacturing structures and processes, quickly brought the United States prominence in world economic activity by the late 1800s. The expanding internal market of the United States provided the demand that stimulated many of the developments. Americans achieved **economies of scale** by building larger factories for increased output. **Horizontal integration** occurred when financiers bought up several producers of the same product, giving the new owners a large share of the total production of goods and enabling them to set market prices. **Vertical integration** took this process further by uniting the producers of inputs with the manufacturer in a single corporate structure. As an example of these processes, Andrew Carnegie built larger steel mills to gain economies of scale. He then bought out other steelmakers in a time of economic recession in the 1870s, to integrate the steel industry horizontally. In the next phase of his vertical integration scheme, he purchased both coal and iron mines that provided the raw materials and the heavy engineering corporations that used the steel. By 1900, Carnegie's United States Steel Corporation dominated the industry. As another example, Henry Ford took the concept of the factory **production line** to a new level, applying it to the assembly of a wide range of components in the production of automobiles. His company produced thousands of a limited number of models for a growing market. His methods, called **Fordism**, were widely adopted by other industries and in other countries. The mechanization of production was another feature that made the United States the world leader in output of manufactured goods by the mid-1900s.

Canada Emerges

Not until the early 1900s did Canada begin the process of becoming a major industrial country. British rule that suppressed internal development, as well as a lack of capital and expertise, hindered the creation and expansion of Canada's global connections. After the British North America Act of 1867, Canada became a dominion within the British Empire with increased responsibility for its own affairs. At that stage, Canadian leaders began to integrate their vast country by building transcontinental railroads to encourage the settlement of the prairie grasslands and the Pacific coast. Manufacturing industries, including wood processing and steelmaking, were established behind tariff walls to protect them from American competition in particular.

Political Geography

Indigenous Peoples and Federal Governments

The stability and wealth of North America create the perception for many that there are few, if any, geopolitical issues. Despite very strong economies and the presence of stable governments in both Canada and the United States, many people in both countries feel lacking in a political voice, unable to compete for jobs or obtain employment, especially since the Great Recession of 2008, or they feel marginalized by mainstream society.

Native American peoples are arguably the most spatially segregated community in the United States, and many Native Americans feel abandoned by the federal government. Indigenous Americans are primarily concentrated on federally sequestered reservation lands that are often arid and unsuitable for agricultural productivity. Native American peoples within the United States therefore are among the country's most materially impoverished communities. Complicating the reservation lifestyle was an administration structure imposed by the federal government, rather than state or local governments. By the 1970s, some 50 percent of Native Americans were unemployed, and 90 percent were on welfare.

Attempts were made in the later 1900s to reduce the poverty imposed on Native Americans who had often been forced to live on marginal and largely unwanted lands. Native American communities with natural resources attempted to capitalize on the economic value of desired commodities such as fresh water and timber, while others filed lawsuits seeking compensation for being placed on resource-deficient land. Lucrative deals with governments and water projects, however, did little to change the economic structure of the Native American reservations.

New enterprises brought limited satisfaction. Some Native Americans developed tourist facilities on their reservations, but others chose to forgo such opportunities—often to avoid copying the commercial practices of European Americans. Casino gambling, legalized on Native American reservation lands in 1998, is now a significant source of income for many Native American groups such as the Ojibwa in Michigan. Among the casino-related challenges are managing the influx of capital, lack of high-wage jobs, combating gambling addiction, avoiding an increasing dependence on a single employment and revenue stream, and numerous social ills that often accompany a gaming-based culture. While casino-related development on reservations remains controversial, such enterprises may encourage some Native American groups to engage with the wider U.S. economy. Native American peoples began to increase their proportion of the U.S. population in the 1970s for the first time since census records began in 1790.

In 1973, the Canadian government opened the possibility of negotiating indigenous land claims with organizations representing native peoples. Until that date, little had been done in much of Canada to implement treaties negotiated with native peoples in the 1800s. Today there are several areas throughout Canada where indigenous communities have local or regional governmental control. The largest area, Nunavut ("Land of the People" in the Inuit language), became a new territory with its own elected government in 1999, although it will remain subject to federal control. In 2012, Nunavut relied on the Canadian federal government for over 90 percent of its income, demonstrating the economic challenges presented in this new province. Some current geopolitical issues facing Nunavut and other indigenous peoples of Canada are complex because of overlapping land claims—resolving these claims involves the often-opposing interests of the federal government and provincial governments.

North America and Geopolitics

The United States wields a great degree of political influence throughout all world regions primarily due to the size of its economy and military presence in almost every location on Earth. The United States and Canada participate in a number of intraregional and interregional diplomatic, economic, and military alliances. Both the United States and Canada, along with Japan, Germany, the United Kingdom, France, Italy, and Russia, are part of the **Group of Eight (G8)**, an economic discussion forum of the world's eight most materially wealthy countries. The United States wields significant influence in the United Nations (UN), the North Atlantic Treaty Organization (NATO), the World Bank, the International Monetary Fund (IMF), the Organization of American States (OAS), and the World Trade

Organization (WTO), among many others. The UN, World Bank, IMF, and OAS are all headquartered in the United States.

The United States enjoys a distinctive role as the host country of the UN, with the international headquarters based in New York City. The **UN Security Council** is the most powerful branch of the organization, with the purpose of the Security Council being the "maintenance of international peace and security." The United States (along with China, France, the United Kingdom, and Russia) is one of five countries granted a permanent seat on the Security Council. These permanent members of the Security Council have the power to veto council decisions. The decisions of the Security Council are binding for the entire UN General Assembly.

Economic Geography

The United States of America is the world's most developed country in terms of economic prosperity, size, and influence. The United States maintains the largest total GNI of any country in the world, although Switzerland, Japan, and some Scandinavian countries exceed its GNI per capita (Figure 11.11a, b). The United

States had the world's largest aggregate economy of any single country in 2012. Canada, with a population of only 10 percent the size of the United States, ranked eighth in the world. However, when adjusted for purchasing power parity, Canada's aggregate GNI PPP is not among the world's top 10 countries.

The international economic influence of the United States began to significantly expand after World War II during the economic recoveries in Europe and East Asia. Investment growth and influence diffused into many developing countries during the Cold War. After 1950, the United States established a huge lead in the initial development and use of computers. Although Western Europe and Japan now challenge this economic and technological superiority, no other single country is close to rivaling the total GNI PPP of the United States or the size of its internal market for products.

In 1988, the United States and Canada established the United States–Canada Free Trade Agreement, which led to the **North American Free Trade Agreement (NAFTA)** in 1994 by adding Mexico to the arrangement. The agreement between the United States and Canada continues to thrive under NAFTA, despite a series of disputes over individual items.

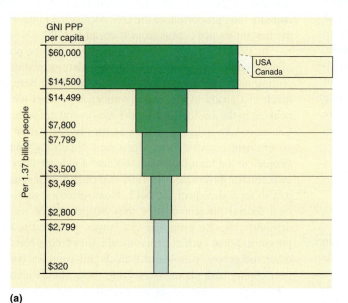

(a)

Figure 11.11 **North America: country incomes and well-being compared.** (a) GNI PPP figures for each country. (b) A comparison of the ownership of consumer goods and access to piped water and energy. The figures for the United States and Canada are among the world's highest. *Sources (a): Data (for 2012) from* World Development Indicators, *World Bank, and Population Reference Bureau; (b) Data (for 2012) from* World Development Indicators, *World Bank 2012.*

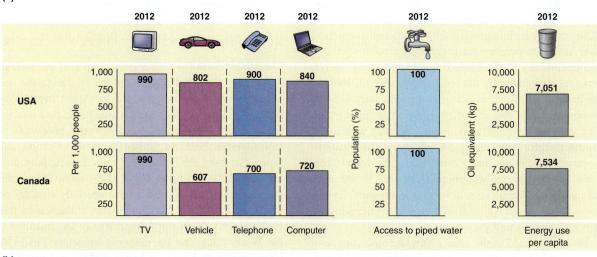

(b)

The Changing U.S. Economy

Major federal investments in transport infrastructure stimulated much of the spread of economic growth. Beginning in the 1850s, transcontinental railroads such as the Union Pacific Railroad were financed by federal and state governments granting lands along the routes to the railroads, which sold them and encouraged homesteading. Many communities, from small towns to major urban centers such as Chicago (Figure 11.12) blossomed along the railroad lines as people migrated away from the northeastern United States into the Midwest and Great Plains, following the promise of cheap land and superior agricultural environments.

In the late 1800s, the value of U.S. manufactured goods exceeded the value of commercial farm products for the first time. Agriculture became industrialized with the increasing use of mechanization and chemicals, while markets for crops and livestock products were linked to the growing railroad network. The railroads reached their greatest extent across the United States by 1916, with over 228,000 track-miles scattered over the nation. By comparison, there are only approximately 116,000 track-miles in existence in the United States today, a direct reflection of changing and improved transportation conditions such as the advent of the automobile and airplane.

Until 1950, manufacturing was the primary engine fueling the expansion of the U.S. economy. These industries were less tied to sources of raw materials than the 1800s industries, and many were market-based or assembly industries where the best locations were central to a range of component producers or large markets (Figure 11.13).

While products diversified and became more technologically sophisticated, production and management techniques developed. American-based multinational corporations took their products to the world and opened factories in many countries in Europe and other continents.

U.S. manufacturing started to decline around the decade of the 1970s, and in 2012 it employs fewer than 15 percent of the work force, as compared to nearly 40 percent in 1950. Products were even more diverse, with some heavy industry and producer units moving to countries with lower labor costs. The United States is a world leader in applying improved technology to manufacturing and service industries. An increasing range of high tech goods is produced in the industrial areas of Silicon Valley in California, metropolitan Boston, metropolitan Washington, D.C. (primarily Fairfax County, Virginia), the Research Triangle region in North Carolina, and areas of the Pacific Northwest especially near the cities of Seattle and Tacoma.

Retail, service, education, medical, and information technology accounted for approximately 70 percent of U.S. economic activity in 2012. Among numerous service industries experiencing growth were investment advice and wireless telecommunications services. Tourism and leisure-based services also became a leading part of many local economies. In 2010, over 55 million tourists from other countries visited the United States, whereas 57 million traveled from the United States to other countries. Additionally, many millions of Americans were tourists in their own country, with Orlando, Florida, being the primary tourist destination with attractions such as Walt Disney World and Universal Studios.

Large U.S.-based corporations increasingly replaced higher-paid service jobs in the United States with less-expensive employees working in offices in India (see Chapter 6). Millions of educated, English-speaking workers in India are hired by big U.S. companies to fill back-office positions such as customer service call centers, accounting and tax services, computer software design, and technology and medical research. The average college graduate working in such a call center makes around US$200 per month, whereas a college graduate in the United States would demand an average starting salary of around US$30,000 per year. White-collar workers in technology-related positions in the United States are increasingly facing greater competition for fewer U.S.-based positions. Critics of this trend say corporations are exporting U.S. economic growth to India, hampering the U.S. economy. Advocates of outsourcing argue that

Figure 11.12 **Freight railroad yard on the outskirts of Chicago, with the downtown skyline in the distance.** Chicago is North America's greatest hub of railroad activity, with over 600 freight trains originating, terminating, or passing through the region daily. *Photo: © Chris Valle/Flickr/Getty Images RF.*

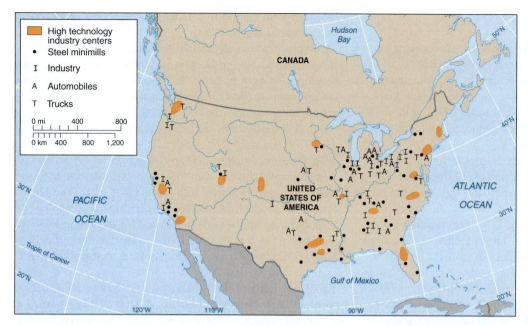

Figure 11.13 **United States: distribution of some major manufacturing industries.** The continuing concentration of steelmaking and vehicle manufacturing in the Manufacturing Belt contrasts with the widespread nature of high-tech industry centers. Steel production in minimills has become more widely distributed since 1970.

it makes U.S. companies stronger, and is a way of supplying young, skilled workers to meet the needs of the U.S. economy without encouraging further immigration into the country.

Starting in 2008, the U.S. economy experienced a dramatic shift as many banks declared bankruptcy due in part to lax regulation and high-risk loan defaults. Stock prices fell, the nearly nationwide real estate bubble deflated, and the U.S. economy shrank. Many linked economies around the world experienced similar economic crises. The U.S. economy had exhibited signs of stabilization and a very slow rebound by the fall of 2012, though progress was challenged by continued high unemployment and underemployment.

Canadian Economic Development

Canadian economic development generally followed that of the United States, though often with a lag time of several years. By 2012, Canada had combined a continuing emphasis on its natural resource base with an affluent, high-tech society. It had one of the highest proportions of trade per capita in the world, and Canada's GNI per capita rivaled that of many European countries.

Through the 1800s and 1900s, Canada's economy depended mainly on primary products such as grain, timber, and minerals. Canada remains a major world producer of newsprint, wood pulp, and timber, and it is one of the world's leading exporters of minerals, wheat, and barley. Canada produces 30 percent of the world's newsprint, manufactured primarily along the lower St. Lawrence River valley. Most timber output comes from the great west coast forests of British Columbia. Canada is one of the world's top producers of minerals that include coal, oil, and natural gas from Alberta, iron ore from Labrador and Québec,

uranium from Ontario and Saskatchewan, and nickel from Ontario and Manitoba. Agriculture remains significant in the Prairie Provinces, southern Ontario, and the specialized fruit-growing districts of British Columbia, where wine production is gaining international attention.

Industrialization in Canada was a phenomenon that commenced far later than in the United States. Beginning before World War II and developing rapidly during the 1940s, the production of aluminum, vehicles, and consumer goods became important as Canada developed import-substitution industries to supply its own markets and avoided total economic domination by the United States. Coincidentally, American investment in Canada through placement of automobile manufacturing plants, for example, allowed Canada's economy to graduate from one based on agriculture to one with a more diverse base overall. The city of Hamilton, located at the western end of Lake Ontario near Toronto, became a steelmaking center. Montréal, Toronto, and Vancouver became centers of financial services and a wide range of commercial enterprises. The signing of NAFTA facilitated much greater trade between the two countries, resulting in further U.S. investment in Canadian industries.

Canada lagged behind its neighbor through the early 1900s, partly because of restrictions resulting from its colonial ties to Britain, and also due to the smaller size of its home market for goods and the protection of its own industries. It gradually became more closely intertwined with the U.S. economy and dependent upon it for financing and markets for its mineral and agricultural outputs. In the post–World War II era, rich stocks of metal ores in Canada's north, along with coal, oil, and natural gas in Alberta and hydroelectric power generation in northern Québec, ignited new geographic directions of development in the country and provided the basis for growing Canadian affluence. Canada's centers of manufacturing production around Toronto and Vancouver, and to a lesser extent around Montréal and in Alberta, now rival individual centers in the United States in the size and diversity of output. Canadians enjoy material living standards almost equal to those in the United States. Some Americans prefer the Canadian way of life, and live across the border from major cities such as Detroit, Michigan, and Buffalo, New York. The 2012 United Nations human development and gender-related development indices (HDI and GDI) both ranked Canada in the top five among the countries of the world.

11.4 Geographic Diversity

The United States

The United States of America (Figure 11.14) dominates the midlatitudes of continental North America, with Canada to the north and Mexico to the south. Alaska and Hawaii are separated from the conterminous 48 states by Canada and the Pacific Ocean respectively.

Metropolitan areas of the United States house the majority of the country's transportation and communications connections, manufacturing facilities, corporate headquarters, and financial and business services.

The U.S. market economy, which is the world's largest, significantly influences economic activity across the globe. The country remains a global leader in high technology and is the world's entrepreneurial leader. The United States was the world's leader in raising venture capital in 2012 with more than US$31.3 billion raised overall.

Problems of Affluence

The country's material wealth does not reach all within its borders. Those with advanced educations or multiple income families continue to gain greater material wealth, while citizens who are located in impoverished regions or lack access to educational opportunities struggle to acquire the basic necessities for nutrition, shelter, and health care (see Figure 11.11).

The gap between the materially wealthy and the materially poor widened from the 1970s to the present day in the United States. Some of the increasing inequality may be explained by the following factors: (1) Lightly regulated labor markets react to world conditions by depressing wages for unskilled jobs in order to compete with poorer countries. (2) Increasingly higher salaries are paid to skilled workers as demand from service industries and professions rises. (3) Lower-paying service jobs replaced relatively higher-paying manufacturing jobs, while technological advances increased the need for more highly skilled and specialized workers. (4) Increasing numbers of lesser-skilled immigrants are filling low-paying service jobs and have become a growing portion of the impoverished residents of the United States.

The materially poor areas of many inner cities have lower-quality buildings and limited access to good education and health care, due to the inability of small jurisdictions with predominantly poor populations to support such services. The contrasts produced by **uneven development** are illustrated by the example of Boston, Massachusetts: Some of the best public secondary education in the country is found in the suburbs, whereas some of the worst is in Boston's inner city—both ends of the spectrum in the same metropolitan area.

Figure 11.14 The United States of America: the 50 states and major cities.

The income and material wealth gaps are often linked to perceived ethnic and racial differences. High income-earning people have many choices with respect to where they live. Many congregate with those of similar socioeconomic standing in trendy downtown areas or established suburban neighborhoods. They may live close to shopping, ethnic restaurants, cinemas, places of worship, or recreation facilities. Less-affluent groups in U.S. society have fewer opportunities for choosing where they live, and often become segregated into communities that more-affluent groups avoid. Such segregated groups often occupy inner-city areas that contain high proportions of African-American, Latin American, or recent immigrant communities, most of whom have poor access to high-quality education and job opportunities.

Other problems of affluence result from the environmental impacts of intensive use and extraction of resources. The United States, with just under 5 percent of the world's total population, consumed over 40 percent of the world's oil production in 2012 and also accounted for over 33 percent of the world's garbage. Such disproportionate consumption often makes the United States the target of international protests.

In the early 1970s, the United States passed stringent environmental legislation to improve its air and water quality. Following the rise of the environmental movement in the United States and related legislative changes, air and water quality standards improved substantially. Such regulations, however, sometimes led to the relocation of the most polluting industries to more materially impoverished countries such as China, where lax environmental restrictions combined with a cheap labor force to producegreater profits for companies.

Urban Landscapes

Urbanization within the United States produces distinctive landscapes through building types and styles, the differentiation of land uses, and the distribution of groups of people. Human variation in the landscape includes the materially wealthy and the materially poor, African-Americans, Hispanic Americans, European Americans, Asian Americans, and many others. American cities grew from colonial ports and inland market centers to industrial societies by the late 1800s. Further evolution in the 1900s and 2000s has promoted the rise of commercial metropolitan centers with well-defined suburbs surrounding the perimeter of many major cities in the United States and Canada.

Industrial and Commercial Cities In the later 1800s and early 1900s, many cities in the eastern and central United States expanded into large conurbations of housing, factories, shops, and offices with the rapid growth of manufacturing industry and the railroad network. The larger factories and mills required many workers, who were accommodated in surrounding housing units.

By 1920, American cities often evolved into a **concentric pattern of urban zones** around the downtowns (Figure 11.15), also known as the central business districts (CBDs). Immediately adjacent to the CBD is the industrial/low-income housing ring, which typically contains poor people and high crime rates. As one moves farther away from the downtown, you generally encounter areas of progressively better housing and standards

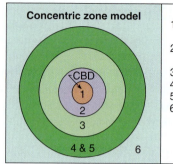

Concentric zone model

1. Central business district (CBD)
2. Wholesale, light manufacturing
3. Lower-class residential
4. Medium-class residential
5. Higher-class residential
6. Residential suburbs

(a)

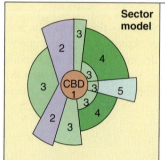

Sector model

1. Central business district (CBD)
2. Wholesale, light manufacturing
3. Lower-class residential
4. Medium-class residential
5. Higher-class residential

(b)

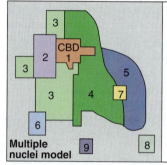

Multiple nuclei model

1. Central business district (CBD)
2. Wholesale, light manufacturing
3. Lower-class residential
4. Medium-class residential
5. Higher-class residential
6. Heavy manufacturing
7. Outlying business district
8. Residential suburb
9. Industrial

(c)

Figure 11.15 **United States: urban landscapes.** (a) The concentric zone model, based on a 1922 study of Chicago. (b) The sector model, based on 1930s census information, emphasizing the impact of railroads and highways on the location of land uses. (c) The multiple nuclei model of 1945 depicts a large central business district and smaller peripheral business district. *Source: Data from Harris et al., 1945.*

of living, from middle-class to upper-class societies and commuter zones (in which people sometimes will drive from 45 minutes to two hours from their "bedroom communities" to a work destination downtown!).

An explosion of house and road building opened up huge suburban areas of single-family homes for those with rising incomes after World War II, as the United States dominated the world economy. New federal roads that linked major cities, particularly those established by the Eisenhower Interstate System in 1956, made greater mobility possible and also encouraged suburban housing and commercial development.

These trends gave rise to a new form of city that developed from the monocentric (single center) before 1940 to the multicentric pattern of today. The CBD became more specialized around financial businesses, as many of its shopping and industrial functions moved to the suburbs. Competition for space elevated land prices, which precipitated the construction of tall office buildings primarily in each city's commercial center. New highways, often elevated or in tunnels, circled the CBD.

Beyond this sector of intense commercial activity, the older pre-1940 suburbs, crossed by railroad yards and older highways with commercial strip developments, were often taken over by expanding African-American, Hispanic, or other ethnic groups. Such groups either did not have the resources to move to newer suburban homes, faced discrimination by financial institutions or home sellers when they tried to do so, or preferred to stay in an area that was familiar to them. The neighborhood concentrations of African-Americans in northern cities did not appear until influxes of African-Americans from the U.S. South combined with the suburb-bound movements of the European American population of northern cities in the 1950s.

The changing face of U.S. cities since World War II resulted in multiple migrations of people within the cities. The growth of the suburbs involved the movement of wealthier families out of the older inner-city areas to new suburban single-family homes. This trend was labeled "white flight," since it resulted in low-income African-Americans and other minority groups being left in the older areas that lost many services and jobs to the growing suburbs. Property values in the older areas declined dramatically, leading to degradation and even abandonment.

Postindustrial Cities Expanding suburban growth through the 1970s created several smaller suburban cities around the primary urban center. Such suburban developments contained significant commercial space and service industries. The term **edge city** was coined for new exurban developments relying primarily on car and truck transport and secondarily on air travel. Edge cities include large developments of shopping malls, offices, and warehouses located on the edge of major metropolitan areas (Figure 11.16). Since the 1970s, there has been a phenomenon of movement back to the city center by affluent groups in a process known as gentrification.

Gentrification Gentrification is the movement of higher-income residents and business owners into materially poor areas of inner cities, leading to the physical and economic improvement of property and potentially dramatic changes in the quality of life in the districts affected.

The availability of inexpensive property, coupled with the proximity to in-town jobs, attracted many suburbanites back into inner-city areas starting in the 1990s. The proximity of inner-city neighborhoods to a wealth of urban amenities including leisure-time facilities and theaters, symphony halls, opera houses, museums, restaurants, improved parks and waterfront areas, and major sports facilities, fueled dramatic urban renewal and inner-city growth for many locations through the present day. The primary drawbacks consisted of higher crime rates in these revitalized neighborhoods—spilling over from nearby "ghetto" sectors—and less funding for public schools. Crime rates have decreased in many inner-city areas, resulting from changes in

Figure 11.16 Tyson's Corners, Virginia, an example of an edge city in the suburbs of Washington, D.C. *Photo: © Glowimages/ Getty Images RF.*

police and city government attitudes as well as changes brought about by the overall gentrification process. As a result, the quality of schools has improved in some inner-city areas. However, it remains common for younger affluent families to move out to suburban locations for better school systems, or to send their children to private schools.

The movement of high-income people into city centers often causes the displacement of more materially impoverished occupants of the urban zone that is undergoing the gentrification process. As money moves into previously poor neighborhoods, old buildings are renovated or replaced with new structures. As a result, the cost of commercial and residential property in the gentrified areas escalates dramatically. Housing opportunities for the materially poor, who can no longer afford to live in the gentrified neighborhoods, are often not part of the process. Proponents argue that the process dramatically increases the economic and social well-being and the standards of the gentrified neighborhoods, along with those of the CBD.

Many cities are affected by gentrification, but its impact is modest in relation to the total built environment of each inner city. Gentrification usually occupies several blocks that constitute a small percentage of the land area of a CBD or low-income sector. Examples of gentrification are found prominently in Baltimore, Boston, Philadelphia, Pittsburgh, Washington, D.C., New York City, Atlanta, and San Francisco.

Regions of the United States

The large areal extent of the United States makes it a country of many internal geographic subregions that developed from the interaction of people with the natural environments and resources of the land (Figure 11.17).

New England Manufacturing-related industries, including textiles and paper-based products, dominated New England's economy from the mid-1800s until World War II. From then through the present day, the economy has shifted to one marked by education and research (the metropolitan Boston area has more colleges and/or universities than any other city in the

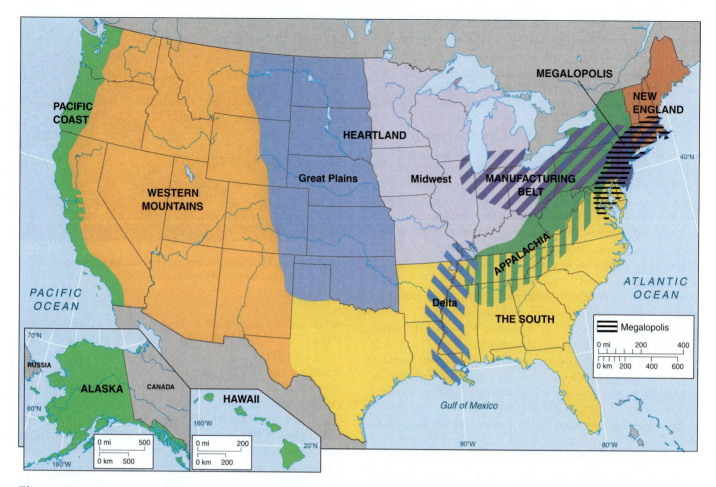

Figure 11.17 **United States: subregions.** The regions reflect differences in natural environments and patterns of human occupation. The overlap of regions in the northeast resulted from early concentrations of people and changing economic and social emphases.

world!). These large numbers of higher education institutions combined their research innovations with the local availability of entrepreneurial capital to create a post–World War II boom of economic development. This is sometimes referred to as the "Massachusetts Miracle" because of the numbers of new jobs created. New England's financial and professional services continue to grow today, and its natural environment of rocky and sandy coasts, wooded hills, interior mountains, and well-preserved early American history appeals to many Americans as a place to vacation or as a high-quality living environment.

Megalopolis The Megalopolis region stretches from the greater Washington, D.C., area (by some definitions, as far south as Richmond, Virginia) at the southwest end of its long urban extent, to Boston at the northeastern end. The urban chain of the Megalopolis includes the huge metropolitan areas of Baltimore, Philadelphia, and New York City/Newark, as well as many smaller urban centers, positioned between metropolitan Washington, D.C., and Boston. Approximately 54 million people—nearly one-sixth of the total population of the United States—live in this region. The major cities of the region grew out of the country's initial port, commercial, and transportation hubs that were based along the Atlantic coastline for trade purposes with Europe.

Despite the migration of some corporation headquarters out of the Megalopolis cities to other parts of the United States, this region still dominates the American economy, federal government, and the modern newspaper and magazine industry. While other groups of cities within the United States have been identified as new versions of the megalopolis phenomenon, none match this region in terms of its numbers of people or its economic and political impact.

Manufacturing Belt From the early 1800s through the mid-1900s, the area stretching from Boston on the East Coast through New York and Philadelphia westward to Pittsburgh, Cleveland, Detroit, and Chicago, became so dominated by manufacturing industries that it was termed the "Manufacturing Belt." This region produced over two-thirds of U.S. manufactured products until challenged by the West Coast and southern states during and after World War II.

Although the old Manufacturing Belt has declined in importance in the last several decades, service industries continue to take on increasing significance in this region. Chicago became a major center for financial services arising out of its futures trading in agricultural products. Pittsburgh, once considered the industrial furnace of the United States, underwent a dramatic transformation during the 1970s and 1980s, and

Figure 11.18 Pittsburgh, Pennsylvania, built at the confluence of the Allegheny and Monongahela rivers. This photo shows the modern skyline of a once highly industrial city that today is a regional center of services and financial activities. Note the inclined railway car in the foreground of the photograph, a popular tourist attraction for those visiting the area. *Photo: © Jeff Greenberg/Peter Arnold/Getty Images.*

in 2012 is one of the best examples of urban renewal in the United States (Figure 11.18). Pittsburgh played a prominent role in the growth of the country through the development of an intense concentration of iron and steel production and related heavy industry. In addition to being known for its significant industrial output, the city became infamous for its pollution. During the city's renaissance of the 1970s and 1980s, many manufacturing jobs were replaced with service positions. Many factories closed to make way for Fortune 500 companies. The city emerged as an international center for medical research and health care. Pollution-filled air and the glow from the iron and steel blast furnaces have been replaced with a dramatically modern and clean skyline.

Appalachia The region of the Appalachian Mountains extends from the middle part of the Manufacturing Belt in Pennsylvania, deep into the U.S. South. Eastern Kentucky, West Virginia, western Virginia, and eastern Tennessee form central Appalachia and remain one of the poorest parts of the United States. Although coal mining continues to be somewhat important, it is subject to significant price fluctuations that produce inconsistent regional conditions. A measure of economic diversity was introduced into southern Appalachia with the production of cheap electrical power through the **Tennessee Valley Authority** from the late 1930s. This project dammed over 95 percent of the rivers of the region, creating hydroelectric resources along with a stable supply of water and opportunities for recreation. Major aluminum companies and federal facilities, such as the Oak Ridge atomic laboratories and the Huntsville rocket center, were joined by many other manufacturing concerns. Cities such as Asheville, Knoxville, Chattanooga, and Huntsville expanded, developing a wide range of service industries. Contrasts of material

wealth, infrastructure, health care, and educational opportunities remain between the urban centers and the surrounding, more materially impoverished rural mountain communities.

U.S. Heartland: Midwest and Great Plains The lowland area between the Appalachians and Rockies that is north of the Ohio River and the Ozark Mountains is a dominant agricultural region of the country, producing major cash crops such as corn, soybeans, wheat, hay, and dairy products (Figure 11.19). Cities such as Chicago, Detroit, Cleveland, and Cincinnati stand amid the world's most productive farming region and are home to a wide variety of natural tourism opportunities (Figure 11.20). The Great Plains, a subregion of the Midwest, include North and South Dakota, Nebraska, Kansas, Oklahoma, and parts of Montana, Wyoming, Colorado, and northern Texas. While the midwestern United States is dominated by corn and soybean production, the Great Plains is overwhelmingly devoted to wheat farming.

Agribusiness—the close commercial linking of inputs to farming, farm activities, and the processing and marketing of farm products—made American agriculture an integral part of the overall economy. World markets for the trade of wheat and grain opened, especially in the former Soviet Union republics of Russia and Kazakhstan.

Despite the high and increasing farm productivity of this region, major problems arose. Many farmers had to obtain off-farm jobs, rent some of their land to other farmers, or sell their lands to banks and other financial businesses in order to make loan payments and cover their costs. Much land that had resided in family farms was sold to large farm corporations. Problems of declining population and few alternative occupations that had plagued other rural regions for decades extended to America's richest farming region during the 1980s. The resulting out-migration of people from the Great Plains to other regions of North America continues to this day, especially as younger people are leaving the farm lifestyle in search of better jobs and opportunities elsewhere.

The South The American South extends from the Atlantic coastal plain westward to the Mississippi River and beyond into eastern Texas, encompassing all areas south of this transitional border. While tourism is the staple of this region's economy today (Figure 11.21), its common cultural identity stems from colonial migration patterns and agricultural practices, as well as states' membership in the Confederacy during the Civil War. Most of the South at that time was dominated by the plantation economy, which was established in colonial times. All of the southern states suffered after the Civil War, experiencing decades of poverty at a time when the northern states made huge strides in industrial expansion and garnered significant material wealth. Black slaves were "freed" by the outcome of the Civil War, but soon became debt slaves of the agricultural lands on which they had been working.

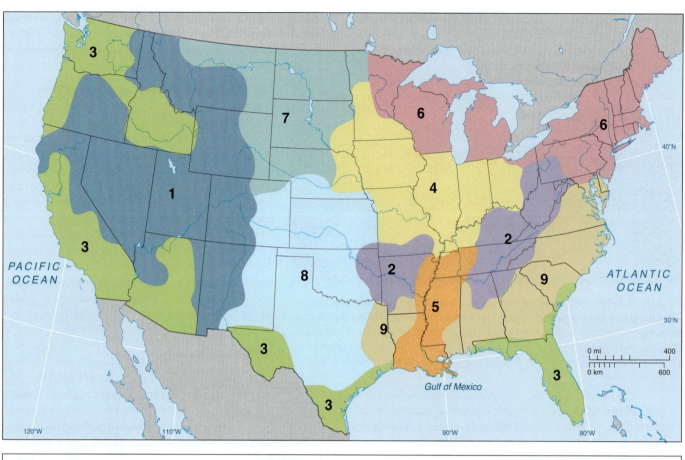

U.S. Farm Resource Regions

Basin and Range
1 Largest share of nonfamily farms, smallest share of U.S. cropland. 4 percent of farms, 4 percent of value of production, 4 percent of cropland.
Cattle, wheat, and sorghum farms.

Eastern Uplands
2 Most small farms of any region. 15 percent of farms, 5 percent of production value, and 6 percent of cropland.
Part-time cattle, tobacco, and poultry farms.

Fruitful Rim
3 Largest share of large and very large family and nonfamily farms. 10 percent of farms, 22 percent of production value, 8 percent of cropland.
Fruit, vegetable, nursery, and cotton farms.

Heartland
4 Most farms (22 percent), highest value of production (23 percent), and most cropland (27 percent).
Cash grain and cattle farms.

Mississippi Portal
5 Higher proportions of both small and larger farms than elsewhere. 5 percent of farms, 4 percent of value, 5 percent of cropland.
Cotton, rice, poultry, and hog farms.

Northern Crescent
6 Most populous region. 15 percent of farms, 15 percent of production value, 9 percent of cropland.
Dairy, general crop, and cash grain farms.

Northern Great Plains
7 Largest farms and smallest population. 5 percent of farms, 6 percent of production value, 17 percent of cropland.
Wheat, cattle, and sheep farms.

Prairie Gateway
8 Second in wheat, oat, barley, rice, and cotton production. 13 percent of farms, 12 percent of production value, 17 percent of cropland.
Cattle, wheat, sorghum, cotton, and rice farms.

Southern Seaboard
9 Mix of small and larger farms. 11 percent of farms, 9 percent of production value, 6 percent of cropland.
Part-time cattle, general field crop, and poultry farms.

Figure 11.19 **Agricultural regions of the United States.** The corn belt (Heartland Region) is the most productive and versatile sector, where corn and soybeans are grown either for direct sale or for livestock feed. Dairying is important in the northern regions of the United States, whereas ranching and grazing dominate the western and southwestern parts of the country. *Source: Data from De Blij, 2012.*

The African-American descendants of the former slaves did not gain full political and social rights for more than 100 years until the successes of the civil rights movement in the 1960s. This development coincided with the building of the interstate highway system that opened up much of the South to new economic opportunities, and the rapid expansion of cities such as Miami, the urbanizing area from Charlotte to Atlanta, New Orleans, Houston, Dallas–Fort Worth, Memphis, and Louisville. Both manufacturing and service industries moved into the area, spurred by the advent of air conditioning and the lower wages of southern workers combined with the absence of unions. Its changing fortunes were highlighted by local promoters who took up the term "Sun Belt" to refer to the U.S. South with its mild winters and sunnier climate. Walt Disney purchased several thousand acres of land just southwest of the small town of Orlando, Florida, in the late 1960s. The resulting opening of Disney World in 1971 creating a tremendous tourist economy in central Florida and saw Orlando's metro population increase from 50,000 in 1965 to over 3 million by 2012. Additional tourist destinations such as Universal Studios and a plethora of hotels and time-share resorts subsequently were built in this region, resulting in tremendous economic gains in the area over the last few decades (Figure 11.21).

Figure 11.20 **Recreation in northern Michigan.** Hiking in the mountains of Michigan's Upper Peninsula. Scenes such as these are repeated throughout the United States in a wide variety of parks and recreation venues. *Photo: © Per Breiehagen/The Image Bank/Getty Images.*

Aside from Florida, widespread poverty still exists in the South today. The Delta lowlands, extending up the Mississippi River valley from Louisiana and Mississippi through Arkansas and into southeastern Missouri, was a major area of cotton production until the early 1900s. The boll weevil plague in the early 1900s wiped out nearly 90 percent of the cotton crops, forcing people to migrate to other locations for work. Today, the region remains materially poor with relatively less infrastructure and fewer educational opportunities.

Manufacturing and service industries have thrived in the areas around Greenville, South Carolina. Production of Michelin tires (from France), BMW cars (from Germany), Lucas automotive electronics (from Britain), and Hitachi electronics (from Japan) provides jobs for many residents and demonstrates active

Figure 11.21 **Sunset over the Gulf of Mexico at Clearwater Beach, Florida.** Tourist destinations such as these bring substantial revenues to the local economy, especially in the winter months when many people migrate to escape harsh climates. *Photo: © Nicholas Pitt/Photodisc/Getty Images RF.*

foreign investment in U.S. labor. In Texas, manufacturing paralleled the accumulation of oil wealth and services development around the Houston and Dallas–Fort Worth regions.

Hurricanes are a threat to this region in the summer months from July to late September. Several strong hurricanes made landfall in southern states along the U.S. Gulf Coast in 2004 and 2005. The costliest storm in U.S. history, Hurricane Katrina, struck southeast Louisiana and southern Mississippi in August 2005, leaving more than 1,800 people dead and tens of billions of dollars in damages.

Western Mountains The western one-third of the United States is a mountainous region interspersed with large valley systems. The highest ranges are the Rockies on the east and the Sierra Nevada of California and the Cascades on the west, separated by broad plateaus and areas of basin and range. The mountains block most of the humid westerly winds flowing into the region from the Pacific Ocean, resulting in vast arid rain shadow areas throughout the interior and creating the great deserts of North America. Settlement focuses on towns that grew up as markets based on cattle ranching, mining, and/or division headquarters on the transcontinental railroads. A large proportion of the land between such centers is still sparsely settled, and most of the land in this region is owned by the federal government, including more than 70 percent of the state of Nevada.

Most of the water available in the region falls on the mountains as snow and runs off in swollen streams during the spring season. Many federally funded irrigation projects manage this water supply, with the largest group of projects along the Colorado River in the south and the Columbia River in the north. Rapidly growing metropolitan Las Vegas continues to capitalize on its gaming and entertainment industries, sunny and dry climate, and proximity to national parks and government military bases, which serve as large employers. The Las Vegas metropolitan area enjoys a healthy business climate and strong tourism industry that ranges from gaming to spectacular nearby scenery (Figure 11.22). The metropolitan region's economic growth slowed in recent years as a result of the U.S. and global economic crisis.

Hispanic people constitute a large proportion of the population in this region. Defined in part by the Rio Grande, the border with Mexico is largely a mountainous desert where few people live outside the border towns.

Pacific Coast The American Pacific coast has recently become a second national core, close in economic importance to that of the region between Boston and Washington, D.C. Settlement increased after the arrival of transcontinental railroads in the 1870s and 1880s. National defense needs for World War II in the Pacific placed manufacturing and military centers on the West Coast. Puget Sound in the north and San Diego in the south became major naval centers. Seattle grew in part as a manufacturing center for military and commercial aircraft. By 2012, Seattle housed many software firms, including the corporate headquarters of Microsoft, along with major Internet retail centers such as Amazon.com. Boeing has its largest airplane manufacturing plant near Seattle, due to mild climates and the need for skilled workers that the region provides.

Figure 11.22 **A view of the Hoover Dam from the pedestrian deck of the newly completed Pat Tillman Bridge, located just south of the dam on the Colorado River.** The bridge is a four-lane highway that was constructed to improve traffic flow on U.S. Highway 93 and offer tourists a more panoramic view of the region. *Photo: © Siegfried Layda/Photographer's Choice/Getty Images.*

Important to economic growth in the southern half of California was the supply of water to its largely arid area. Federal and state government funds supported huge water storage and distribution projects that were built throughout the 1900s. However, substantial droughts during the last 20 years have strained the region's limited water supplies. It is likely that farmers, who use 85 percent of the water to produce less than one-tenth of the state's economic output, will have to pay more for this resource and manage it a lot more carefully. At present, farmers pay only half the cost of delivering water to them, while city dwellers pay 20 times as much as the farmers for the same resource.

Arizona, California, and Nevada compete for fresh water from the Colorado River basin. These states have signed an agreement that limits each to an allocated percentage of fresh water from the river. Arizona's Central Arizona Project delivers approximately 1.5 million acre-feet of Colorado River water per year to several Arizona counties. Las Vegas is one of the fastest-growing metropolitan areas in the United States, and its water needs increase significantly each year as both the population and the number of tourists visiting the region continue to grow (Figure 11.23). Increasing water demand in the Las Vegas Valley encroaches on the supply of water downstream in southern

Both San Francisco and Los Angeles attracted financial service businesses after World War II, and by the mid-1990s high-tech jobs, a more diversified range of service industries, and increased overseas exports created thriving economies in each. The metropolitan Los Angeles economy today is greater than the economy of South Korea, and the city now contains a thriving motion-picture industry, a major intermodal port, and more computer software jobs than Silicon Valley, the best-known high-tech center situated just south of San Francisco. California has the world's eighth-largest economy and is America's most productive farming state, with the Central Valley leading the country in the production of lettuce, carrots, grapes, tomatoes, and other major cash crops.

(a)

(b)

Figure 11.23 **Growth and development in the desert.** Rapid growth and development in metropolitan Las Vegas expands across the desert valley floor. A new golf course community (a) and several new subdivisions (b) required large amounts of water from very limited resources. Nevada, Arizona, and California compete for fresh water from the Colorado River. *Photos: (a), (b) © Joseph P. Dymond.*

California. As southern California's water needs increase, officials are looking to alternative sources in the water-scarce region such as seawater desalination.

The West Coast developed a new economic importance through an increasing volume of business with Asia and other Pacific countries. Seattle, Portland, and Los Angeles are major world ports where manufactured goods from container ships are transferred to railroads for shipment across the country to their various destinations. The West Coast attracts most of the growing Asian investment in the United States. Asian investment, particularly Japanese and Chinese, finances car design centers, media industry corporations, and leisure industries.

California's proximity to Mexico and the state's trade links with the Latin America region continue to attract Hispanic peoples from south of the U.S. border. Southern California has the highest proportion of Hispanic Americans in the country. Los Angeles doubled its middle-class Hispanic population since 1980 to more than half a million people, and Hispanic businesses in the city have doubled since 1993.

Alaska and Hawaii In 1959, Alaska and Hawaii became the 49th and 50th states, respectively, added to the United States (Figure 11.24). They contribute to the extreme variety

of environments and resources within the United States, and each has its own issues.

Alaska is a huge area of northern land that was bought from Russia in 1867. High mountains, ice fields, and glaciers mark its southern coast. Inland, the Yukon lowlands have long and bitterly cold winters, and the Brooks Range and North Slope leading down to the Arctic Ocean are even colder. Any potential for commercial farming is severely limited by climatic conditions. The Yukon lowlands widen westward to the Bering Sea, but there is little mining or manufactured output to trade in that direction. The dramatic mountain scenery and abundant wildlife widespread in the state do support a growing tourist industry.

Alaska's Pacific location and proximity to Russia make it a strategic site for U.S. military bases, and the U.S. government owns most of Alaska's land area. Alaska also contains vast deposits of natural resources, including fuels and metal ores. The discovery of oil on the North Slope forced the state and federal governments to develop policies for Alaska's native peoples who live within that area. Residents of the state receive payments from the state government from taxes on oil production. The oil discoveries, and especially the 1989 oil spill from the Exxon *Valdez* tanker, made Alaska a focus of environmental

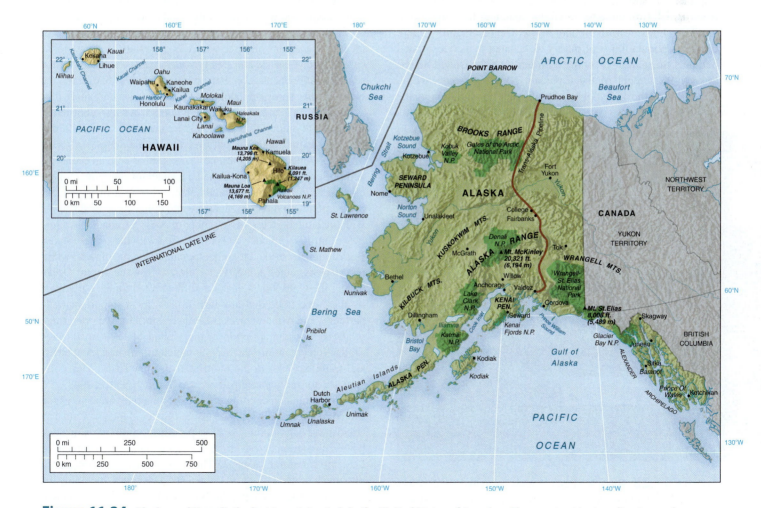

Figure 11.24 **Alaska and Hawaii: the last two states to join the United States of America.** They contrast in size, climates, and resources. The much smaller Hawaiian Islands have twice the population of Alaska.

concern. Although controversy remains concerning the prevention of drilling in the Arctic National Wildlife Refuge on the north coast east of Prudhoe Bay, oil extraction is thriving on other parts of Alaska's North Slope.

Hawaii has been of profound importance to the United States as a naval port and as a source of agricultural products such as pineapples and sugar. Containing a larger population than Alaska, the state has a major international tourist industry that is the backbone of its economy today (Figure 11.25). Hawaii has recently experienced economic challenges with high unemployment, and its tourist industry suffered from cutbacks on travel by the Japanese. Sugarcane production decreased slightly as Asian competition and changing U.S. consumption habits reduced the market. Hawaii also suffers from a poor business climate due to many regulations, poor schools, a high union presence, and its geographic isolation from globally connected business systems. However, coastal areas of the state attract wealthy retirees from the mainland United States and other parts of the world, creating dramatic disparities in real estate values and material wealth.

Canada

The land area of Canada extends laterally over 3,000 miles from British Columbia to Newfoundland, and from the southern U.S. border north to the Arctic Circle (Figure 11.26). Canada's population is primarily distributed in a concentrated belt near the country's southern border with the United States, as over 90 percent of the Canadian population lives within 100 miles of the U.S. border. The linear concentration of the

Figure 11.25 Hawaiian Luau. Tourists visiting the Polynesian Cultural Center on the island of Oahu watch as center staff prepare a roasted pig during a Luau ceremony. *Photo: © Joseph P. Dymond.*

Canadian populace along the border suggests the likelihood that Canadians interact frequently with cities and systems in the United States, and for warmer climates as well. Although Canadians do interact with urban systems across their border, the people operate internally within domestic subregions and on a national level through their federal government and national identity. The distinctive subregional groups within Canada's borders include the people of the Atlantic Provinces, the French-speaking peoples of Québec, the multicultural population of Ontario—especially Toronto—and the scattered farming populations and industrial cities of the Prairie Provinces. Additional subregions include Vancouver and the British Columbia coast, where people look increasingly toward Asia for economic reasons, and the indigenous Canadians, or First Nations, of the northlands. The Canadian provinces were created a century after the U.S. states, are generally much larger, and have greater internal political control than U.S. states do with respect to the federal government.

The geographic proximity of Canadian cities and commercial development to the United States, combined with the extensive exposure to entertainment and news media based across the border in the United States, have powerful influences on modern Canadian life that blur the clarity of a cohesive Canadian national identity. However, the unique population distribution of Canadians, combined with living in the shadow of the United States, actually appears to provide a strong incentive to form a national identity. Canadians disagree with U.S. nationals on many political and social issues, and this provides many Canadians with a sense of national cohesion. Perceptions of some U.S. citizens and other peoples across the globe that Canada is nothing more than a "satellite" or "subsidiary" of the United States also contributes to the desire among Canadians to demonstrate to the world that they are separate and different from the United States.

The United States continues to be seen as both an economic and political partner with Canada. Canadians enjoy many benefits of affluence and national security because of their close integration with the U.S. economy and defenses. Canada, along with the United States, is a member of NATO. Canada has a history of military alliance and cooperation with the United States. Both countries worked together to maintain a line of radar warning stations facing the Soviet Union across the Arctic Ocean during the Cold War. Canada and the United States participate in open free trade and are each other's most important trade partners. The two countries first entered into formal open-trade relations with the adoption of the United States–Canada Free Trade Agreement of 1987–1988, which eliminated most tariffs and trade barriers. Both countries then joined with Mexico to sign and formally participate in NAFTA in 1993 to 1994.

Political and economic ties are not the only areas in which Canadians cooperate with the United States. Much of Canada's television programming is U.S. produced, although Canadian television stations are restricted to showing only 30 percent of U.S. programming at most. The two countries are integrated into many of the same professional sports leagues, including the National Hockey League, the National Basketball Association, and the professional baseball leagues.

Figure 11.26 Canada: regions, provinces, territories, and cities. The Canadian provinces have greater powers than the individual states of the United States. Their geographic distribution in a line north of the U.S. border makes communications difficult among different parts of the country. The red lines indicate the division of regions within Canada.

Although Canada and the United States enjoy one of the most stable political boundaries in the world, there are environmental, social, and political issues that create occasional friction between the two countries. Among the issues that create tension between the two allies is a strong Canadian objection to acid rain, which is primarily produced from U.S.-generated industrial pollution. Canadians also have willingly accepted more stringent gun-related legislation from their government, as compared to the United States. There is also a perception that the United States is exceptionally crime-ridden: many Canadians joke that everyone in the United States is walking around armed with a gun. From the United States' perspective, the U.S. government takes issue with Canada's attitudes toward marijuana, its relationship with Cuba, and the potential access international terrorists might gain to U.S.

territory via the relatively open border between the two countries. However, such issues do not create enough friction to jeopardize strong relations between the two countries. The strong economic aspects of the geographic juxtaposition of Canada and the United States, the shared media and trade, political cooperation, and generally friendly attitudes the residents of the two countries have toward one another far outweigh any degrading attitudes or friction emanating from either country.

Canadian City Landscapes

Québec City is the only Canadian city with an historic core containing buildings dating back to the 1700s. Most cities from Toronto westward were built almost entirely in the 1900s as centers of agricultural commerce (e.g., Calgary, Edmonton, Regina). Other major differences between Canadian and American cities include the level of planning involved and the relationships of government units within metropolitan centers.

Toronto is the largest Canadian city and the center of Canadian financial services and manufacturing industries, as well as the provincial capital of Ontario, Canada's wealthiest province. When Toronto began to extend suburbs into surrounding jurisdictions in the 1950s, the province of Ontario required that these jurisdictions and the city plan cooperatively to integrate metropolitan services and regional road construction. Further growth in metropolitan Toronto was encouraged around hubs outside downtown, including North York and the area adjacent to the international airport. A major project was also undertaken to redevelop the city's waterfront, eliminating a major freeway in the process of gentrification.

Toronto welcomed the growing immigrant groups that moved into older inner suburbs: Italian, Greek, Portuguese, and Chinese districts form distinctive ethnic enclaves. The city capitalizes on the increasing diversity of its residents as restaurants, stores, art galleries, and festivals reflecting the vast international heritage of the citizens continue to multiply throughout the city. Locals and tourists alike crowd ethnic business establishments, helping the diversifying economy of the metropolitan region. Toronto retains a busy downtown and surrounding older suburbs, and has the lowest homicide rate of large North American cities. The city also contains good public transportation systems with ever-increasing mobility, and controlled urban sprawl.

Regions of Canada

Canada has a federal government that links the 10 provinces and three territories together. This arrangement was decided by Britain in the 1867 Act of Confederation, uniting the different parts of what became modern Canada in the face of a perceived military threat from the United States after its Civil War ended in 1865. In 1982, Canada formally ended its legal ties to Britain, although it remained within the Commonwealth of Nations. Canada next faced the challenge of determining the constitutional roles of its federal and provincial governments. Several attempts to reconcile the wishes of the people of Québec with those of other provinces highlighted the problems of a federal constitution in which the provinces have, in many ways, greater powers than the central government. This situation contrasts with that in the United States, where the Constitution was generated internally nearly a century earlier. U.S. states are generally smaller in size than Canadian provinces, and they lost powers to the federal government during the 1900s. In terms of tourism and development, Canada is a popular destination for tourists in both the summer and winter climates, enjoying spectacular scenery from Québec to British Columbia (Figures 11.27 and 11.28).

Atlantic Provinces The eastern coast of Canada was settled first by the French and British. It contains the small, hilly Atlantic Provinces of Newfoundland (which has jurisdiction over the almost uninhabited Labrador), Nova Scotia, Prince Edward Island, and New Brunswick. The small-scale economy of these

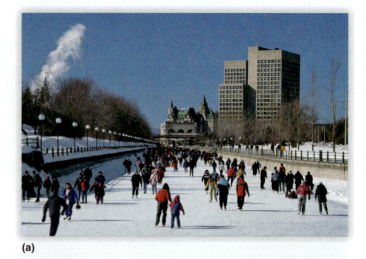

(a)

(b)

Figure 11.27 **Seasons in Canada.** (a) Ice-skating on the world's largest and longest public skating rink, the Rideau Canal in Ottawa, Ontario. Note the Parliament building in the background. (b) Contrasted with the above winter scene, summertime in Canada means enjoying the natural features of the region, such as this lake in southern Québec. *Photos: (a) © Cheryl Forbes/Lonely Planet Images/ Getty Images; (b) © Flickr Open/Getty Images RF.*

areas, based in fishing and farming, was augmented locally by mining and manufacturing, and by the naval base at Halifax. These provinces remain the primary recipients of federal regional aid due to a meager to poor overall economy today.

In the 1980s and early 1990s, the region was hit badly by declining fish stocks on the Grand Banks. Some 30,000 fishers and fish plant workers lost their jobs as cod stocks virtually disappeared. Unemployment in Newfoundland rose to over 20 percent, and new industries were not attracted by government efforts to revive the region. Relief prospects, such as pumping oil from the offshore undersea oil fields, would be very costly and environmentally risky. The development of nickel and cobalt mining at Voisey Bay, Labrador, which began construction in 2002, is expected to be in production by 2014 and could bring wealth to the owners and possibly to a relatively small population

Figure 11.28 The spectacular natural scenery of Moraine Lake against the backdrop of the Rocky Mountains in Banff National Park, Alberta. *Photo: © Ryan McVay/Lifesize/Getty RF.*

of local workers. It is significant on a world scale, however, that the cobalt from Voisey Bay replaces the falling output from the previously dominant mines in the Democratic Republic of Congo.

Nova Scotia, Prince Edward Island, and, to a lesser degree, New Brunswick are developing tourist industries in an attempt to replace lost components of the economy. Nova Scotia has a well-structured and geographically comprehensive tourist industry covering all parts of the province. New jobs in the region may be found in associated retail, hotel, and restaurant services.

Québec The province of Québec was settled in the late 1600s, shortly after the establishment of the first Atlantic coast settlements by French people, who developed a distinctive type of long-lot system of land division that maximized land ownership along either side of the St. Lawrence River. Québec became part of British Canada following General James Wolfe's defeat of the French army under Montcalm at Québec City in 1759. The French settlers and their descendant Québecois resented living under their conquerors. The economics of their culture were barely above subsistence, although traders bringing furs from far inland supported a merchant class during the 1800s. The combination of French language and Roman Catholic religion forged a strong loyalty that shifted to political activism in the 1900s as a reaction against Anglo-Canadian control.

As a result of political lobbying and clashes with the provincial government, French was accepted as an equal national language in the mid-1900s. Québec gained other concessions from the remainder of Canada, despite not acknowledging English as an alternative language within its borders—street signs are exclusively in French throughout this province. Québec looks beyond Canada to other French-speaking countries and takes a leading role in developing a new global French technical language, rather than simply accepting and adopting English words.

The province of Québec covers a large area, extending northward to include most of the peninsula east of Hudson Bay (Figure 11.29). The majority of the population lives along the St. Lawrence River estuary in the southern parts of the province. The two largest cities, Québec City and Montréal, have a range of manufacturing and service industries that place them among the world's great cities (Figure 11.30). A series of industrial towns use local hydroelectricity to power timber industries, pulp and paper mills, and aluminum refineries along one of the world's main shipping lanes. With global market production of wood pulp extending to tropical areas, the prices of these commodities—and hence, the well-being of Québec's major industry—have fluctuated since the late 1980s. Expansion of global demand in Asia, Europe, and North America does not always balance the production output of this province.

North of the St. Lawrence River estuary, the land is bleak, eroded by former ice sheets and covered by coniferous forest, lakes, and tundra. The rocks contain large mineral resources that are gradually being exploited as world markets and transport facilities become available. One of the major potential resources is hydroelectricity, and Québec is investing heavily in new facilities to export power to the United States. Québec

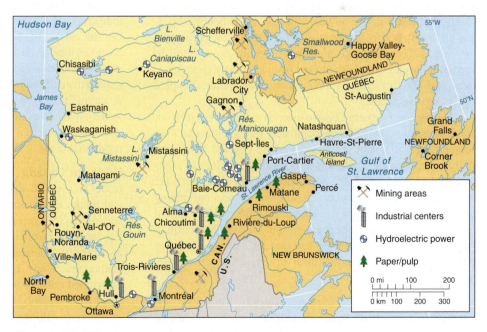

Figure 11.29 Québec Province, Canada: major cities, mining areas, industrial centers, and hydroelectricity sites. Nearly all economic activity crowds into the St. Lawrence River lowlands.

Figure 11.30 Québec City. The provincial capital, Québec City is situated at the confluence of the St. Lawrence and St. Charles rivers. The waterfront area is the Lower Town. The Upper Town is built on a bluff with the imposing Château Frontenac hotel dominating the cityscape. Québec City is designated a World Heritage City by the United Nations. *Photo: © Joseph P. Dymond.*

alone produces 25 percent of Canada's manufacturing output, making it critical to Canada's overall economy.

Ontario Southern Ontario is situated between three of the Great Lakes: Huron, Ontario, and Erie. It has a relatively mild climate, modified by a maritime influence from the very large lakes and from being situated farther south than most other regions of Canada. Soils and climate suitable for pasture, grain crops, and corn made this the most attractive part of Canada for immigrant farmers in the 1800s. The city of Toronto, Canada's largest metropolitian center, arose in the 1800s and then significantly developed in the middle of the 1900s, when the opening of the St. Lawrence Seaway turned it into an ocean port city. The French-language policies of Québec drove English-speaking businesspeople westward from its former rival, Montréal, and Toronto's role as the economic and financial capital of Ontario expanded.

In contrast, northern Ontario extends to the shores of Hudson Bay, but it is very sparsely populated. Its ancient rocks were scraped bare of soil by ice sheets and are covered today partially by coniferous forest. Local deposition of clays in meltwater lakes provided usable soils, though growing seasons are short. Most of the settlements in northern Ontario are mining towns, including the huge complex around the nickel mines of Sudbury. Settlements in the areas west of Sudbury and north of Lakes Huron and Superior are few and far between, and they constitute a major gap in the linear east-west Canadian settlement pattern.

Prairie Provinces The Prairie Provinces of Manitoba, Saskatchewan, and Alberta were settled following the building of the Canadian Pacific and Canadian National railroads in the late 1800s. In the early 1900s, they became major wheat-growing areas. The crop was planted in the spring and ripened in the short growing season, which was as little as 90 days on the northern margin of the plowed area. Toward the west, there was insufficient moisture for wheat, and cattle ranching took over. The area is essentially a northward extension of the Great Plains in the United States. The ground rises westward in a series of steps, further reducing the growing season and precipitation. Many of the agricultural districts had out-migrations of people since 1950 as a result of the mechanization of farming and the closing of marginal farms.

Winnipeg, Regina, Edmonton, Saskatoon, and Calgary were railroad towns that became grain and meat markets, and three of them became provincial capitals. After 1950, the discovery of coal, oil, and natural gas in Alberta brought extractive and manufacturing industry wealth and more people to that province. The large distances east and west to Canadian ports raised the price of exports and led to some of the main energy markets being southward in the United States.

West Coast British Columbia is situated along Canada's Pacific Coast and primarily consists of the rugged terrain of the Canadian Rockies and Pacific Coast Mountains. It is a province of geographically concentrated economic activity, including pockets of mining, fruit growing, and tourism in small lowland areas that are separated by high mountains. Most of the people of the province live in the Vancouver area, extending westward to Vancouver Island.

The people of British Columbia are both physically and mentally distanced from the rest of Canada. Some of them act more British than those who actually live in Britain, with elaborate rituals of afternoon tea common in the city of Victoria. Others emphasize the Pacific connections of Vancouver, which has an increasing population of East Asian peoples and growing economic ties to Asia. Asian migrants brought money to metropolitan Vancouver, facilitating the growth of many metropolitan industries. The expanding Asian community further diversifies the culture of the city. In some Vancouver neighborhoods, however, tension developed between some Canadians of British descent and Asian migrants. Many Asians living in Vancouver garnered wealth in business endeavors in Hong Kong, and tore down traditional dwellings in parts of Vancouver, erecting large Asian-style homes in their place. This frustrated long-time residents who preferred the traditional architectural styles of the residential areas of the city.

Northern Canada The northern parts of all the provinces from Québec to British Columbia end in the barren rock, trees, and tundra that characterize northern Canada. The Northwest and Yukon Territories long formed the largest part of Canada—a federally controlled zone where few people live permanently apart from American Indian and Inuit (Eskimo) groups of Native Americans. In 1999, Nunavut became a new territory designated entirely for the Inuit people and carved out of the Northwest Territories.

Mining settlements produce a range of metallic ores, and there are few communications in this wilderness area apart from those linking mines to their respective markets. In the late 1990s a new diamond mine was opened near Yellowknife,

Northwest Territories, and it now produces over 3 percent of the world's diamonds. The diamond mine needs to address local community concerns about caribou herd migration paths and threats to hunting and fishing if it is to establish a lasting impact within the local community. On a global scale, the mine, together with others outside South Africa, could reduce the South African De Beers company's domination of global diamond markets. The future of this region is not bright, as mining activities and federal subsidy incomes decline. A possible alternative income is from tourism, especially in the form of fishing and hunting expeditions and resort lodges situated on the major lakes of the area such as Great Bear Lake and Great Slave Lake.

11.5 Contemporary Geographic Issues

Immigration

The contemporary immigration picture for the United States is extremely dynamic. Immigrant communities from all world regions are occupying space in the country's metropolitan centers, such as the growing Ethiopian communities in Los Angeles and Washington, D.C., and in smaller cities and towns throughout as exhibited by the dramatic growth in the Hispanic communities of western Michigan. Although distinctive ethnic spatial patterns are evident as detailed in Figure 11.31, rapid growth in some communities creates new ethnic maps almost daily. While many highly educated people migrate to the United States to fill employment demands in high-tech and computer-related industries, medical research, and numerous fields within the physical sciences, more significant numbers come with no material wealth or education in search of a life different from the environment in their country of origin.

Over half of the African-Americans in the United States live in the South in both rural and urban areas. The majority of the remaining African-American population is clustered in northern and western metropolitan centers. The African-American percentage of the total U.S. population is diminishing as the percentage of Asian Americans and especially Hispanic Americans increases.

Asian immigrants are currently concentrated along the West Coast of the United States, especially in the larger urban centers. There are also significant concentrations of Asian Americans in parts of the upper Midwest and in large cities in the eastern United States. The climate and ecosystems of the Gulf Coast region attracted and now house a significant population of people who have emigrated from Vietnam. Vietnamese Americans living along the Gulf Coast engage in agricultural and fishing practices similar to those of Vietnam.

The U.S. Census Bureau 2012 estimates reported a total population of 313.5 million people with 202.8 million of European origin, 38.7 million (12 percent) African-American, 2.5 million (0.7 percent) Native American, and about 15.1 million (4 percent) Asian. Some 49.1 million (14 percent) were Hispanics, an ethnic designation based mainly on Latin American heritage and the Spanish language.

Hispanic peoples in the United States were the country's largest and fastest-growing minority group in 2012. Hispanic Americans are located mainly in the states bordering Latin America, including Florida, Texas, New Mexico, Arizona, and California; most large cities of the northeastern part of the United States; and in the expansive urban reaches of metropolitan Chicago.

People of European ancestry in the United States have the country's lowest birth rates. Without immigrant communities from Latin America and other world regions, the United States might experience a population decline as in some European countries. Young immigrant communities with relatively higher birth rates maintain a replacement population for jobs, economic growth, and the infusion of money into the tax base for infrastructure expenditures.

Immigration into the United States was one of the country's most controversial issues in 2012. While it is important to understand that most people of Latin American heritage in the United States pay taxes, legally reside within the U.S. borders, and directly contribute to the U.S. economy, the number of illegal immigrants was on the rise. Increasing numbers of illegal immigrants, or "undocumented aliens" (estimated by the Pew Hispanic Center to be as high as 11 million in 2012) mainly of Hispanic origin, live and work in the United States.

Millions of legal and illegal Hispanic immigrants in the United States are laborers in commercial agriculture, construction, landscaping, housekeeping, and fast-food service positions. Although their wages are relatively low by U.S. standards, they may earn more per week working in the United States then they do for a month in their country of origin. Many send a significant percentage of their wages, or **remittances**, to relatives living in Mexico, El Salvador, Guatemala, Honduras, and Nicaragua—where such receipts may be the most significant foreign revenue earned (as is the case for El Salvador).

Increasing public dialogue concerning immigration and undocumented workers led to a series of significant events from the spring of 2006 through the fall of 2012. Millions of Hispanic Americans rallied and staged "sit outs" from work in U.S. cities to assert their importance to the U.S. economy and to voice their concerns over existing and proposed immigration laws. The United States in 2008 increased the number of National Guard troops stationed along the U.S.-Mexico border and fortified structural barriers in populated border corridors.

Another controversial trend surrounding Hispanic immigrants is the day laborer phenomenon. Day laborers are workers lacking formal contractual relationships with employers who operate in the informal economy. Homebuilders, for example, might pick up groups of workers (who may be waiting in front of a convenience store in the hopes of encountering work) and take them to a building site, where they will be paid to work for that day. Day laborers do not enter into any type of formal or contractual position, pay taxes, or receive any benefits. This practice became very common in many U.S. cities in recent years. Controversy is generated by those asserting the laborers should be paying

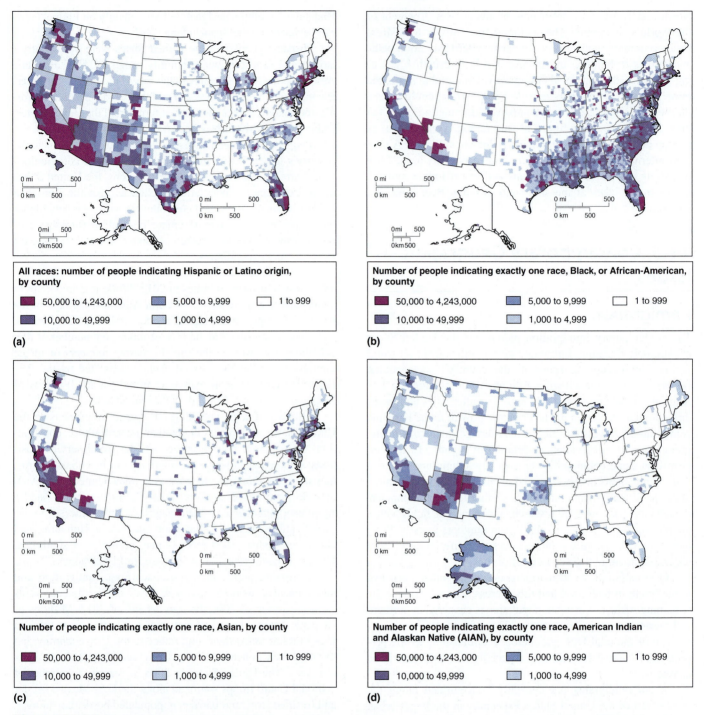

Figure 11.31 **United States: ethnic distributions.** (a) Primary concentrations of Latin Americans include California and the Southwest, Florida, Chicago, and the cities of Megalopolis. (b) Primary African-American concentrations include California, the South from east Texas through North Carolina, Florida, and major cities of the Great Lakes and Megalopolis. (c) Primary concentrations of Asian Americans include the Pacific coast, Chicago, and the cities of Megalopolis. (d) Primary concentrations of Native American and Alaskan Native peoples include Alaska, California, Arizona, New Mexico, and Oklahoma. *Source: Data from U.S. Census Bureau, Census 2000 Redistricting Data (PL 94-171) Summary File. Cartography: Population Division U.S. Census Bureau; American Factfinder at factfinder.census.gov provides census data and mapping tools.*

taxes, and by neighborhood residents and business owners who are concerned by large groups of laborers who "hang out" in front of their businesses or in their neighborhoods while waiting for prospective employers to pick them up. Fairfax County, Virginia, in suburban Washington, D.C., used local funds to build a day laborer center where the

workers could wait to be picked up for employment. Some residents of the county did not approve of their tax dollars being used to support what they see as an illegal practice, while other county residents felt it was the perfect solution.

Controversy among voters, business owners, and government officials focuses on the responsibility of the government in terms of resolving issues related to illegal immigration. Members of the domestic Latin American community and many employers would like to see the U.S. government grant amnesty to illegal Hispanic peoples within the U.S. borders. Others propose temporary "guest worker" arrangements where undocumented immigrants, in conjunction with secured employment, could obtain work permits allowing their residence for a stated period (with an end date). Those against the presence of illegal immigrants within the United States would like the government to track them down, deport them, and dramatically strengthen border security in order to prevent their return.

North American Free Trade Agreement (NAFTA)

The formal implementation of the North American Free Trade Agreement (NAFTA) in 1994 created one of the world's largest trading blocs. The United States, Canada, and Mexico moved toward an extensive elimination of trade barriers to increase the economic activity between them and strengthen their economic and political positions at the global scale. The three countries tied by NAFTA are now each other's largest trading partners.

NAFTA has many proponents and critics in each of the three countries (Table 11.3). Similar living standards in the United States and Canada contrast sharply with poverty conditions common to millions in Mexico. Proponents of NAFTA claim it creates jobs, strengthens and expands business and industry, diversifies the economies of each of the three country partners, enhances foreign revenues, and fortifies each country's ability to compete in a global trade arena. Those opposing NAFTA assert the agreement results in the exportation of jobs and revenues from Canada and the United States to Mexico, the perpetuation of harsh labor conditions for Mexican workers, the degradation of the natural environment in all countries, and the granting of nearly supreme powers to big business and industry at the expense of workers, consumers, and those living in the shadow of production facilities.

Political leaders from the three NAFTA partners (Mexican President Carlos Salinas de Gortari, 1988–1994; U.S. President George H. W. Bush, 1988–1992, and, later, U.S. President Bill Clinton, 1993–2001; and Canadian President Brian Mulroney, 1984–1993) signed the 1992 trade pact. The respective legislative bodies each then passed the agreement, which was formally implemented in 1994. The agreement effectively removed most tariffs immediately, and established a timeline by which all tariffs would be eliminated within 15 years of implementation.

The United States led efforts to push NAFTA forward in hopes of creating a trade bloc to compete with the European Union. The majority of NAFTA proponents, who pushed through an accelerated legislative process in the United States, were mostly representatives of large corporations. Labor unions, human rights groups, and environmentalists were strongly united against the rapid development of a trade agreement between the three countries and hoped to stop the fast-track legislation by

Table 11.3	Debate: The effects of NAFTA	
Supporters of NAFTA	**Critics of NAFTA**
Opened new markets for the three countries.	Primarily opened new markets for Mexico and Canada.
Competition of lower-priced goods produced in Mexico forces down the price of goods produced in the United States and Canada; thus, U.S. and Canadian consumers "win" with lower-priced goods.	Low labor costs in Mexico and lower-priced goods in the United States and Canada cause U.S. and Canadian companies to move their production operations to Mexico; thus, thousands of jobs are lost ("exported") to Mexico, hurting employment in the United States and Canada.
Strengthens the global economic weight of the three countries and makes them better able to compete with the EU and other trade blocs and countries of the world.	Significant economic disparity exists between the affluent United States and Canada and the relatively materially impoverished Mexico, placing Mexico at a disadvantage and creating more of a service role for Mexico to Canada and especially the United States, rather than truly making it an equal trade partner and stronger international economic player.
Promotes democracy and political stability in Mexico and strengthens the Mexican economy, thus ensuring greater stability for North America.	Perceptions by some Mexicans of heightened economic disparity in their country due to NAFTA result in political instability such as the Zapatista uprising.
Creates thousands of jobs in Mexico. Cities and towns in northern Mexico, where the majority of NAFTA-related production (*maquila*) takes place, enjoy much higher living standards and higher rates of employment than most other parts of the country.	Cultural distinctions are blurred in all three countries. U.S. culture may overpower parts of Canada and especially northern Mexico. Fast food is replacing traditional food; U.S. holiday celebrations are replacing traditional celebrations. Areas in the U.S. Southwest are developing a watered-down culture that is a mixture of U.S. and Mexican elements. The increased use of the Spanish language in parts of the United States, especially areas in the Southwest, increases tension with some English-speaking residents.
Strengthens Mexican environmental conditions through environmental side agreements negotiated along with the primary trade agreement, resulting in a healthier environment for Mexico and especially the border region; reverses environmental damage on the U.S. side of the border.	Side agreements negotiated with NAFTA fall far short of strengthening environmental laws in Mexico due to lax enforcement. The wording of NAFTA facilitates environmental abuse by companies in all three countries as NAFTA protects the companies' rights to free trade over the rights of people living in areas polluted by factories and other production facilities.
Forces the equal treatment of corporations in the three countries.	Corporations are too powerful under NAFTA.

arguing that they were largely excluded from the negotiations. Environmental critics of NAFTA claim that many of the key negotiators who rapidly pushed NAFTA through the U.S. Congress were among the United States' biggest polluters.

Proponents of NAFTA assert that the trade agreement forces the equal treatment of corporations and industry throughout the region. Business leaders claim that the package promotes fair and equal business opportunities for all companies in the three countries. Corporate leaders also assert that equal treatment reduces the risks of governmental interference in free market flows. Mexico has a long history of protectionism, or imposing restrictions, taxes, and quotas on goods and services produced outside of Mexico, and nationalization of industry (government takeover and control of business and industry, thus removing private ownership). Business proponents claim that protectionist policies are curtailed under NAFTA. Business leaders firmly state that capitalism thrives best in NAFTA-created conditions in the region, and this promotes and fortifies economic and political stability for all three countries.

Strong opposition to NAFTA emanates from labor unions in the United States and advocates of human and worker's rights in Mexico. Union leaders argue that the agreement exports thousands of jobs from the United States to Mexico. Working conditions in production facilities in Mexico are dramatically different from those in the United States. Mexican wages are, on average, one-eighth to one-tenth of U.S. wages, and benefits and other labor costs are nonexistent or significantly lower in Mexico. Laws restricting or regulating environmental pollution from production facilities, and those concerning labor safety, are fewer and less frequently enforced in Mexico. U.S.-based companies can manufacture products in Mexico with significantly lower labor costs and with far fewer environmental or labor restrictions. U.S. labor leaders say these factors are far too attractive for U.S. companies to forgo, and they claim numerous companies are shifting production to Mexico and therefore replacing U.S. workers with Mexican ones.

The Mexican Border Industrial Program, first implemented in 1965, slowly opened the door for foreign firms to establish production facilities in Mexico. The idea of the program was to stimulate growth in northern Mexico (and thus relieve some pressure on the then rapidly growing metropolitan area of Mexico City), create jobs, and infuse capital into a region that had operated under protectionist and nationalist economic policies for decades. The program created a manufacturing zone across northern Mexico known as the *maquila* zone (the factories are referred to as **maquiladoras**). The idea was that foreign firms, such as U.S. companies, could import raw materials or parts, pay Mexican workers low wages to refine or assemble them into finished goods, take the goods back across the border (to the United States for a U.S. firm) relatively tax free, then sell them to consumers.

The controversy surrounding NAFTA is far from over. Further trade barriers have been reduced or eliminated in the past few years in accordance with the progression plan of the agreement. Activist groups both for and against NAFTA continue to lobby the respective governments to promote their views on the future of the agreement.

The Athabasca Tar Sands

Buried under Canada's boreal forest in northeast Alberta is one of the world's largest reserves of oil, in a deposit known as the Athabasca Tar Sands, named after the river that bisects this region. According to a 2003 estimate, this location has the capacity to produce 174.5 billion barrels of oil. These are extracted by injecting hot water into a well that liquefies the oil for pumping. To extract the oil at these locations, oil producers remove the sand in large, open-pit mines. The sand is rinsed with hot water to separate the oil, and then the sand and wastewater are stored in retention ponds at the site. The process of extracting oil from the sand is expensive. It takes 2 tons of sand to produce one barrel of crude oil. Great Canadian Oil Sands opened the first large-scale mine in 1967, but growth was slow until 2000 because the global cost of a barrel of oil was too low to make oil sands profitable. After 2000, the price of oil began to climb, and investment in oil sands became worthwhile in the eyes of investors.

Oil sand mining has a large impact on the environment. Forests must be cleared for both open-pit and *in situ* mining. Pit mines can grow to more than 80 meters depth, as massive trucks remove up to 720,000 tons of sand every day. As of 2012, roughly 256 square miles of land had been disturbed for oil sand mining. Both water and air pollution could pose a health hazard, although an independent panel of experts found no definite connection between the mines and specific illnesses as of 2012. Because it takes energy to mine and separate oil from the sands, oil sands extraction releases more greenhouse gases than other forms of oil production. For all of these reasons, some groups have labeled the oil sands an environmental menace. On the other hand, they offer a stable source of energy and economic growth. The Athabasca oil sands are the largest segment of the economy in Alberta, composing just over 30 percent of the gross domestic product. In 2012, the Alberta government received more than US$3 billion in royalties from oil companies. It is estimated that the region contains enough oil to produce 2.5 million barrels per day for the next 187 years.

Desalination as a Groundwater Alternative in the 21st Century

In the past 30 years, Tampa, Florida, has experienced a tremendous population increase, as both retirees and job-seekers have migrated into the "Sunshine State." With these elevated population levels came greater needs for groundwater resources—in the year 2012 alone, Tampa and the surrounding metropolitan areas were using an average of over 100 millions gallons of water per day. Because this was perceived as unsustainable, leaders in the region examined alternative methods of acquiring fresh water for their citizens to use, easing the pressure on the natural aquifer systems. **Desalination**, the process of converting seawater (salt water) into fresh water, had been utilized in the Middle East in recent years, and the costs of building and operating a desalination plant in the Tampa region seemed to be justifiable, given the water needs of the region. From 2001 to 2003, a seawater desalination plant was built on Tampa Bay just south of the city of Tampa, situated next to a power plant (see Figure 11.3). Desalination works best when water temperatures are warm (i.e., in the 90- to 110°F range). However, the costs of heating seawater were greater than the benefit unless the desalination plant was located next to a power-generating facility that already was using seawater to cool its turbines. The seawater that flowed out from the power generation plant was already heated to the optimal temperature (about 110°F), and the desalination plant siphoned this water into its facility. It uses a process known as reverse osmosis to filter out the molecule-sized sodium chloride particles and other debris in the seawater, with the end result that 1 gallon of seawater is converted into a half-gallon of pure, fresh water and a half-gallon of twice-as-salty brine. The brine is pumped out into the Gulf of Mexico, where it is flushed back into the water system and reaches background levels of salinity within 100 meters of the pipeline, minimizing environmental impact. In 2012, the Tampa Bay Desalination Plant was producing over 25 million gallons of fresh water, and a second plant located near St. Petersburg was under construction. Seawater desalination greatly reduced the pressure on the local aquifer system, providing a sustainable alternative to groundwater withdrawal in the region.

Geography at Work

Geography and the National Geographic Society

Kaitlin Yarnall (Figure 11.32) is a research cartographer with the National Geographic Society in Washington, D.C. Her work revolves almost completely around geography. Kaitlin researches and edits projects produced by the Maps Department of the Society. These products are varied and include maps published within *National Geographic*, reference maps of various countries and regions, and special print and online projects that cover topics ranging from climate change to language diversity loss. Along with gathering the latest datasets, staying abreast of current trends, and fact-checking all material placed on maps, research often involves working with experts and consultants, which insures that the maps produced are accurate and include the latest data and trends.

For the past two years, Kaitlin has been a research editor of *Earth-Pulse*, a special issue of *National Geographic* magazine. This publication focuses on global trends that are affecting the planet and its inhabitants. Trends include demographic growth, increased global labor migration, rise in fuel consumption, deforestation, and India's growing middle class. This publication is geography-based and is used by geography students and professors, as all topics are viewed through a spatial lens and presented through maps and graphics. Kaitlin has a bachelor's degree in geography from Humboldt State University and a master's degree in geography from The George Washington University. Her

Figure 11.32 Kaitlin Yarnall is a research cartographer for the Maps Department of the National Geographic Society (NGS). Kaitlin's work at NGS includes serving as a research editor of *EarthPulse*, a special issue of *National Geographic* magazine. *Photo: Courtesy Kaitlin Yarnall.*

geography background has allowed her to create and edit content that will be useful to future geographers and will communicate to the public the importance of trends facing our planet.

Glossary

A

Aborigines (9; 272). Indigenous people of Australia.

absolute location (1; 5). Location of a place on Earth's surface as defined by latitude and longitude or by distance in kilometers (miles) from another place.

acid deposition (2; 45). Dry or wet acidic deposition of acidic material from the atmosphere, often resulting from sulfur and nitrate gases and particles emitted into the air from coal combustion in power plants.

afforestation (9; 280). The replanting of large areas of forest.

African Union (8; 260). The organization that combines all African countries since 2001. Formerly the Organization of African Unity (OAU).

agglomeration economies (2; 55). The total economies achieved by a production unit because of a large number of related economic activities in the same area.

agribusiness (2; 58). The large-scale commercialization of agriculture that places farming within the broader context of inputs of seeds, fertilizer, machinery, and so on, and of outputs of processing, marketing, and distribution.

ALBA. *See* **Bolivarian Alternative for the Americas (ALBA).**

alluvial fan (6; 166). A fan- or cone-shaped river deposit, often formed where a stream issues from a mountain gorge into an open plain.

alluvial soils (4; 116). Soil materials deposited in valley floors by annual river floods.

alluvium (6; 165). Deposits of rivers in their flood plains, often composed of mud, silt, sand, or gravel.

Altiplano (10; 294). High plateau in the Andes Mountains of Peru and Bolivia.

altitudinal zonation (10; 298). Zones of climate, vegetation, and crops that change with height above sea level.

animism (8; 234). Traditional religious beliefs based on the worship of natural phenomena and the belief in spirits separable from bodies.

Antarctic Peninsula (9; 269). The northernmost part of the mainland of Antarctica.

Antarctic Treaty (9; 284). Signed in 1961 by 39 countries, provided a basis for nonmilitary scientific cooperation, environmental safeguards, and international control of Antarctica.

Antarctic Treaty System (ATS) (9; 284). Body that, in effect, "governs" Antarctica composed of signatories of the Antarctic Treaty. The ATS encourages scientific research and exchange among interested signatory countries.

apartheid (8; 236). The separation of people of different ethnic groups, affecting housing, education, and jobs. It was government policy in South Africa until 1994.

Arab League (7; 206). Organization created in 1945 to encourage the united action of Arab countries for their mutual benefit.

artesian wells (9; 270). Wells drilled into rocks where water flows to the surface without pumping.

atmosphere-ocean environment (1; 7). The combination of atmosphere and oceans in which movements that cause weather and climate are controlled by solar energy.

atoll (9; 268). A coral island with a central lagoon.

B

badlands topography (6; 166). Closely spaced networks of deep gullies often carved by occasional streams in soft sediments unprotected by vegetation cover. They destroy the usefulness of the land.

barrier reef (9; 268). A coral reef structure surrounding an island and separated from it by a lagoon. The Great Barrier Reef lies off Queensland, Australia.

Benelux (2; 67). An acronym for **Be**lgium, the **Ne**therlands, and **Lux**embourg. The term was coined in recognition of the close working relationship these countries have with one another.

Berlin Conference (8; 235). Conference held by the European nations in 1884 that divided the continent of Africa among themselves, with no political representation from anyone in Africa.

biome (1; 14). A world-scale ecosystem type such as tropical rainforest or savanna grassland.

black earth soils (chernozems) (3; 79). Highly fertile soil type in which organic matter accumulates near the surface, commonly beneath temperate grassland communities.

Black Triangle (2; 45). A heavily polluted industrial area straddling the Polish, Czech, and German borders.

Bolivarian Alternative for the Americas (ALBA) (10; 305). A political and economic pact signed by Venezuela, Bolivia, and Cuba in the early 2000s with the focus of excluding the United States.

(British) East India Company (6; 173). A British company that traded with and conquered much of the Indian subcontinent. After an 1857 mutiny in India, the British government took political control.

British Indian Empire (6; 173). Established after 1857 on the Indian subcontinent, including Ceylon and later extended to Burma. Lasted until independence and partition in 1947.

British North America Act (11; 339). Act that established independence for Canada from Britain on July 1, 1867.

brown earth soils (3; 80). Fertile soil type in which plant matter replenishes nutrients in the upper layers, commonly forming beneath temperate deciduous forest.

Buddhism (6; 171). A religion that began in South Asia but became the major religion of East Asia. Followers of Buddhism have a greater social openness than do followers of Hinduism and accommodate other philosophies and religions such as Confucianism and Shinto.

C

Canadian Shield (11; 332). The ancient rocks of Northern Canada, the oldest in North America, which contain many mineral deposits.

capital city (1; 27). The city in which central government functions are concentrated; sometimes the largest city, but often one that is specially designed for the purpose.

capitalist system (1; 30). *See* **free-market (capitalist) system**.

cartel (7; 208). An organization that coordinates the interests of producers (such as OPEC).

Caribbean Community Common Market (CARICOM) (10; 306). Formed in 1973 by 13 former British colonies to provide special entry to U.S. markets.

caste (6; 171). The basis of social class divisions in South Asia.

caste order (6; 171). A social class system associated with Hinduism that is based on the supremacy of Aryan peoples. The Aryan castes include priests (Brahmans), warriors, and other Aryan people; non-Aryan castes include cultivators, craftspeople, and untouchables.

central planning (3; 93). The Soviet Union practice in which the government decided how many goods and services were needed by society, and gave instructions for their production almost without cost considerations.

Christianity (7; 199). A religion that developed out of Judaism, based on the belief that God came to Earth in the form of Jesus of Nazareth. Main religion of the Western world.

ciclovía (10; 319). Bogotá, Colombia, city government program in which several major streets are closed to automobile traffic on Sundays and holidays and open for families to bicycle around the city.

class (1; 21). Hierarchies within societies created by an emphasis on such criteria as ethnicity, race, religion, material wealth, education, perceived birthright, and other social characteristics.

climate (1; 10). The long-term atmospheric conditions of a place.

collectivization (4; 129). The transformation of rural life in Communist countries such as China and the Soviet Union in which individual farmers were grouped in cooperatives that took ownership of their land and labor.

colonialism (2; 54). The system by which one country extends its political control to another territory to improve local conditions and/or economically exploit the human beings and natural resources of the subordinate territory.

command economy (3; 93). The Soviet practice whereby the government ran the economy, owned all industries, and set quotas, favoring heavy industry over the production of consumer goods.

Common Market of the South (MERCO-SUR) (10; 306). Trading group established in 1991 among Argentina, Brazil, Paraguay, and Uruguay.

Commonwealth of Independent States (CIS) (3; 78). A political and economic organization created in 1991 by 11 republics of the Soviet Union. Members included Russia, Belarus, Ukraine, Moldova, Armenia, Azerbaijan, Kazakhstan, Kyrgyzstan, Tajikistan, Turkmenistan, and Uzbekistan. Georgia became the twelfth member in 1993. The CIS coordinates relations between member countries, including issues involving economics, foreign policy, and defense matters.

Communauté Financierè Africaine (CFA) (8; 244). Economic links between France and its former colonies in Africa. From independence until 1994, it involved links to the French franc, but this arrangement ended with the devaluation of the CFA franc.

commune (4; 129). The organization that controls rural life in some Communist countries, including agriculture, industry, trade, education, local militia, and family life.

communism (2; 53). A system in which the workers govern and collectively own the means of economic production. When spelled with a capital *C*, *Communism* refers not to the system as it was originally envisioned, but rather to the totalitarian systems adopted in countries such as the Soviet Union and China, where small, elite groups rule or ruled under the guise of communism.

concentration (2; 58). Agricultural production carried out on fewer and larger farms and limited to smaller areas of higher productivity.

concentric pattern of urban zones (11; 346). A pattern or urban geography with a central business district surrounded by a hierarchy of residential zones.

Confucius (Kong Fuzi) (4; 128). Chinese administrator who established a system of efficient and humane political and social institutions that became the basis of procedures and ways of life in much of East Asia.

coniferous (11; 334). Forest of needle-leaf trees commonly occurring in areas with long winters or poor, sandy soils.

continental temperate climate (2; 43). The climates of continental interiors in mid-latitudes. They are often drier and have greater extremes of summer and winter temperatures than the climates of coastal areas in these latitudes.

continentality (3; 79). Especially cold winters and hot summers resulting from locations on landmasses that are far from the moderating effects of large water bodies such as oceans.

convergent plate boundary (1; 8). In plate tectonics, where two plates move toward each other, causing earthquakes, volcanic activity, and mountain-building as they meet.

copra (9; 282). Dried white meat that lines the inside of a coconut shell.

coral atoll (fringing reef) (9; 268). A coral reef along a coast without a lagoon.

core country (1; 34). A materially wealthy country that plays a major part in controlling world economic processes.

countermigration (11; 337). Movement of people returning to regions they once left, as in the return of African Americans from northern U.S. cities to the South.

country (1; 26). A self-governing political unit having sovereignty within its borders and recognition by other countries.

crony capitalism (5; 155). The economic system in which entrepreneurs gain advantages through close links to government officials and ministers.

crude birth rate (1; 24). The number of live births per 1,000 of the population in a year.

crude death rate (1; 24). The number of deaths per 1,000 of the population in a year.

cultural geography (1; 18). The study of spatial variations in cultural features such as material traits, social structures, languages, or belief systems.

Cultural Revolution (4; 129). The attempt by Mao Zedong between 1966 and 1976 to change the basis of Chinese society. The disruption held back the country's economic development.

cultural rights (1; 36). The rights to protect one's cultural traditions.

D

Daoism (4; 128). The teachings of Chinese philosopher Laozi, who disliked the organized and hierarchical social system of Confucius and advocated a return to local, village-based communities with little outside interference.

deciduous (11; 334). Plants with broad leaves that mostly lose their foliage in cold or dry seasons.

decolonization (2; 73). The process by which mainly European countries enabled their colonies to become independent countries.

deindustrialization (2; 55). A rapid fall in manufacturing employment and the abandonment of factories in a once-important industrial region.

democratic centralism (2; 53). The practice of sole governance by the Communist Party, the political party of the working class, because it is believed that only the Communist Party is the true representative of the people.

demography (1; 24). The study of human populations in terms of numbers, density, growth or decline, and migrations from place to place.

deposition (1; 9). The dropping of particles of rocks carried by rivers, wind, or glaciers when they stop flowing, blowing, or melt, respectively.

desalination plant (7; 210. 11; 362). A mechanism that extracts fresh water from seawater by evaporation and condensation.

desert (8; 233). An area that does not support vegetation, often associated with an arid climate.

desert climate (7; 195). Area where evaporation rates greatly exceed precipitation through most of the year.

desertification (1; 15. 8; 233). Processes that destroy the productive capacity of an area of land.

development (1; 34). The process by which human societies improve their quality of life, including economic, cultural, political, and environmental aspects.

devolution (2; 70). The process by which local peoples desire less rule by their national governments and seek greater authority in governing themselves.

diaspora (7; 199). The scattering of a people, such as the Jews and Chinese, to other countries.

distance (1; 6). The space between places measured in kilometers (miles), travel time, or travel cost.

divergent plate boundary (1; 8). In plate tectonics, a zone between two plates that are moving apart, often marked by an ocean ridge.

diversified economy (7; 216). An economy in which manufacturing is more important than primary products and where there is a variety of manufactured products and a growing service sector.

E

Earth-surface environments (1; 7). Where atmosphere–ocean environments interact with solid earth environments to produce relief features such as river or glacial valleys and coastal cliffs and beaches.

East African Community (EAC) (8; 246). A grouping of Kenya, Tanzania, and Uganda, launched in 1967 and relaunched in 2000.

Eastern Antarctic Shield (9; 268). Landform containing rocks over 4 billion years old.

Economic Community of West African States (ECOWAS) (8; 244). Organization of Western African countries, founded in 1975 and involved in peacekeeping, a consultative parliament (2002), free trade and movement of labor, and banking.

economic geography (1; 28). The study of the spatial aspects of material wealth and poverty, the use of resources, and the production of goods.

economic leakage (10; 308). Foreign ownership and control of industry, such as tourism in many parts of Latin America, where the revenue gathered by the industry flows to the foreign owners and foreign employees and largely bypasses local communities and thus "leaks" from the local economy.

economies of scale (11; 340). Increased productivity gained by building larger factories and gaining access to larger markets.

ecosystem (1; 14). The total environment of a community of plants and animals, including heat, light, and nutrient supplies.

ecosystem environments (1; 7). The environments of living organisms where weather, mineral nutrients, and solar energy combine.

edge city (11; 347). A post-industrial city, usually in the suburbs of a major city, containing significant commercial space and service industries.

El Niño Southern Oscillation (10; 295). The full name of the phenomenon known as El Niño, in which the warm equatorial waters push back the cold Peruvian current off the western coast of tropical South America, producing local fish kills and wider world climatic effects.

emerging countries (1; 35). In the early 2000s, a group of developing countries with economic growth that is beginning to challenge wealthier countries.

encomienda **system** (10; 304). The Spanish colonial system employed in Latin America in which lands were allotted to Spaniards who were responsible for exploiting their wealth and who had jurisdiction over the native peoples.

entrepôt (4; 136). A port that collects goods from several countries to trade with the wider world and distributes imports to its immediate area or hinterland. Examples include Hong Kong and Singapore.

erosion (1; 9). The wearing away of rocks at Earth's surface by running water, moving ice, the wind, and the sea to form valleys, cliffs, and other landforms.

ersatz capitalism (5; 155). The economic system that implies an inferior substitute for locally based economic development when the capital, skills, and management are all imported to produce goods for export.

estuary (2; 42). A wide river mouth that experiences changes in tidal water level and quality.

ethnic cleansing (2; 72). The process by which one ethnic group removes other ethnic groups from a territory either through expulsion, extermination, or forced assimilation. (*See also* **genocide**.)

ethnic group (1; 21). A cultural group whose members are defined by such characteristics as common origin (real or imaginary), religion, language, customs, or physical features.

ethnic religion (1; 20). A religion that is linked to a particular ethnic group, such as Judaism and Hinduism.

ethnocentrism (1; 21). Judging other ethnic groups harshly because they have different traits and practices.

European Union (EU) (2; 67). Name adopted by the European Community in 1993, suggesting both an expansion to other European countries following the end of the Cold War and the possibility of a future closer political federation.

evapotranspiration (10; 324). The combination of evaporation from water and transpiration through plant leaves that produces humidity in the air.

extensification (2; 58). In agriculture, the production of fewer livestock or crops from the same area.

F

favela (10; 325). A shantytown in a Brazilian city.

First Nations (11; 339). Indigenous or native Canadians.

five-year plan (3; 93). A comprehensive economic planning scope in the former Soviet Union. These plans were followed from 1928 until 1991.

fjord (2; 42). A formerly glaciated valley that was flooded with ocean water after the sea level rose in the post-glacial age.

Fordism (11; 341). The application of production lines in the assembly of a wide range of components in various manufacturing industries based on the production system established by Henry Ford for the automobile industry.

formal economy (1; 35). The economic sector in which workers have recognized or licensed jobs, receive agreed-upon wages, and pay taxes.

free-market (capitalist) system (1; 30). The economic system that is based on competition and pricing of goods determined by the market. It is the basis of capitalism, but market "freedom" is often reduced by government and the actions of major producers.

Free Trade Area of the Americas (FTAA) (10; 306). An attempt by the United States and several Latin American countries to create the world's largest trading group.

free trade zone (10; 321). An area within a country where components can be imported without tariffs for assembly with a view toward exporting the finished goods.

friction of distance (1; 6). The relative difficulty of moving from one place to another, which increases with kilometers (miles), cost, or travel time.

G

gender (1; 22). The cultural implications of being male or female, with particular reference to the inequalities suffered by females in human societies.

genocide (2; 53). The systematic extermination of an ethnic group, nation, or racial or religious group.

gentrification (2; 51). The movement of higher-income groups to occupy and improve residences in older and poorer parts of cities.

geographic inertia (2; 55). Once capital investments are made in factories and infrastructure that give a region agglomeration economies, production will continue there for a period of years after other areas emerge with lower production costs.

geographic information system (GIS) (1; 5). The computer-based combination of maps, data, and often satellite images that is a foundation for geographic analysis.

geography (1; 4). The study of spatial patterns in the human and physical world; where and how the human and natural features of Earth's surfaces are distributed, are related to each other, and change over time.

glasnost (3; 93). The Soviet Union policy of the late 1980s designed to create greater openness and exchange of information.

global warming (1; 17). The process by which average temperatures in Earth's atmosphere rise over a period of several decades or centuries, leading to the melting of ice masses and a rising sea level.

globalization (1; 7). The growing interconnectedness of the world's peoples and the integration of economies, technologies, and some aspects of cultures.

Gondwanaland (5; 143. 9; 266). The former huge continent from which the southern continents and Indian peninsula broke away.

governance (1; 27). The coordination and regulation of human activities at different levels of geographic scale, often outside the powers of sovereign countries.

Great Australian Desert (9; 279). In the Australian interior, this desert lies west of the sparsely populated farming region and covers most of the remainder of the continent.

Great Barrier Reef (9; 267). The Earth's largest continuous chain of coral located off the northeastern coast of Australia.

Great Dividing Range (9; 267). Australia's one significant mountain chain running along the continent's east coast.

Great Leap Forward (4; 129). The attempt by the Chinese Communist government in the late 1950s to increase the pace of industrialization. It failed by ignoring food production at a time when bad weather brought poor harvests and famine.

Green Revolution (5; 154). The result of introducing high-yielding strains of wheat and rice. The outputs of commercial farms in South and East Asia increased by several times, but the costs of seeds, fertilizers, and pesticides were too high for smaller farmers.

greenhouse effect (1; 10). The natural process of heating Earth's atmosphere. Solar rays of short wavelengths reach Earth's surface and are absorbed and reradiated as long wavelength (heat). This radiation is partially absorbed by and heats the lower atmosphere, which contains water vapor and carbon gases. When humans add to the carbon gases, they enhance this process, raising temperatures above "natural" levels.

gross domestic product (GDP) (1; 29). The total value of goods and services produced within a country in a year. Often expressed as GDP per capita, when the total GDP is divided by the country's population.

gross national income (GNI) (1; 29). The total value of goods and services produced within a country in a year, together with income from labor and capital working abroad, minus deductions for payments to those living abroad.

Group of Eight (G8) (11; 341). An economic discussion forum consisting of the world's eight most materially wealthy countries; it includes the United States of America, Canada, Japan, Germany, the United Kingdom, France, Italy, and Russia.

guest worker (2; 73). A foreigner who has permission to reside in a country to work but is not a citizen of that country. From the German word *gasterbeiter.*

gulag (3; 106). Short for the Russian name *Glavnoe upravlenie ispravitel no-trudovykh lagerei* (Main Directorate for Corrective Labor Camps), a collection of prison camps in the former Soviet Union where criminals and those who opposed the Communist government were sent to perform hard labor as punishment.

H

Hamas (7; 220). Founded in 1987 as an offshoot of the Muslim Brotherhood, it is an Islamic resistance movement whose goal is to establish a Palestinian state on the current site of Israel, the West Bank, and the Gaza Strip.

Han Chinese (4; 119). The largest group (94 percent) of people in China. An ethnic grouping based on the administrative culture spread by the Chinese empire in the AD 200s and 300s.

heartland (3; 99). The area of a country that contains a large percentage of the country's population, economic activity, and political influence.

Hezbollah (7; 221). Inspired by the Iranian Revolution of 1979, it was founded in 1982 by Shia Muslims as a resistance movement to the Israeli invasion of Lebanon. It has close connections and support from Syria and Iran.

Hinduism (6; 171). A religion of South Asia, observed mainly in India, that includes the worship of many gods. Related to varied historic experiences and is associated with the caste system.

hinterland (3; 99). The areas of a country that lie outside the heartland. The hinterland usually has a relatively small percentage of a country's population, economic activity, and political influence compared to the heartland, though it may be well endowed with natural resources.

HIV/AIDS (8; 251). Disease affecting the immune system that is particularly prevalent in Southern Africa. It is regarded as a long-term pandemic that can be alleviated, but not cured, by drug treatments.

Homestead Act of 1862 (11; 339). The U.S. act of 1862 that provided land cheaply or freely to families settling the U.S.West.

horizontal integration (11; 340). The combining of the producers of the same product in a single corporation to create economies of scale and the control of prices.

household responsibility system (4; 130). The replacement for communes in rural China after 1976, returning ownership and decision-making rights to individuals and groups that could sell surpluses in open markets.

human development (1; 34). A broader view of development that focuses on people rather than economic change; regarded as a means to the end of enabling people to enlarge their capabilities so that they can enjoy the richness of being human.

human development index (HDI) (1; 34). A measure of human development based on income, life expectancy, adult literacy, and infant mortality.

human geography (1; 4). The study of geographic aspects of human activities, often with a population, political, economic, cultural, or social focus.

human rights (1; 35). The rights that should be part of normal human experience, including justice, a decent standard of living, personal security, and freedom of thought and speech.

hybrid religion (8; 234). A religion that combines ideals, traditions, and customs from two or more separate religions into a single religion.

hydrology (11; 333). The surface water drainage of a region, including lakes and rivers.

I

imperialism (2; 54). The practice of extending the rule of an empire over foreign lands.

import substitution (1; 34). A policy in which countries develop manufacturing industries to fulfill internal market demands, often protected by high tariffs to exclude foreign competition.

import-substitution manufacturing (9; 279). Government protection and encouragement of domestic industries through tariffs and restrictions on certain imported goods.

indigenous people (1; 27). The first inhabitants of an area or those present when the area is taken over by another group.

Industrial Revolution (2; 54). The period of the late 1700s and early 1800s when increasingly complicated machines and chemical processes, fueled by inanimate power sources such as water and coal, replaced traditional ways of making goods by hand with simple tools. The mass production of goods resulted, as did the need for raw materials. The Industrial Revolution began in England and then spread to other areas of Europe and the world.

infant mortality (1; 24). The number of deaths per 1,000 live births in the first year of life.

informal economy (1; 35). The economic sector in which workers act outside the formal sector.

intensification (2; 58). In agriculture, the increased output of crops or livestock per area unit of land.

intergovernmental organization (IGO) (1; 27). A group of country governments working together (e.g., the United Nations, the European Union).

irredentism (2; 53). The desire to gain control over lost territories or territories perceived to belong rightfully to a group; associated with nationalism.

Islam (7; 200). A religion of northern Africa and Southwestern Asia, and parts of South and East Asia, based on the teachings of Muhammad as recorded in the Qu'ran. *Islam* means "submission to the will of God."

isthmus (10; 313). A naturally occurring land bridge.

J

Jainism (6; 171). A religion, mainly in India, that involves a nonviolent code and has laws against harming animals (thus forbidding farming).

Judaism (7; 199). The religion of the Jewish people who worship Yahweh as the creator and lawgiver.

K

kibbutz (pl. kibbutzim) (7; 219). A communal farming village in Israel, the social and spiritual basis of the new Israeli nation after 1948, but now the home of fewer than 5 percent of the Israeli population.

Kyoto Protocol (1; 18). Statement adopted by the United Nations Convention on Climate Change in 1997 setting targets for reducing primary greenhouse gases.

L

language (1; 18). The means of communication among people by speaking, writing, and signing.

laterite (8; 233). Soils that form in seasonal tropical conditions. The cementation of clay and iron minerals form an impenetrable layer that is used as building blocks.

latitude (1; 5). The distance of a place north or south of the equator measured in degrees.

lingua franca (5; 156). A language that aids communication among people of varied languages, often at first used for trade, but later for varied purposes.

localization (1; 7). The geographic differentiation of places among and within countries.

loess (2; 44. 11; 333). Fine-grained and fertile soils developed from windblown deposits.

longitude (1; 6). The distance of a place east or west of 0° longitude measured in degrees.

Lost Decade (10; 306). In the 1980s, country economies in Latin America stalled because of high interest rates and debts.

M

mallee (9; 270). Type of Australian vegetation formed of eucalyptus shrubs that grow into dense thickets of many close-spaced stems.

Maoris (9; 272). Indigenous people of New Zealand.

map (1; 4). The representation of the features of Earth's surface on paper at varying scales.

maquiladora (10; 312. 11; 362). Mexican government program that encourages foreign-owned factories to be sited in Mexico by not charging import duties on raw materials for assembly.

market gardening (truck farming) (2; 58). The commercial production of high-cash-value, especially fruit and vegetable crops such as table grapes, raisins, oranges, grapefruit, apples, and lettuce.

marsupial (9; 270). A mammal that raises its young in a pouch instead of a womb. Mainly found in Australia, these include the kangaroo and koala.

medina (7; 203). The crowded streets of older sections of Arab towns in Northern Africa and Southwestern Asia.

megacity (1; 24). A city with a population of 10 million inhabitants or greater.

megalopolis (4; 118. 11; 337). An expanded urbanized area that includes several metropolitan areas with over a million people and dominates the economy of surrounding areas. First identified in the northeastern United States, covering the area between Boston and Washington, D.C.

Melanesian (9; 272). One of three broad categories of inhabitants of South Pacific oceanic islands, so named by Westerners because of their darker skin.

meridian of longitude (1; 6). An imaginary line joining places of the same longitude on Earth's surface.

microclimate (10; 295). The climate of a small area such as a valley; may also include the climate of a single plant.

Micronesian (9; 272). One of three broad categories of inhabitants of South Pacific oceanic islands; so named by Westerners because they inhabited "small islands."

midlatitude cyclone (9; 268). Cyclonic storm that occurs primarily in the midlatitudes.

migration (1; 25). The long-term movement of people into (in-migration, immigration) or out of (out-migration, emigration) a place.

Ministry of Economy, Trade, and Industry (METI) (4; 124). The Japanese government body responsible for assistance and advice to industry, including the export marketing office.

monotheism (7; 199). A religion based on belief in a single god.

Monroe Doctrine (10; 305). An 1823 declaration asserting U.S. rights to control activity in the Americas over those of countries outside the region. In the Monroe Doctrine, the United States vowed to resist any intervention from outside countries in Latin American affairs.

monsoon (9; 268). Seasonal summer rain.

monsoon tropical climatic environment (5; 144). A tropical climatic environment in which wind shifts between summer and winter bring heavy rains from oceanic air in the summer and dry winds of interior continental air in winter.

Mughal (Mogul) dynasty (6; 172). Turkish invaders of India from Persia in the AD 1500s who conquered most of the region and left a heritage of magnificent buildings such as the Taj Mahal.

multinational corporation (MNC) (1; 31). A corporation that makes goods and provides services in several countries, but directs operations from headquarters in one country.

Muslims (7; 200). "Those who submit to Allah." Followers of Islam.

N

nation (1; 26). A group of people who share a common identity, a sense of unity, and a desire for self-governance.

nationalism (1; 27). A pride in one's national identity and the belief that one's national interests are more important than all other interests (e.g. personal, local, regional, global) as well as other nations' interests.

nation-state (2; 53). The linking of a separate and distinct people (nation) and a politically organized territory with its own sovereign government (state).

nation-state ideal (2; 53). The belief that each people (nation) must have its own country (state) in order to be free and govern itself as it desires.

Native Americans (11; 339). Indigenous people who inhabited the Americas before the European arrival, including Amerinds and Inuits (Eskimos).

natural hazard (1; 14). A natural event, such as a volcanic eruption, earthquake, tornado, hurricane, or flood, that interrupts human activities by causing extensive damage and deaths.

natural resource (1; 14). Material present in the natural environment and recognized by humans as of practical worth (minerals, soils, water, building stones, timber).

New Partnership for Africa's Development (NEPAD) (8; 260). A South African organization dedicated to African development through African-based efforts.

New Zealand film industry (9; 281). A recent impetus to New Zealand's tourism industry. The immensely popular *Lord of the Rings* trilogy showcased the country's spectacular beauty, thereby attracting visitors from around the globe.

Nile Waters Agreement (7; 210). Agreement between Egypt and Sudan in 1959 to share Nile River waters, with 70 percent allocated to Egypt.

nongovernmental organization (NGO) (1; 27). Groups of people who act outside government and major commercial agencies, mainly in advocacy roles such as delivering aid and lobbying for particular causes.

nonrenewable resource (1; 14). A natural resource that is used up once it is extracted; for example, coal or metallic minerals.

North American Free Trade Agreement (NAFTA) (10; 305. 11; 342). An economic agreement among Canada, Mexico, and the United States signed in 1994.

North American plate (11; 332). A tectonic plate covering most of North America, Central America, and Greenland.

North Atlantic Treaty Organization (NATO) (2; 53). A military alliance of non-Communist European countries and the United States, founded in 1949 to counter the military threat of the Soviet Union. In recent years, former Communist countries have joined the alliance, and Russia has formed a partnership with NATO.

northern coniferous forest (3; 80). A forest composed of coniferous trees (firs, pines, cedars) common in the northern parts of temperate continental interior climatic environments.

O

oceanic temperate climate (2; 43). The climate of the western margins of midlatitude continents, in which cool, moist air from the ocean brings precipitation and moderate air temperatures.

offshore financial centers (1; 32). Places of low tax rates, few regulations, and secrecy where individuals and businesses move their financial assets to save money. They are typically small islands in the Caribbean and the South Pacific, though some of them are mainland countries like Switzerland, Luxembourg, Costa Rica, and Kuwait.

offshoring (1; 32). The shifting of a job to another country, often overseas as suggested by the term.

Organization of Petroleum Exporting Countries (OPEC) (7; 207). Established in 1960 to further the interests of oil and gas producers throughout the world, often in materially poorer countries and often to resist the overriding power of multinational oil corporations.

Organization of the Islamic Cooperation (OIC) (7; 206). Originally named the Organization of the Islamic Conference, it was established in 1970 by foreign ministers of Muslim countries throughout the world. It has 57 members but advances individual countries' interests rather than pursuing a common agenda. The current name was adopted in 2011.

orographic (9; 268). Mountain-related.

orographic lifting (1; 10). Hilly areas, usually facing oceanic moisture sources, cause uplift of air and enhanced precipitation levels.

outsourcing (1; 32). A business or government agency contracting with a company to produce a good or perform a service that it once produced or performed for itself. It occurs both inside and outside of countries.

ozone hole (9; 285). A thinning of the Earth's protective ozone layer formed when small quantities of chlorine gases penetrate Antarctica's atmosphere from lower latitudes.

P

Palestine Liberation Organization (PLO) (7; 206). An organization that promotes the reestablishment of a country of Palestine. It has a secular, left-wing political basis.

Pan-Arab country (7; 206). The idea of Arab countries joining to form a single country. Egypt and Syria were joined for a short while in the 1960s, but few attempts have been made to achieve this end since.

pandemic (8; 251). A disease occurring over a wide area and affecting a high proportion of the population.

parallel of latitude (1; 5). An imaginary circle joining places of the same latitude on Earth's surface.

perestroika (3; 93). The Soviet Union policy of the late 1980s designed to reconstruct the political and economic structure of the country so that it could compete in the capitalist world economic system.

peripheral country (1; 34). A materially poor country that is dependent on the world economy and materially wealthy countries.

permafrost (3; 81). Permanently frozen ground extending several hundred meters below the surface in Siberia and northern Canada. In summer, water in the surface active layer, 50 to 100 cm deep, melts.

petrodollars (10; 305). Dollars invested by Middle East oil producers during high oil prices in the 1970s.

physical geography (1; 4). The study of geographic aspects of natural environments.

place (1; 4). A point or area on Earth's surface having a geographic character defined by what it looks like, what people do there, and how they feel about it.

planned economy (2; 53). The Communist practice of the government, rather than the free market, deciding what goods and services need to be produced within a country.

podzol soil (3; 81). Soils of low fertility in which plant nutrients are removed by water passing through. Commonly develop beneath temperate coniferous forest and on sandy soils.

polar climate (1; 13). Climate typical of the polar regions; extremely cold all year.

political geography (1; 26). The study of how governments and political movements influence the human and physical geography of the world and its resources.

political rights (1; 35). The rights to vote and participate in one's own government.

Polynesian (9; 272). One of three broad categories of inhabitants of South Pacific oceanic islands; so named by Westerners because they were perceived as inhabiting "many islands."

population density (1; 23). The number of people per given area.

population distribution (1; 23). The spread of people in a region, incorporating areas of high, medium, and low density.

population doubling time (1; 25). The number of years taken to double the population of a country.

post-Panamax (10; 315). The government of Panama's expansion of the Panama Canal in the early 2000s to enable larger ships to pass through it.

primary sector (1; 30). The sector of an economy that produces output from natural sources, including mining, forestry, fishing, and farming.

primate city (1; 24). A city that contains a large proportion of the urban population of a country, often several times the population of the second city.

producer goods (2; 55). Industrial goods used by other industries to make consumer goods.

producer services (2; 57). Service industries that are involved in the output, including-market research, advertising, accountancy, legal, banking, and insurance industries.

production line (11; 341). The system of manufacturing in which components are made and assembled into the final product in a sequence of factory-based processes.

productive capacity (2; 55). The total amount of goods a country's industries can produce during a given period.

productivity (2; 57). The measure of the amount of product generated or work completed per hour of labor.

protectionism (1; 31. 10; 306). Governmental policies that focus on government-controlled industry and the protection of domestic products through tariffs, quotas, and red tape.

purchasing power parity (PPP) (1; 29). The measure of GNI or GDP that is based on internal country costs of living rather than external exchange rates related to the U.S. dollar.

Q

quaternary sector (1; 30). The sector of an economy that specializes in producer services, including financial services and information services.

Qu'ran (7; 200). The holy book of Islam.

R

race (1; 21). A biologic stock of people with similar physical characteristics, or a group of people united by a community of interests.

rain shadow (10; 295). Low rainfall in an area to the lee of a mountain range, where winds warm and get drier as they descend after flowing across the mountains.

region (1; 6). An area of Earth's surface with physical and human characteristics that distinguish it from other places.

regional geography (1; 6). The study of different regions at Earth's surface in their country and global contexts.

relative distance (1; 6). The direction and distance of a place relative to others, often affected by factors that slow or increase contacts among people.

relief (1; 9). The physical height and slope of the land, as in hills, mountains, and valleys.

religion (1; 18). An organized system of practices that seeks to explain our purpose on Earth and may include a set of values and/or worship of a divine being.

remittance (10; 305. 11; 359). Funds sent home by workers in foreign countries.

renewable resource (1; 14). A resource that is replaced by natural processes at a rate that is faster than its usage.

reservations (11; 339). Government-designated lands set aside for Native American people, found in the central and northwestern parts of the United States.

responsible growth (1; 35). A term originating in the 2002 World Summit on Sustainable Development. The summit urged development policies that linked economic growth, environmental sustainability, and social equity.

rift valley (8; 230). A deep valley caused by the rocks of Earth's crust arching and cracking to let down a section of crust to form the valley floor.

rural area (1; 24). Land outside urban areas, often having an economic emphasis on farming, mining, and/or forestry. Such areas may dominate population distribution in materially poorer countries.

Russification (3; 81). Policies directed at making non-Russians into Russians by encouraging or forcing non-Russians to adopt Russian cultural characteristics such as the Russian language.

S

Sahel (8; 233). The zone immediately to the south of the Sahara Desert in Africa that suffers droughts as the arid area expands by natural or human-induced actions.

salinization (7; 198). The process by which soils become unproductive because of an accumulation of alkaline salts near the surface. Often associated with poorly managed irrigation systems in arid areas.

savanna (8; 233). A tropical ecosystem that in areas of seasonal rainfall; dominated by grasses and home to large herbivores and carnivores. Many such areas show signs of burning by humans to restrict tree growth.

secondary sector (1; 30). The sector of an economy that changes the raw materials from the primary sector into useful products, thus increasing their value, as in chewing gum or parts for airplanes.

sedimentation (10; 299). The deposition of rock debris, including in offshore areas, where too much fine debris may kill coral reefs.

separatism (2; 70). The desire by an ethnic group for independence, as in the case of the Basque group on the French–Spanish border.

shantytown (8; 254). An unplanned residential sector of urban areas in poorer countries. Housing is often built of any materials that come to hand and does not have electricity, water, or waste disposal.

shatter belt (6; 187). A zone between distinctive cultures and political groups that experiences conflict and a slowing of development.

Shia Muslims (7; 200). Also known as Shiites. Muslims who are partisans of the imam Ali (not acceptable to Sunni Muslims) and look to his return. They make up 90 percent of the Iranian population and 60 percent of Iraquis.

Shinto (4; 123). The traditional religion of Japan built on ancient myths and customs that promote national interests and identity.

Sikhism (6; 171). A Hindu-related religion with a strict code of conduct. Its temple kitchens provide food for all.

social rights (1; 36). The rights to have a job and earn a living with basic material standards.

soil (1; 10). Weathered rock material that develops by the actions of water, animals, and plants into a basis for plant growth.

solid earth environment (1; 7). The solid planet Earth, incorporating core, mantle, and crust, in which movements of solid rock are caused by interior heat energy.

South Pacific Forum (9; 280). Established in 1971, links 13 regional countries; Australia, New Zealand, Papua New Guinea, the Solomon Islands, the Cook Islands, Fiji, Kiribati, Nauru, Niue, Tonga, Tuvalu, Vanuatu, and Western Samoa. Its aim is to develop regional political cooperation.

Southern African Development Community (SADC) (8; 248). Established by countries in Southern Africa that were opposed to South Africa's apartheid policy in order to organize alternative trade outlets. Now the community encourages trade among the constituent countries, including South Africa.

Southern Ocean (9; 268). The fourth largest of the Earth's five oceans.

spatial view (1; 4). The geographic view that focuses on differences among places.

specialization (2; 58). The concentration on fewer commercial products within a farming region.

state (2; 53). The country or division of a country within a federal government.

state socialism (2; 53). The Communist Party actively running the political, social, and economic activities of the people.

steppe climate (semiarid) (7; 195). Conditions in which evaporation rates exceed precipitation during periods of the year.

steppe grasslands (3; 79). Temperate grasslands typical of the transition between forest and arid areas of temperate continental interior climatic environments.

subduction (1; 8). In plate tectonics, where two converging plates clash and one is pushed beneath the other, causing earthquakes and volcanoes.

subsistence agriculture (8; 243). The growing of crops to satisfy local needs.

subtropical rainy climate (4; 115). A climate found on east coasts characterized by hot, rainy summers and cool winters. Such areas experience tropical cyclones (hurricanes, typhoons).

subtropical winter rain (Mediterranean climate) (2; 43). A climate that is characterized by summer drought and winter rains.

Sunni Muslims (7; 200). Also known as Sunnites. Traditional, conservative followers Islam forming the majority in most Muslim countries. Includes strict sects such as the Wahabi.

supranationalism (2; 69). The idea that differing nations can cooperate so closely for their mutual benefit that they can share the same government, economy (including currency), social policies, and even military.

sustainable forestry (9; 280). The application of principles of sustainability to reforestation.

sustainable human development (1; 34). A level of development in which resources are exploited at a rate that is sustainable for future generations.

T

tectonic plate (1; 7). A large block of Earth's crust and underlying rocks approximately 100 km thick and up to several thousand kilometers across. Earth's interior heat causes plates to move apart and crash together, forming major relief features including ocean basins, mountain systems, and continental areas.

temperate climate (1; 12). The climates of mid-latitudes, in which there are summer–winter temperature contrasts without the extremes of lengthy very hot or cold periods.

temperature inversion (10; 299). Naturally occurring periods during the winter months when cold, dense air remains "trapped" at the surface under warmer air for several days or even weeks. The cold air is usually trapped by mountains or some related physical barrier. Temperature inversions may produce stagnant air, which may complicate pollution problems in urban areas.

Tennessee Valley Authority (11; 349). Established by the U.S. government in 1933 to stimulate economic growth in southern Appalachia through the damming of rivers to improve transportation, flood control, and electricity costs.

tertiary sector (1; 30). The sector of an economy concerned with the distribution of goods and services, including trade, professions, and government employment.

Thar Desert (6; 166). Arid region extending from Afghanistan through Pakistan into westernmost India.

total fertility rate (1; 24). The number of births per woman in her childbearing years.

Transantarctic Mountains (9; 268). One of the Earth's longest continuous mountain chains and a continuation of the Andes Mountains; divide Antarctica into East and West Antarctica.

transform plate margin (9; 267). Area where plates slide horizontally along one another.

triangular trade pattern (8; 234). Pattern of trade in the 1700s that involved Britain, West Africa, and the Americas, driven by the demand for slave labor in North and Central America.

tropical climate (1; 10). Climatic environment typical of the tropical zone and having high temperatures all year.

tropical rain forest (8; 232). A forest ecosystem characterized by a great variety of species that are dependent on high temperatures and rainfall.

tsunami (5; 162). A huge wave generated by an earthquake that changes the shape of the ocean floor suddenly. The wave travels fast across the ocean and causes destruction on entering shallow coastal waters and confined valleys.

tundra (3; 81). Ecosystem type occurring in cold polar environments and consisting of mosses, grasses, and low shrubs.

typhoon (5; 144. 9; 268). A tropical storm of hurricane type experienced in Southeast and East Asia.

U

UN Security Council (11; 342). One of the principal divisions of the United Nations. The purpose of the Security Council is the maintenance of international peace and security. The United States, China, France, the United Kingdom, and Russia are the five countries granted permanent seats on the Security Council. The permanent members of the Council have the power to veto Council decisions. The decisions of the Security Council are binding for the UN General Assembly.

uneven development (11; 345). The increase in the gap between poor and wealthy regions in a country and the shifting locations of economic growth and decline over time seen by Marxists as an outcome of capitalism.

universalizing religion (1; 20). A religion that seeks to be global in its application, such as Islam and Christianity.

urban area (1; 24). An area with high densities of people, buildings, transportation linkages, and human activities of a high economic, political, and cultural order. Urban areas dominate population distribution in materially wealthy countries.

vertical integration (11; 340). The combining of producers of raw materials, manufacturers that process the materials, and those that assemble the products in a single corporation to achieve economies of scale.

Virgin Lands Campaign (3; 95). A Soviet agricultural campaign begun in the 1950s. It promoted farming in lands where it had never taken place before, primarily in lands that were very marginal because the soil was poor or not enough water or heat was present to grow crops. Much of the land was in Siberia and Central Asia, especially in the Kazakh Republic.

Wallace Line (9; 270). Line drawn by botanist Alfred Russell Wallace in the mid-1800s marking the edge of plate tectonic action that forced Indonesia's eastern islands against its western islands and brought Australasian plants and animals with them.

weathering (1; 9). The action of atmospheric forces (through water circulation and temperature changes) on rocks at Earth's surface that breaks the rocks into fragments, particles, and dissolved chemicals.

Wellington Agreement (9; 284). Agreement that banned commercial mining activities and introduced environmental protection regulations to Antarctica.

white Australia policy (9; 272). Informal policy that encouraged the acceptance of European immigrants and discouraged immigration from neighboring Asian countries.

world languages (1; 19). *See* **lingua franca**.

world region (1; 36). This text recognizes nine world regions, each of which includes a number of countries linked by cultural, political, economic, and environmental conditions.

xenophobia (3; 98). A fear of foreigners.

Index